Lecture Notes in Networks and Systems 1418

The series "Lecture Notes in Networks and Systems" publishes the latest developments in Networks and Systems—quickly, informally and with high quality. Original research reported in proceedings and post-proceedings represents the core of LNNS.

Volumes published in LNNS embrace all aspects and subfields of, as well as new challenges in, Networks and Systems.

The series contains proceedings and edited volumes in systems and networks, spanning the areas of Cyber-Physical Systems, Autonomous Systems, Sensor Networks, Control Systems, Energy Systems, Automotive Systems, Biological Systems, Vehicular Networking and Connected Vehicles, Aerospace Systems, Automation, Manufacturing, Smart Grids, Nonlinear Systems, Power Systems, Robotics, Social Systems, Economic Systems and other. Of particular value to both the contributors and the readership are the short publication timeframe and the world-wide distribution and exposure which enable both a wide and rapid dissemination of research output.

The series covers the theory, applications, and perspectives on the state of the art and future developments relevant to systems and networks, decision making, control, complex processes and related areas, as embedded in the fields of interdisciplinary and applied sciences, engineering, computer science, physics, economics, social, and life sciences, as well as the paradigms and methodologies behind them.

Indexed by SCOPUS, EI Compendex, INSPEC, WTI Frankfurt eG, zbMATH, SCImago.

All books published in the series are submitted for consideration in Web of Science.

For proposals from Asia please contact Aninda Bose (aninda.bose@springer.com).

Ivan Ostroumov · Karen Marais ·
Maksym Zaliskyi
Editors

Advances in Civil Aviation Systems Development

International Workshop of ACASD on Joint Event in West Lafayette, USA and Kyiv, Ukraine on March 18–20, 2025

Springer

Editors
Ivan Ostroumov
Air Navigation Systems Department
National Aviation University
Kyiv, Ukraine

Karen Marais
School of Aeronautics
Purdue University
West Lafayette, IN, USA

Maksym Zaliskyi
Telecommunication and Radioelectronic Systems
National Aviation University
Kyiv, Ukraine

ISSN 2367-3370 ISSN 2367-3389 (electronic)
Lecture Notes in Networks and Systems
ISBN 978-3-031-91991-6 ISBN 978-3-031-91992-3 (eBook)
https://doi.org/10.1007/978-3-031-91992-3

This Springer imprint is published by the registered company Springer Nature Switzerland AG
The registered company address is: Gewerbestrasse 11, 6330 Cham, Switzerland

Preface

This volume contains the papers presented at the third International Workshop on Advances in Civil Aviation Systems Development (ACASD 2025) which is a joint event of the School of Aeronautics and Astronautics of Purdue University (IN, USA) and National Aviation University (Kyiv, Ukraine) during March 18–20, 2025. ACASD 2025 received 47 research submissions from different countries; among them 37 papers were accepted for presentation. There are contributors from Ukraine, USA, Poland, Ireland, UK, Germany, France, Latvia, and Spain.

ACASD was formed to bring together outstanding researchers and practitioners in the field of civil aviation systems including avionics systems, communication, navigation and surveillance systems, aerospace systems reliability development, and engineering from all over the world to share their experience and expertise. ACASD plays a unique role in the representation of new innovative technologies in the aerospace domain to researchers and industry practitioners to share complex problems and their solutions. The workshop covers major research areas focused on advances in civil avionics system design, modern trends in communication, navigation, and surveillance systems development, human factor and safety system analysis, civil avionics system maintenance questions, and air traffic management systems. Workshop includes following tracks: Aerodynamics, air transportation, management and economics, interference in global navigation satellite system, aircraft noise, communication systems for civil aviation systems, surveillance data processing in civil aviation, methods of operational efficiency improvement for civil aviation systems, sensors in civil aviation, human factor in civil aviation, and unmanned aircraft systems.

Special attention during ACASD workshop was paid to Impact of the Russian-Ukrainian war on Global Air Transportation System and Recovery of Ukrainian Civil Aviation. Experts in civil aviation from all over the world are brought together to discuss new challenges in the global air transportation system related to closing and limiting the use of wide volumes of airspaces caused by the war in Ukraine. Development of ground air navigation infrastructure, recovery of aviation equipment from long periods of not use, and all other topics related to the impact of war in civil aviation.

Workshop was in a hybrid format. Most reports are presented in-person mode.

The organizers also wish to thank our publisher, Springer, for their acceptance to publish the proceedings and for their kind assistance and cooperation during the editing process. The organizing committee wishes to thank CMT Research Conference Management System as it is a wonderful tool for easy organization and compilation of conference documents.

All our team invites you to visit the ACASD workshop website https://www.acasd.nau.edu.ua where you can find all the relevant information about this exciting aviation event.

March 2025

Ivan Ostroumov
Karen Marais
Maksym Zaliskyi

About this Book

This book includes high-quality research papers presented at 3rd International Workshop on Advances in Civil Aviation Systems Development (ACASD 2025), which was joint event of School of Aeronautics and Astronautics of Purdue university (IN, USA) and National Aviation University (Kyiv, Ukraine). This book presents original results of a scholarly study of unique research teams and market leaders on the development in civil aviation systems and its application. The book topics include major research areas focused on advances in air transportation, interference in global navigation satellite system, aircraft noise, communication systems for civil aviation systems, surveillance data processing, methods of operational efficiency improvement, sensors in civil aviation, human factor, and unmanned aircraft systems. Book will be useful for scholars and professionals in the civil aviation domain.

Organization

The 3rd International Workshop on Advances in Civil Aviation Systems Development (ACASD 2025) was joint event of School of Aeronautics and Astronautics of Purdue university (IN, USA) and National Aviation University (Kyiv, Ukraine).

General Chairs

Ivan Ostroumov	Department of Air Navigation Systems, National Aviation University, Ukraine
Karen Marais	School of Aeronautics and Astronautics, Purdue University, West Lafayette, USA
Maksym Zaliskyi	Department of Telecommunication and Radioelectronic Systems, National Aviation University, Ukraine

Program Committee Chairs

Vitalii Larin	National Aviation University, Kyiv, Ukraine
Olha Sushchenko	National Aviation University, Kyiv, Ukraine
Fulya Aybek Cetek	Eskisehir Technical University, Eskisehir, Turkey
Sergiy Gnatyuk	National Aviation University, Kyiv, Ukraine
Neelakshi Majumdar	University of Arkansas, USA
Caner Ozcan	Karabuk University, Engineering Faculty, Karabuk, Turkey
Viktoriia Ivannikova	Dublin City University Business School, Ireland
Oleksandr Zaporozhets	Lukasiewicz Institute of Aviation, Warsaw, Poland
Valeriu Vitan	Eurocontrol, Brussels, Belgium
Tetana Shmelova	National Aviation University, Kyiv, Ukraine
Oleksii Holubnychyi	National Aviation University, Kyiv, Ukraine
Nina Rizun	Gdansk University of Technology, Poland
Vladimir Pavlikov	National Aerospace University "KhAI", Kharkiv, Ukraine
Simeon Zhyla	National Aerospace University "KhAI", Kharkiv, Ukraine
Kostiantyn Dergachov	National Aerospace University "KhAI", Kharkiv, Ukraine

Leonid Emelyanov	Institute of Ionosphere NAS and MES, Kharkiv, Ukraine
Roman Odarchenko	National Aviation University, Kyiv, Ukraine
Samiksha Shukla	CHRIST University, Bangalore, India

Sponsoring Organizations

ACASD 2025 could not been possible without support of National Aviation University, Ukraine (https://www.nau.edu.ua) and Springer International Publishing AG, Germany (https://www.springer.com). Organizers and all participants of International Workshop on Advances in Civil Aviation Systems Development express a lot of thanks.

Contents

Interference in Global Navigation Satellite System

Impact or Russian-Ukrainian war on Global Air Transportation System and Recovery of Ukrainian Civil Aviation

Aircraft Noise

Communication Systems for Civil Aviation Systems

Sensors in Civil Aviation

Human Factor in Civil Aviation

Unmanned Aircraft Systems

Contributors

Maksym Alpert National Technical University of Ukraine "Igor Sikorsky Kyiv Polytechnic Institute", Kyiv, Ukraine

Sofiia Alpert National Aviation University, Kyiv, Ukraine

Veniamin Antonov National Aviation University, Kyiv, Ukraine

Yunifa Miftachul Arif Universitas Islam Negeri Maulana Malik Ibrahim Malang, Malang, Indonesia

Maryna Arkhyrei National Aviation University, Kyiv, Ukraine

Yuliya Averyanova National Aviation University, Kyiv, Ukraine

Oleksandr Avramenko The National Defence University of Ukraine Named After Ivan Cherniakhovskyi, Kyiv, Ukraine

Oleksandra Bankova National Aviation University, Kyiv, Ukraine

Yurii Bezkorovainyi National Aviation University, Kyiv, Ukraine

Nataliia Bilak National Aviation University, Kyiv, Ukraine

Bohdan Blazhei National Aviation University, Kyiv, Ukraine

Oleksandr Bondarev National Aviation University, Kyiv, Ukraine

Ihor Bovdui Anatolii Pidhornyi Institute of Mechanical Engineering Problems of the National Academy of Sciences of Ukraine, Kharkiv, Ukraine

Iryna Bychkova National Aerospace University KhAI, Kharkiv, Ukraine

Natali Caccioppoli EUROCONTROL Innovation Hub, Bretigny sur Orge, France

Kostiantyn Cherednichenko National Aviation University, Kyiv, Ukraine

Oksana Chumak National Aviation University, Kyiv, Ukraine

Oleksii Chuzha National Aviation University, Kyiv, Ukraine

Kostiantyn Dergachov National Aerospace University KhAI, Kharkiv, Ukraine

Peter Dischler ISGlobal, University of Barcelona, Barcelona, Spain

Anastasiia Dmytruk National Aviation University, Kyiv, Ukraine

Vladyslav Dubinin National Aerospace University KhAI, Kharkiv, Ukraine

David Duchet EUROCONTROL Innovation Hub, Bretigny sur Orge, France

Valentyn Dyptan The National Defence University of Ukraine Named After Ivan Cherniakhovskyi, Kyiv, Ukraine

Sergiy Gnatyuk National Aviation University, Kyiv, Ukraine

Andriy Goncharenko National Aviation University, Kyiv, Ukraine

Andrii Grekhov National Aviation University, Kyiv, Ukraine

Yaroslav Gritsev National Aviation University, Kyiv, Ukraine

Andreii Hnashuk Private Enterprise "Secret-Service", Kyiv, Ukraine

Oleksii Holubnychyi National Aviation University, Kyiv, Ukraine

Yurii Hryshchenko National Aviation University, Kyiv, Ukraine

Oleksii Hurtovyi National Aerospace University KhAI, Kharkiv, Ukraine

Ludvig Ilnitskiy National Aviation University, Kyiv, Ukraine

Olga Ivanets National Aviation University, Kyiv, Ukraine

Viktoriia Ivannikova Dublin City University, Dublin 9, Ireland

Viktoriia Ivannikova Dublin City University, Dublin, Ireland

Oleg Ivashchuk National Aviation University, Kyiv, Ukraine

Fedir Katushonok National Aviation University, Kyiv, Ukraine

Kateryna Kazhan National Aviation University, Kyiv, Ukraine

Volodymyr Kharchenko National Aviation University, Kyiv, Ukraine

Rimvidas Khrashchevskyi National Aviation University, Kyiv, Ukraine

Volodymyr Klymchuk National Aviation University, Kyiv, Ukraine

Anzhelika Kokarieva National Aviation University, Kyiv, Ukraine

Denys Kolesnichenko National Aviation University, Kyiv, Ukraine

Volodymyr Kolotusha National Aviation University, Kyiv, Ukraine

Georgiy Konakhovych National Aviation University, Kyiv, Ukraine

Vasyl Kondratiuk National Aviation University, Kyiv, Ukraine

Valeriy Konin National Aviation University, Kyiv, Ukraine

Lesia Konoplianyk National Aviation University, Kyiv, Ukraine

Valentina Konovaliuk National Aviation University, Kyiv, Ukraine

Pavel Krents Antonov Company, Kyiv, Ukraine

Andrii Krupko National Aviation University, Kyiv, Ukraine

Dmytro Krytskyi National Aerospace University KhAI, Kharkiv, Ukraine

Oleksandr Kutsenko National Aviation University, Kyiv, Ukraine

Nataliia Kuzmenko National Aviation University, Kyiv, Ukraine

Borys Kuznetsov Anatolii Pidhornyi Institute of Mechanical Engineering Problems of the National Academy of Sciences of Ukraine, Kharkiv, Ukraine

Olena Kyzymchuk ITM Technische Universitat Dresden, Dresden, Germany

Nataliia Ladohubets National Aviation University, Kyiv, Ukraine

Vitalii Larin National Aviation University, Kyiv, Ukraine

Anzhela Lelechenko National Aviation University, Kyiv, Ukraine

Elvira Luzik National Aviation University, Kyiv, Ukraine

Vitalii Makarenko National Aviation University, Kyiv, Ukraine

Tetiana Maliutenko National Aviation University, Kyiv, Ukraine

Illia Marienkov National Aviation University, Kyiv, Ukraine

Kristina Marintseva Kingston University London, London, UK

Tatyana Nikitina Educational Scientific Professional Pedagogical Institute Ukrainian Engineering Pedagogical Academy, Kharkiv, Ukraine

Pavlo Openko The National Defence University of Ukraine Named After Ivan Cherniakhovskyi, Kyiv, Ukraine

Valerii Orlianskyi National Aviation University, Kyiv, Ukraine

Ivan Ostroumov National Aviation University, Kyiv, Ukraine

Eugene Ovdiyuk National Aerospace University KhAI, Kharkiv, Ukraine

Yuliia Petrova National Aviation University, Kyiv, Ukraine

Zarina Poberezhna National Aviation University, Kyiv, Ukraine

Artem Polozhentsev National Aviation University, Kyiv, Ukraine

Igor Prokopenko National Aviation University, Kyiv, Ukraine

Dmytro Prosvirin Antonov Company, Kyiv, Ukraine

Irina Pryhodko National Aviation University, Kyiv, Ukraine

Denys Puhach National Aerospace University KhAI, Kharkiv, Ukraine

Serhii Pustovyi LLC "Adron-DAM", Kyiv, Ukraine

Valentyna Semychenko National Aviation University, Kyiv, Ukraine

Olga Shcherbyna National Aviation University, Kyiv, Ukraine

Tetiana Shmelova National Aviation University, Kyiv, Ukraine

Leonid Sibruk National Aviation University, Kyiv, Ukraine

Yuliya Sikirda Ukrainian State Flight Academy, Kropyvnytskyi, Ukraine

Stepan Simchenko National Aviation University, Kyiv, Ukraine

Serhii Smerichevskyi National Aviation University, Kyiv, Ukraine

Petro Snitsarenko National Defence University of Ukraine Named After Ivan Cherniakhovskyi, Kyiv, Ukraine

Olena Sokolova National Aviation University, Kyiv, Ukraine

Oleksandr Solomentsev National Aviation University, Kyiv, Ukraine

Olha Sushchenko National Aviation University, Kyiv, Ukraine

Oleksii Sushych National Aviation University, Kyiv, Ukraine

Serhii Sydorenko National Aviation University, Kyiv, Ukraine

Viktoria Sydorenko National Aviation University, Kyiv, Ukraine

Kateryna Synylo National Aviation University, Kyiv, Ukraine

Olena Tachinina National Aviation University, Kyiv, Ukraine

Serhii Tarasiuk National Aviation University, Kyiv, Ukraine

Daria Tkachuk National Aviation University, Kyiv, Ukraine

Vadim Tokarev National Aviation University, Kyiv, Ukraine

Oleksandr Truhan Zhytomyr Institute PrJSC "Higher Educational Institution" "Interregional Academy of Personnel Management (IAPM)", Zhytomyr, Ukraine

Nina Volianska Taras Shevchenko National University of Kyiv, Kyiv, Ukraine

Roman Voliansky National Technical University of Ukraine "Igor Sikorsky Kyiv Polytechnic Institute", Kyiv, Ukraine

Petro Yablonsky National Defence University of Ukraine Named After Ivan Cherniakhovskyi, Kyiv, Ukraine

Maxim Yatsko SmartLynx Airlines "Mazrudas", Marupes novads, Latvia

Anatolii Yaremenko National Aerospace University KhAI, Kharkiv, Ukraine

Gennadiy Yun Private Enterprise "Secret-Service", Kyiv, Ukraine

Oleksandr Zadorozhnyi National Aviation University, Kyiv, Ukraine

Ihor Zakutynskyi National Aviation University, Kyiv, Ukraine

Maksym Zaliskyi National Aviation University, Kyiv, Ukraine

Oleksandr Zaporozhets Łukasiewicz Research Network – Institute of Aviation, Warsaw, Poland

Oleksandr Zaporozhets Institute of Aviation, Warsaw, Poland

Oleksander Zhdanov National Aviation University, Kyiv, Ukraine

Yevheniia Znakovska National Aviation University, Kyiv, Ukraine; Interregional Academy of Personnel Management, Kyiv, Ukraine

Oleksiy Zuiev National Aviation University, Kyiv, Ukraine

About the Editors

Dr. Ivan Ostroumov is a full professor of the Air Navigation Systems Department at National Aviation University of Ukraine. He graduated from National Aviation University (Master degree in Engineering of maintenance and services of airspace systems) in 2006. In 2009, he received Ph.D degree in navigation and air traffic control. He received D.Sc. degree in engineering with research "Methodological foundation of integrated aircraft positioning by multiple navigational aids under risk condition" in 2020. In 2013, he completed a traineeship in the Navigation department of Eurocontrol (Brussels). Secondment in Institute of Air Navigation (Eurocontrol, Luxembourg, 2015). In 2017/2018, he was a visiting scholar in the School of Aeronautics and Astronautics at Purdue University (IN, US) with a fulbright grant for research "An investigation and development of alternative methods of positioning and navigation for air transport". He participated in several research and development projects under the Ministry of Education and Science of Ukraine. Ivan is the head of bachelors educational and professional program "Systems of Air Navigation Services" which is fully compatible with ICAO ATSEP Training program.

Dr. Karen Marais' research focuses on developing ways to deliver value through reliability, safety, and sustainability. She is the author or co-author of several technical publications, including 24 journal papers, two reports for the National Academies, and two book chapters. She received an NSF CAREER award in 2014. Dr. Karen Marais has worked in engineering for two decades. Dr. Marais holds a B.Eng. in electrical and electronic engineering from the University of Stellenbosch and a B.Sc. in mathematics from the University of South Africa. She also holds a master's degree in space-based radar from MIT. She received her Ph.D. from the Department of Aeronautics and Astronautics at MIT in 2005. Prior to graduate school, she worked in South Africa as an electronic engineer.

Dr. Maksym Zaliskyi is a full professor of the Department of Telecommunication and Radioelectronic Systems at the National Aviation University. He received his B.Sc. and M.Sc. degrees in radio engineering from the National Aviation University, Ukraine, in 2005 and 2007, respectively. He obtained a Ph.D. and D.Sc. degrees in aircraft equipment operation from the National Aviation University, in 2012 and 2021, respectively. His areas of interest include data processing, equipment operation, and reliability analysis. He participated in several research and development projects under the Ministry of Education and Science of Ukraine. Maksym is the head of Ph.D. educational and scientific program "Telecommunications and Radio Engineering". A total number of research publications is 280.

Aerodynamics

Analysis of Measuring Errors During Aerodynamic Research in Wind Tunnel

Oleksander Zhdanov, Olha Sushchenko(✉), and Valerii Orlianskyi

National Aviation University, Liubomyra Huzara Ave., 1, Kyiv 03058, Ukraine
sushoa@ukr.net

Abstract. This paper deals with estimating errors in the measurement systems during experimental tests of unmanned aerial vehicles in the wind tunnel. The instrumentation and technique of the calibration for measuring pressure parameters are represented. Features of the measurement in the wind tunnel are characterized. The transformation function of the measuring channel was proposed. The estimation of high-velocity head measurement errors has been carried out. Relative and absolute errors of velocity head were estimated. Errors in measuring airflow speed have been estimated. It is shown that the velocity of the airflow is the result of an indirect nonlinear measurement. Errors in determining coefficients of aerodynamic forces and moments are calculated and represented. The appropriate graphical dependencies are represented. Errors of calibrating coefficients during the experiment are calculated, Instrumental errors in measuring aerodynamic coefficients are calculated. The obtained results can be useful for tests of the unmanned aerial vehicles of the wide class.

Keywords: Wind Tunnel · Experimental Research · Calibration · Pressure Measurement · Airflow Speed · Aerodynamic Coefficients · Error Models

1 Introduction

Nowadays, unmanned aerial vehicles (UAVs) are widespread in many industries [1, 2]. During developing UAV, it is possible to improve its aerodynamic characteristics [3, 4]. The important stage of such an approach is researching the aerodynamic characteristics of the UAV in the wind tunnel [5–7]. This process is accompanied by measuring head velocities, airflow speeds, and aerodynamic coefficients of forces and moments [8]. It is important to create error models to assess the quality of measuring processes and to improve calibration processes used during tests in the wind tunnel. The most important topic is the calibration of pressure measurement channels. This requires improved instrumentation and technique for measuring pressure. The paper's aim is to study measuring errors in the aerodynamic characteristics of UAVs during tests in the wind tunnel.

The paper is organized as follows: In Sect. 2, we consider the process of calibrating pressure measurement channels. Section 3 deals with estimating high-velocity head measurement errors, including relative and absolute ones. The estimations of systematic errors in airflow speed measurements are given in Sect. 4. The estimations of errors in determining coefficients of aerodynamic forces and moments are represented in Sect. 5.

I. Ostroumov et al. (Eds.): ACASD 2025, LNNS 1418, pp. 3–19, 2025.
https://doi.org/10.1007/978-3-031-91992-3_1

2 Literature Review and Problem Statement

Many scientific articles and papers represent features of aerodynamic tests. For example, article [3] studies the aerodynamics of a vortex wing of an unmanned aerial vehicle, including difficult situations such as achieving large and closed angles of attack. Features of vortex generators used for improving the aerodynamic characteristics of UAVs are described in [4, 5]. Tests on aerodynamic characteristics of wings with leading-edge protuberances were represented in publications [6–8]. The characteristics of tools and methodologies taking into consideration conceptual aerodynamics for aircraft design are described in [9, 10].

The effect of sinusoidal protrusions on the leading edge of nominally two-dimensional airfoils was investigated experimentally in a series of wind tunnel tests [6]. In [11], the effect of triangular vortex generators placed on the upper surface of the airfoil on the aerodynamic coefficients under turbulent flow conditions was investigated. These vortex generators were used as passive flow control devices. The tests were carried out using a wind tunnel.

The basics of the experimental dynamic including the last achievements in this area are given in the book [12].

Features of experiments in the wind tunnel are described in books [13–15] in detail. Nevertheless, some aspects of experimental tests in the wind tunnel require further research. This paper is connected with researching the measurement system of the wind tunnel. The main goal of the paper is the analysis of measuring errors during experimental tests in the wind tunnel. For this, the procedure of calibrating measurement pressure channels has been developed and described.

The basic stages of the experiment in the wind tunnel researched in this paper are given in publications [16, 17]. To achieve the set problem, the approaches represented in [18–20] have been used.

3 Calibration of Measurement Pressure Channels

Calibration of the pressure measurement channel is performed at the beginning of the research to reduce the measurement channel's conversion errors generally. The calibration results are the coefficients of the approximating polynomial characteristic of the channel transformation and an estimate of possible measurement errors. The coefficients of the approximating polynomial are used during experiments [9–11].

The pressure measurement channel includes an MPXV5004G pressure sensor and an analog-to-digital conversion (ADC) of the interface board card. The main sources of pressure measurement errors are:

- the constant component of the output signal of pressure sensors of the MPXV5004G type, which is in the range from 1 V to 1.25 V and is a systematic additive component of the error (stable during the experiment, if the temperature fluctuations do not exceed ±5°C of the calibration temperature);
- pressure conversion sensitivity error S (multiplicative component) and nonlinearity of the MPXV5004G type pressure sensor conversion characteristic – a systematic error not exceeding 2.5% of the maximum deviation of the output signal;

- noise of the output signal of MPXV5004G type pressure sensors, a random component that does not exceed 5 mV (established experimentally);
- ADC zero offset is a systematic additive component that is set during ADC calibration (its level does not exceed one lower digit of the maximum value of the set measurement range);
- non-linearity of the ADC conversion, which does not exceed the level of the unit of the lowest bit of the source code – a systematic error;
- quantization noise, a random component, the root mean square value of $\sigma = Q/\sqrt{12}$, where Q is the value of the unit of the lowest bit of the source code.

When the PCI-1747 U card is operated in the range from −5 to 5 V, the value of the unit of the least significant bit of the source code Q will be equal to $Q = 10/2^{16} = 152$ μV, and $\sigma = 14$ μV. At the same time, the component of pressure measurement error during analog-to-digital conversion will be $\sigma_p = \sigma/S = 5 \cdot 10^{-3}$ mm H_2O, which is significantly less than random conversion errors of MPXV5004G pressure sensors (output signal noise level up to 5 mV). Therefore, analog-to-digital conversion errors can be neglected and the main sources of pressure measurement errors are the conversion errors of pressure sensors. To reduce the impact of all systematic error components on the result of pressure measurement, calibration of the measurement channel as a whole is performed [11–14]. The scheme according to which the calibration of the pressure measurement channel of the system is carried out is shown in Fig. 1. The MKV-250 micromanometer with a limit of absolute pressure measurement error of 0.02 mm of water column (0.2 Pa) was chosen as the reference. Calibration is performed by applying the reference pressure, which is formed relative to atmospheric pressure, to the MKV-250 micromanometer and simultaneously to the pressure sensor through the direct input of the sensor (input P+) using a single pneumatic line.

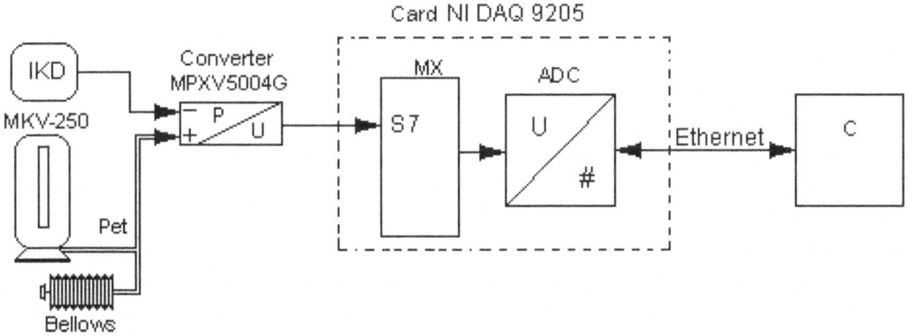

Fig. 1. The structural diagram of the calibrating channel of pressure measurement.

At the same time, atmospheric pressure is applied to the inverse input of the pressure sensor P-. The reference pressure, formed with the help of a bellows, has a positive sign, that is, it exceeds the atmospheric pressure. This is because during the experiment, the working air pressure at the inlet, supplied to the inverse inlet of the P- pressure sensor, is expected to be negative about atmospheric pressure. Thus, the effects of air pressure on the sensor's diaphragm during research and in the calibration mode coincide in direction [15, 16].

Manometer readings are recorded and entered into the computer. The output signals of the pressure sensor, which are proportional to the pressure difference at the direct and inverse inputs of the sensor, are measured by the IBS and entered into the computer. Calibration is performed by cyclically increasing and decreasing the reference pressure from zero to 120 mm H_2O (1.2 kPa).

Measurements of the output signals at each value of the reference pressure are performed multiple times (at least 5000) and subjected to statistical processing to reduce the random error component caused by the noise of the output signal of the pressure sensors and the quantization noise of the ADC. An experimental study of the nature of the noise component showed that the probability density of the distribution of voltage levels of the output signal of pressure sensors approaches the normal distribution law. Thus, if the maximum noise level does not exceed 5 mV, then its root mean square value $\sigma \approx 1.6$ mV will be reduced by more than 70 times as a result of averaging 5000 readings and will be $\sigma_{av} \approx 0.02$ mV, which is equivalent to a pressure of 0.002 mm H_2O. The level of such an error is much smaller than the error due to the temperature instability of the pressure sensor, the nonlinearity of the conversion, and the hysteresis of the characteristic, which is within 2.5% of the maximum value of the converted pressure and is several mm H_2O. The level of such an error is also less than the error of pressure measurement by the MKV-250 micromanometer by more than 10 times. Therefore, the influence of the random component of the error due to the pressure sensors' noise on the measuring channels' calibration results can be neglected.

As a result of the calibration, the transformation function of the measuring channel is created, which is then approximated by a polynomial according to the formula [17]:

$$P_j = \sum_{i=0}^{n} a_i \cdot pU_{sj}^{i}, \tag{1}$$

where a_i are the coefficients of the polynomial; U_{sj}^{i} are the output voltages of the pressure sensor; n is the degree of the polynomial.

The use of the function of the transformation of measurement channels (1) takes into account almost all components of systematic error, such as additive, multiplicative, and error of nonlinearity of the characteristics of the transformation of measurement channels [16, 17]. When choosing the order of the polynomial n, it should be taken into account that the additive and multiplicative components of sensor conversion errors are already compensated by the first-degree polynomial [20–22]. Reducing the nonlinearity error requires the use of higher-order polynomials. However, if a pressure measurement error correction is used outside the calibration range (i.e., greater than 1.2 kPa), then as a result of such a correction, the pressure conversion error can increase significantly. In this regard, the approximation of the first degree ($n = 1$) was chosen.

The technique for calibration of measuring channels provides [15, 16]:

- control of the calibration process in manual or semi-automatic modes, when the operator receives a prompt to set the necessary pressure level at the input of the pressure sensor using the MKV-250 micromanometer;
- control of the calibration process by the instantaneous values of voltages and pressures in each channel, which are presented in the form of tables and a multi-channel graph on the user interface;
- registration (recording in a file) of reference pressure values at sensor inputs;
- multiple measurements of reference pressures on each measuring channel and registration (recording in a file) of measurement information;
- primary statistical processing of measurement information;
- polynomial approximation of transmission characteristics of pressure sensors;
- estimation of approximation errors for each channel: the root mean square value of the pressure measurement error is calculated using the approximated transformation characteristic and reference pressure values;
- registration of the coefficients of polynomials of the approximating characteristic, which will be used during the next stage in model research modes.

The registered data (calibration results) are used to form the calibration protocol, which contains the polynomial coefficient and accuracy characteristics [23–25].

The experimental results of the calibration have showed sufficient accuracy of the pressure measurement, which is at the level of $\sigma < 0.035$ mm H_2O (0.35 Pa) for the root mean square value (rms) of the approximation error. The maximum value of the pressure measurement error does not exceed $\Delta = 0.067$ mm H_2O (0.7 Pa) Thus, with a range of 120 mm H_2O, the value of the relative combined error of pressure measurement will not exceed the value of $\gamma < 0.06\%$.

4 Estimation of High-Speed Velocity Measurement Errors

The velocity head in the operating part of the wind tunnel can be determined by the expression [15]:

$$q = \zeta_{apr}\left(P_f - P_r\right), \tag{2}$$

where $\Delta P_q = P_f - P_r$ is the difference between the pressure at the front point and the static pressure receiving points of the air pressure receiver; ζ_{apr} is the pressure calibration factor of the air pressure receiver.

Since the calibration of the equipment was performed for the measuring channel, the results of the measurement of pressure differences can be considered the results of direct measurements. The systematic error of determining the pressure difference as a result of direct measurement does not exceed the calibration error [26–28].

During the experiment, the speed of the airflow was between 5 and 28 m/s. At the same time, the pressure sensor performed pressure measurements in the range from 1.5 to 50 mm of water column (15–500 Pa). The maximum value of the absolute error of determining the pressure drop on the air pressure receiver should be considered $\theta_{\Delta P_q} = 0.7$ Pa.

Let's represent expression (2) in the form [15]:

$$q = \zeta_{ps} \Delta \mathrm{P}_q. \tag{3}$$

Velocity head according to expressions (2) and (3) is the result of indirect measurement. To determine its error, we prologarithmize and differentiate the expression and proceed to finite differences [16]:

$$\gamma_q = \frac{\theta_q}{q} = \frac{\theta_{\zeta_{\mathrm{ps}}}}{\zeta_{ps}} + \frac{\theta_{\Delta \mathrm{P}_q}}{\Delta \mathrm{P}_q}, \tag{4}$$

where γ_q is a relative error of the velocity head; θ_i is the absolute error of the corresponding argument.

Blowing was carried out under pressure in the prechamber $\Delta \mathrm{Pq} = 15 \div 500$ Pa, while the pressure calibration coefficient was $\zeta_{ps} = 1.07 \div 1.058$. Therefore, the error of the pressure sensor coefficient ζ_{ps} used in the experiments does not exceed 0.002. Then the value of the combined relative error according to (4) γ_q and absolute error θ_q will acquire the following values:

$$\gamma_q = \frac{0.002}{1.07} + \frac{0.7}{15} \approx 0.049; \theta_q = \gamma_q q = 0.049 \cdot p15 = 0.73 \text{ [Pa]}, \tag{5}$$

$$\gamma_q = \frac{0.002}{1.058} + \frac{0.7}{500} \approx 0.0034; \theta_q = \gamma_q q = 0.0034 \cdot p500 = 1.7 \ [Pa]. \tag{6}$$

Thus, the relative cumulative error of high-speed pressure measurement will not exceed $\gamma < 4.9\%$ at a speed of 5 m/s, $\gamma < 1.6\%$ at a speed of 9 m/s (ΔPq =50 Pa), $\gamma < 0.57\%$ at a speed of 18 m/s (ΔPq =190 Pa), and will decrease to $\gamma < 0.34\%$ with an increase in speed up to 29 m/s.

5 Estimation of Systematic Errors of Airflow Speed Measurement

The speed of the airflow based on the known value of the dynamic pressure is determined according to the expression [16]:

$$V = \sqrt{2q/\rho}, \tag{7}$$

where q is the dynamic pressure (velocity pressure); ρ is the air density.

In the wind tunnel, we do not measure the dynamic pressure. We measure the pressure difference between the prechamber and the atmosphere, which is related to the dynamic pressure by the ratio [16]:

$$q = \xi_{ps}\left(P_{atm} - P_{prc}\right). \tag{8}$$

Since the wind tunnel is a direct-acting air pipe, the air expands during the movement of air through the pre-chamber and the nozzle. We will assume that the expansion occurs according to the adiabatic law. Then the ratio between the air density in the prechamber and in the atmosphere used in expression (8) can be written:

$$\frac{P_{prc}}{P_{atm}} = \frac{\rho_{prc}^{k}}{\rho_{atm}^{k}}, \tag{9}$$

where ρ_{prc}, ρ_{atm} are air densities in the prechamber and in the atmosphere, respectively; $k = c_v/c_p = 1.4$ is the adiabatic exponent for air under near-normal conditions.

The sum of static pressure, velocity pressure, and hydraulic losses in the section from the input device to the working part is equal to atmospheric pressure:

$$P_{prc} = P_{atm} - q/\mu_q. \tag{10}$$

Substituting expression (10) into (9) and solving it relatively ρ_{prc}, we will obtain:

$$\rho_{prc} = \rho_{atm}\left(1 - \frac{P_{atm} - P_{prc}}{P_{atm}}\right)^{1/1.4}. \tag{11}$$

Finally, the expression (7) using expressions (11) can be rewritten in the form:

$$V = \sqrt{\frac{2\mu_q\left(P_{prc} - P_{atm}\right)}{\rho_{atm}\left(1 - \frac{\left(P_{prc} - P_{atm}\right)}{P_{atm}}\right)^{1/\kappa}}}. \tag{12}$$

According to expression (12), the velocity of the airflow is the result of an indirect nonlinear measurement. The air density in the prechamber is the result of an indirect non-linear measurement. Atmospheric air density, taking into account humidity, is calculated according to the expressions given in [18]. Since these expressions are quite complicated from the point of view of analytical analysis of errors, we will calculate the influence coefficients of the density measurement result on the measurement errors of the arguments according to the methodology given in [19]:

$$\begin{aligned} \frac{\partial\rho}{\partial P_{atm}} &= +0.0016\left[\left(\text{kg/m}^3\right)/(1\text{mm Hg})\right]; \\ \frac{\partial\rho}{\partial t_{atm}} &= -0.0045\left[\left(\text{kg/m}^3\right)/(1^\circ\text{C})\right]; \\ \frac{\partial\rho}{\partial v_{atm}} &= -0.000075\left[\left(\text{kg/m}^3\right)/(1\%)\right]. \end{aligned} \tag{13}$$

Expressions (13) are obtained by calculating the air density for weather conditions according to the algorithm used in the measuring system of the wind tunnel. Considering that the specified errors are independent, the total error of the air density measurement is determined according to the expression given below. Since the coefficient of influence of air humidity on the density is small, in the calculation we will use the largest error value of the psychrometer (7%) [15]:

$$\Theta_{\rho_{atm}} = \sqrt{\left(\Theta_{atm}\frac{\partial\rho}{\partial P_{atm}}\right)^2 + \left(\Theta_{t_{atm}}\frac{\partial\rho}{\partial t_{atm}}\right)^2 + \left(\Theta_{v_{atm}}\frac{\partial\rho}{\partial v_{atm}}\right)^2} =$$
$$= \sqrt{(1 \cdot p0.0016\ )^2 + (0.2 \cdot p0.0045)^2 + (7 \cdot p0.000075)^2} = 0.0018\ \left[kg/\mathrm{m}^3\right]. \quad (14)$$

Denote:

$$Q = 1 - \frac{P_{atm} - P_{prc}}{P_{atm}}. \quad (15)$$

Then the expression (12), taking into consideration the relation (15) becomes:

$$V = \sqrt{\frac{2\mu_q P_{atm} - P_{prc}}{\rho_{atm} Q^{1/\kappa}}}. \quad (16)$$

Let us perform an error analysis for expression (15). The systematic error Θ_Q is completely defined by the expression:

$$\Theta_Q = \Theta\left(\frac{(P_{atm} - P_{prc})}{P_{atm}}\right). \quad (17)$$

After differentiating the expression (17), we will obtain:

$$\delta_Q = \frac{\Theta_Q}{Q} = \frac{\Theta P_{atm-P_{prc}}}{(P_{atm} - P_{prc})} + \frac{\Theta P_{atm.}}{P_{atm}}. \quad (18)$$

The value (18) is dependent on the value of the pressure drop in the prechamber and the atmospheric pressure. To simplify the analysis, we will assume that the change in atmospheric pressure is not significant and is equal to the standard value P_{atm} of 760 mm. mercury (or 101,360 Pa). Taking into account that $\Theta_{P_{atm}-P_{prc}} = 0.7$ Pa, we will obtain:

$$\frac{\Theta_Q}{Q} = \frac{\Theta P_{atm-P_{prc}}}{P_{atm} - P_{prc}} + \frac{1}{P_{atm}}. \quad (19)$$

Let us write down the expression for the measurement error of the airflow speed in the working part of the wind tunnel. To do this, we differentiate the expression (16) taking into account the relation (19):

$$\frac{\Theta_V}{V} = \frac{1}{2}\left[\left(0.01 + 2\frac{\Theta P_{atm-P_{prc}}}{P_{atm} - P_{prc}}\right) + \frac{\Theta P_{atm-P_{prc}}}{P_{atm} - P_{prc}} + \frac{\Theta_{\rho_{\mathrm{atm}}}}{\rho_{atm}} + \frac{1}{1.4}\left(\frac{\Theta P_{atm} - P_{prc}}{P_{atm} - P_{prc}} + \frac{\Theta P_{atm}}{P_{atm}}\right)\right]. \quad (20)$$

Expression (20) depends on three factors such as the pressure difference between the atmosphere and the prechamber, the density of the atmospheric air, and the atmospheric pressure. We will assume that the main error is determined by the change in the pressure difference between the atmosphere and the prechamber, and consider the rest as additional errors. Then, summing similar terms and opening the parentheses, we obtain for standard atmospheric conditions:

$$\left(\frac{\Theta_V}{V}\right)_{\text{bas}} = \delta_{V_{\text{bas}}} = 0.0062 + \frac{1.768}{P_{atm} - P_{prc}}, \tag{21}$$

and the corrections due to the deviation of atmospheric conditions from the standard ones are determined by the expressions:

$$\Delta\delta_V(\rho) = \frac{0.0009}{\rho_{atm}} - 0.00073, \tag{22}$$

$$\Delta\delta_V(P_{atm}) = \frac{1}{P_{atm}} - 0.00094. \tag{23}$$

Graphical dependencies of the main and additional systematic errors of speed measurement based on expressions (14), (16), (19), (21)–(23) are shown in Figs. 2 and 3.

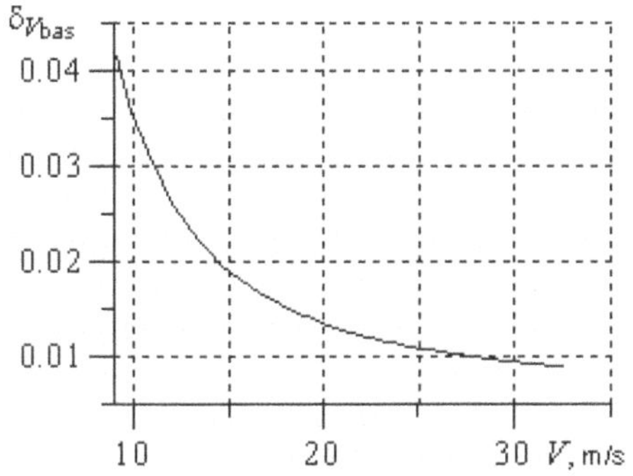

Fig. 2. Basic speed measurement error.

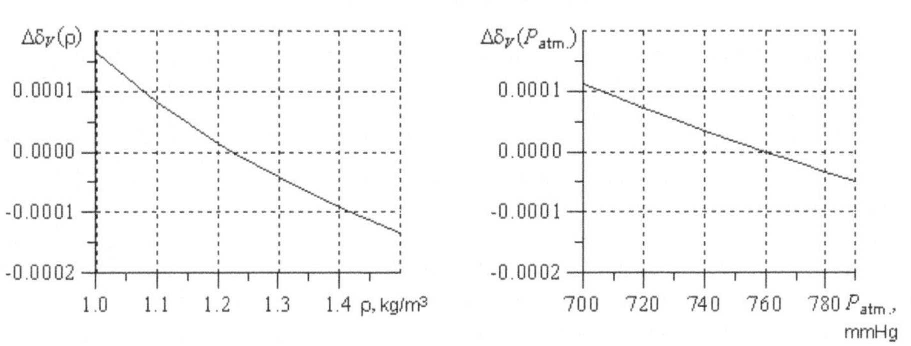

Fig. 3. Additional speed measurement errors due to changes in density and barometric pressure.

As can be seen from the comparison of graphs represented in Figs. 2 and 3, the main speed measurement error is at least 50 times greater than the maximum values of the additional ones.

Therefore, these corrections can be neglected and the systematic error of speed measurement can be calculated using expression (21). For example, for airflow speed $V = 5\ m/s$ ($P_{atm} - P_{prc} = 15$ Pa), the systematic error of speed measurement will not exceed $\delta V \leq 11\%$, and for $V = 28 m/s$ ($P_{atm} - P_{prc} = 450$ Pa) – $\delta V \leq 1\ \%$.

In expressions (21)–(23), the pressure difference between the atmosphere and the prechamber is measured in Pa, the air density ρ_{atm} in kg/m^3, and the atmospheric pressure P_{atm} in mm of mercury. The full systematic relative error of the airflow velocity measurement in the wind tunnel will be written in the form:

$$\delta_V = \delta_{V_{bas}} + \Delta\delta_V(\rho) + \Delta\delta_V(P_{atm}). \qquad (24)$$

Corrections from additional errors are added taking into account their signs. The systematic absolute error is determined in the usual way:

$$\Theta_V = \delta_V V[\mathrm{m/s}]. \qquad (25)$$

Equations (24) and (25) allow to estimate relative errors of the measurement system.

6 Estimation of Errors in Determining Coefficients of Aerodynamic Forces and Moments

Errors in the results of determining the coefficients of aerodynamic forces and moments are divided into systematic and random [29–31]. It is believed that the systematic components of the errors such as calibration, environmental, and inherent errors of the transducers are completely determined by the instrumental errors of the measuring equipment [32]. The random components are due to the scattering of experimental data due to the instability of the experimental environment. To reduce the instrumental components of the error in determining the coefficients of aerodynamic forces and moments, calibration of aerodynamic weights is performed at the beginning of a series of experimental studies.

The weighing experiment system provides the measurement of the loads on the research object in terms of 3 components Y_1, Y_2, and X_a with the help of strain transducers installed on wind tunnel scales. Signals of strain transducers are amplified by SKAIME-SMJ strain amplifiers in the range of ±5 V. Such an approach ensures the efficient operation of the system. The range of measured values of weight loads on component Y_1 achieves 20 N, on components Y_2, X_a – up to 10 N.

Calibration of 3-component scales was performed for the entire measuring channel as a whole. With the help of a special loading device, known loads with a margin of error of 0.01% were set on individual components of the scales, which were created by sample loads. Blowout models were installed on a scale stand with transverse and longitudinal bases of 350 × 450 mm with blowouts in the range of angles of attack −5 ÷ 32 degrees. Accordingly, the calibration of the load balancers was performed for the specified bases.

Calibration was performed according to the technique [20], as a multifactorial experiment using an incomplete factorial scheme with randomization of factor values [21].

The transformation functions for each of the three measurement channels are represented by a linear regression model:

$$F_i = k_i U_i + \sum k_{ji} U_j, \tag{26}$$

where F_i is the load on the ith component of the scales; k_i is the conversion factor for the ith component; U_i is the output signal on the ith component; k_{ji} is the coefficient of influence of the jth component, which affects the change in the output signal of the ith component when the load on the jth component changes; U_j is the output signal on the jth component.

The errors of reproduction of the reference loads by the measuring channels of the corresponding weight components are the values calculated with the help of expression (26) are given in Table 1.

Table 1. Calibration errors of measuring channels of scales.

Component	Y_1	Y_2	X_a
Range of calibration, kg	2	1	1
Maximum relative error Δ, %	0.258	0.489	0.538
Maximum error, kg	0.005	0.005	0.005
Relative error, %	0.12	0.23	0.28
Error (rms), kg	0.0024	0.0023	0.0028

To estimate the instrumental error of measuring the coefficients of aerodynamic forces and moments, it is necessary to know the geometric parameters of the model, the system of decomposition of forces, and load values, to perform the calculation of the traditional aerodynamic components Y_b and X_b in a semi-connected coordinate system. The latter coincides with the weight or pipe coordinate system with zero flow bevels in the working part of the pipe and zero corrections to the angle of attack due to the influence of airflow boundaries.

The coefficients of aerodynamic forces and moments are calculated according to the following equations (to simplify the writing, we omit the notation of the dependence of aerodynamic coefficients and components of weight loads on the angle of attack α):

$$C_x(\alpha) = a_{x,1}C_X;\ C_y(\alpha) = a_{y,1}C_{Y1} + a_{y,2}C_{Y2};\ C_{mz1}(\alpha) = a_{mz1,1}C_{Y2}, \tag{27}$$

$$C_{mz0}(\alpha) = a_{mz0,1}C_X + a_{mz0,2}C_{Y1} + a_{mz0,3}C_{Y2};\ C_{mz0}(\alpha) = a_{mz0,1}C_X, \tag{28}$$

$$C_X = \frac{x_a - x_{ap}}{q};\ C_{Y_1} = \frac{kY_1 - Y_{1p}}{q};\ C_{Y_2} = \frac{kY_2 - Y_{2p}}{q}, \tag{29}$$

where C_X, C_{Y_1}, C_{Y_2} are the coefficients of aerodynamic loads on the weight components; q is the speed pressure; $a_{j,\,i}$ are coefficients have the following values:

$$a_{x,1} = a_{y,1} = a_{y,2} = \frac{1}{S}; \tag{30}$$

$$a_{mz1,1} = -\frac{lx \cos \alpha}{S_{mz}}; \tag{31}$$

$$a_{mz0,1} = -\frac{x_{t1} \sin \varphi_m}{SL_{mz}}; a_{mz0,2} = -\frac{x_{t1} \cos \varphi_m}{SL_{mz}}; a_{mz0,3} = -\frac{l_x \cos \alpha + x_{t1} \cos \varphi_m}{SL_{mz}}, \tag{32}$$

where $\varphi_m = k\alpha + \varphi^{\text{o}}$ is the position angle of the model in the experiment; $S = 0.06125$ m^2 is the characteristic plane of the model; $l_x = 0.35$ m is the suspension base along the X axis; $x_{t1} = 0.04375$ are coordinates of the conditional center of mass in the semi-coupled coordinate system; $L_{mz} = 0.125$ m the characteristic size, k is the coefficient of position of the model (k = 1 corresponds to the wind tunnels, $k = -1$ corresponds to the flight), $\varphi^o = 0$ is an angle of installation of the model, α is an angle of attack.

Expressions (27)–(29) are convenient for analyzing the characteristics of quantities calculated during multiple measurements with a given angle of model installation, as well as for analyzing measurement errors. Indeed, coefficients $a_{j,\,i}$ are compositions of constant parameters of the model, which are measured once and remain unchanged throughout the experiment, and the value of the angle of attack of the model installation α, which is measured once for each position of the model.

The experience of operating the wind tunnel showed that the random components of the instrumental errors of the scales and subsystems for measuring the high-speed head are much smaller than the calibration errors (which determine their full instrumental errors). Therefore, the limit values of the measurement errors of the coefficients of the aerodynamic loads on the weight components can be calculated based on the values of the instrumental errors of the measurement of the loads and the velocity head for the result of indirect measurement as follows.

Table 1 shows the values of the instrumental errors of the weight components, and formulae (5), and (6) allow you to calculate the errors of the high-speed pressure measurement subsystem, which are $\gamma < 4.9\%$ at a speed of 5 m/s, $\gamma < 1.6\%$ at a speed of 9 m/s (ΔPq = 50 Pa), $\gamma < 0.57\%$ at a speed of 18 m/s (ΔPq =190 Pa) and $\gamma < 0.34\%$ at a speed of 28 m/s (ΔPq = 450 Pa). Expanding expressions (27)–(32) and taking into account that k = ±1, we will obtain the coefficient of aerodynamic loads on the weight components:

$$C_F = \frac{\pm F - F_n}{q}. \tag{33}$$

If we assume that the instrumental measurement errors of each weight load are independent, then the value of the measurement error of the aerodynamic loads $\pm F - F_n$ will be:

$$\Delta_{\sum F} = \sqrt{\Delta_F^2 + \Delta_{F_n}^2}. \tag{34}$$

Given the same measurement range for each component in (34), we get the relative errors:

$$\gamma_{\sum F} = \sqrt{\gamma_F^2 + \gamma_{F_n}^2} = \gamma_F\sqrt{2}. \tag{35}$$

Let's use expression (33) and take partial derivatives:

$$\frac{\Delta_{C_F}}{C_F} = \frac{\Delta_F}{F} + \frac{\Delta_q}{q} = \gamma_{\sum F} + \gamma_q. \tag{36}$$

Let us determine the maximum systematic error for the coefficient C_x. The maximum relative error of measuring loads on the X_a component, based on expressions (35)–(36) the calibration results given in Table 1, will be:

$$\gamma_{\sum X_a} = \sqrt{\gamma_{X_a}^2 + \gamma_{X_p}^2} = \gamma_{X_a}\sqrt{2} = 0.75\%. \tag{37}$$

The maximum relative error of measuring the aerodynamic coefficient (37), taking into account the errors of measuring the high-speed head at air flow speeds of 9 m/s, 18 m/s, and 28 m/s will be:

$$\gamma_{\sum X_a} = \sqrt{\gamma_{X_a}^2 + \gamma_{X_{ap}}^2} = \gamma_{X_a}\sqrt{2} = 0.75\%, \tag{38}$$

$$\gamma_{C_{XV=18}} = \gamma_{\sum X_a} + \gamma_q = 0.75 + 0.57 = 1.32\%, \tag{39}$$

$$\gamma_{C_{XV=18}} = \gamma_{\sum X_a} + \gamma_q = 0.75 + 0.34 = 1.09\%. \tag{40}$$

The maximum absolute value of the instrumental error for the aerodynamic coefficient C_x, based on the maximum range of loads of 1 kg on the X_a component and the values of the velocity pressure $q = 50$ Pa ($V = 9$ m/s), $q = 190$ Pa ($V = 18$ m/s) and $q = 450$ Pa ($V = 27$ m/s), respectively, will be:

$$\Delta C_{X_{\max} V=9} = 0.0235\frac{1}{Sq} = 0.0235\frac{1}{0.061 \cdot p50} = 0.0077, \tag{41}$$

$$\Delta C_{X_{\max} V=18} = 0.0132\frac{1}{Sq} = 0.0132\frac{1}{0.061 \cdot p190} = 0.0011, \tag{42}$$

$$\Delta C_{X_{\max} V=28} = 0.0235\frac{1}{Sq} = 0.0235\frac{1}{0.061 \cdot p450} = 0.0004. \tag{43}$$

Similarly, according to (38)–(43) absolute and relative measurement errors of other aerodynamic coefficients are determined, which are represented in Table 2. The relative and absolute errors of the aerodynamic coefficients, calculated from the root mean square deviation of the measured value from the reference value of the loads during calibration of the scales, are also given there.

Table 2 shows that the maximum lateral errors are generally less than 2.5% of the calibration range and rms of the relative error is within 1%.

Table 2. Instrumental errors in measuring aerodynamic coefficients.

Aerodynamic coefficient	C_x	C_Y	C_{mz0}	C_{mzl}
For airflow speed m/s (q = 50 pa)				
Maximum relative error γ, %	2.35	2.37	2.12	1.96
Maximum absolute error Δ	0.0077	0.008	0.007	0.0064
Relative error (rms) $\Delta_{.}$, %	1.012	0.98	0.93	0.52
Absolute error (rms) γ	0.0026	0.0033	0.0029	0.0024
For airflow speed 18 m/s (q = 190 pa)				
Maximum relative error, %	1.32	1.34	1.09	0.93
Maximum absolute error Δ	0.0011	0.0012	0.0011	0.0008
Relative error (rms) $\Delta_{.}$, %	0.656	0.712	0.69	0.385
Absolute error (rms) $\gamma_{.}$	0.0004	0.0005	0.0005	0.0003
For airflow speed 28 m/s (q = 450 pa)				
Maximum relative error γ, %	1.09	1.11	0.856	0.74
Maximum absolute error Δ	0.0004	0.0004	0.0007	0.00025
Relative error (rms) $\Delta_{.}$, %	0.436	0.485	0.408	0.321
Absolute error (rms) γ				0.0001

The magnitude of the absolute error depends on the geometric dimensions of the blowing model. To reduce the absolute instrumental errors, it is necessary to increase the dimensions of the purge model to realize the aerodynamic loads corresponding to the full range of loads during calibration, or to reduce the range of loads during the calibration of strain gauges.

At the same time, relative errors can be reduced, and absolute errors will decrease in inverse proportion to the increase in characteristic geometric dimensions. In the considered situation, the aerodynamic loads on the model are several times smaller than the calibration range of the weights by components. Therefore, it can be confidently stated that the actual absolute errors in determining the aerodynamic coefficients are several times smaller than those given in Table 2.

7 Conclusions

The instrumentation and technique of the pressure measurement channel are represented. Expressions for estimation of speed head have been represented. Mathematic descriptions of systematic errors of airflow speed measurement errors based on such three factors as the pressure difference between the atmosphere and the prechamber, the density of the atmospheric air, and the atmospheric pressure are proposed.

The appropriate graphical dependences are represented. Estimation of errors in determining aerodynamic coefficients has been carried out. Calibration errors of measuring channels of scales are given. Instrumental errors in measuring aerodynamic coefficients have been calculated.

In future work, it is proposed to implement research relative to the full experimental test of UAVs aerodynamic characteristics in the wind tunnel, including features of the measurement procedure.

References

1. Larin, V., Solomentsev, O., Zaliskyi, M., Shcherban, A., Averyanova, Y., Ostroumov, I.: Prediction of the final discharge of the UAV battery based on fuzzy logic estimation of information and influencing parameters. In: 3rd KhPI Week on Advanced Technology (KhPI Week), pp. 44–49. IEEE, Kharkiv (2022). https://doi.org/10.1109/KhPIWeek57572.2022.9916490
2. Ostroumov, I., Kuzmenko, N., Bezkorovainyi, Y., Averyanova, Y., Larin, V.: Relative navigation for vehicle formation movement. In: 3rd KhPI Week on Advanced Technology (KhPI Week), pp. 10–13. IEEE, Kharkiv (2022). https://doi.org/10.1109/KhPIWeek57572.2022.9916414
3. Udartsev, E.P., Alekseenko, S.I., Zhdanov, O.O.: Unsteady aerodynamics of a vortex jet wing of an unmanned aerial vehicle at large and closed angles of attack. Electron. Contr. Syst. **4**(46), 40–45 (2015) http://nbuv.gov.ua/UJRN/etsu_2015_4_8
4. Lin, C.J.: Review of research on low-profile vortex generators to control boundary-layer separation. Prog. Aerosp. Sci. **38**, 389–420 (2002). https://doi.org/10.1016/S0376-0421(02)00010-6
5. Delinero, J.S., Leo, J.M., Camocardi, M.E.: Vortex generators effect on low Reynolds number airfoils in turbulent flow. Int. J. Aerodyn. **2**(1), 1–14 (2020). https://doi.org/10.1504/IJAD.2012.046539
6. Johari, H., Henoch, C., Custodio, D., Levshin, A.: Effects of leading-edge protuberances on airfoil performance. AIAA J. **45**(11), 2634–2642 (2007). https://doi.org/10.2514/1.28497
7. Custodio, D., Henoch, C.W., Johari, H.: Aerodynamic characteristics of finite span wings with leading-edge protuberances. AIAA J. **53**(7), 1878–1893 (2015). https://doi.org/10.2514/1.J053568
8. Malipeddi, A.K., Mahmoudnejad, N., Hoffmann, K.A.: Numerical analysis of effects of leading-edge protuberances on aircraft wing performance. J. Aircr. **49**(5), 1336–1344 (2012). https://doi.org/10.2514/1.C031670
9. Divekar, P.S., Ekbote, T.: Design of aerodynamic of an airplane wings. IJSDR. **4**, 59–64 (2019) https://www.ijsdr.org/viewpaperforall.php?paper=IJSDR1910011
10. Salem, K.A., Palaia, G., Vittorio, C., Vincenzo, B.: Tools and methodologies for box-wing aircraft conceptual aerodynamic design and aeromechanic analysis. Mech. Indus. **22**(39), 1–19 (2021). https://doi.org/10.1051/meca/2021037
11. Delnero, J.S., Maranon, J., Di Leo, J.M., Camocardi, M.E.: Vortex generators effect on low Reynolds number airfoils in turbulent flow. Int. J. Aerodyn. **2**(1), 1–14 (2012). https://doi.org/10.1504/IJAD.2012.046539
12. Chanetz, B., Delery, J., Gilliéron, P., Gnemmi, P., Gowree, E.R., Perrier, P.: Experimental Aerodynamics. Springer, Cham (2022)
13. Britcher, C., Landman, D.: Wind Tunnel Test Techniques. Academic Press, Cambridge (2024)
14. Barlow, J.B., Rat, W.H.: Low-Speed Wind Tunnel Testing. Wiley, London (1999)
15. Aftab, S.M., Razak, N.A., Mohd Rafie, A.S., Ahmad, K.A.: Mimicking the humpback whale: an aerodynamic perspective. Prog. Aerosp. Sci. **84**, 48–69 (2016). https://doi.org/10.1016/j.paerosci.2016.03.002

16. Zhdanov, O., Orlianskyi, V., Sushchenko, O.: Researching influence of vortex generators on aircraft aerodynamic characteristics. In: Ostroumov, I., Zaliskyi, M. (eds.) Proceedings of the 2nd International Workshop on Advances in Civil Aviation Systems Development. ACASD 2024, Lecture Notes in Networks and Systems, vol. 992, pp. 410–422. Springer, Cham (2024). https://doi.org/10.1007/978-3-031-60196-5_30
17. Sushchenko, O., Bezkorovainyi, Y.: Design of robust aircraft control system. In: Ostroumov, I., Zaliskyi, M. (eds.) Proceedings of the 1st International Workshop on Advances in Civil Aviation Systems Development. ACASD 2023, Lecture Notes in Networks and Systems, vol. 736, pp. 17–29. Springer, Cham (2023). https://doi.org/10.1007/978-3-031-38082-2_2
18. Sundararajan, D.: Digital Signal Processing: an Introduction. Springer, Cham (2024)
19. Tayor, J. R.: An Introduction to Error Analysis: the Study of Uncertainties in Physical Measurements. Mill Valley, CA (2022)
20. Buoaccorsi, J.P.: Measurement Error: Models, Methods and Applications. CRC Press, Boca Raton (2010)
21. Ivashchuk, O., et al.: A configuration analysis of Ukrainian flight routes network. In: 16th International Conference on the Experience of Designing and Application of CAD Systems, pp. 6–10. IEEE, Lviv (2021). https://doi.org/10.1109/CADSM52681.2021.9385263
22. Bezkorovainyi, Y.N., Sushchenko, O.A.: Improvement of UAV positioning by information of inertial sensors. In: 5th International Conference on Methods and Systems of Navigation and Motion Control (MSNMC), pp. 123–126. IEEE, Kyiv (2018). https://doi.org/10.1109/MSNMC.2018.8576307
23. Solomentsev, O., et al.: Efficiency analysis of current repair procedures for aviation radio equipment. In: Ostroumov, I., Zaliskyi, M. (eds.) Proceedings of the 2nd International Workshop on Advances in Civil Aviation Systems Development. ACASD 2024, Lecture Notes in Networks and Systems, vol. 992, pp. 281–295. Springer, Cham (2024). https://doi.org/10.1007/978-3-031-60196-5_21
24. Chikovani, V., Sushchenko, O., Tsiruk, H.: Redundant information processing techniques comparison for differential vibratory gyroscope. East. Eur. J. Enterp. Technol. **4**(7), 45–52 (2016). https://doi.org/10.15587/1729-4061.2016.75206
25. Voliansky, R., Sadovoi, A., Volianska, N.: Interval model of the piezoelectric drive. In: 14th International Conference on Advanced Trends in Radioelecrtronics, Telecommunications and Computer Engineering (TCSET), pp. 1–6. IEEE, Lviv-Slavske (2018). https://doi.org/10.1109/TCSET.2018.8336211
26. Voliansky, R., Krasnoshapka, N., Statsenko, O., Shramko, I., Sadovoi, O., Dwiyanto, F.A.: The interval perturbed motion of the generalized nonlinear dynamical plants. In: 4th International Conference on Modern Electrical and Energy System (MEES), pp. 1–6. IEEE, Kremenchuk (2022). https://doi.org/10.1109/MEES58014.2022.10005720
27. Voliansky, R., Sadovoi, O., Sergienko, O., Zhelinskyi, M., Statsenko, O., Volianska, N.: Interval modeling and simulation of duffing pendulum. In: 4th KhPI Week on Advanced Technology (KhPIWeek), pp. 1–6. IEEE, Kharkiv (2023). https://doi.org/10.1109/KhPIWeek61412.2023.10312997
28. Osadchy, S.I., Zozulya, V.A.: Optimal filtering of hexapod acceleration data obtained under action of electromagnetic interference. In: 3rd International Conference Methods and Systems of Navigation and Motion Control (MSNMC), pp. 21–23. IEEE, Kyiv (2014). https://doi.org/10.1109/MSNMC.2014.6979719
29. Osadchy, S.I., Zozulya, V.A., Rudiuk, G.I.: The dynamics of 3-dimensional micro-mechanic sensor of angle motions of a robot-hexapod. In: 8th IEEE International Conference on Intelligent Data Acquisition and Advanced Computing Systems: Technology and Applications (IDAACS), pp. 908–912. IEEE, Warsaw (2015). https://doi.org/10.1109/IDAACS.2015.7341435

30. Sushchenko, O.A., Bezkorovainyi, Y.N., Novytska, N.D.: Theoretical and experimental assessments of accuracy of nonorthogonal MEMS sensor arrays. East. Eur. J. **3**(9), 40–49 (2018). https://doi.org/10.15587/1729-4061.2018.131945
31. Fuller, W.A.: Measurement Error Models. Wiley, London (2006)
32. Holubnychyi, O., Zaliskyi, M., Ostroumov, I., Sushchenko, O., Solomentsev, O., Averyanova, Y.: Self-organization technique with a norm transformation based filtering for sustainable Infocommunications within CNS/ATM systems. In: Ostroumov, I., Zaliskyi, M. (eds.) Proceedings of the 2nd International Workshop on Advances in Civil Aviation Systems Development. ACASD 2024, Lecture Notes in Networks and Systems, vol. 992, pp. 262–278. Springer, Cham (2024). https://doi.org/10.1007/978-3-031-60196-5_20

Air Transportation

Advanced IoT Sensor Network for Continuous Cargo Condition Surveillance During the Flight

Dmytro Prosvirin[1](✉), Volodymyr Kharchenko[2], and Pavel Krents[1]

[1] Antonov Company, Akademika Tupoleva Str., 1, Kyiv 03062, Ukraine
dmytro.prosvirin@antonov-airlines.aero

[2] National Aviation University, Liubomyra Huzara Ave., 1, Kyiv 03058, Ukraine

Abstract. This paper dwells upon network for continuous cargo condition surveillance during the flight. Advanced IoT sensor network for continuous cargo condition surveillance during the flight addresses flight data and cargo surveillance by Airlines. The core objective of this is the automated processing of flight information dispatched from the aircraft in real-time (24/7/365) for the purpose to enhance airborne safety, and exchange of information between the aircraft crew for operational purposes and ground personnel to improve the performance and financial success of freight transportation of Airlines. For this purpose, an onboard interaction link of short burst data (SBD) from the aircraft and circuit switched data (CSD) a communication link to the plane is set up alongside the available use of satellite navigation system (GPS). As part of the Horizon Europe project, tools aimed at enhancing the security of an air cargo monitoring system will be developed. These tools will leverage onboard indicators, information traffic analysis, and anomaly detection to improve system security. AI-based tools created by consortium specialists will aid in detecting abnormal behaviour of aircraft units, anomalous network activity.

Keywords: IoT Sensor · Surveillance · Network

1 Introduction

In accordance with the requirements and recommendations of ICAO, EASA and the State Aviation Administration of Ukraine, airlines must have a fully functioning flight safety management system (the requirements for the system are set out in Annex 19 of the Chicago Convention and in the Flight Safety Management Manual Doc. 9859) [1], the full functioning of the SMS requires a system for identifying hazardous factors, which is based on a reactive, proactive and predictive method, as well as a risk assessment system. The predictive method is primarily based on the analysis of flight data through a direct aircraft monitoring system.

To ensure a high level of safety and profitability of aircraft flight operation, as well as compliance with the requirements of national and international aviation regulations, civil aviation airlines use modern information technologies. With their help, operational management of the airline is carried out through the flight control centre, based on

I. Ostroumov et al. (Eds.): ACASD 2025, LNNS 1418, pp. 23–32, 2025.
https://doi.org/10.1007/978-3-031-91992-3_2

risk management, planning, optimization and continuous improvement of technological processes.

A characteristic feature of modern aircraft is the ever-increasing complexity of aircraft systems and the growth of functions performed automatically by systems. At the same time, operational monitoring tools are being developed that receive, process and record large amounts of information. The function of recording flight information and its ground processing for the purpose of monitoring the condition of aircraft systems is called the Aircraft Health Monitoring System in the ARINC-624 document, Section 8, as part of the Onboard Maintenance System [2].

Operational control means ensure both the localization of the failure that has occurred and the registration of information characterizing the parameters of the functioning of the aircraft systems, the actions of the pilots, the spatial position of the aircraft, etc.

Information generated by operational control means is used both by the aircraft crew during the flight and by ground technical personnel for the prompt elimination of problems that have arisen.

The most important factor is the timeliness of the provision of information both for taking measures to promptly eliminate the failure and for the implementation of preventive measures, including the ability to promptly inform the crew about an abnormal situation and deviations from the specified (acceptable) conditions. For this, both the results of monitoring centralized control systems and the records of operational and emergency recorders are used. The obligation to post-flight process information is determined by the rules of airlines and national requirements.

In order to efficiently transmit information from centralized control systems, international practice currently uses standard communication facilities that automatically transmit information. This function is implemented on the An-148-100 aircraft via the BSTO (On board system of technical service).

Recently, On-Board Video Surveillance Systems (OBVS) have been introduced as additional means of operational control, which are an additional source of information not recorded by other means of operational control.

An effective means of transmitting information are satellite channels with minimal tariffs for data transmission. They transmit information from the aircraft's operational control facilities in real time to the airline's Flight Control Centre for analysis and prompt decision-making, if necessary. First of all, such facilities are used for aircraft performing long flights.

Advanced IoT sensor network for continuous cargo condition surveillance during the flight This paper dwells upon network for continuous cargo condition surveillance during the flight. Advanced IoT sensor network for continuous cargo condition surveillance during the flight addresses flight data and cargo surveillance by Airlines. The core objective of this is the automated processing of flight information dispatched from the aircraft in real-time (24/7/365) for the purpose to enhance airborne safety, and exchange of information between the aircraft crew for operational purposes and ground personnel to improve the performance and financial success of freight transportation of Airlines. For this purpose an onboard Interaction link of short burst data (SBD) from the aircraft and circuit switched data (CSD) a communication link to the plane is set up alongside the available use of satellite navigation system (GPS). The scheme of interaction between on-board equipment and ground station is given in Fig. 1.

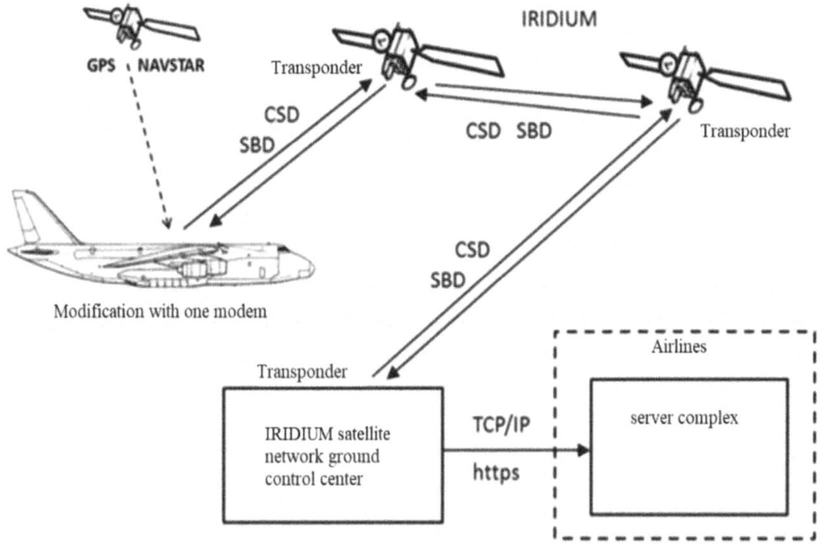

Fig. 1. The Scheme of Interaction between on-Board Equipment and Ground Station.

Advanced IoT sensor network is a tool for:

- facilitate the efficient handling and evaluation of flight data, including parametric, voice, and video recordings from the onboard flight data recorder, for routine fleet operations and the analysis of aviation accidents.
- accomplishment of the 4D surveillance function.
- automating the management of the aircraft's "Electronic Form" to control the lifespan of the airframe and engines, and analyse its technical condition, production quality, and repair records.

The general requirements for the ground-based software and hardware complex are as follows:

- availability of a mechanism for receiving in real-time mode on a permanent basis via satellite communication channels the required volume of specified types of information from the airline aircraft;
- availability of a mechanism for receiving the full volume of flight information from the aircraft after the flight, including when it is outside the Airlines base, via Internet channels;
- availability of a mechanism for the exchange of information on requests between aircraft and the Airlines via satellite communication channels in sufficient volume provided by the IRIDIUM (global satellite communication) system (at the first stage 9.6 kbit/sec and at the second stage 100 kbit/sec) and Internet channels;

- availability of a mechanism for unambiguous interpretation of the data collected by the onboard system and registration transmitted to the Airlines for storage, processing, distribution and analysis;
- ensuring the possibility of storing the entire volume of all types of flight data (parametric, voice and video) received from the Airline's fleet equipped with onboard flight information and recording equipment and working with it in real-time for at least 3 months;
- ensuring the possibility of archiving the entire volume of flight data (parametric, voice and video) received from the aircraft;
- provision of the required information security level for storage of the Airlines technical flight data, access to it and work with it;
- availability of automated standard services for work with flight information in accordance with the needs of existing user groups;
- availability of settings of standard automated services of work with flight information, which will provide the necessary completeness of information services to users for the performance of their functional duties and solution of production tasks;
- the structure of information provision to the user and user interfaces for access to existing flight information resources and services should be intuitive;
- the possibility of subsequent expansion of functional capabilities without structural changes in the software and hardware components of the complex.

The main principles of ground-based software and hardware complex creation are:

- use of generally accepted and widely used standards of information structuring and service provision;
- unification of formats and protocols of information exchange;
- compatibility with the existing Airlines hardware and software complex;
- ensuring operation on the basis of the local area network of the Airlines high degree of scalability of software and hardware;
- use of effective methods of protection against unauthorized access to information resources and compliance with the information security policy of the enterprise.

2 Use Case Architecture (Cargo Monitoring)

There several cyber-risks should be taken into account during the design system architecture.

Hidden attack: in this case a random IP addresses are sent by the attacker in large quantities which causes a ruckus in the data operation centre. It becomes difficult to differentiate between the legitimate and illegitimate IPs, since the data operation centre is not capable of doing that.

Mirroring attack: it is a case where the victim's IP address is shared with random unknown sources. It is a situation of serious concern, as the victim's information is share illegally.

Imitation attack: here, a legitimate user's identity is illegally used by the attacker by using his/her IP address and portraying themselves in their place. If there, is no proper detection, this attack can cause grave damage both to the victim and the company.

Physical risks: IoT involves hardware systems, thus, in case of any malfunction like power glitch, over-heating, etc., the object might harm the user which cannot be controlled by IoT.

Software/Network security risks: there are in proper software upgradations of IoT applications, which put it at a great software security risk. Illegal/illegitimate modifications or corruption of software can put various clients at risk, and especially those dependent upon a particular API (Application Programming Interface). Furthermore, the IoT devices use internet for their connectivity to other devices, this can cause bandwidth corruption leading to communication loss, losses in data, technical errors and other failures [3].

Cargo monitoring system are designed to track and register a certain set of parameters. Architecture diagram of the system is given in Fig. 2. The system consists of the following main software and hardware elements:

- a computer with a recording device and a parameter indicator;
- three digital pressure, humidity and temperature sensors;
- digital three-axis overload sensor;
- a special program for monitoring parameters, which is installed on a laptop.

Case 1 Architecture Diagram includes AIS (Aircraft information system), electronic flight bag (EFB), IoT sensors installed on the cargo (S11, SA12, S13), communication unit, data transfer standard for aircraft avionics (ARINC429), and aircraft condition monitoring system (ACMS) [4].

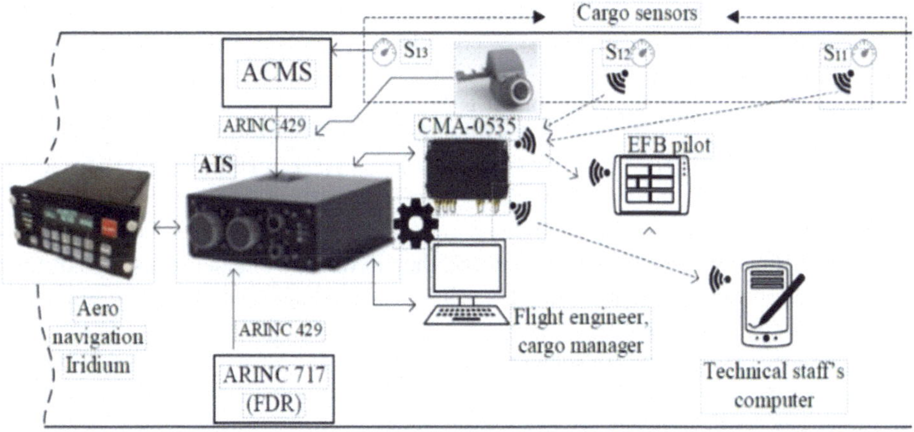

Fig. 2. Use case 1 architecture diagram (Wireless approach with using IoT sensors).

AIS providing registration of flight information from Flight data recorder (FDR) express processing and automatic output of information via the INMARSAT/IRRIDIUM channel during the flight and through block to ground personnel after the flight.

Communication unit CMA-0535 provides wireless communication on the aircraft for crew members and wireless output of information after the flight to the technical staff's computer, as well as in 4G/LTE format to the airport server (if available) for further transmission via the Internet to the airline.

Sensors S11, SA12, S13 are providing information about temperature, pressure, humidity in cargo compartment to AIS for further delivering info to the ground control centre, to the customer and to the cargo monitoring specialist onboard. The last one is checking whether the cargo is transporting in proper condition, as per customer requirements.

The components appearing in this use case are described in Table 1.

Table 1. Components of the use case aviation.

Component Name	Description
Cargo sensor	Digital pressure, humidity and temperature sensor
IP5	A computer with a recording device and a parameter indicator
Control panel	For reset or/and turn off command
Laptop	For visual monitoring parameters

3 Storyline Flight Certification and Safety Testing of Flight Systems

Under current regulations, safety must be considered not only as part of the aircraft type certification in accordance with RTCA DO-326 and EUROCAE ED-202, but also as part of ongoing airworthiness certification. It is proposed that all IoT (Internet of Things) devices on board an aircraft should be certified for security, generally in accordance with RTCA DO-178 and EUROCAE ED-12 for software and RTCA DO-274 and EUROCAE ED-80 for hardware.

In order for a new aircraft to be allowed to fly and transport cargo, especially passengers (so that the company can make a profit), the aircraft must undergo a series of tests and certification of all flight systems. Therefore, certification is indirectly an important factor potentially affecting airline profits.

The security assessment for integrated and complex aerospace systems will be provided via a risk assessment methodology that will use integrated modelling tools to determine a systemic cybersecurity risk assessment.

To meet the requirements of the European Union Aviation Safety Agency (EASA) and Federal Aviation Administration (FAA), the development of Avionics software will be done in Safety Critical Application Development Environment (SCADE) [5] therefore checking for compliance with DO-178C to be done optional [6], the third year of the AN-124 retrofit program after implementation of flight monitoring system—two different satellite communication channels—INMARSAT (standard, based on the HD-710 unit) and IRIDIUM) use of effective methods of protection against unauthorized access to information resources and compliance with the enterprise's information security policy.

Hazard evaluation, along with identification of irregularities and improper use detection within elements (sensors) and network, will be carried out to identify hazardous conditions at both the component and system levels as part of the newly financed European grand. This process will uncover detecting weaknesses in permission management to the network elements and surveillance the generation of short-term or long-term users. It will also track typical and atypical behaviours, test incident responses at both hardware and software levels (revoking user rights, securing logs, quarantining clusters, notifying responsible personnel, setting timeframes, issuing recommendations for rule or condition violations during incident resolution, and logging actions during incident management). Noncompliance in the Aviation Application scenarios are given in Table 2.

Table 2. Noncompliance in the Aviation Application scenario.

Data Protection Violation	Description
Supply chain attack	Unmaintained, unpatched (or un-patchable) firmware could be vulnerable due to exploits
Irregularities in functioning	Detection of irregularities in functioning In the system and allocation of threat levels to each sensorsю
Workflow disruption	The connection of the sensors to the recording unit is interrupted
Unauthorized endpoint	The security system has been manipulated, so that it sends out strong signals, which can be picked up and make the plane with the precious cargo easy to locate (and could be physically attacked)
Immediate alarm mechanism	The cargo endures high or low temperatures for an unreasonably long time, which could be mitigated during the flight

There will be different steps of implementation during the project run-time. The phases of the implementations are:

1. Cabled method without satellite communication;
2. Cabled method with satellite communication;
3. WLAN method with satellite communication;
4. WLAN method with satellite communication.

Security risks Aligned with the ongoing deployment

- Supply chain attack;
- Irregularities in functioning;
- Workflow disruption;
- Unauthorized endpoint;
- Immediate alarm mechanism.

4 TELEMETRY Approach

As part of the Horizon Europe project, tools aimed at enhancing the security of an air cargo monitoring system will be developed. These tools will leverage onboard indicators, information traffic analysis, and anomaly detection to improve system security. AI-based tools created by consortium specialists will aid in detecting abnormal behavior of aircraft units, anomalous network activity. This process will enable monitoring of typical and atypical system behavior and ensure an effective response to incidents involving vulnerabilities at both hardware and software levels (e.g., revoking user rights, ensuring secure logging, quarantining clusters, notifying responsible personnel, setting timeframes, issuing recommendations for resolving incident violations, and logging actions taken during incident resolution) [7].

Specific tools will be tested to address the defined aviation storylines. These tools can be used individually or in combination with others to address particular storylines.

The tools will be installed on a separate laptop onboard the aircraft, connected to the airplane's sensor network. This laptop will host all necessary tools for both wired and wireless approaches and will also record raw data for evaluation purposes. This setup enables the evaluation of tool events in relation to real-world events.

We consider several evaluation scenarios: Rogue Device, Device operation anomaly, and Authorization control threat.

Rogue Device. The security system has been manipulated, so that it sends out strong signals, which can be picked up and make the airplane with the precious cargo easy to locate (and could be physically attacked).

Device Operation Anomaly. Unusual network activity is observed, and a risk profile is allocated to each sensors. If a Minimum/Maximum value is surpassed, an alert will be sent to the user.

Authorization Control Threat. Reviewing the cybersecurity defenses of IT systems will facilitate the detection of weaknesses in user access regulation for the system elements and regulate the creation of short term or long-term users with varied access permissions and sets of permissions, and also allow to monitor their typical or unusual activities. Evaluating the effectiveness of incident management process at the HW and SW level will be done (revoking user permissions, securing log files, isolating clusters for containment, alerting relevant personnel, configuring time schedules, offering recommendations for addressing rule violations in incident handling, tracking actions performed during incident handling, etc.).

Test of the system has been evaluated based on different scenarios:

- Scenario 1: Rogue device. Covertly placed devices within the company network can independently monitor network activity by exploiting known vulnerabilities. Anomaly detection in network traffic will be performed, and a risk profile will be assigned to the sensor. In case the threshold value is exceeded, the user will be notified.
- Scenario 2: Device operation anomaly. The system detects abnormal operational changes using anomaly detection, informs the user of deviations in network traffic and classifies each device with a risk profile. If a threshold value is exceeded, the operator is promptly alerted.
- Scenario 3: Access control risk. Multiple entry points to data provide users with tools to obtain permissions from various policies, causing an increase in effective permissions, which collectively increasing the associated threat. A comprehensive

tool for testing permission control systems will empower the network administrator to enhance decision-making raise the quality and minimize incident response times.

Evaluation Strategy for Scenario 1 is given in Table 3.

Table 3. Evaluation strategy scenario 1.

Instrument 1—Misuse Detection ML Tool	Instrument 2—Distributed Ledger Technology based Data Sharing	Instrument 3—Platform
1. Detect anomaly	2. Record the detected activity	3. Display message for operator

5 Conclusions

A defining characteristic of modern aircraft is the increasing complexity of aircraft systems and the growing number of functions automatically managed by these systems. Simultaneously, operational control tools are being developed to receive, process, and record vast amounts of data. The function of recording flight information and processing it on the ground to monitor the condition of aircraft systems is outlined in Section 8 of the ARINC-624 document. This function is referred to as the "Aircraft Surveillance System" within the Onboard Maintenance System.

As a part of the recently funded Horizon Europe project, TELEMETRY advanced IoT sensor network for continuous cargo condition surveillance during the flight will be designed. Once this system passed all checks including cybersecurity testing it will be wildly used on the aircraft for cargo monitoring during the flight and on the ground especially for transportation sensitive cargo such as satellite.

References

1. Safety Management Manual (SMM). Doc 9859, ICAO (2013)
2. Standard Design Guidance for Onboard Maintenance System. ARINC 624 (1993)
3. Taylor, S., Jaatun, M., Mc Gibney, A., Seidl, R., Hrynchenko, P., Prosvirin, D., Mancilla, R.: A Framework Addressing Challenges in Cybersecurity Testing of IoT Ecosystems and Components. In: Proceedings of the 9th International Conference on Internet of Things, Big Data and Security, pp. 226–234 (2024). https://doi.org/10.5220/0012676300003705
4. Digital Information Transfer System (DITS), Part 1, Functional Description, Electrical Interfaces, Label Assignments and Word Formats. ARINC429P1-19, 429P1-19 (2019).
5. Prosvirin, D., Kharchenko, V.: Hierarchical-correlation method for designing of an adaptive neural flight control system in compliance with European and US standards. In: Ostroumov, I., Zaliskyi, M. (eds.) Proceedings of the International Workshop on Advances in Civil Aviation Systems Development. Lecture Notes in Networks and Systems, vol. 992, pp. 86–97. Springer, Cham (2024). https://doi.org/10.1007/978-3-031-60196-5

6. Software Considerations in Airborne Systems and Equipment Certification: DO-178C. RTCA, Inc., Washington, DC (2011)
7. Prosvirin, D., Kharchenko, V.: Frameworks for testing and monitoring in IoT systems. CEUR Workshop Proc. **3895**, 160–169 (2024)

Simulation Modelling for Supply Chain Optimization with a Logistics Hub Based at an Airport Complex

Kostiantyn Cherednichenko[1](✉), Viktoriia Ivannikova[2], and Olena Sokolova[1]

[1] National Aviation University, Liubomyra Huzara Ave., 1, Kyiv 03058, Ukraine
cherednichenko.kostya@gmail.com
[2] Dublin City University, Glasnevin, Dublin 9, Ireland

Abstract. Efficient supply chain management is a key factor in the competitiveness of modern transportation companies, especially those operating airport-based logistics hubs. This study applies simulation modelling to analyze and optimize cargo transportation processes within a supply chain network, integrating a central hub at Warsaw Chopin Airport and regional distribution centers in Poland. The optimization focused on adjusting the number of transport vehicles, improving route planning, and maximizing fleet utilization. As a result, the number of vehicles was reduced from 28 to 15, while the average load factor increased from 0.1 to 0.79, significantly improving transport resource efficiency. Additionally, simulation identified critical inefficiencies in vehicle utilization and optimizing transport routes to minimize empty trips and waiting times. Implementation of optimization strategies led to a reduction in operational costs and enhanced the responsiveness of the logistics network to fluctuating demand. The results demonstrate that dynamic routing, adaptive fleet allocation, and intelligent cargo distribution significantly enhance the performance, cost-effectiveness, and reliability of an airport-integrated logistics system. The study highlights the benefits of simulation-based decision-making in optimizing logistics networks, reducing transportation costs, and increasing overall supply chain resilience.

Keywords: Simulation Modelling · Supply Chain · Airport · Transport Hub · Optimization

1 Introduction

Nowadays efficient supply chain management is one of the most critical factors for ensuring the competitiveness and sustainability of transportation companies [1, 2]. The increasing complexity of logistics operations, coupled with the need for faster and more cost-effective cargo transportation, has led to a growing interest in the adoption of advanced modelling and optimization techniques. As supply chains expand and integrate with multimodal transport systems, companies face significant challenges related to fleet management, cargo flow coordination, warehouse operations, and overall supply chain resilience [3, 4].

I. Ostroumov et al. (Eds.): ACASD 2025, LNNS 1418, pp. 33–44, 2025.
https://doi.org/10.1007/978-3-031-91992-3_3

One of the key components of modern transportation systems is the integration of logistics centres with major airport hubs. Airports serve as strategic nodes within global and regional transportation systems, facilitating the efficient movement of goods across different modes of transport, including air, road, and rail [5]. The presence of a transport hub at an airport complex enables companies to streamline cargo handling processes, reduce delivery times, and optimize inventory management [6]. However, despite these advantages, companies often encounter issues such as inefficient vehicle scheduling, unbalanced resource utilization, bottlenecks in cargo processing, and increased operational costs due to poor supply chain coordination.

Efficient management of logistics processes and transportation systems is essential for ensuring the stability and development of modern supply chain networks. Given the increasing complexity of supply chains, researchers have explored various methods for optimizing logistics operations, leveraging mathematical modeling, simulation approaches, and digital transformation strategies. These studies highlight the necessity of improving repair procedures, enhancing navigation systems, optimizing infrastructure, and integrating advanced technologies into logistics networks.

One of the critical areas of research focuses on optimizing repair procedures for transportation infrastructure to enhance system reliability and reduce operational costs. Recent studies analyze the efficiency of current maintenance processes for aviation radio equipment, proposing optimization strategies to streamline technical servicing [7]. An importance of risk assessment in multimodal transportation systems has been emphasized, with models developed to evaluate safety threats from both external environmental factors and internal technical failures [8]. Alongside maintenance improvements, researchers have also explored mathematical models and navigation systems to enhance transport efficiency. Studies have introduced predictive models based on variable-structure oscillators to analyze transport process dynamics. Moreover, advanced algorithms have been developed to calculate optimal UAV operating parameters, facilitating their integration into logistics networks and improving autonomous transport capabilities [9, 10].

The development of transport infrastructure is another essential aspect of logistics system optimization. Research assessing completed transportation interchanges in Kyiv provides insight into how infrastructure changes influence road capacity and congestion levels. These findings reinforce the need for strategic planning and integration of infrastructure projects within broader supply chain networks to ensure seamless operations [11]. In addition to physical infrastructure, information and communication technologies (ICTs) play a crucial role in modern logistics. Recent advancements in self-organization techniques and signal filtering methods for air traffic management systems have contributed to greater efficiency in real-time operations. Digital transformation frameworks have been proposed to guide logistics companies in adopting IoT, Big Data, and AI-driven solutions for optimizing supply chain performance [12, 13].

Ensuring logistics safety and mitigating technological risks remains a major research priority. Studies have examined the impact of electromagnetic interference on logistics operations and proposed strategies to minimize the negative effects of magnetic fields on electronic transport systems [14].

Beyond risk mitigation, researchers have emphasized simulation-based optimization as a key strategy for improving supply chain efficiency. Simulation techniques have been widely applied to logistics processes, offering algorithmic approaches to minimize costs, enhance route planning, and optimize fleet management. Several studies extend simulation methodologies to dynamic supply chain environments, demonstrating their effectiveness in adapting logistics networks to fluctuating demand and external disruptions [15, 16]. Complementing these simulation approaches, researchers have examined the optimization of aviation-related infrastructure, proposing a methodology for substantiating the development of aviation radio equipment repair centers. This work is crucial for ensuring uninterrupted logistics operations in the aviation sector, where system reliability is paramount [17].

The role of simulation modeling extends beyond traditional logistics operations to specialized sectors, such as biofuel supply chains and aviation transport. Simulation-based studies of ethanol transportation, for example, have demonstrated their effectiveness in refining delivery strategies, reducing transportation costs, and improving overall resource efficiency [18]. As logistics systems become more reliant on autonomous and unmanned transport solutions, external environmental factors such as weather conditions must be accounted for. Research on wind-related hazards has led to the development of predictive algorithms that assess meteorological risks and enable real-time adjustments to unmanned aerial vehicles and automated transport operations, significantly enhancing operational resilience [19, 20].

Recent advancements in virtual supply chain optimization further expand the scope of logistics modeling. Simulation-based optimization using virtual platforms has been explored as a means of structuring supply chains and improving decision-making processes. The implementation of digital twin technologies has proven particularly effective in modeling complex logistics networks and testing various operational scenarios before real-world deployment [21, 22].

To address those challenges, this study develops a simulation-based framework for optimizing supply chain networks with a logistics center integrated into an airport hub. A digital model of a transportation network is developed to analyze logistics flows and evaluate different operational scenarios. The objectives of this research are:

- to model the supply chain network and visualize logistics operations within an airport-integrated logistics system;
- to determine the optimal number of transport vehicles required for maintaining cost-efficient and reliable operations.

The case study for the research is the logistics network of Freight Transport Partner, the Ukrainian transportation company, centered at Warsaw Chopin Airport, with regional warehouses strategically positioned in Wroclaw, Lublin, Rzeszow, Warsaw, Krakow, Poznan, and Gdansk [23].

The simulation model aims to evaluate the effectiveness of different supply chain configurations, identify areas for improvement, and provide actionable insights for optimizing logistics operations. The findings of this research will contribute to the development of more efficient and resilient logistics management practices, particularly in the context of transportation networks that rely on airport transport hub as key distribution points.

2 Research Methodology

This study employs a simulation-based approach to optimize the supply chain network with an airport-integrated logistics center. The model is developed using AnyLogic software and incorporates Geographic Information System (GIS) technologies to simulate transportation flows, warehouse operations, and vehicle allocation strategies [24, 25]. The methodology consists of several key stages, ensuring a comprehensive assessment of logistics performance and optimization scenarios.

The first step involved studying and validating the process of forming a batch cargo order at the regional warehouses of the transportation company. For this purpose, a specialized agent was created in the AnyLogic software environment to represent the order and define its interaction with the distribution warehouse located at the airport. The model assumes that the warehouses regularly generate supply requests, transitioning from normal operation mode to awaiting cargo. The order formation occurs at a specified frequency, after which the request is transmitted to the distribution center, where the processing begins (Fig. 1).

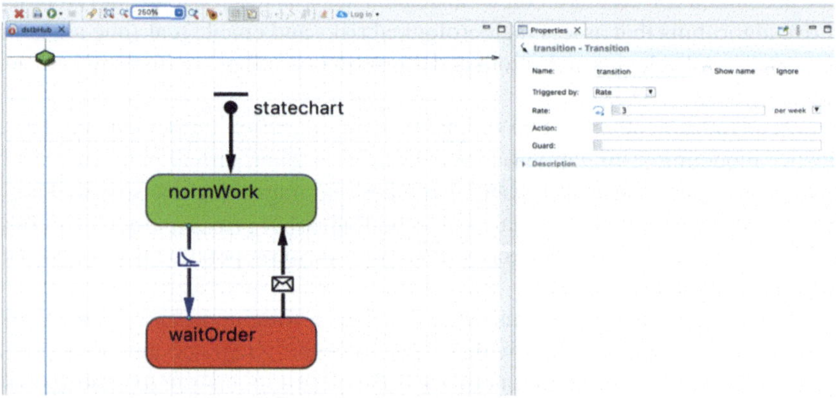

Fig. 1. State transition logic for regional warehouse functioning from normal operation to order waiting.

The operational process of the regional warehouses within the transportation company follows a structured sequence that ensures efficient management of cargo flow and order processing. Initially, the regional hub operates in its default state, where it remains ready to receive and process orders. This phase is referred to as the initial operating state.

Once the hub begins functioning under normal conditions, it transitions to the "normal work" state, during which orders are processed, and cargo is received, stored, or dispatched based on predefined schedules. However, when the hub reaches a point where there are no incoming orders or all current orders have been fulfilled, it enters a transition phase controlled by a specific parameter. This parameter regulates the shift from "normal work" to "order waiting", ensuring that the hub does not remain operational without active tasks, thereby optimizing resource usage.

In the "order waiting" state, the transportation hub remains idle until a message confirming the receipt of cargo is received. Instead of activating in response to a new order request, the system monitors the status of inbound shipments and triggers the state transition only after the cargo has arrived and been successfully processed. The transition between states is determined by a specific control parameter, which receives the cargo arrival message and initiates the return to the "normal work" state (Fig. 2).

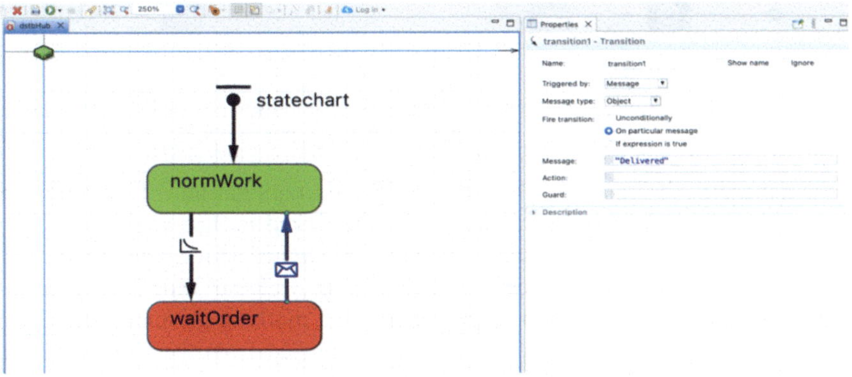

Fig. 2. State transition logic for regional warehouse functioning from order waiting to normal operation.

This cyclical process ensures that the regional warehouses efficiently manage incoming and outgoing shipments, minimizing unnecessary downtime and enhancing the adaptability and responsiveness of the logistics system.

This logic allows the model to reflect the real cycle of the logistics system: from the creation of order to its fulfilment and subsequent transition to the waiting period before the next request.

The next step is modelling the routing process and cargo flow management within the logistics model. The primary goal is to organize efficient movement of goods between the central hub at the airport and regional warehouses. For this purpose, transport routes are created in the model, taking into account optimal distances, vehicle capacity constraints, and delivery time parameters.

The delivery process is automated through the implementation of transport management algorithms, which determine the sequence and priority of shipments. Freight trucks receive transportation tasks, move along designated routes, complete unloading, and return to the starting point. This approach ensures effective transport utilization, minimizes idle time, and optimizes resource usage (Fig. 3).

The process begins with the "*trucks*" block, which defines the available fleet of transport vehicles. When a new order arrives from a regional hub, it enters the system through the "*enter*" block, where it is registered for processing. The "*resource task start*" block initiates resource allocation by assigning a specific truck to handle the cargo. Once assigned, the truck moves to the "*loading*" block, where the cargo batch is prepared for shipment. The loading process takes between 1 and 3 h, depending on the cargo volume and operational conditions. After loading, the truck proceeds to the "*delivery*" block,

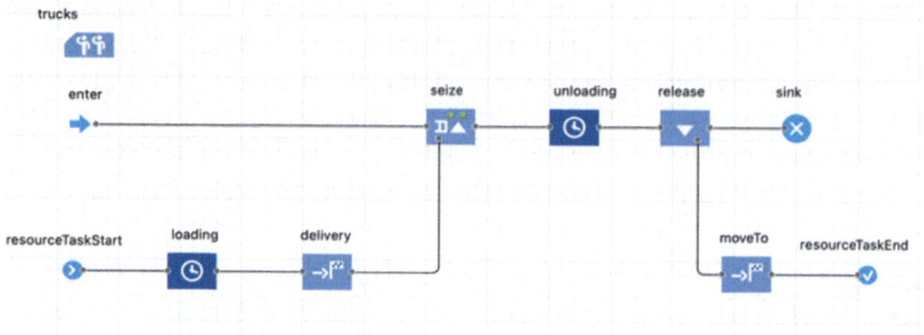

Fig. 3. Operational logic of the central hub of the transportation company at the airport.

where it transports the cargo batch to the designated regional hub. To ensure proper tracking and control, the truck and cargo order are linked in the "*seize*" block. Upon arrival at the regional hub, the truck enters the "*unloading*" block, where the cargo is offloaded. Once the unloading process is complete, the "release" block disconnects the truck resource from the completed cargo batch order, marking the successful delivery of goods. At this stage, the cargo batch order is removed from the system through a dedicated block, while the truck is directed to the "*move to*" block, returning it to a designated parking area at the central hub. Finally, the "*resource task end*" block signals the completion of the transportation cycle, allowing the truck to be reassigned for a new delivery.

At the final stage, the efficiency of the transportation system is analysed, including vehicle utilization rates, fuel consumption, and order processing speed. The data obtained enables the evaluation of various logistics scenarios and the optimization of delivery processes, minimizing delays and enhancing the efficiency of the supply chain network. The 2D simulation of the developed model is presented in Fig. 4.

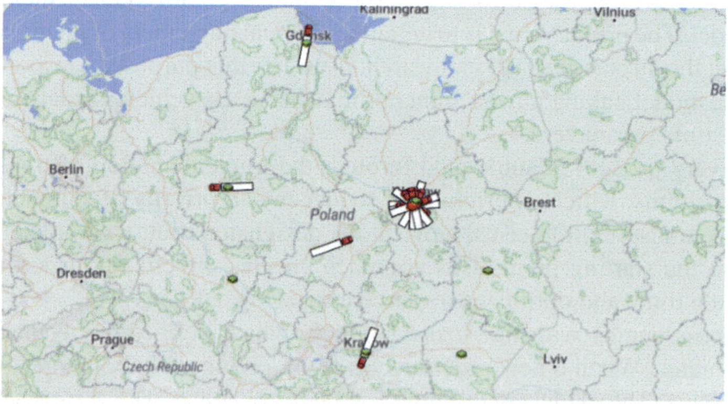

Fig. 4. The 2D simulation of the supply chain operation of the transportation company based at Warsaw Chopin Airport.

The model begins with the "*dstbHub_location*" element, which defines the geographical locations of the regional hubs. This element provides spatial positioning for the logistics network, allowing the system to determine optimal transport routes and evaluate distance-based constraints. The "*dstbHubs*" agent is responsible for simulating the operations of regional hubs, including cargo reception, temporary storage, and dispatching processes. These hubs act as intermediate points between the airport-based logistics center and final delivery locations. At the core of the simulation, the "*airportHub*" agent models the functioning of the central hub located at the airport, where cargo shipments are received, sorted, and dispatched to regional hubs. This agent ensures that airport logistics operations are efficiently coordinated with the overall supply chain network. The "*trucks*" agent is responsible for simulating truck operations, including cargo transportation between the central hub and regional distribution points. It incorporates real-world parameters such as vehicle loading capacity, travel time, and routing efficiency to optimize fleet utilization. To regulate the availability of transport resources, the "*numberTrucks*" parameter defines the total number of trucks in the fleet. This parameter plays a crucial role in capacity planning, allowing the simulation to analyze how fleet size impacts delivery efficiency, cost-effectiveness, and resource utilization.

A simulation of the supply chain's annual operation revealed that transport vehicle resource utilization is low, averaging approximately 10% (Fig. 5).

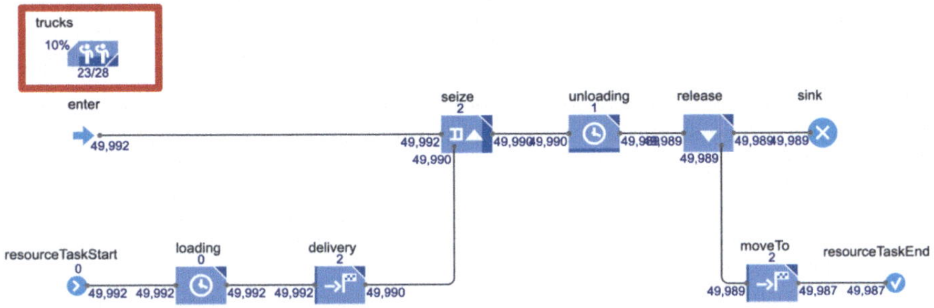

Fig. 5. Simulation of trucks utilization in the supply chain of the transportation company.

Thus, the current logistics process was identified as inefficient, due to high operational costs and poor utilization of transport resources. Based on results it makes sense to conduct an optimization experiment aimed at improving resources utilization, reducing operational expenses, and enhancing the overall efficiency of the logistics network.

3 Results

As part of the optimization experiment, the number of trucks and their assignment to routes were optimized to improve resources utilization efficiency and reduce operational costs. The initial analysis revealed that the average vehicle load factor was only about 10%, which was deemed inefficient for the long-term functioning of the system.

To optimize operations, the vehicle load factor was set at 0.9, reducing the number of empty trips and increasing fleet utilization. The optimization process was conducted by varying the number of trucks within a range of 1–50, with a step of 1. The objective function was defined as maximizing the utilization of transport resources at the airport hub. This was achieved using the standard *utilization()* function, which calculates the average resource block utilization.

To ensure realistic simulation conditions, the simulation runtime was set to 365 days, allowing for a full annual cycle of the transportation company's operations. However, to prevent risks associated with accidents, vehicle damage, or cargo loss, an upper utilization limit was established. The optimal value was set at 0.85 (85%), reflecting a balance between operational efficiency and transportation safety.

The optimization experiment interface is designed to facilitate the analysis and improvement of transport resource utilization in the supply chain simulation. The transport resource optimization function serves as the core mechanism for adjusting fleet operations to enhance efficiency. It allows for the dynamic allocation of trucks based on demand, minimizing idle time and optimizing cargo distribution [26, 27].

The optimization parameters define the key variables influencing the experiment, such as the number of trucks, load capacity, and delivery schedules. These parameters are adjusted to test different scenarios and determine the most efficient operational strategy. The simulation time for the optimization experiment sets the duration of the experiment, typically covering an annual operational cycle (365 days). This enables the evaluation of long-term impacts on fleet efficiency, cost reduction, and delivery performance. The constraints on transport resource utilization ensure that optimization remains within realistic operational limits. These constraints include maximum vehicle load capacity (set to 0.85 to prevent overloading), travel time restrictions, and maintenance scheduling, all of which contribute to maintaining a balance between efficiency and reliability. The optimization experiment interface is presented in Fig. 6.

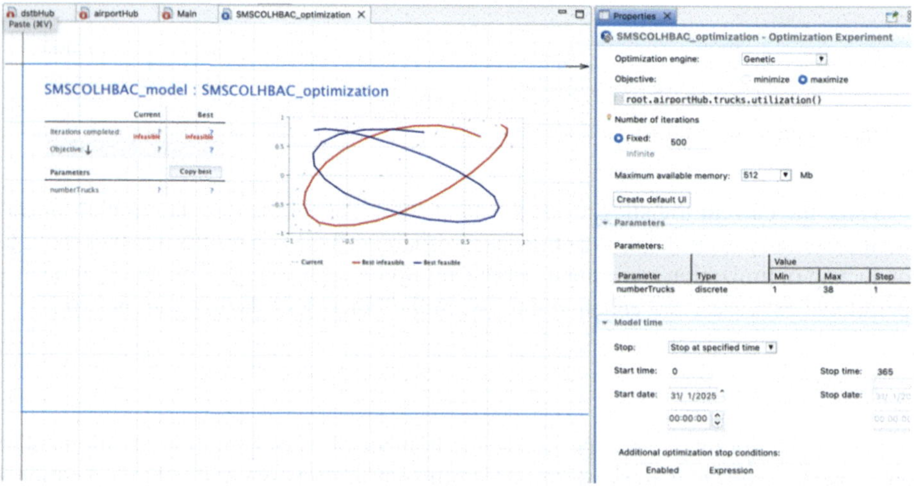

Fig. 6. Optimization experiment interface.

Results of simulation are given in Figs. 7 and 8. Obtained results indicate that optimizing the number of transport vehicles and their load capacity significantly enhanced the efficiency of the logistics system.

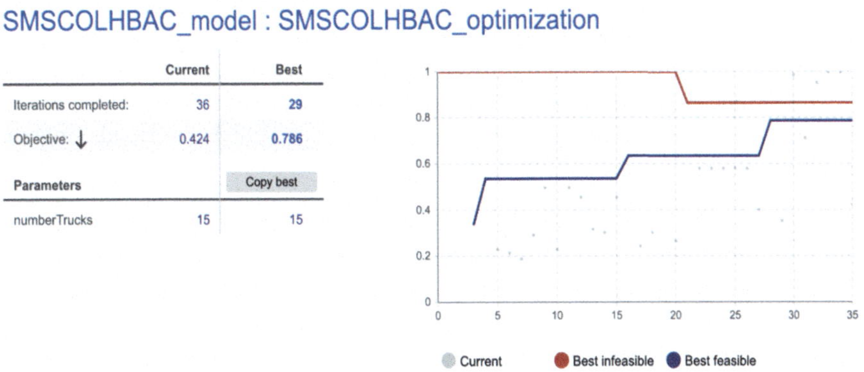

Fig. 7. Simulation of the optimization experiment for enhancing transport vehicle utilization in the transportation company.

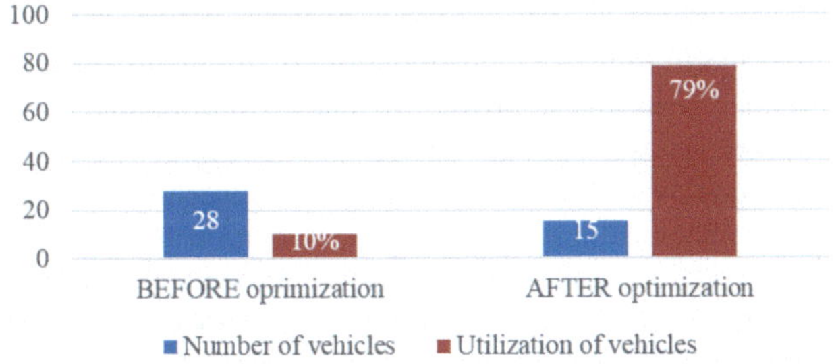

Fig. 8. Comparison of transport vehicle resource utilization before and after the optimization experiment.

The reduction in the number of transport vehicles from 28 to 15 units resulted in lower operation costs, while the increase in the average utilization factor from 0.1 to 0.79 contributed to better usage of transport resources without the risk of overloading.

4 Conclusions

Presented results demonstrate the importance of simulation modeling as a key decision-support tool for optimizing supply chain operations within a transportation company based at an airport logistics hub. The initial analysis of the system revealed inefficiencies, including low transport resource utilization ($\approx$10%), high operation costs, and an excessive number of empty or idle trips. These factors negatively affected the overall efficiency of the logistics network, leading to increased expenses and suboptimal resources allocation.

To address these issues, an optimization experiment was conducted, focusing on adjusting the number of transport vehicles, improving route planning, and maximizing fleet utilization. As a result, the number of transport vehicles was reduced from 28 to 15, while the average vehicles utilization rate increased from 0.1 to 0.79. These adjustments allowed for a more balanced distribution of cargo flows, ensuring a reduction in unnecessary transportation costs and an overall improvement in system performance. The optimization process was based on adjusting the number of vehicles within a range of 1–50 to determine the optimal workload distribution while maintaining an upper safety limit of 0.85 (85%) to prevent operational risks such as vehicle overloads, increased maintenance requirements, or delays.

The results confirm that dynamic route planning, fleet adjustment, and intelligent cargo distribution significantly enhance the performance and reliability of a supply chain that integrates an airport logistics hub. The approach demonstrated in this paper not only reduces operation costs but also improves service quality by ensuring faster and more reliable deliveries. Furthermore, the integration of adaptive resource allocation and real-time monitoring within logistics operations opens new opportunities for scalability, automation, and increased resilience to fluctuating demand.

However, despite the improvements achieved through optimization, the current model has certain limitations that can be addressed in future studies. One of the main drawbacks is the lack of batch cargo formation strategies for multi-destination deliveries. The current simulation assumes that each cargo batch is delivered to a single destination, whereas in real-world logistics, consolidation of shipments for multiple locations can significantly reduce costs and improve efficiency. Additionally, the model does not fully account for uncertainties in demand fluctuations, last-mile delivery constraints, and external disruptions such as weather conditions or infrastructure limitations.

Future research will focus on expanding the simulation framework by incorporating real-time data integration, machine learning-based demand forecasting, and multimodal transport solutions. Additionally, testing the developed approach in real-world case studies with larger datasets and different geographical conditions will allow for further refinement and validation of the proposed optimization strategies, ensuring their practical applicability in modern logistics networks. Implementing cargo consolidation mechanisms and multi-destination routing will be a crucial next step in improving the adaptability and economic efficiency of logistics operations.

References

1. Oliskevych, M., Danchuk, V., Pelo, R., Nikipchuk, S., Brytkovskyi, V., Sorokivskyi, O.: Coordination of material and information flows in intercity logistics systems. In: Logistics Systems: Technological and Economic Aspects of Efficiency, p. 10.15587/978-617-7319-66-4.ch3. Privat Company Technology Center, Kharkiv (2022)
2. Namazov, M., Matsiuk, V., Bulgakova, I., Nikolaienko, I., Vernyhora, R.: Agent-based simulation model of multimodal ironore concentrate transportation. Naukovij zurnal Tehnika ta energetika. **14**(1), 46–56 (2023). https://doi.org/10.31548/machinery/1.2023.46
3. Chayka-Petegyrych, L.: Multimodal and intermodal cargo transportation in the system of international transport logistics. Herald UNU. Int. Econ. Relat. World Econ. **33**, 114–117 (2020). https://doi.org/10.32782/2413-9971/2020-33-41

4. Trost, M., Claus, T., Herrmann, F.: Master production scheduling with consideration of utilization-dependent exhaustion and capacity load. Sustain. For. **15**(8), 6816 (2023). https://doi.org/10.3390/su15086816
5. Sokolova, O., Ivannikova, V., Cherednichenko, K.: Fundamentals of a sustainable supply chain designing with the involvement of air transport. In: 7th International Conference on Actual Problems of Unmanned Aerial Vehicles Development (APUAVD), pp. 140–145. IEEE, Kyiv (2024). https://doi.org/10.1109/apuavd64488.2024.10765912
6. Sokolova, O., Cherednichenko, K., Ivannikova, V.: Mathematical model of airport aviation security. In: Prentkovskis, O., et al. (eds.) TRANSBALTICA XIII: Transportation Science and Technology. TRANSBALTICA 2022. Lecture Notes in Intelligent Transportation and Infrastructure, vol. 1, pp. 773–781. Springer, Cham (2023). https://doi.org/10.1007/978-3-031-25863-3_75
7. Solomentsev, O., Zaliskyi, M., Holubnychyi, O., Ostroumov, I., Sushchenko, O., Bezkorovainyi, Y.: Efficiency analysis of current repair procedures for aviation radio equipment. In: Ostroumov, I., Zaliskyi, M. (eds.) Proceedings of the 2nd International Workshop on Advances in Civil Aviation Systems Development. ACASD 2024, Lecture Notes in Networks and Systems, vol. 992, pp. 281–295. Springer, Cham (2024). https://doi.org/10.1007/978-3-031-60196-5_21
8. Cherednichenko, K., Ivannikova, V., Sokolova, O., Yanchuk, M.: Model of transport safety assessment in multimodal transportation systems. Transport. **38**(4), 204–213 (2023). https://doi.org/10.3846/transport.2023.20865
9. Voliansky, R., Ostroumov, I., Sushchenko, O., Averyanova, Y., Solomentsev, O., Holubnychyi, O., Zaliskyi, M.: Variable-structure interval-based duffing oscillator. In: 42nd International Conference on Electronics and Nanotechnology (ELNANO), pp. 581–586. IEEE, Kyiv (2024). https://doi.org/10.1109/ELNANO63394.2024.10756860
10. Teterin, S.: Algorithm for calculating optimal parameters contributing to UAV operation efficiency. In: 7th International Conference on Actual Problems of Unmanned Aerial Vehicles Development (APUAVD), pp. 304–307. IEEE, Kyiv (2024). https://doi.org/10.1109/apuavd64488.2024.10765853
11. Bespalov, D., Tarasiuk, V., Osetrin, M., Dorosh, M., Pryimachenko, O.: An analysis of the transport impact of finish building on-ramps and off-ramps from the Dnipro embankment to the Darnytskyi Bridge in Kyiv. Arch. Stud. **10**(2), 172–190 (2024). https://doi.org/10.56318/as/2.2024.172
12. Holubnychyi, O., Zaliskyi, M., Ostroumov, I., Sushchenko, O., Solomentsev, O., Averyanova, Y.: Self-organization technique with a norm transformation based filtering for sustainable infocommunications within CNS/ATM systems. In: Ostroumov, I., Zaliskyi, M. (eds.) Proceedings of the 2nd International Workshop on Advances in Civil Aviation Systems Development. ACASD 2024, Lecture Notes in Networks and Systems, vol. 992, pp. 262–278. Springer, Cham (2024). https://doi.org/10.1007/978-3-031-60196-5_20
13. Gulmez, S., Denktas-Sakar, G.: Digital transformation framework for logistics service providers. In: Handbook of Digital Innovation, Transformation, and Sustainable Development in a Post-Pandemic Era, pp. 60–77. CRC Press (2024). https://doi.org/10.1201/9781003438748-4
14. Kuznetsov, B., Bovdui, I., Nikitina, T., Voliansky, R., Sushchenko, O., Bezkorovainyi, Y., Averyanova, Y., Ostroumov, I., Solomentsev, O., Holubnychyi, O., Zaliskyi, M., Cherednichenko, K., Sokolova, O., Ivannikova, V.: Efficiency of overhead power lines magnetic field combined electromagnetic silence. In: Fifth International Conference on Green Energy, Environment, and Sustainable Development (GEESD 2024) Proc. SPIE, vol. 13279, 1327907 (2024). https://doi.org/10.1117/12.3043769
15. Brinza, G.: Simulation and optimization in supply chain. Proc. Econ. Finan. **3**, 635–641 (2012). https://doi.org/10.1016/s2212-5671(12)00207-9

16. Raoui, H., Oudani, M., Alaoui, A.: Optimization/simulation in the supply chain context. In: Proceedings of the International Conference on Learning and Optimization Algorithms: Theory and Applications (LOPAL '18), pp. 1–7. APC (2018). https://doi.org/10.1145/3230905.3230965
17. Zaliskyi, M., Solomentsev, O., Holubnychyi, O., Ostroumov, I., Sushchenko, O., Averyanova, Y., Bezkorovainyi, Y., Cherednichenko, K., Sokolova, O., Ivannikova, V., Voliansky, R., Kuznetsov, B., Bovdui, I., Nikitina, T.: Methodology for substantiating the infrastructure of aviation radio equipment repair centers. CEUR Workshop Proc. **3732**, 136–148 (2024) https://ceur-ws.org/Vol-3732/paper11.pdf
18. Kim, S., Choi, Y., Kim, S.: Simulation modeling in supply chain management research of ethanol: A review. Energies. **16**(21), 7429 (2023). https://doi.org/10.3390/en16217429
19. Averyanova, Y., Zaliskyi, M., Solomentsev, O., Holubnychyi, O., Ostroumov, I., Sushchenko, O., Bezkorovainyi, Y., Cherednichenko, K., Sokolova, O., Ivannikova, V., Voliansky, R., Kuznetsov, B., Bovdui, I., Nikitina, T.: Algorithm of wind-related hazards prediction for UAS flight and urban operations based on meteorological data fusion. CEUR Workshop Proc. **3895**, 132–142 (2024) https://ceur-ws.org/Vol-3895/paper11.pdf
20. Ostroumov, I., Ivannikova, V., Zaliskyi, M., Kuzmenko, N.: Impact analysis of Russian-Ukrainian war on airspace. J. Air Transp. Manag. **124**, 102742 (2025). https://doi.org/10.1016/j.jairtraman.2025.102742
21. Liu, Z., Shirakashi, R., Kamiebisu, R., Nishi, T., Matsuda, M.: Simulation-based optimization using virtual supply chain structured by the configuration platform. IFAC-PapersOnLine. **56**(2), 7840–7845 (2023). https://doi.org/10.1016/j.ifacol.2023.10.1145
22. Sushchenko, O., Bezkorovainyi, Y., Solomentsev, O., Zaliskyi, M., Holubnychyi, O., Ostroumov, I.: Algorithm of determining errors of gimballed inertial navigation system. In: Gervasi, O., Murgante, B., Garau, C., Taniar, D.C., Rocha, A.M.A., Faginas Lago, M.N. (eds.) Computational Science and Its Applications ICCSA 2024 Lecture Notes in Computer Science, vol. 14816, pp. 206–218. Springer, Cham (2024). https://doi.org/10.1007/978-3-031-65223-3_14
23. Freight Transport Partner. https://ftpua.com/index-en.php#delivery_methods. Last accessed 18 Jan 2025
24. Wolniewicz, L.: Railway transport system modelling approach for robustness analysis. Transport. **39**(4), 287–301 (2024). https://doi.org/10.3846/transport.2024.22877
25. Nouh, A., Elkasaby, E., Wageh, O.: Dynamic approach for investigating delivery system effects on infrastructure project performance. Eng. Constr. Archit. Manag. **31**, 4791–4808 (2023). https://doi.org/10.1108/ecam-10-2022-1022
26. Karmaker, C.L., Aziz, R., Ahmed, T., Misbauddin, S.M., Moktadir, M.A.: Impact of industry 4.0 technologies on sustainable supply chain performance: the mediating role of green supply chain management practices and circular economy. J. Clean. Prod. **419**, 138249 (2023). https://doi.org/10.1016/j.jclepro.2023.138249
27. Abdelaziz, S., Munawaroh, M.: Mitigating supply chain vulnerabilities: a bibliometric analysis of sustainable logistics for resilience and risk management. Int. J. Automot. Sci. Technol. **8**(4), 544–588 (2024). https://doi.org/10.30939/ijastech.1554338

Theory of Subjective Preferences Entropy in Application to Air Transportation Technologies: Rational Recombination of Resources

Andriy Goncharenko(✉)

National Aviation University, Liubomyra Huzara Ave., 1, Kyiv 03058, Ukraine
andrii.honcharenko@npp.kai.edu.ua

Abstract. The paper deals with the combined problem of the subjective analysis grounded upon the important parameters of the air transportation technologies resources recombination. It is considered the time of the aircraft flight to the point of connection for two aircrafts. Starting with the delay time for one of the aircraft, the problem was simplified to the simultaneous problem setting in order to distinguish the rational element of the objectively existing optimal values (aircraft flight speeds, thus, covered distances) determination. The criterion of the optimization was chosen as an economical one (minimal flight cost). The solution has been obtained with the help of the parametric relation between the criterion (the minimal cost) and the optimized value of the distance flown by the aircraft. Fortunately, due to this, the parametric approach yielded the sought solution. Summarized aspects of human factor, safety system analysis, internal active participant vs. external contemplation observer, theorem vs. axiom/postulate, Weber-Fechner Law as well as Jacob Bernoulli Law are discussed. Numerical simulation is conducted, the necessary diagrams are plotted.

Keywords: Human Factor · Safety System Analysis · Transportation Technologies · Resources Recombination · Internal Participant · External Observer · Optimization · Aircraft Speed · Minimal Cost

1 Introduction

1.1 A Literature Survey

Human Factor and Safety System Analysis are important for civil aviation development. Taking into account the system of the individuals' subjective preferences distribution is an essential problem [1].

The theory of subjective entropy is grounded upon Jaynes' entropy principle [2–4] in the formal mathematical notation but has a specific meaning. Aviation transport systems' issues could be considered at the angle of both narrow specialist approaches [5–7] and tourism management endeavors enforcements [8–10] dealing with environmental sustainability, mega-sport influences, and public policy implications. Other human factor's

I. Ostroumov et al. (Eds.): ACASD 2025, LNNS 1418, pp. 45–55, 2025.
https://doi.org/10.1007/978-3-031-91992-3_4

issues important here are the leisure travel, vulnerability to hazards, and information technologies on air transportation [11–13].

In any case, such approaches might use the entropy elements as in the references of [14–16] where internal participant's attitudes are considered vs. external observer's objective speculations characterizations are distinguished, individuals' preferences entropy is applied to conditional effectiveness optimizations, and uncertainty hybrid models are proposed.

Then, in [17–19] there are some attempts to develop and apply the entropy concept to the intelligent aircraft control, using airworthiness analogy, and maintenance technologies issues. Which has a severe impact upon safety system analysis.

According to [20–22] human factor is realized at the conditional optimization of aircraft maintenance preference uncertainty, control and optimization for air transportation systems as well as airline profit maximization.

This, in turn, opens new horizons for prospective transport technologies developments, for instance, widely depicted in the various works likewise of [23–25] dedicated to the multi-objective optimization for economic outcomes on socio-economic segments, stakeholder theory analysis, and transport security and supply chains. The questions raised in [26–28] emphasize the impacts in the aviation sector, inter-city air mobility, and airline behavior simulation importance.

Some aspects of the touched problematics have already been contoured out. Those deal with the Internal active participant vs. External contemplation observer [14], approaches to Conflict Theory based upon the subjective preferences distribution differences. Theoretically, it is an important issue to make persistent attempts to find at least the regularities in relations between objective vs. subjective phenomena.

This requires sometimes a renovation to the analytical endeavors and hypotheses review. It is one of the goals of the presented research. In fact, the theory development needs critical contemplations and the prolific speculations drive the authors to review the foundations of Theorems vs. Axioms/Postulates.

In addition to Discrete vs. Continuous alternatives preferences considerations, there is a good potential in a fresh sight to the well-known human factor provisions influencing the safety moments in the generally accepted forms. It is worth mentioning here the main law of the psychophysics, Weber-Fechner Law [15]. Similar entropy approaches allow rediscovering the law of the subjective value by Jacob Bernoulli [15] as well.

The traditional type entropy of the Shannon's view is laid down into the foundation of the multiple uncertainty descriptions as in reference [16].

1.2 Problem Formulation

The problem of the presented study could be formulated as to combine human factor and safety system analysis objective characteristics with the individuals' subjective preferences through the entropy approaches for civil aviation technologies development.

And the aim of the research is to ensure, based upon [1], a steady development of the theory reckoning with the uncertainties of the operational situations in air transport technological processes.

2 Main Content

2.1 Theoretical Background

It is proposed to consider several theoretical aspects.

An active element of the safety system inevitable bares a human factor. In such context, it is proposed to consider the options that could be chosen (in subjective analysis theory they are called 'alternatives'). However, the subjective choice is based upon some objective parameters.

Distance Optimization. For example, in the problem with the delay, it is a specific consideration.

The Delay Case (Only One Aircraft Delay). Taking into account that in the simplest case just only one aircraft delay time is considered, one can come to such a basic form as a linear combination.

That is:

$$t_1 = t_2 + t_{z2}, \tag{1}$$

where t_1 is the flight time to the connection point of the first aircraft; t_2 is the flight time to that point of by the second aircraft; t_{z_2} is the time of the delay of the second aircraft compared to the first one.

Then from (1)

$$|AC| = \frac{|AB| + v_2 t_{z2}}{v_1 + v_2} v_1, \tag{2}$$

where $|AC|$ is the first aircraft flown distance; $|AB|$ is the distance between the departure airports of both aircraft; v_1 and v_2 are the aircraft speeds in respect.

Now, it can be set a problem of $|AC|$ optimization.

Delay Case Conditional Optimization (Parametric Solution). In such circumstances the optimization criteria have to be chosen.

For the beginning it is proposed to consider a without delay case:

$$t_{z_2} = 0. \tag{3}$$

Optimization criterion is chosen as the minimum flight cost for the first aircraft:

$$S_1(v_1, v_2) = s_1^{(1)} \frac{|AB|}{v_1 + v_2} + k_1 v_1^{n_1} \frac{|AB|}{v_1 + v_2}, \tag{4}$$

where $S_1(v_1, v_2)$ is the cost of the flight of the first aircraft, it is implied here that the cost is the function of the motion speeds for the two aircraft correspondingly; $s_1^{(1)}$ is the cost an hour of the first aircraft during the flight that does not depend upon the speed of the aircraft; k_1 is the coefficient of proportionality of the first aircraft hourly flight cost dependent upon the aircraft speed; n_1 is the power index for the proportionality of the first aircraft hourly flight cost dependence upon the aircraft speed.

If the first aircraft optimal speed: $v_{1_{opt}}$, is hard (or impossible) to obtain from (4) based upon the first order conditions for the extremum existence, the optimal solution can be found when taking into account the parametric form of (4) and (2) on condition of (3) dependence.

Such parametric approach yields

$$S_1(|AC|, v_2) = s_1^{(1)} \frac{|AB| - |AC|}{v_2} + k_1 \left(\frac{|AC|}{|AB| - |AC|} v_2 \right)^{n_1} \frac{|AB| - |AC|}{v_2}, \quad (5)$$

Formula (5) delivers the sought optimal solution.

Human Factor and Safety System Analysis. Subjective Analysis Theory of Individuals' Preferences [1] is a powerful tool for researching in the field of active systems.

Human factor implies an active system.

Elements of Subjective Analysis (Entropy of Preferences). From [1]

$$\Phi_\pi = H_\pi + \beta\varepsilon + \gamma n, \quad (6)$$

where Φ_π is the objective functional of the unknown so far individual subjective preferences functions: π, given upon the set of the attainable (achievable) alternatives; H_π is the subjective entropy (measure of the subjective uncertainty of the individual subjective preferences functions: π, also a functional of the preferences); β is the cognitive parameter that evaluates the weight of the subjective effectiveness functions, sometimes it could be reckoned with as a Lagrange uncertainty multiplier; ε is the subjective effectiveness function, also depending upon the individual subjective preferences functions: π; γ is the cognitive parameter that evaluates the weight of the normalizing conditions for the individual subjective preferences functions: π; n is the normalizing condition of the individual subjective preferences functions: π.

Then, developing from (6)

$$\Phi_\pi = H_\pi + \beta\varepsilon(\pi_i, F_i) + \gamma n, \quad (7)$$

where $\varepsilon(\pi_i, F_i)$ is the subjective effectiveness function of the own, relevant to the attainable alternatives, arguments; F_i is an i-th alternative effectiveness function.

Subjective entropy:

$$H_\pi = -\sum_{i=1}^{N} \pi_i \, ln \, \pi_i, \quad (8)$$

where N is the number of the considered alternatives.

Subjective effectiveness function:

$$\varepsilon(\pi_i, F_i) = \sum_{i=1}^{N} \pi_i F_i. \quad (9)$$

Normalizing condition:

$$\sum_{i=1}^{N} \pi_i = 1. \quad (10)$$

The necessary extremum existence conditions:

$$\frac{\partial \Phi_\pi}{\partial \pi_i} = 0. \tag{11}$$

Thus, (11) for (6)–(10) gives

$$\pi_i = \frac{e^{\beta F_i}}{\sum_{j=1}^{N} e^{\beta F_j}}. \tag{12}$$

And

$$\pi_i(x) = \frac{e^{\beta F_i(x)}}{\sum_{j=1}^{N} e^{\beta F_j(x)}}. \tag{13}$$

Formula (13) symbolizes functional dependence upon some independent variable of x in an explicit view compared to (12).

Operational alternatives of air transportation technologies can be treated in the Stimulus and Perception manner in accordance with the main law of psychophysics by Weber-Fechner.

Weber-Fechner Law (Entropy Principle). Consider stimuluses and their perceptions.

That is instead of, [1]:

$$\Phi_\pi = -\sum_{i=1}^{N} \pi_i \, ln \, \pi_i + \beta \sum_{i=1}^{N} \pi_i F_i + \gamma \left(\sum_{i=1}^{N} \pi_i - 1 \right). \tag{14}$$

One can consider [15]:

$$\Phi_S = -\sum_{i=1}^{N} S_i \, ln \, S_i + \beta \sum_{i=1}^{N} S_i p_i + \gamma \left(\sum_{i=1}^{N} S_i - 1 \right), \tag{15}$$

where S_i are the stimuluses; and p_i are the perceptions.

The entropy approach likewise (6)–(14) gives for (15)

$$\frac{\partial \Phi_S}{\partial S_i} = 0. \tag{16}$$

Then

$$\frac{\partial \Phi_S}{\partial S_i} = -ln \, S_i - 1 + \beta p_i + \gamma. \tag{17}$$

For every two stimuluses

$$-ln \, S_1 + \beta p_1 = 1 - \gamma = -ln \, S_2 + \beta p_2. \tag{18}$$

Which means for the threshold stimulus and the corresponding threshold perception:

$$S_0 \text{ and } p_0; \tag{19}$$

and any other stimulus and its corresponding perception:

$$S \text{ and } p, \tag{20}$$

there must be

$$-ln\, S_0 - 1 + \beta p_0 + \gamma = -ln\, S - 1 + \beta p + \gamma. \tag{21}$$

And

$$ln\, S - ln\, S_0 = \beta p - \beta p_0. \tag{22}$$

Thus, (6)–(22) leads to the well-known Weber-Fechner law (the main law of psychophysics).

Indeed, from (22)

$$p - p_0 = \frac{1}{\beta} ln \frac{S}{S_0}. \tag{23}$$

And (23) is the Weber-Fechner law.

Its traditional notation is

$$p - p_0 = klog \frac{S}{S_0}, \tag{24}$$

where k is the coefficient of proportionality.

Weber-Fechner Law (Differential Principle). Initially the main law of psychophysics was derived based upon a differential principle.

Supposed dependence is

$$\frac{dp}{dS} S = k = p'_S S. \tag{25}$$

Solving the differential equation

$$dp = k \frac{dS}{S} \tag{26}$$

it is possible to find

$$p = klnS + C, \tag{27}$$

where C is the constant of integration which can be found from the initial conditions:

$$p|_{S=S_0} = p(S = S_0) = p_0. \tag{28}$$

Then

$$p_0 - klnS_0 = C. \tag{29}$$

And

$$p - p_0 = k(ln\, S - ln\, S_0). \tag{30}$$

Finally, (25)–(30) gives (23) or (24).

The law of the subjective value by Jacob Bernoulli can be obtained in the same mathematical manner [15]. That is, alternatives of the air transport technologies might be considered in a multi-alternative problem setting in combination with the objective parameters' optimization.

2.2 Simulation

In order to demonstrate the procedure by (1)–(5) it is proposed to conduct a numerical experimentation.

The data needed and accepted for the calculations are as $v_1 = 600\ldots1000$, the speed range for the first aircraft $s_1^{(1)} = 5000$ knots in conditional units, let us say $/h:

$$k_1 = 1.5 \cdot p10^{-5}; \quad n_1 = 3; \quad |AB| = 10 \cdot p10^3 \text{ [km]}. \tag{31}$$

The results of the simulation are shown in Figs. 1 and 2.

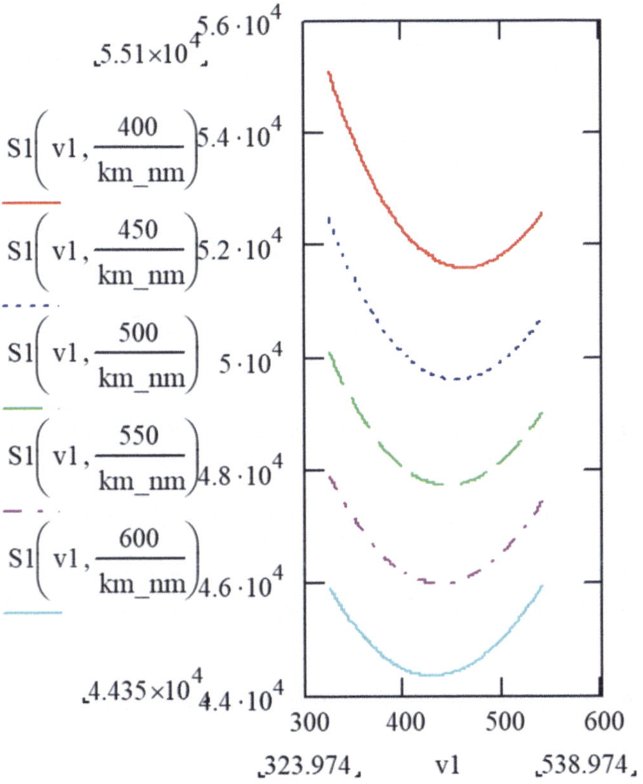

Fig. 1. Dependence of flight costs upon aircraft speeds.

In Fig. 1, the aircraft speed is in knots. The cost is in $. In Fig. 2, the aircraft speed is in knots. The distance is in nm.

3 Discussions

As it is visible from Fig. 1 the cost of flight for the first aircraft has minimal (delivered by optimal speed/distance) values in accordance to the supposed parameters (31). In the given consideration the first aircraft this is the airplane having no delay time. Moreover, for the basic problem setting presented and simulated herewith, there is no delay,

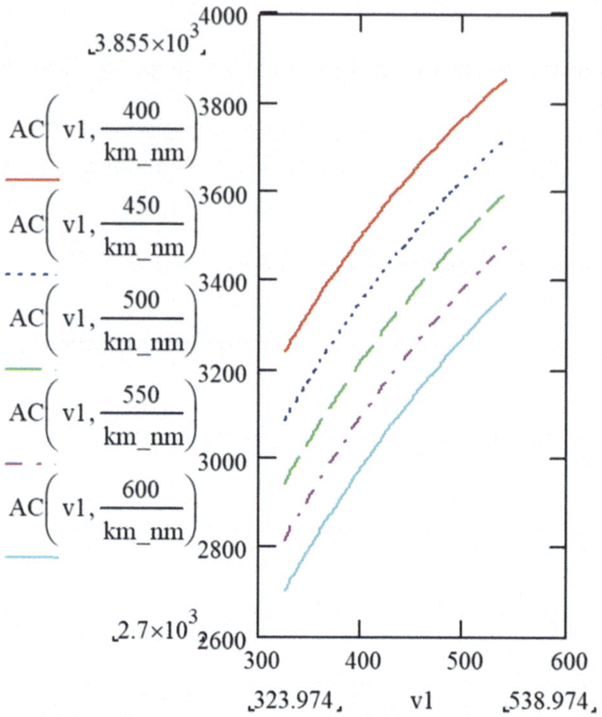

Fig. 2. Dependence of flight distance upon aircraft speeds.

according to the supposed and accepted simplification of (3). That means that "the first" aircraft is just one of the two (conventional enumeration).

The costs were calculated based upon the idea of the two hourly components (4) depending upon the aircraft speed and independent upon the speed of flight. Thus, deliberately omitting for the perceptional ease the obvious mathematical transformations, the concept embodied in formula (4) leads to the desirable objective function (5). To that end, the corresponding distance flown by the first aircraft will be optimal (1)–(5) (see Fig. 2).

Different speeds of both aircraft could be considered as the alternatives of the air transport technological process. Thus, the human factor undoubtably presents and plays sometimes the crucial role in the air transportation technological operational situations. The safety system analysis is recommended, therefore, to conduct with taking into account the subjective entropy provisions (6)–(30).

Internal active participant vs. external contemplation observer [14] will change their roles in respect to the corresponding aircraft operation.

Pseudo-entropy hybrid relative combined function [16] is used to model the conflict of the individuals' subjective preferences.

Theorem vs. axiom/postulate theoretical aspects is in distinguishing the provision accepted as obvious for axioms and the acceptance of a non-obvious hypothetical idea (postulate) leading to some certain plausible results although.

The statements proven based upon both axioms and postulates appear theorems. In such context, the subjective entropy conditional optimality principle should be dealt with.

The Weber-Fechner Law, the main law of the psychophysics as well as the law of the subjective value by Jacob Bernoulli [15] are derived in result of such postulates acceptance.

4 Conclusions

Human factor and safety system analysis are to be rationally dealt with as the elements of subjective analysis. In such understanding every alternative or each time chosen options for aviation system operation are being treated based upon the individuals' effectiveness preferences functions uncertainty/entropy conditional optimization.

The presented herewith problem setting shows untrivial solution that is invisible in the less sophisticated elaborations. This might be reckoned with as an attempt to develop the general theory in terms of air transportation technologies resources reasonable recombination (economic criterion/criteria maximum/minimum in regards with the air transportation technological parameter/parameters, optimized value/values).

The development of the entropy theory allows combinations of objectively existing options and subjectively preferred alternatives.

In further research it is recommended to consider dependences with time intervals.

Also, it is promising to set a number of variational problems.

References

1. Kasianov, V.: Subjective entropy of preferences. Subjective analysis: monograph. Institute of Aviation Scientific Publications, Warsaw, Poland (2013)
2. Jaynes, E.T.: Information theory and statistical mechanics. Phys. Rev. **106**(4), 620–630 (1957). https://doi.org/10.1103/PhysRev.106.620
3. Jaynes, E.T.: Information theory and statistical mechanics. II. Phys. Rev. **108**(2), 171–190 (1957). https://doi.org/10.1103/PhysRev.108.171
4. Jaynes, E.T.: On the rationale of maximum-entropy methods. Proc. IEEE. **70**, 939–952 (1982). https://doi.org/10.1109/PROC.1982.12425
5. Sushchenko, O., et al.: Algorithm of determining errors of gimballed inertial navigation system. In: Gervasi, O., et al. (eds.) ICCSA 2024. LNCS, vol. 14816, pp. 206–218. Springer, Cham (2024). https://doi.org/10.1007/978-3-031-65223-3_14
6. Kazak, V., et al.: Reconfigurable flight control system design for aircraft when unexpected problem occurs during flight. In: 2nd International Conference on Methods and Systems of Navigation and Motion Control (MSNMC), pp. 112–115. IEEE, Kyiv (2012). https://doi.org/10.1109/MSNMC.2012.6475105
7. Solomentsev, O., Zaliskyi, M., Shcherbyna, O., Kozhokhina, O.: Sequential procedure of changepoint analysis during operational data processing. In: Microwave Theory and Techniques in Wireless Communications (MTTW), pp. 168–171. IEEE, Riga (2020). https://doi.org/10.1109/MTTW51045.2020.9245068
8. Demeter, C., Fechner, D., Dolnicar, S.: Progress in field experimentation for environmentally sustainable tourism—a knowledge map and research agenda. Tour. Manag. **94**, 104633 (2023). https://doi.org/10.1016/j.tourman.2022.104633

9. Fourie, J., Santana-Gallego, M.: The impact of mega-sport events on tourist arrivals. Tour. Manag. **32**, 1364–1370 (2011). https://doi.org/10.1016/j.tourman.2011.01.011
10. Aguinis, H., et al.: The why, how, and what of public policy implications of tourism and hospitality research. Tour. Manag. **97**, 104720 (2023). https://doi.org/10.1016/j.tourman.2023.104720
11. Laroche, P.C.S.J., et al.: The role of holiday styles in shaping the carbon footprint of leisure travel within the European Union. Tour. Manag. **94**, 104630 (2023). https://doi.org/10.1016/j.tourman.2022.104630
12. Alvarez, S., Bahja, F., Fyall, A.: A framework to identify destination vulnerability to hazards. Tour. Manag. **90**, 104469 (2022). https://doi.org/10.1016/j.tourman.2021.104469
13. Hansman, R.: The impact of information technologies on air transportation. In: 43rd AIAA Aerospace Sciences Meeting and Exhibit - Meeting Papers (2005). 10.2514/6.2005-1
14. Goncharenko, A.: The ant colony probabilistic model equivalency to the options uncertainty extremized one. In: 10th International Conference on Advanced Computer Information Technologies (ACIT), pp. 541–544. IEEE, Deggendorf (2020). https://doi.org/10.1109/ACIT49673.2020.9209008
15. Goncharenko, A.: Optimal price choice through buyers' preferences entropy. In: 10th International Conference on Advanced Computer Information Technologies (ACIT), pp. 537–540. IEEE, Deggendorf (2020). https://doi.org/10.1109/ACIT49673.2020.9208892
16. Goncharenko, A.: Multi-optional hybridization for UAV maintenance purposes. In: 5th International Conference on Actual Problems of Unmanned Aerial Vehicles Developments (APUAVD), pp. 48–51. IEEE, Kyiv (2019). https://doi.org/10.1109/APUAVD47061.2019.8943902
17. Goncharenko, A.: Several models of artificial intelligence elements for aircraft control. In: 4th International Conference on Methods and Systems of Navigation and Motion Control (MSNMC), pp. 224–227. IEEE, Kyiv (2016). https://doi.org/10.1109/MSNMC.2016.7783148
18. Goncharenko, A.: Airworthiness support measures analogy to the prospective roundabouts alternatives: theoretical aspects. J. Adv. Transp. **2018**, 1–7 (2018). https://doi.org/10.1155/2018/9370597
19. Goncharenko, A.: An entropy model of the aircraft gas turbine engine blades restoration method choice. In: International Conference on Advanced Computer Information Technologies (ACIT), pp. 2–5. CEUR, České Budějovice (2018)
20. Goncharenko, A.: Development of a theoretical approach to the conditional optimization of aircraft maintenance preference uncertainty. Aviation. **22**(2), 40–44 (2018). https://doi.org/10.3846/aviation.2018.5929
21. Balakrishnan, H.: Control and optimization algorithms for air transportation systems. Annu. Rev. Control. **41**, 39–46 (2016). https://doi.org/10.1016/j.arcontrol.2016.04.019
22. Dray, L., Doyme, K., Schafer, A.W.: Airline profit maximisation, cost pass-through, and scarcity rents in capacity-constrained aviation systems. J. Trans. Econ. Policy (JTEP). **54**(4), 244–266 (2020)
23. Balcells, C.C., Krueger, R., Bierlaire, M.: Multi-objective optimization of activity-travel policies for epidemic control: balancing health and economic outcomes on socio-economic segments. Trans. Res. Interdiscip. Persp. **27**, 101183 (2024). https://doi.org/10.1016/j.trip.2024.101183
24. Ceulemans, E., et al.: Synchromodal transport vs. conventional hinterland transport: a stakeholder theory analysis. Transp. Rev. **45**(1), 1–25 (2024). https://doi.org/10.1080/01441647.2024.2403471
25. De Liso, N., Zamparini, L.: Innovation, transport security and supply chains: a review. Transp. Rev. **42**(6), 725–738 (2022). https://doi.org/10.1080/01441647.2022.2105415

26. Burbidge, R., Paling, C., Dunk, R.M.: A systematic review of adaption to climate change impacts in the aviation sector. Transp. Rev. **44**(1), 8–33 (2023). https://doi.org/10.1080/01441647.2023.2220917
27. Ehlers, T., et al.: On key parameters for the attractiveness of on-demand inter-city air mobility. In: 33rd Congress of the International Council of the Aeronautical Sciences (ICAS), Stockholm, Sweden (2022)
28. Doyme, K., et al.: Simulating airline behavior: application for the Australian domestic market. Transp. Res. Rec. **2673**(2), 104–112 (2019). https://doi.org/10.1177/0361198119826533

Management and Economics

Modeling of Integration Business Processes for Aviation Enterprise in the Context of Sustainable Development

Zarina Poberezhna and Maksym Zaliskyi(✉)

National Aviation University, Liubomyra Huzara Ave., 1, Kyiv 03058, Ukraine
mzaliskyi@kai.edu.ua

Abstract. The paper examines the content and process of modeling of the integration business processes of an aviation enterprise in the context of sustainable development. It is determined that integration is the process of combining various elements, systems, or structures to create a single, coordinated mechanism that ensures increased efficiency. The author's definition of the integration business processes of the aviation enterprise is presented as coordinated actions aimed at combining the functional units, operations, and resources into a single system to achieve strategic goals. The substantive characteristics of integration business processes in the environment of aviation enterprises are determined. It is proposed that modeling the integration business processes of aviation enterprises should be considered, such as creating mathematical, information, or conceptual models that reflect the interaction of various business processes to optimize their efficiency and coordination. It has been substantiated that the importance of integration business processes modeling for aviation enterprises is manifested in the following aspects: optimization of resources, improvement of coordination, forecasting and risk management, improvement of customer service, innovation, and adaptation to changes. A system of expedient methods for integration business processes modeling for aviation enterprises has been formed. An algorithm for modeling the integration business processes of aviation enterprise is presented, which allows to increase the efficiency of the aviation enterprise, optimizing all operations and minimizing risks and costs.

Keywords: Integration Business Processes · Aviation Enterprise · Aviation Industry · Sustainable Development · Business Process Modeling · Operation · Repair · Data Processing

1 Introduction

1.1 Introduction to the Problem

The integration development of aviation enterprises is of particular importance in the context of the current challenges facing the global economy, society, and ecosystem [1, 2]. The growth of international trade, tourism, and migration requires effective and

I. Ostroumov et al. (Eds.): ACASD 2025, LNNS 1418, pp. 59–73, 2025.
https://doi.org/10.1007/978-3-031-91992-3_5

continuous interaction between aviation market participants. At the same time, the aviation industry is under the influence of a number of crises, such as economic turbulence, disruptions in global supply chains, cyber threats, and the impact of scientific and technological progress [3]. Climate change is putting additional pressure on the aviation industry, forcing enterprises to implement environmentally friendly technologies and optimize operational processes [4].

Integration development allows aviation enterprises to join forces to increase their resilience to risks and modern challenges. Sharing innovations, coordinating routes, and standardizing processes help to increase efficiency and reduce costs [5]. In addition, integration helps to improve the quality of passenger service by providing convenient flight connections, transparency and accessibility of information, and more reliable logistics. In the context of globalization transformations, this helps to strengthen international relations and economic cooperation. The formation of integration business processes of aviation enterprises is an important element of their adaptation to modern market conditions and ensuring sustainable development. Integration business processes help to optimize the use of resources, reduce costs, improve the quality of services, and ensure the long-term sustainability of airlines. The introduction of modern digital technologies to automate operations and ensure effective communication between different departments is particularly important. Automation of business processes has a positive effect on the indicators of the aviation enterprise's activity and increases the need for their assessment in terms of various areas of activity (purchasing, finance, personnel, sales, and marketing) [6].

In the context of sustainable development, it is important to take into account environmental aspects, such as reducing carbon dioxide emissions and introducing energy-efficient technologies [7, 8]. Conservation and sustainable management of ecosystems is a necessary condition for ensuring sustainable development. Integration of ecosystem services into business management, decision-making, and policy development helps to preserve valuable natural resources and ensure long-term economic and social development [9].

Modeling the development of integration business processes of aviation enterprises allows to analyze existing processes, identify their weaknesses, and offer effective solutions for integrating various functions and departments [10, 11]. This helps to achieve high synergy between the operational, financial, and marketing areas of the aviation enterprise. Using modern software tools and analytical methods allows to optimize resource management and improve the accuracy of forecasting and strategic decision-making [12, 13]. Modeling is especially important to ensure continuity of operations and minimize risks in crises [14, 15]. Integration of processes such as maintenance and repair, automation and digitalization of key operations, and interaction with customers and partners can reduce costs and improve the quality of service for delivery [16, 17].

1.2 Motivation

In order to optimize the use of resources, reduce costs, improve the quality of services, and ensure long-term sustainability, aviation enterprises need to develop effective models of integration business process, which will increase the efficiency of operations and adaptation to crisis phenomena.

1.3 Contribution

The key contribution of this study is the developed methodology for modeling integration business processes with the aim of optimizing the main tasks of resource management, increasing the accuracy of forecasting, and making strategic decisions in the context of sustainable development, which provides the aviation enterprise with a high level of competitive advantages in the market, increases its profitability, and significantly simplifies the work of employees of all levels.

1.4 The Organization of the Paper

The paper contains six sections. The first section describes the introduction. The second section highlights the key findings of the literature review. In the third section, the main features and methods for modeling of integration business processes of the aviation enterprises are considered. The fourth section presents an example of calculation and modeling. The fifth section is the conclusion. Finally, the sixth section outlines the future scope.

2 Literature Review and Problem Statement

In modern research, there is a tendency towards an interdisciplinary approach that takes into account both theoretical aspects and practical experience of successful integration. At the same time, the lack of common standards and universal models for integration modeling creates additional challenges. This necessitates an in-depth analysis and development of adapted solutions for different aviation enterprises, which emphasize the complexity and multidimensionality of this scientific problem.

In today's environment, integration is the process of combining various elements, systems, or structures to create a single, coordinated mechanism that ensures increased efficiency. It covers the interaction between different levels of management, departments, business processes, and partners aimed at achieving a synergistic effect. Integration is becoming a key tool for enterprises to adapt to a rapidly changing market environment, where globalization, digitalization, and increased competition require maximum coherence [18].

Publications [19, 20] note that integration includes the introduction of common standards, technologies, and procedures that ensure effective communication, transparency, and interaction. An important aspect is the harmonization of environmental, social, and economic goals, which is in line with the principles of sustainable development. Thus, integration becomes the basis for creating competitive advantages and long-term success for enterprises in the modern world.

Paper [21] defines integration as a process of organizational changes aimed at creating a single economic mechanism. The purpose of this mechanism is to achieve a common strategic goal by combining efforts in its implementation, strengthening interaction, coordinating business processes, and harmonizing the use of the resource potential of participants.

The authors of the paper [22] believe that the integration of existing processes with new digital processes contributes to their automation and information exchange between the main divisions.

Publications [23, 24] have determined that modeling of integration processes helps to identify inefficient areas, predict risks, and develop strategies to minimize them. In addition, it supports the introduction of new technologies and innovations, increasing the overall productivity and competitiveness of the enterprise in a dynamic market.

Paper [25] emphasizes the value of the business process management (BPM) methodology in improving them, which is important for decision-making, and proposes a model of adaptive process improvement focused on value creation in complex organizational environments.

Thus, by integrating business processes, aviation enterprises become more flexible and able to respond quickly to changes in demand, improve communication between different parts and departments of the airline, ensure efficient operation, increase its competitiveness, and attract investment.

The aim of this research is to generalize the theoretical foundations and develop methodological approaches to modeling the integration business processes of aviation enterprises in the context of sustainable development.

3 Materials and Methods

Integration plays a key role in the development of aviation enterprise's business processes, providing coordinated actions aimed at combining functional units, operations, and resources of the enterprise into a single system to achieve strategic goals. They cover the interaction between various components of the business, such as flight management, maintenance, cargo logistics, passenger services, and financial management.

Integration allows businesses to adapt to changes in the external environment, introduce innovative technologies, optimize resource use, and improve interaction with partners and customers.

The substantive characteristics of integration business processes in the environment of aviation enterprises are summarized in Table 1.

Thus, the integration business processes of aviation enterprises contribute to the achievement of strategic goals through the introduction of modern technologies, and process automation, which allows to increase labor productivity and service quality. Modeling the aviation enterprise's integration business processes is the process of creating mathematical, information, or conceptual models that reflect the interaction of various business processes of the enterprise to optimize their efficiency and coordination. This allows to predict the results of changes in the system, analyze existing problems, identify opportunities for improvement, and optimize the use of resources.

An enterprise's stronger competitiveness requires ensuring the growth of its sales of products or services. However, even a rise in sales volumes can produce meaningful results only if it is accompanied by positive outcomes in the enterprise's profits, and cost of sales [26].

The importance of modeling the integration business processes of aviation enterprise is manifested in the following aspects [27, 28]:

Table 1. Content characteristics of integration business processes in enterprise environment.

Characteristic	Content
The purpose of integration	Achievement of strategic goals by combining the activities of departments and improving coordination between them
Main components	Flight management, maintenance, cargo logistics, passenger services, and financial management
Types of integration	Integration of internal processes (between business units) and integration with external partners and suppliers
Tools and technologies	Use of IT systems for process automation, modern methods of forecasting and optimization (for example, ERP systems)
Competitive advantages	Reduced costs, improved service quality, increased productivity, and flexibility in responding to changes
Risks and challenges	Technical problems in integrating new systems, problems with coordination between departments, and high initial implementation costs
Attracting innovation	Adoption of new technologies, such as digitalization, automation and the use of drones for monitoring and delivery
Resource management	Optimize the use of the fleet, personnel, and infrastructure, reduce maintenance and fuel costs
Social and environmental aspects	Improving safety, reducing environmental impact, and improving working conditions for staff
Interaction with customers	Improving passenger service, reducing delays, and increasing customer satisfaction

- Optimization of resources. Modeling allows for the efficient allocation of resources (personnel, technical equipment, and aircraft fleet), which reduces costs and increases the productivity of the enterprise.
- Improved coordination. Models help to ensure coordinated interaction between different departments of the enterprise, which increases the efficiency of decision-making and reduces the risk of errors.
- Forecasting and risk management. Modeling helps to predict possible problems or crises in business processes, which makes it possible to develop strategies to prevent or minimize them.
- Improving customer service. Modeling customer interaction processes allows to improve the quality of service, reduce waiting time, and increase passenger satisfaction.
- Innovation and adaptation to change. Integration models allow airlines to adapt more quickly to changes in the market environment, technological innovations, or regulatory changes, which helps them maintain a competitive advantage.

There is a wide range of tools and approaches to modeling the integration business processes of aviation enterprises, which makes it possible to take into account various aspects of its operations. Each approach has its advantages and is used depending on the specifics of the processes being modeled. The main tools include mathematical models, business process management systems, data optimization and analysis methods, and modern information technologies such as Enterprise resource planning (ERP) systems.

The introduction of digital tools contributes to increased competitiveness, customer satisfaction, and profitability, so it is a necessary step to ensure their sustainable development [29, 30].

These tools allow not only to accurately reproduce existing processes but also to predict their possible changes and consequences.

The choice of a particular approach depends on the goals set and the level of complexity of the integration processes that need to be improved. The development of complex models requires flexibility in the choice of methods, which allows to adapt to changing conditions and achieve high efficiency. Thus, the use of different approaches enables aviation enterprises to optimize their business processes, increasing their overall competitiveness.

The system of methods for modeling individual integration business processes of aviation enterprises is presented in Table 2.

Assessing the effectiveness of business process transformation using information technology, as well as determining the level of efficiency of individual components of business processes can be used during the making management decisions on analyzing, preventing, and countering real and potential threats in a dynamic business environment [31].

In today's environment, aviation enterprises face the need to optimize their operations to increase efficiency and reduce costs. For this purpose, it is important to develop a common approach to the step-by-step process of modeling integration business processes.

This approach facilitates the automation and integration of processes, which increases the speed of data processing and the accuracy of tasks. At the same time, it significantly reduces the risks associated with the human factor and malfunctions of individual systems.

The development of a unified, step-by-step approach will ensure the sustainability and transparency of all aviation enterprise's processes, which is important for ensuring its competitiveness in the market.

The algorithm for modeling the integration business processes of the aviation enterprise is shown in Fig. 1.

Thus, the proposed algorithm for modeling the integration business processes of the aviation enterprise covers several key stages, starting with the analysis of current processes and the definition of integration goals. An important part is the development of a map of integrated processes, which allows to accurately determine the interaction between different departments of the aviation enterprise. The next step is to evaluate the technologies and tools for automation and integration that will ensure the smooth operation of the systems.

Table 2. Methods of modeling individual integration processes of aviation enterprises.

Integration business process	Modeling methods	Description of methods	Advantages
Flight and route management	Mathematical modeling	Use of models to predict the load, routes, time, and resources used in flights	Improves planning accuracy, optimizes resource utilization, and reduces delays
Fleet maintenance	Discrete-stochastic modeling	Modeling the operation of technical processes using random variables, which accounts the probability of breakdowns and the need for maintenance	It allows to take into account uncertainty and risks, increases the reliability of fleet maintenance
Cargo and passenger logistics	Business process modeling (BPM)	Using BPM to describe, analyze, and optimize processes related to transportation	It increases logistics efficiency, reduces cargo delivery time, and improves passenger service
Financial management	Decision support systems (DSS)	Use of specialized systems to analyze financial data and support decisions on budget, expenditures, and investments	It helps to make informed decisions on financing and reduces financial risk
Personnel management	Human resources modeling	Use of models to forecast staffing needs, their optimal distribution, and training.	Increases staff efficiency and optimizes training and development costs
Marketing and customer service	Data mining systems	Using big data analysis methods to identify patterns and trends in customer behavior	It improves customer experience, personalizes service, and increases loyalty
Procurement and supply management	Supply chain modeling	Modeling of supply flows, including the purchase, storage, and delivery of materials and spare parts for aviation companies	Reduces supply costs, optimizes warehouses, and improves the availability of necessary materials

(*continued*)

Table 2. (*continued*)

Integration business process	Modeling methods	Description of methods	Advantages
Information technology and cybersecurity	Modeling threats and risks	Assessment and modeling of cybersecurity risks and threats to enterprise information systems	It protects data, reduces the risk of information leakage, and improves the security of internal systems
Development strategy management	Strategic modeling (SWOT, PESTLE)	Use of models to analyze the strategic direction of enterprise development, identify opportunities and threats in the market	It improves the efficiency and competitiveness of the enterprise and allows to respond effectively to changes in the market

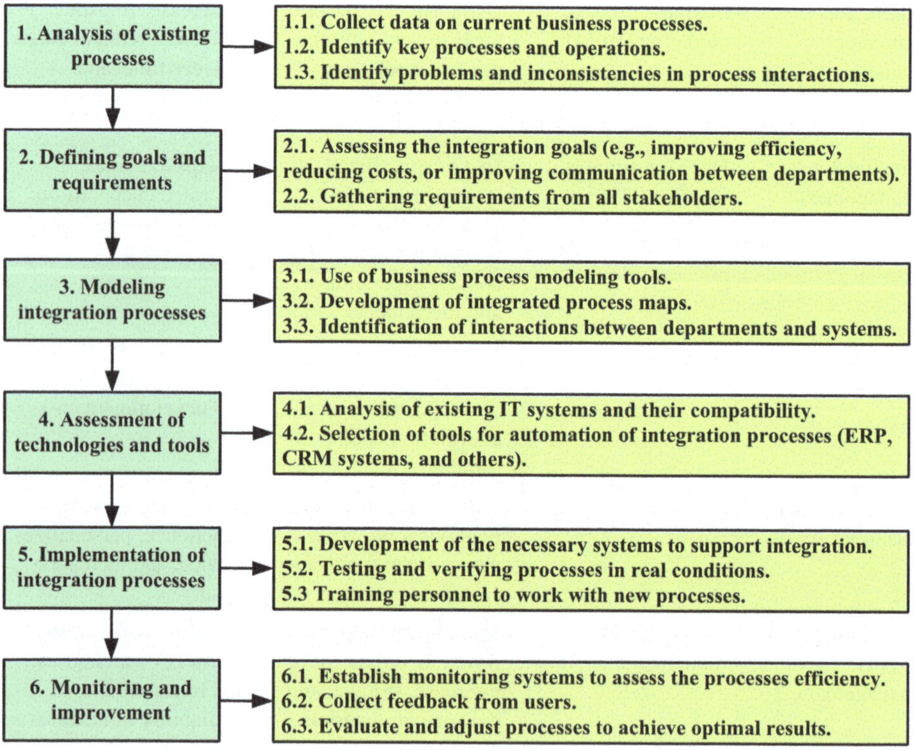

Fig. 1. Algorithm for modeling the integration business processes of aviation enterprise.

After new processes are implemented, it is important to continuously monitor their effectiveness and make adjustments to achieve better results. In general, such a scheme can significantly improve the efficiency of the aviation enterprise, optimizing all operations and minimizing risks.

4 Results and Discussions

Let us consider the features of modeling business processes of the aviation enterprise. As the studied business process, we will choose the maintenance and repair process, which is important from the point of view of providing equipment reliability.

The implementation of the maintenance and repair process has the most significant costs among all the processes of aircraft equipment operation. This is explained by the fact that technological procedures are performed by personnel, it is necessary to have a stock of spare parts, measuring equipment, and other resource support.

The maintenance and repair business process includes the following procedures:

1. Monitoring equipment performance indicators.
2. Data collection.
3. Data processing.
4. Making a decision on the occurrence of equipment failure.
5. Making a decision on preventive maintenance implementation.
6. Performing repairs by replacing an equipment component using the spare parts stock.
7. Performing repairs through adjustments, regulations, and restoration work.
8. Performing maintenance.
9. Using equipment for functional purposes.

The structural diagram of the interconnection of the procedures of the maintenance and repair business process is shown in Fig. 2.

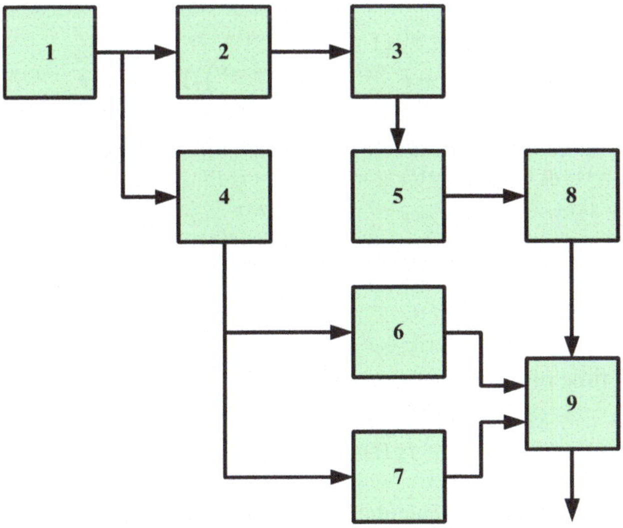

Fig. 2. Business process of maintenance and repair.

The numbers in Fig. 2 indicate the numbers of the procedures of the maintenance and repair business process given above.

Discrepancies may occur during the execution of certain procedures. The moment of occurrence of the discrepancy is random and is characterized by the probability density function $f_i(t)$, where i is a number of procedures. We assumed that the business process has a general discrepancy if the efficiency of maintenance and repair performance has significantly decreased. The decrease in efficiency will also occur at a random point in time and be described by its PDF $f_\Sigma(t)$. Let's consider the features of finding this PDF.

Let the probability density function $f_i(t)$ is exponential, therefore

$$f_i(t) = \lambda_i e^{-\lambda_i t}, \lambda_i > 0, \tag{1}$$

where λ_i is the rate of discrepancy occurrence.

The probabilities of discrepancy occurrence and absence for the i-th procedure of business process

$$q_i(t) = \int_0^t f_i(t)dt = 1 - e^{-\lambda_i t}. \tag{2}$$

$$p_i(t) = \int_t^\infty f_i(t)dt = e^{-\lambda_i t}. \tag{3}$$

Then the probability of conformity of the entire business process as a whole will be equal to

$$p_\Sigma(t) = p_1(t)(1 - (1 - p_2(t)p_3(t)p_5(t)p_8(t)) \times (1 - p_4(t)(1 - q_6(t)q_7(t))))p_9(t). \tag{4}$$

Taking into account (1)–(3), we will get

$$\begin{aligned} p_\Sigma(t) &= e^{-\lambda_1 t}\left(1 - \left(1 - e^{-(\lambda_2+\lambda_3+\lambda_5+\lambda_8)t}\right)\right. \\ &\times \left(1 - e^{-\lambda_4 t}\left(e^{-\lambda_6 t} + e^{-\lambda_7 t} - e^{-(\lambda_6+\lambda_7)t}\right)\right)\Big)e^{-\lambda_9 t} = e^{-(\lambda_1+\lambda_9)t} \\ &\times\left(1 - \left(1 - e^{-(\lambda_2+\lambda_3+\lambda_5+\lambda_8)t}\right)\left(1 - e^{-(\lambda_4+\lambda_6)t} - e^{-(\lambda_4+\lambda_7)t} + e^{-(\lambda_4+\lambda_6+\lambda_7)t}\right)\right) \\ &= e^{-(\lambda_1+\lambda_9)t}\left(e^{-(\lambda_4+\lambda_6)t} + e^{-(\lambda_4+\lambda_7)t} - e^{-(\lambda_4+\lambda_6+\lambda_7)t} + e^{-(\lambda_2+\lambda_3+\lambda_5+\lambda_8)t}\right. \\ &\left. - e^{-(\lambda_2+\lambda_3+\lambda_5+\lambda_8+\lambda_4+\lambda_6)t} - e^{-(\lambda_2+\lambda_3+\lambda_5+\lambda_8+\lambda_4+\lambda_7)t} + e^{-(\lambda_2+\lambda_3+\lambda_5+\lambda_8+\lambda_4+\lambda_6+\lambda_7)t}\right) \\ &= e^{-(\lambda_1+\lambda_4+\lambda_6+\lambda_9)t} + e^{-(\lambda_1+\lambda_4+\lambda_7+\lambda_9)t} - e^{-(\lambda_1+\lambda_4+\lambda_6+\lambda_7+\lambda_9)t} \\ &+ e^{-(\lambda_1+\lambda_2+\lambda_3+\lambda_5+\lambda_8+\lambda_9)t} - e^{-(\lambda_1+\lambda_2+\lambda_3+\lambda_4+\lambda_5+\lambda_6+\lambda_8+\lambda_9)t} \\ &- e^{-(\lambda_1+\lambda_2+\lambda_3+\lambda_4+\lambda_5+\lambda_7+\lambda_8+\lambda_9)t} + e^{-\sum_{i=1}^{9}\lambda_i t}. \end{aligned} \tag{5}$$

The PDF of time of discrepancy occurrence

$$f_\Sigma(t) = -\frac{dp_\Sigma(t)}{dt}. \tag{6}$$

We performed modeling and calculation. The initial data for calculation and modeling are: $\lambda_1 = 0.001$, $\lambda_2 = 0.004$, $\lambda_3 = 0.002$, $\lambda_4 = 0.001$, $\lambda_5 = 0.003$, $\lambda_6 = 0.001$, $\lambda_7 = 0.002$, $\lambda_8 = 0.0015$, $\lambda_9 = 0.0008$. The number of iterations is 1000.

The modeling includes the following steps:

- Obtaining exponentially distributed random variables that characterize the moment of occurrence of the discrepancy in each of the business process procedures.
- Developing a model for functional diagram of the business process.
- Constructing a distribution histogram.
- Comparing the histogram with analytical equations.
- Finding the expected value and standard deviation.

Figure 3 shows the simulation results in the form of a random process of discrepancies occurrence. A comparison of the simulation results with analytical equations is presented in Fig. 4. We can see that visually the simulation results are consistent with the given formulas (4)–(6). For the final decision on consistency, the chi-square criterion was used, which confirmed the coincidence of the data with the distribution with a confidence probability of 0.05.

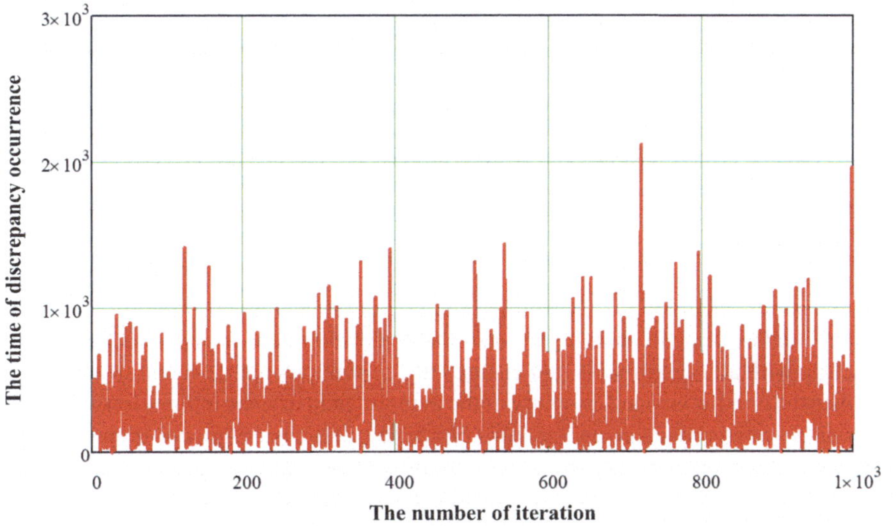

Fig. 3. The example of discrepancy occurrence simulation.

The expected values and standard deviations of time moment of discrepancy occurrence are 299.6, 270.9, respectively, for simulation results and 306.2, 271.8, respectively, for analytical calculations.

The presented approach to analysis and modeling can be applied to any business process of the aviation enterprise.

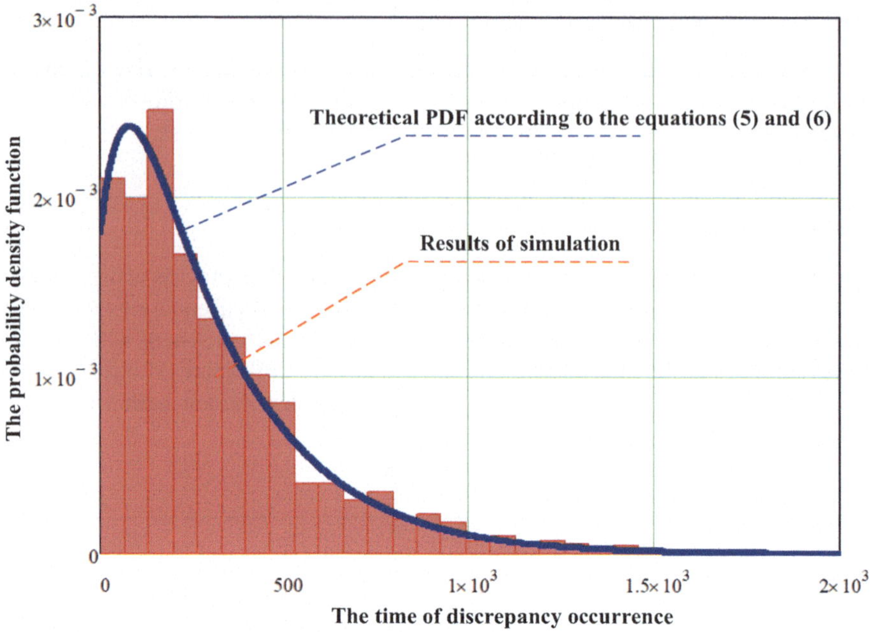

Fig. 4. The comparison of the simulation results and analytical equations.

5 Conclusions

Thus, it can be concluded that modeling the integration business processes of an airline company is a key stage in ensuring its sustainable development, as it allows for effective interaction with all departments and ensures the continuous operation of all operations.

Given the complexity and diversity of the aviation industry, the integration of business processes helps to reduce costs, increase efficiency, and ensure greater transparency in the work of the aviation enterprise. Automation of processes is also important, as it allows for a quick response to changes and market demands, and ensures the accuracy of tasks.

Since airlines operate in a highly competitive environment, effective modeling of integration business processes helps to reduce the time for operations and the number of errors. This, in turn, improves the quality of customer service, which is an important factor in ensuring competitiveness in the market. That is why modeling of integration business processes is a key tool for ensuring the sustainable development of an airline company and helps to increase its efficiency, competitiveness, and ability to adapt to changes. This allows the aviation enterprise to be more resilient to economic challenges and provide a high level of service quality to its customers.

The paper considers the example of the analysis of the business process of maintenance and repair of aviation equipment. For this business process, the possibilities of discrepancies during the execution of individual procedures were analyzed. To quantitatively assess the efficiency of the business process, analytical equations for statistical characteristics of the process of discrepancy occurrence were obtained. To verify the obtained formulas, the imitation model of the business process was developed. The consistency of the analytical equations and the modeling results confirmed the correctness of the research.

The results of the paper can be used when improving the structure of the aviation enterprise in terms of the use of information technologies for data processing and the automation of its activities.

6 Future Scope

Future research is related to:

- development of an imitation model of business processes of the aviation enterprise;
- analysis of the efficiency of the business process carried out;
- development of recommendations for the implementation of data processing technologies in the activities of the aviation enterprise;
- minimization of costs for the equipment operation of aviation enterprise;
- optimization of the organizational structure of an aviation enterprise.

References

1. Acemoglu, D., Laibson, D., List, J.A.: Economics. Pearson, London (2018)
2. Kabashkin, I., Misnevs, B., Zervina, O.: Artificial intelligence in aviation: new professionals for new technologies. Appl. Sci. **13**(11660), 1–33 (2023). https://doi.org/10.3390/app132111660
3. Averyanova, Y., et al.: Turbulence detection and classification algorithm using data from AWR. In: IEEE 2nd Ukrainian Microwave Week (UkrMW), pp. 518–522. IEEE, Ukraine (2022). https://doi.org/10.1109/UkrMW58013.2022.10037172
4. Zaporozhets, A., Babak, V., Isaienko, V., Babikova, K.: Analysis of the air pollution monitoring system in Ukraine. In: Babak, V., Isaienko, V., Zaporozhets, A. (eds.) Systems, Decision and Control in Energy I. Studies in Systems, Decision and Control, vol. 298, pp. 85–110. Springer, Cham (2020). https://doi.org/10.1007/978-3-030-48583-2_6
5. Raza, A., Ulansky, V.: Through-life maintenance cost of digital avionics. Appl. Sci. **11**(2), 715 (2021). https://doi.org/10.3390/app11020715
6. Poberezhna, Z.: Formation of a system for optimizing business processes of aviation enterprises based on their automation. In: Ostroumov, I., Zaliskyi, M. (eds.) Proceedings of the 2nd International Workshop on Advances in Civil Aviation Systems Development. ACASD 2024, Lecture Notes in Networks and Systems, vol. 992, pp. 325–338. Springer, Cham (2024). https://doi.org/10.1007/978-3-031-60196-5_24
7. Korba, P., Sekelova, I., Koscakova, M., Behunova, A.: Passengers' knowledge and attitudes toward green initiatives in aviation. Sustain. For. **15**(7), 6187 (2023). https://doi.org/10.3390/su15076187
8. Calvet, L.: Towards environmentally sustainable aviation: a review on operational optimization. Future Transp. **4**(2), 518–547 (2024). https://doi.org/10.3390/futuretransp4020025
9. Smerichevska, S., Poberezhna, Z., Mykhalchenko, O., Shtyk, Y., Pokanevych, Y.: Modeling and evaluation of organizational and economic support for sustainable development of transport enterprises: innovative and ecological aspects. Finan. Credit Activ. Prob. Theor. Pract. **4**(51), 218–229 (2023). https://doi.org/10.55643/fcaptp.4.51.2023.4121

10. Andre, J.: Industry 4.0: Paradoxes and Conflicts. Wiley, New York (2019)
11. Tachinina, O.M., Lysenko, O.I., Ponomarenko, S.O., Alekseeva, I.V.: Conceptual proposals for the creation of a fully reusable light-class aerospace system in Ukraine. In: IEEE 6th International Conference on Methods and Systems of Navigation and Motion Control, pp. 85–88. IEEE, Kyiv (2020). https://doi.org/10.1109/MSNMC50359.2020.9255504
12. Ostroumov, I.V., Kuzmenko, N.S.: Accuracy assessment of aircraft positioning by multiple radio navigational aids. Telecomm. Radio Eng. **77**(8), 705–715 (2018). https://doi.org/10.1615/TelecomRadEng.v77.i8.40
13. Ivashchuk, O., et al.: A configuration analysis of Ukrainian flight routes network. In: 16th International Conference on the Experience of Designing and Application of CAD Systems, pp. 6–10. IEEE, Lviv (2021). https://doi.org/10.1109/CADSM52681.2021.9385263
14. Solomentsev, O.V., Zaliskyi, M.Y., Zuiev, O.V., Asanov, M.M.: Data processing in exploitation system of unmanned aerial vehicles radioelectronic equipment. In: 2nd International Conference Actual Problems of Unmanned Air Vehicles Developments, pp. 77–80. IEEE, Kyiv (2013). https://doi.org/10.1109/APUAVD.2013.6705288
15. Ostroumov, I.V., Kuzmenko, N.S.: Accuracy estimation of alternative positioning in navigation. In: 4th International Conference Methods and Systems of Navigation and Motion Control, pp. 291–294. IEEE, Kyiv (2016). https://doi.org/10.1109/MSNMC.2016.7783164
16. Solomentsev, O., Zaliskyi, M., Nemyrovets, Y., Asanov, M.: Signal processing in case of radio equipment technical state deterioration. In: Signal Processing Symposium 2015 (SPS 2015), pp. 1–5. IEEE, Debe (2015). https://doi.org/10.1109/SPS.2015.7168312
17. Solomentsev, O., Zaliskyi, M., Herasymenko, T., Petrova, Y.: Data processing method for deterioration detection during radio equipment operation. In: IEEE Microwave Theory and Techniques in Wireless Communications (MTTW), pp. 1–4. IEEE, Riga (2019). https://doi.org/10.1109/MTTW.2019.8897232
18. Vergal, K.Y.: Theoretical and multiple approaches to managing the integration processes of enterprises. Theor. Pract. Aspect. Econ. Intell. Prop. **1**(10), 189–193 (2014)
19. Kondratiuk, V.V.: Integration processes: concepts, approaches, foreign experience. Scientific works of Kirovograd National Technical University. Econ. Sci. **12**(2), 215–219 (2007)
20. Grinchenko, Y.L., Nienno, I.M.: Innovative Forms of Integration in High-tech Industries. Innovative Economy: Theoretical and Practical Aspects. Aldi-plus, Kherson (2019)
21. Mitsenko, N.G.: Integration processes in economic theories. Entrepren. Trade. **23**, 42–49 (2018)
22. Goonetillake, J.F., Ren, G., Li, H.: An integration of business processes and information management for improving the efficiency and reliability of infrastructure. Appl. Sci. **13**(24), 12974 (2023). https://doi.org/10.3390/app132412974
23. Melnyk, O.G., Dvulit, Z.P., Zlotnik, M.L., Malynovska, Y.B.: Features of the application of a systematic approach to modeling business processes at the enterprise. Manag. Entrepren. Ukraine Stag. Format. Prob. Dev. **1**(1), 46–54 (2019)
24. Shmatkovska, T., Dzyamulych, M., Stashchuk, O.: Features of business process modeling in the context of the formation of the digital economy. Econ. Soc. **26**, 1–4 (2021). https://doi.org/10.32782/2524-0072/2021-26-66
25. Teixeira, A.R., Ferreira, J.V., Ramos, A.L.: Optimization of business processes through BPM methodology: a case study on data analysis and performance improvement. Information. **15**(11), 724 (2024). https://doi.org/10.3390/info15110724
26. Poberezhna, Z.: Comprehensive assessment of the airlines' competitiveness. Econ. Ann. XXI. **167**(9–10), 32–36 (2017). https://doi.org/10.21003/ea.V167-07
27. Kononenko, Z.A., Sharavara, R.I., Yakovenko, T.I.: Business process modeling as a component of enterprise management. Econ. Soc. **62**, 1–8 (2024). https://doi.org/10.32782/2524-0072/2024-62-134

28. Paranytsia, N.V., Paranytsia, S.P., Bulichov, O.S.: Methodology of business process modeling. Econ. State. **3**, 59–62 (2022). https://doi.org/10.32702/2306-6806.2022.3.59
29. Ostroumov, I., Ivannikova, V., Kuzmenko, N., Zaliskyi, M.: Impact analysis of Russian-Ukrainian war on airspace. J. Air Transp. Manag. **124**, 102742 (2025). https://doi.org/10.1016/j.jairtraman.2025.102742
30. Smerichevskyi, S., Poberezhna, Z., Kryvovyazyuk, I., Ivanenko, L., Malnov, D.: Formation of principles of a customer-oriented approach by transport enterprises in conditions of sustainable development. In: E3S Web Conference. Innovations in Construction and Smart Building Technologies for Comfortable, Energy Efficient and Sustainable Lifestyle (ICSBT 2024), vol. 534, pp. 1–12 (2024). https://doi.org/10.1051/e3sconf/202453401022
31. Poberezhna, Z., Truhan, O.: Strategic diagnostics of business processes of enterprises using information technologies in the context of modern challenges. In: 14th International Conference on Advanced Computer Information Technologies, pp. 455–459. Ceske Budejovice, IEEE (2024). https://doi.org/10.1109/ACIT62333.2024.10712609

Experimental Study of the Method for Cyber Incidents Management in Aviation Critical Infrastructure

Sergiy Gnatyuk(✉), Viktoria Sydorenko, Artem Polozhentsev, and Serhii Sydorenko

National Aviation University, Liubomyra Huzara Ave., 1, Kyiv 03058, Ukraine
serhii.hnatiuk@npp.kai.edu.ua

Abstract. Ensuring the security of civil aviation infrastructure (CAI) is critical in a world where cyber threats are increasingly targeting critical systems, potentially impacting global aviation safety, operational continuity, and regulatory compliance. CAI includes essential sectors such as air traffic control, airport operations and communication networks. Disruptions to these systems can have far-reaching consequences, requiring effective approaches to identifying, assessing and managing cyber threats. Building on their previously developed universal method for managing IT threats in Critical Information Infrastructures (CII), the authors present an adapted method specifically tailored to the unique challenges of CAI. This method integrates the STRIDE threat identification model with the TODIM multi-criteria decision-making approach to enable systematic threat assessment and prioritization. The method also introduces a criticality variable to account for the varying importance of CAI components. Experimental validation across three CAI sub-sectors – communication networks, Air Traffic Control (ATC) systems, and Passenger Data Management Systems (PDMS) – demonstrated the method's adaptability and effectiveness. The results identified Denial of Service (DoS) and Information Disclosure as critical threats, providing a robust framework for prioritizing risks and enhancing the cybersecurity of CAI.

Keywords: Civil Aviation Infrastructure · Cybersecurity · Cyber Threat Prioritization · STRIDE Model · TODIM method · Aviation Cybersecurity Risks · ATC · PDMS · Critical Information Infrastructure

1 Introduction

In an era of rapid technological advancement, critical infrastructure is increasingly reliant on digital systems to ensure efficiency, connectivity and operational reliability. Civil aviation infrastructure (CAI), a cornerstone of global and national transport networks, is no exception. Comprising air traffic control systems, communications networks, airport operations and passenger data management, CAI is a highly interconnected and interdependent system that is essential for the seamless functioning of the aviation industry. However, this reliance on digital technologies has made CAI particularly vulnerable to cyber threats, which pose risks to operational continuity, security and compliance with international standards [1].

I. Ostroumov et al. (Eds.): ACASD 2025, LNNS 1418, pp. 74–91, 2025.
https://doi.org/10.1007/978-3-031-91992-3_6

Effective cybersecurity management for CAI requires addressing a number of challenges, including the complexity of aviation systems, the need for real-time operations, and the critical importance of ensuring passenger safety [2]. Existing methodologies for managing IT threats to critical information infrastructure (CII), while valuable, often do not fully address these unique challenges. Frameworks such as STRIDE and TODIM provide systematic approaches to threat modelling and decision making, but their direct application to CII requires modifications to address the specific requirements of the sector.

This paper builds on previous research in IT threat management for CII by adapting a universal methodology to the specific needs of CAI. The proposed approach combines the structured threat identification capabilities of the STRIDE model with the multi-criteria decision flexibility of the TODIM method. In addition, it incorporates a criticality variable to account for the hierarchical importance of different CAI components, allowing for a more accurate assessment of cyber risks. By aligning an established methodology with the specific needs of CAI, this research seeks to advance cybersecurity management in the aviation sector and provide a foundation for mitigating the risks posed by evolving cyber threats.

2 Literature Review

In their previous articles [3, 4], the authors conducted a detailed analysis of methodologies for managing IT incidents in CII. The methodologies compared included STRIDE [5], NIST SP 800-30 [6], ISO/IEC 27005 [7], OCTAVE [8], and COBIT [9]. These methodologies were evaluated using specific criteria designed to measure their effectiveness in addressing IT threats within CII. The criteria used for the comparison were:

- Ease of Use (EU): How easily the methodology can be applied in practice.
- Comprehensiveness (CM): The extent to which the methodology addresses all aspects of IT incident and threat management.
- Integration with Other Systems (IS): The ability of the methodology to integrate seamlessly with existing security tools and frameworks.
- Critical Infrastructure Focus (CI): Whether the methodology accounts for the unique characteristics and needs of critical infrastructure sectors.
- Objectivity (OB): The extent to which the methodology reduces subjectivity in decision-making.
- Time to Use (ET): The time and resources required to implement the methodology effectively.

The analysis revealed the following:

- STRIDE stood out as the most effective methodology, balancing comprehensiveness, usability, and integration.
- NIST SP 800-30 offered a detailed framework for risk management but required substantial time and resources, limiting its accessibility for smaller organizations.
- ISO/IEC 27005 aligned well with other ISO standards and provided a structured approach to risk management but posed challenges due to its complexity and resource demands.

- OCTAVE focused on critical asset protection but required extensive staff involvement, complicating its use in larger infrastructures.
- COBIT emphasized IT governance and integration with business objectives but lacked a direct focus on cybersecurity and was resource-intensive.

The analysis demonstrated that STRIDE offers several key advantages, making it the most suitable choice for IT incident management in CII:

- Comprehensive Threat Coverage: STRIDE addresses six critical threat categories – Spoofing, Tampering, Repudiation, Information Disclosure, Denial of Service, and Elevation of Privilege – ensuring all major vulnerabilities are considered.
- Structured and Practical Framework: Its clear methodology supports systematic threat identification and mitigation planning, making it accessible for professionals with varying levels of expertise.
- Effective Integration: STRIDE integrates well with other security tools and practices, enabling organizations to develop cohesive cybersecurity strategies.
- Objective Evaluation: By providing a consistent and structured approach, STRIDE minimizes biases and ensures that threats are evaluated based on clear, measurable criteria.

Building on the prior analysis of IT threat modeling methodologies, the authors further examined decision-making methods essential for prioritizing IT threats within CII. These methods offer structured approaches to evaluate complex problems and determine the most appropriate actions by considering multiple criteria and alternatives.

To assess the suitability of decision-making methods for IT threat prioritization in CII, the authors used the following evaluation criteria:

- Applicability to Critical Infrastructure (CI): The ability of the method to address the specific needs and challenges of CII.
- Flexibility (FL): The capacity of the method to adapt to different decision-making scenarios and incorporate a variety of criteria.
- Scalability (SC): The method's ability to handle decision-making processes for both small-scale and large-scale infrastructure systems.
- Consideration of Risk and Uncertainty (CR): How effectively the method incorporates the assessment of risks and uncertainties in decision-making.
- Ease of Use (EU): The simplicity and practicality of applying the method in real-world scenarios.

The analysis compared three prominent decision-making methods: AHP (Analytic Hierarchical Process) [10], TODIM (Interactive Multicriteria Decision-Making) [11], and TOPSIS (Technique for Order of Preference by Similarity to Ideal Solution) [12]. The findings concluded that TODIM is the most suitable method for IT threat prioritization in CII.

TODIM effectively integrates risk and uncertainty into the decision-making process, a crucial requirement for CII, where threats often involve significant unpredictability. Its flexibility allows for the inclusion of multiple criteria, ensuring a comprehensive evaluation of threats. While it involves complex calculations and subjective judgments, its strengths outweigh these challenges, making it highly effective for managing CII-specific risks.

In contrast, AHP offers an intuitive and scalable approach that balances quantitative and qualitative criteria. However, its inability to adequately address risk and its reliance on subjective weighting reduce its effectiveness for prioritizing threats in dynamic CII environments. Similarly, TOPSIS, while easy to implement and capable of identifying optimal solutions, is less robust in handling risks and uncertainties, which are critical factors in CII decision-making.

The authors conclude that TODIM provides the most balanced and adaptable framework for prioritizing IT threats in CII, as it meets the critical requirements of risk consideration, flexibility, and scalability. By addressing the complexities of CII, TODIM ensures a robust foundation for effective threat prioritization and management.

3 Method for Cyber Incidents Management in Aviation Critical Infrastructure

It is proposed to disclose Method for Cyber Incidents Management in Aviation Critical Infrastructure (CAI) in the following stages:

Stage 1: Identification of Cyber Threats to Civil Aviation Infrastructure.

The identification of cyber threats is a critical stage in the process of managing cybersecurity risks to CAI [13]. The goal of this stage is to recognize potential threats that could compromise the normal operation of CAI. Threat identification can utilize various international frameworks, such as STRIDE, NIST SP 800–30, ISO/IEC 27005, OCTAVE, or COBIT, depending on the unique characteristics of CAI. The identified threats constitute the set of potential cyber threats, referred to as the U_i set (1):

$$U_i = \{U_1, U_2, \ldots, U_n\}, \tag{1}$$

where U_i – the set of potential cyber threats, U_1, U_2, ..., U_n represents the identified potential cyber threats.

Stage 2: Establish Criteria for Assessing Cyber Threats to CAI.

For each identified threat U_i corresponding set of evaluation criteria K is defined:

$$K = \{k_1, k_2, \ldots, k_m\}, \tag{2}$$

where K represents the collection of parameters used to evaluate cyber threats, with, k_1, k_2, ..., k_m – denoting individual evaluation factors.

Each threat U_i is analyzed against the criteria within K to determine its potential impact, significance, and priority for implementing appropriate mitigation measures.

Stage 3. Gather and Standardize Data on Cyber Threats to CAI.

This stage involves the systematic collection, evaluation, and standardization of data related to cyber threats targeting CAI. The goal is to ensure a fair and consistent approach to threat evaluation. Each threat U_i is assessed using the predefined criteria K, such as the likelihood of the threat occurring, its potential impact, and the complexity of its execution.

Each criterion k is assigned a specific weighting factor ω_k, which reflects its importance in the overall assessment. The sum of all weighting factors equals 1, according to (3).

$$\sum\nolimits_{k=1}^{K} \omega_k = 1, \tag{3}$$

where K is overall number of criteria, and ω_k represents the weight assigned to each individual criterion k.

Stage 4. Assign Weights to Cyber Threat Criteria for CAI.

Determining the weighting factors for each criterion used in assessing cyber threats to CAI is a critical step. It enables consideration of the relative importance of various aspects of each threat, ensuring a fair and balanced evaluation process. Each criterion is scored on a predefined scale. The authors recommend using a simple 1-to-5 scale, where 5 indicates the highest level of probability, impact, or complexity, and 1 represents the lowest. This straightforward approach facilitates the assessment process for experts.

For each criterion, the average of the geometric scores provided by experts is calculated using the formula (4):

$$k = \left(\prod\nolimits_{j=1}^{n} \nu k_j\right)^{\frac{1}{n}}, \tag{4}$$

where νk_j is the score assigned to criterion k by expert j, and n is the number of experts.

Following this, a vector of weighting factors is established by normalizing the geometric means of the scores (5):

$$W = (W_1, W_2, \ldots, W_n)^T, \tag{5}$$

where W_i is the weight assigned to criterion k

$$W_{jr} = \frac{W_j}{\sum_{r=1}^{n} W_r}, \tag{6}$$

where W_j is the geometric mean of scores for that criterion.

This process ensures that each criterion's weight reflects its relative importance in assessing cyber threats to CAI while maintaining consistency across the evaluation.

Stage 5. Pairwise Comparisons of Alternative Threats to CAI.

Pairwise comparisons are used to determine the dominance of one threat over another by employing a prospective value function. This function incorporates the weights of the evaluation criteria and the threat ratings for each criterion (7):

$$PV\left(U_i, U_j\right) = \sum\nolimits_{k=1}^{K} W_k f\left(v_{U_i,k}, v_{U_j,k}, \alpha\right), \tag{7}$$

where U_i and U_j are the cyber threats which are compared; k is the evaluation criterion; W_k is the weight of the criterion k; $v_{U_i,k}$ and $v_{U_j,k}$ are the threat ratings by the criterion; α is a parameter that reflects the attitude toward risk.

To ensure an accurate assessment of threats to CAI, a criticality variable C is introduced to reflect the importance of aviation infrastructure and its potential impact. This variable accounts for the diverse levels of significance within the civil aviation sector, such as air traffic control systems, airport operations, and passenger management systems.

The criticality variable C takes values from 1 to 4, corresponding to the infrastructure's criticality level:

- $C = 1$: Infrastructure vital to international aviation safety. Disruption could result in a global aviation crisis.
- $C = 2$: Infrastructure essential to national aviation safety. Disruption could cause significant national delays and operational risks.
- $C = 3$: Infrastructure important for regional operations. Disruption could cause localized aviation disruptions.
- $C = 4$: Infrastructure critical for local airport operations. Disruption could lead to limited, localized issues.

Incorporating the criticality variable C as an additional criterion within the multicriteria analysis provides a more nuanced understanding of the potential impact of cyber threats on different levels of aviation infrastructure. This enhances the precision of threat assessments and informs prioritized management decisions.

Stage 6. Conduct an Integrative Assessment of Alternative Cyber Threats to CAI.

At this stage, the goal is to calculate a comprehensive utility score for each identified cyber threat. This score reflects the combined impact of all criteria and provides a holistic evaluation of each threat's significance. The calculation is performed using the following formula (8):

$$Score(U_i) = \sum_{j \neq i} \sum_{k=1}^{K} Dom\left(U_i, U_j, k\right), \tag{8}$$

where $Score(U_i)$ The overall utility score for cyber threat U_i and U_j, k is the comparison criterion, $Dom(U_i, U_j, k)$ – the prospective value function used to evaluate the dominance of one threat relative to others.

Stage 7. Cyber threat prioritization and decision making for CAI.

In this step, the identified cyber threats must be prioritized, and decisions made on actions to mitigate or minimize them. This is achieved by calculating the relative significance of each threat and ranking them based on their utility scores. The calculation is as follows (9):

$$p(U_i) = \frac{Score(U_i)}{\sum_{i=1}^{n} Score(U_i)}, \tag{9}$$

where $p(U_i)$ is the relative importance of each potential cyber threat U_i, and $Score(U_i)$ is the utility score of cyber threat U_i.

The threats are then ranked from highest to lowest based on their relative importance. Threats with the highest scores are deemed the most critical and require immediate attention. Based on the rankings, specific mitigation measures are determined for each threat. These measures may include:

- Technical: Implementing advanced cybersecurity tools such as intrusion detection systems or encryption.
- Organizational: Enhancing training programs or revising security policies.
- Procedural: Establishing stricter access controls or incident response protocols.

In the civil aviation sector, prioritizing threats ensures that resources are allocated efficiently to address the most significant risks, such as protecting air traffic control systems, securing passenger data, or mitigating denial-of-service attacks (Fig. 1). This structured approach helps maintain the safety, reliability, and integrity of critical aviation operations.

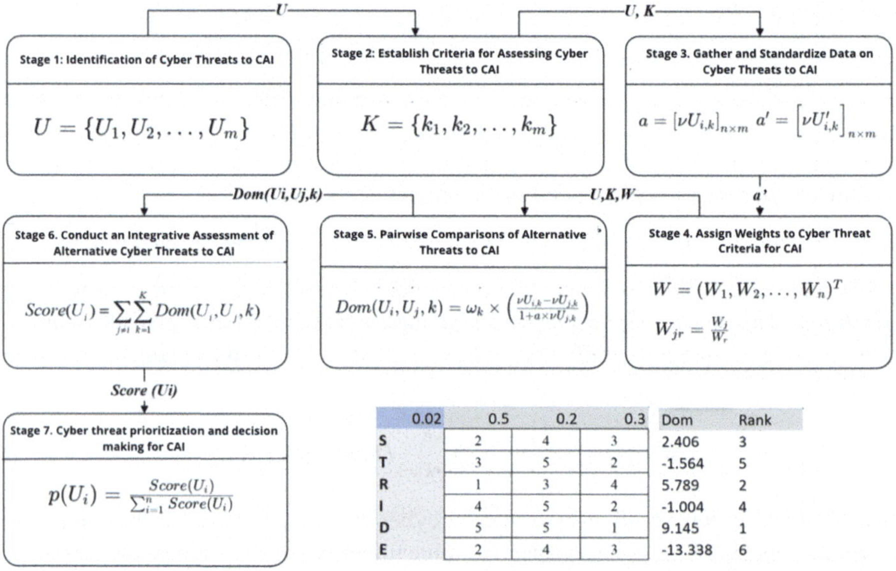

Fig. 1. Flowchart of a method for managing cyber incidents in aviation critical infrastructure.

4 Experimental Study of the Method for Cyber Incidents Management in Aviation Critical Infrastructure

To apply this methodology to the CIA sector, let's focus on the encompass critical infrastructure responsible for ensuring seamless and secure exchange of information between various aviation stakeholders. These networks support essential operations, including real-time communication between air traffic control and pilots, ground handling services, maintenance teams, and airport management systems. Their reliability and security are pivotal for maintaining the integrity, availability, and confidentiality of data, which directly impacts the safety and efficiency of aviation operations [14].

Stage 1: Identification of Cyber Threats to CAI.

Cyber threats to these systems are identified using the STRIDE methodology combined with insights from prior studies. This approach ensures a systematic evaluation of potential vulnerabilities and establishes a comprehensive list of threats that require mitigation to enhance cybersecurity within CAI. The identified threats include the following:

1. *Spoofing*. This threat involves impersonating legitimate systems or entities to gain unauthorized access or interfere with critical aviation operations. Spoofing can undermine the integrity of navigation or communication systems, leading to significant disruptions in operational continuity.
2. *Tampering*. It refers to the unauthorized alteration of data, configurations, or settings within critical aviation systems. Such modifications can compromise system functionality, impact decision-making processes, and pose risks to operational safety.
3. *Repudiation* arises from the lack of mechanisms to verify or trace actions performed within aviation systems. The inability to attribute activities to specific users or entities compromises accountability, which is crucial for ensuring secure operations.
4. *Information disclosure*. This threat entails unauthorized access or leakage of sensitive information related to aviation operations. Exposure of such data undermines confidentiality and poses risks to the security and privacy of passengers, crew, and operations.
5. *Denial of Service*. DoS attacks aim to disrupt the availability of critical aviation systems by overwhelming them with excessive traffic or requests. Such disruptions can delay operations, hinder communication, and lead to significant operational inefficiencies.
6. *Elevation of Privilege*. This threat involves unauthorized escalation of user privileges within aviation systems. Elevation of privilege grants attackers control over critical operations or access to sensitive data, thereby compromising the integrity and security of the systems.

The identification of these threats is guided by their potential impact on the safety, operational continuity, and compliance of civil aviation infrastructure with international standards such as those set by ICAO [15]. The analysis also accounts for the intricate dependencies between interconnected aviation systems, including airports, airlines, ground services, and external vendors, which are essential for maintaining the integrity of aviation operations.

This initial stage provides the groundwork for a deeper analysis of these threats, enabling the development of strategies to assess their impact, prioritize responses, and implement measures to mitigate risks effectively. By understanding the specific challenges facing CAI, this step sets the stage for a systematic and targeted approach to cybersecurity management in civil aviation.

Stage 2: Establish Criteria for Assessing Cyber Threats CAI.

This stage focuses on defining the key criteria necessary to evaluate and prioritize cyber threats specific to CAI. The criteria are essential for conducting a thorough analysis of threats, enabling informed decision-making, and effectively managing cybersecurity risks within CAI systems. For each identified cyber threat U_i, a set of criteria K is defined, incorporating the following key parameters, according to (1), (2):

1. Impact on Safety (IS): this criterion evaluates the extent to which a cyber threat compromises the safety of passengers, crew, and aircraft operations. Safety is the highest priority in aviation. Any threat that could result in harm to individuals, loss of life, or safety incidents must be carefully assessed.
2. Disruption to Operational Continuity (OC): This criterion measures the potential of a cyber threat to disrupt normal aviation operations, such as delays, cancellations, or ground service interruptions. Ensuring uninterrupted operations is vital for the efficiency of aviation systems and customer satisfaction. Threats that cause significant delays or downtime require higher prioritization.
3. Regulatory and Compliance Risk (RCR): This criterion assesses the extent to which a cyber threat could lead to violations of aviation regulations, international standards, or non-compliance with industry best practices. Compliance with aviation-specific regulations (e.g., ICAO Annexes) is crucial for maintaining certifications, avoiding penalties, and ensuring trust in the aviation industry.

Accurate and detailed definition of these evaluation criteria is critical for achieving a comprehensive threat analysis. This approach allows for a balanced assessment of risks, enhancing the prioritization of mitigation efforts and ensuring robust protection for CAI. The number of criteria should be carefully chosen to ensure a detailed yet practical analysis. Standard practice suggests using 3–7 criteria, depending on the complexity of the problem and the availability of reliable data. This ensures that various aspects of the threats are evaluated without overwhelming the decision-making process.

Stage 3. Gather and Standardize Data on Cyber Threats to CAI.

This stage focuses on collecting, evaluating, and normalizing data related to cyber threats specific to CIA. The primary objective is to establish a structured and balanced framework for assessing these threats by determining appropriate weighting factors for each evaluation criterion.

The weighting factors for assessing IT threats to CII should be determined based on their relative importance. The likelihood of a threat occurring was given a high coefficient because it has a significant impact on the risk of the threat being realized. The potential damage from the threat has the highest coefficient because the potential losses from the threat are critical to the functioning of the CII. The complexity of the threat realization received a lower coefficient due to its relatively lower importance compared to other criteria, but it is still important for assessing the technical aspects of protection.

Each criterion k is assigned a weighting factor ω_k, ensuring that the sum of all weights equals 1. This normalization process provides a consistent basis for assessing threats and highlights the sector's priorities, where safety, operational continuity, and compliance are fundamental, establishing the groundwork for an objective and systematic analysis of cyber threats, ensuring that the most critical threats to CIA are accurately identified and prioritized for mitigation, based on (3), as outlined in Table 1.

Stage 4. Assign Weights to Cyber Threat Criteria for CAI.

In this stage, the focus is on determining the relative importance of each criterion used to evaluate cyber threats to CIA. This involves assigning weights to reflect the significance of each criterion in assessing and prioritizing threats.

Table 1. Table of IT threat assessment criteria.

Criterion, k	Description	Weighting index, ω_k
IS	Impact on safety, including risks to passengers, crew, and aviation systems	0.5
OC	Disruption to operational continuity, including delays or service interruptions	0.3
RCR	Risk of regulatory or compliance breaches, such as violations of aviation standards	0.2

A five-point scale is introduced to measure the contribution of each criterion, where 1 indicates the least significant level (e.g., minimal likelihood, low impact, or low disruption), and 5 indicates the highest significance (e.g., high likelihood, severe impact, or major disruption). These values are assigned to each criterion based on its relevance to CIA cybersecurity goals and the unique requirements of the aviation sector.

The weighted criteria are then used to evaluate cyber threats in a pairwise comparison process, helping to identify the most critical threats to CIA. This weighting system not only highlights the importance of individual criteria but also provides a foundation for ranking threats based on their overall impact and risk level. This ranking is instrumental in guiding management decisions and formulating appropriate mitigation strategies for CIA cybersecurity. As outlined in Eqs. (4), (5), and (6), the above scale will be utilized to assess the alternatives based on the defined criteria. The following table presents the evaluation of alternatives according to these criteria (Table 2).

Table 2. Evaluation of alternatives based on defined criteria.

Cyber Threat	IS	OC	RCR
Spoofing	4	3	2
Tampering	5	4	3
Repudiation	2	3	5
Information disclosure	3	4	5
Denial of service	5	5	3
Elevation of privilege	4	2	3

The values are utilized to perform pairwise comparisons of threats, enabling the evaluation of their relative importance and criticality to the CIA. This process identifies the most significant threats that require immediate attention and priority protection measures. Based on the comparison results, an integrative assessment and prioritization of threats are conducted, forming the foundation for informed management decisions on cybersecurity strategies and protective actions for the CIA.

Stage 5. Pairwise Comparisons of Alternative Threats to CAI.

Based on (7), this stage employs the method of pairwise comparisons to evaluate the dominance of each cyber threat over others. This approach enables the assessment of the relative importance and criticality of each threat by comparing them based on specific criteria. The prospective value function is applied in this process, incorporating the weights of the criteria and the scores assigned to each alternative.

1. Inputting Threat Data: Once the criteria-based evaluations for all cyber threats have been completed, the relevant data is input into a custom-built software tool to facilitate automated analysis and computations.
2. Adjusting the Risk Sensitivity Parameter (α)**:** A parameter, denoted as α, is configured to reflect the organization's approach to risk. This parameter, which typically ranges between 0 and 1, is calibrated based on the situation. Lower values reduce the weight of risk factors, while higher values magnify their impact on the analysis.
3. Conducting Pairwise Threat Comparisons: Threats are evaluated relative to one another across all established criteria. For each pair, a comparative dominance score is calculated to indicate which threat poses a greater risk.
4. Calculating Comprehensive Dominance Scores: After completing comparisons for every criterion, an aggregated dominance score is determined for each threat. These scores are used to rank the threats and identify those that require priority action.

This approach provides a detailed and systematic analysis, ensuring an evidence-based process for identifying and prioritizing critical threats to CIA.

Stage 6. Conduct an Integrative Assessment of Alternative Cyber Threats to CAI.

To streamline this process and enhance calculation accuracy, a specialized Cyber Threat Management Methodology software application is utilized at this stage. This software consolidates all input data, pairwise comparisons, and weighting factors to compute the final utility scores for each threat. As outlined in Eq. (8), the prospective values are aggregated to derive a utility score for every threat. The results generated by the cyber threat management software are visualized in Fig. 2, providing a clear and structured representation of the threat evaluation outcomes.

Stage 7. Cyber threat prioritization and decision making for CAI.

According to (9) and based on the analysis performed by the developed method, the threats were ranked according to their total dominance. Let us present the prioritization of threats, where threats with higher values of total dominance should be addressed as the most critical (Table 3).

Figure 3 illustrates the results of the cyber threat assessment for CIA, namely Communication Networks, conducted using the updated method for evaluating cyber threats, as detailed in Table 3.

Cyber Threat	0,5	0,3	0,2	Total Dominance	Rank
Spoofing	4	3	2	-1,945	4
Tampering	5	4	3	3,647	2
Repudiation	2	3	5	-4,154	6
Information disclosure	3	4	5	0,316	3
Denial of Service	5	5	3	5,330	1
Elevation of Privilege	4	2	3	-2,506	5

Fig. 2. Cyber threat management software usage.

Table 3. Cyber threat prioritization for CIA sub-sector communication networks.

Cyber Threat	Dominance level	Priority
Spoofing	5.330	The highest
Tampering	3.647	High
Repudiation	0.316	Average
Information disclosure	−1.945	Medium
Denial of service	−2.506	Low
Elevation of privilege	−4.154	The lowest

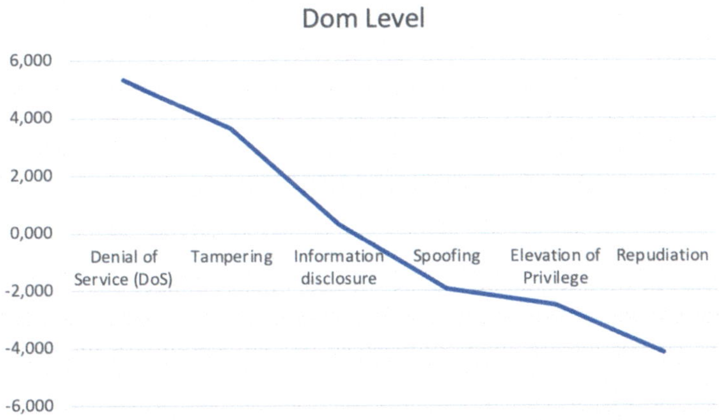

Fig. 3. Cyber threat assessment based on the developed method.

Therefore, based on the results obtained using the specialized software developed, the following recommendations have been proposed to address cyber threats:

1. **Denial of Service (DoS):** The most critical threat that must be addressed as a priority to mitigate the risk of communication network disruptions. DoS attacks can lead to the failure of critical aviation communication channels, including pilot-to-controller interactions and ground services. It is recommended to implement resilient systems

against DoS attacks, including network load balancing, redundancy, and advanced intrusion detection/prevention systems.

2. **Tampering:** Ranked as a significant threat, tampering requires the implementation of robust data integrity controls to prevent unauthorized modifications. Systems that detect unauthorized changes to communication protocols or transmitted data, along with end-to-end encryption, are essential for safeguarding the integrity of aviation communication networks.
3. **Information Disclosure:** This medium-priority threat highlights the need to strengthen data protection mechanisms, especially for sensitive aviation-related information. Data encryption (in transit and at rest), access control measures, and continuous monitoring should be employed to detect and prevent unauthorized disclosures.
4. **Spoofing:** To counter spoofing attacks, which rank as a medium-level threat, it is necessary to enhance authentication procedures. The implementation of multi-factor authentication, secure certificate-based systems, and advanced user identity verification methods are recommended to prevent unauthorized access to critical communication networks.
5. **Elevation of Privilege:** Although this is considered a low-priority threat, measures to prevent privilege escalation must not be overlooked. The principle of least privilege should be implemented across systems, access rights should be regularly reviewed, and privilege escalation attempts must be monitored and blocked using security tools.
6. **Repudiation:** The lowest-ranked threat, repudiation, requires improved logging and auditing systems to ensure accountability within communication networks. Reliable mechanisms for capturing and retaining logs, coupled with regular audits, can effectively detect and address attempts to deny actions or transactions.

5 Verification of the Developed Method for Cyber Incidents Management in Aviation Critical Infrastructure

To validate the effectiveness and adaptability of the proposed method for managing cyber incidents in CAI, the additional experiments focused on two critical sub-sectors of CAI were conducted: air traffic control systems and passenger data management systems. These experiments aimed to evaluate the applicability of the method across different aviation components and to further demonstrate its ability to identify, assess and prioritize cyber threats.

Air traffic control (ATC) [16] systems are the backbone of aviation safety, ensuring the orderly and efficient flow of aircraft through controlled airspace. Given the real-time nature of ATC operations, even minor disruptions can have catastrophic consequences. This experiment aimed to apply the proposed methodology to identify and prioritize cyber threats specific to ATC systems. Using the STRIDE model, potential threats to ATC systems were identified. Key threats included:

1. Spoofing: Impersonation of communication channels between pilots and air traffic controllers, resulting in false or malicious instructions.
2. Tampering: Unauthorized changes to flight plans, radar data or communications system configurations.

3. Repudiation: Inability to trace actions within the system, compromising accountability and incident investigation.
4. Information Disclosure: Exposure of sensitive operational data, such as flight paths or controller instructions, to unauthorized parties.
5. Denial of Service: Overloading ATC communications or radar systems, causing disruption to real-time operations.
6. Elevation of Privilege: Unauthorized access to ATC systems, allowing attackers to alter or disrupt operations.

Using the TODIM method, each identified threat was evaluated against the criteria. Weighting factors were assigned based on expert input, emphasizing safety and operational continuity (Table 4).

Table 4. Evaluation of alternatives based on defined criteria for ATC.

Cyber Threat	IS	OC	RCR
Spoofing	5	4	3
Tampering	5	5	4
Repudiation	3	3	4
Information disclosure	4	3	4
Denial of service	5	5	3
Elevation of privilege	4	4	3

For the evaluation of cyber threats to ATC systems, it was decided to assign the following weights to the criteria, reflecting the critical role of ATC in ensuring the safety and real-time operation of aviation systems:

- Impact on Safety (IS): 0.6 (Safety is paramount in ATC systems);
- Operational Continuity (OC): 0.3 (Maintaining operations is critical);
- Regulatory and Compliance Risk (RCR): 0.1 (Compliance is important but secondary to safety and continuity).

Using specialized software developed by the authors, the following results were obtained (Table 5). The results highlight the most critical threats to the Air Traffic Control (ATC) sub-sector, with Tampering and Denial of Service (DoS) identified as the highest priorities due to their significant impact on safety and operational continuity.

Besides that, the authors conducted an additional experiment to evaluate the effectiveness of the proposed method in another critical sub-sector of CIA – Passenger Data Management Systems (PDMS) [17], which handles sensitive personal and financial information, making them a prime target for cyber attacks.

The aim of this experiment was to apply the proposed method to assess threats specific to PDMS and to prioritize mitigation efforts [18]:

1. Spoofing: Phishing or impersonation of legitimate booking portals to steal passenger credentials.

2. Tampering: Unauthorized modification of passenger data, such as flight details or payment information.
3. Repudiation: Lack of accountability for actions performed within the system, hindering investigation of unauthorized access.
4. Information Disclosure: Leakage of sensitive passenger information, including personal and financial details.
5. Denial of Service: Disruption of system availability, preventing passenger bookings, check-ins, or boarding processes.
6. Elevation of Privilege: Unauthorized escalation of access rights, granting attackers control over sensitive passenger databases.

The evaluation of cyber threats PDMS was conducted using the proposed method, integrating the STRIDE framework and weighted multi-criteria analysis [19] (Table 6).

Table 5. Cyber threat Prioritization for of the CIA sub-sector ATC.

Cyber Threat	Dominance level	Priority
Tampering	4.214	The highest
Denial of service	3.653	High
Spoofing	1.986	Average
Elevation of privilege	−1.328	Medium
Information disclosure	−2.434	Low
Repudiation	−5.747	The lowest

Table 6. Evaluation of alternatives based on defined criteria for PDMS.

Cyber Threat	IS	OC	RCR
Spoofing	3	3	5
Tampering	4	4	4
Repudiation	2	3	5
Information disclosure	3	4	5
Denial of service	4	5	3
Elevation of privilege	3	3	4

To evaluate cyber threats to PDMS, it was decided to assign the weights to the criteria:

- Impact on Safety (IS): 0.3;
- Operational Continuity (OC): 0.4 (Ensuring operational continuity is a key focus);
- Regulatory and Compliance Risk (RCR): 0.3 (Compliance is crucial due to sensitive passenger data).

Following the evaluation of threats to PDMS based on the defined criteria, the specialized software was used to calculate dominance levels and prioritize the identified cyber threats. Prioritization reflects the relative significance of each threat, with higher dominance levels indicating greater criticality. The results highlight the threats requiring immediate attention and mitigation efforts to enhance the cybersecurity of PDMS. The results are summarized in Table 7.

Table 7. Cyber threat Prioritization for of the CIA sub-sector PDMS.

Cyber Threat	Dominance level	Priority
Denial of service	1.855	The highest
Tampering	1.282	High
Information disclosure	1.246	Average
Elevation of privilege	−0.411	Medium
Spoofing	−0.984	Low
Repudiation	−2.677	The lowest

The results of applying the proposed methodology to three sub-sectors of the CIA – communication networks, ATC systems and PDMS – are visualized in the Fig. 4. The dominance levels of identified cyber threats are shown, indicating their relative criticality for each sub-sector of CIA.

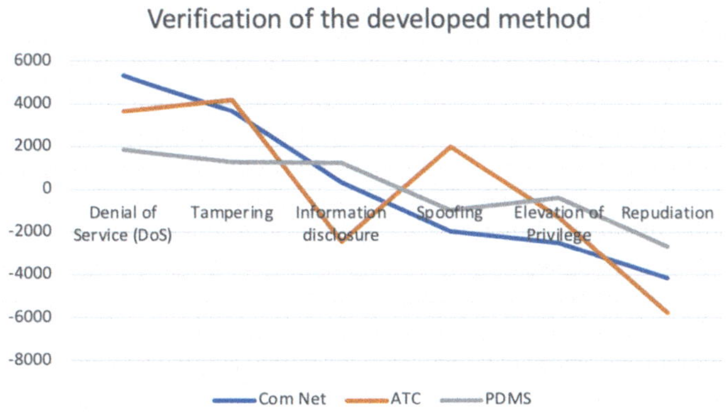

Fig. 4. Validation of cyber threat assessment method across experimental data sets.

6 Conclusions

In conclusion, this study has analyzed cyber threat management for CAI by applying the developed methodology to three critical sub-sectors: Communication Networks, ATC and PDMS. It was found that traditional methods do not fully address the unique challenges of assessing and managing cyber threats to aviation infrastructure, and that a new approach is needed. To address these gaps, the authors developed a methodology that combines the TODIM multi-criteria decision-making methodology with the STRIDE threat modelling approach. This method enables effective identification, assessment and prioritization of threats by considering their likelihood, potential impact and complexity of implementation. The developed method includes several key stages, including threat identification, evaluation criteria determination, data normalization, criteria weighting, pairwise comparison, integrative evaluation, prioritization and decision making. It provides a systematic and efficient framework for improving the cybersecurity of CAI by tailoring its application to the specific characteristics of each sub-sector. Experimental validation in the three sub-sectors demonstrated the adaptability and effectiveness of the method. For communication networks, Denial of Service was identified as the highest priority threat due to its potential to disrupt essential communication processes. For ATC systems, Tampering and DoS were identified as the most critical threats, highlighting the importance of real-time threat mitigation to ensure safety and operational continuity. For PDMS, Information Disclosure and DoS ranked highest, highlighting the need for robust data protection and service reliability.

Overall, the results confirm that the proposed method enables the efficient allocation of resources and implementation of appropriate security measures for CAI. Its ability to prioritize threats based on sub-sector-specific risks ensures a high level of protection for aviation infrastructure, addressing both immediate and long-term cybersecurity challenges.

7 Perspectives for Further Research

In future research, the authors plan to focus on the following areas to enhance the effectiveness of cyber threat management in CIA:

1. Automating Weighting Criteria Assignment: The authors aim to automate the process of assigning weights to evaluation criteria by leveraging advanced AI techniques, such as machine learning and deep learning. This approach will enable dynamic and data-driven determination of criteria importance, improving the consistency, accuracy, and efficiency of methodologies like STRIDE and TODIM.
2. Exploring the Impact of Emerging Technologies: The research will investigate the influence of new technologies, including 5G, quantum computing, and the Internet of Things, on communication networks within aviation infrastructure. The focus will be on identifying potential vulnerabilities and novel threats.
3. Examining Human Factors in Cyber Threat Management: The authors intend to analyze the role of human factors in cybersecurity, emphasizing the development of effective training programs, error reduction strategies, and awareness campaigns for personnel responsible for managing aviation communication systems. This focus aims to foster a more resilient and informed workforce capable of addressing evolving cyber threats.

References

1. Federal Aviation Administration (FAA). New FAA regulations target cybersecurity vulnerabilities in aircraft design. https://www.jonesday.com/en/insights/2024/09/new-faa-regulations-target-cybersecurity-vulnerabilities-in-aircraft-design. Last accessed 29 Dec 2024
2. Federal Aviation Administration (FAA). What a tangled web: Aviation prosperity and cybersecurity risk. https://www.faa.gov/speeches/what-tangled-web-aviation-prosperity-cybersecurity-risk. Last accessed 29 Dec 2024
3. Gnatyuk, S., Sydorenko, V., Polozhentsev, A., Sokolov, V.: Method for managing IT incidents in critical information infrastructure facilities. CEUR Workshop Proc. **3826**, 326–333 (2024)
4. Sydorenko, V.M., Polozhentsev, A.A., Sydorenko, S.Y., Skurativskyi, A.A.: A method for prioritizing IT incidents at critical information infrastructure facilities of the state. Prob. Inform. Manag. **2**(78), 10.18372/2073-4751.78.18967 (2024)
5. STRIDE. Microsoft. The STRIDE Threat Model. https://docs.microsoft.com/en-us/previous-versions/msp-n-p/ee823878(v=pandp.10). Last accessed 29 Dec 2024
6. NIST SP 800-30. National Institute of Standards and Technology (NIST). Guide for Conducting Risk Assessments (NIST Special Publication 800-30 Revision 1). Gaithersburg, MD: U.S. Department of Commerce. https://doi.org/10.6028/NIST.SP.800-30r1. Last accessed 29 Dec 2024
7. ISO/IEC 27005:2018 – Information security risk management. https://www.iso.org/standard/75281.html. Last accessed 29 Dec 2024
8. Alberts, C., Dorofee, A., Stevens, J., Woody, C.: OCTAVE Threat Management Framework. Carnegie Mellon University: Software Engineering Institute. https://resources.sei.cmu.edu/library/asset-view.cfm?assetid=6295. Last accessed 29 Dec 2024
9. COBIT: Framework: Governance and Management Objectives. ISACA, Schaumburg, IL (2019) https://www.isaca.org/resources/cobit. Last accessed 29 Dec 2024
10. Saaty, T.L.: Decision making with the analytic hierarchy process. Int. J. Services Sci. **1**(1), 83 (2008). https://doi.org/10.1504/ijssci.2008.017590
11. Zhou, Z., Xuan, K., Tao, Z., Zhou, L.: Generalized-TODIM method for multi-criteria decision making with basic uncertain information and its application (2021). arXiv preprint arXiv:2104.11597
12. Kumar, S., Joseph, D.: A review on industrial applications of TOPSIS approach. Int. J. Eng. Res. Appl. **8**(1), 50–54 (2018)
13. Bhattacharya, M., Manikandan, S.: Critical infrastructure protection: an overview of civil aviation security. In: Cyber Security and Threats: Concepts, Methodologies, Tools, and Applications, pp. 448–465. IGI Global (2019)
14. International Civil Aviation Organization (ICAO). Aviation cybersecurity strategy. https://www.icao.int/aviationcybersecurity/Pages/default.aspx. Last accessed 29 Dec 2024
15. International Civil Aviation Organization (ICAO). Safety Management Manual (SMM) (Doc 9859), 4 (2018)
16. International Civil Aviation Organization (ICAO): Air Traffic Management (ATM) (n.d.). https://www.icao.int/safety/airnavigation/pages/atm.aspx. Last accessed 29 Dec 2024
17. International Air Transport Association (IATA): Passenger data (n.d.). https://www.iata.org/en/programs/passenger/passenger-facilitation/passenger-data. Last accessed 29 Dec 2024
18. Cybersecurity and Infrastructure Security Agency (CISA). Guide to getting started with a cybersecurity risk assessment. https://www.cisa.gov/sites/default/files/2024-09/24_0828_safecom_guide_getting_started_cybersecurity_assessment_2022_final_508C.pdf. Last accessed 29 Dec 2024
19. Nosal, K., Solecka, K.: Application of AHP method for multi-criteria evaluation of variants of the integration of urban public transport. Transp. Res. Proc. **3**, 269–278 (2014). https://doi.org/10.1016/j.trpro.2014.10.006

Strategic Diagnostics for Optimization of Aviation Enterprises Activity in the Condition of Digital Transformations

Serhii Smerichevskyi[1], Oleksandr Truhan[2], Zarina Poberezhna[1](✉), and Oksana Chumak[1]

[1] National Aviation University, Liubomyra Huzara Ave., 1, Kyiv 03058, Ukraine
zarina.poberezhna@npp.kai.edu.ua

[2] Zhytomyr Institute PrJSC "Higher Educational Institution" "Interregional Academy of Personnel Management (IAPM)", Peremohy Str., 26, Zhytomyr 10003, Ukraine

Abstract. The aim of this study is analysis of the theoretical foundations and proposal methodological recommendations for determining the possibilities of expanding the use of strategic diagnostic tools in the condition of digitalization. It has been established that digital transformations have become one of the key trends in the modern business environment, which significantly affect all aspects of management. The use of digital technologies provides new opportunities for optimizing the activities of aviation enterprises, increasing its efficiency and improving its competitive position. It is determined that strategic diagnostics is a systematic process of analyzing and evaluating internal and external factors that affect the activities of aviation enterprise in order to determine its strategic position, development potential, risks and prospects. The advantages of using digital technologies in the system of strategic diagnostics of aviation enterprises are determined, which are as follows: speed of data collection and processing; accuracy of analysis; automation of business processes; integration and visualization of data; improvement of communication; adaptation to changes; saving resources; support for innovative development. An algorithm for conducting strategic diagnostics for aviation enterprises in the condition of digitalization is proposed, which ensures the integration of traditional methods of analysis with the latest technologies for more efficient management. A modeling of calculating an erroneous decision on the status of a business process is proposed, which makes it possible to determine the probability of an erroneous decision.

Keywords: Strategic Diagnostics · Digital Transformation · Digitalization · Digital Technologies · Aviation Enterprises

1 Introduction

1.1 Introduction to the Problem

The current operating environment of aviation enterprises is characterized by a high level of dynamism, globalization of markets, growing competition, and rapid changes in technology. In such circumstances, effective management of aviation enterprises becomes

I. Ostroumov et al. (Eds.): ACASD 2025, LNNS 1418, pp. 92–104, 2025.
https://doi.org/10.1007/978-3-031-91992-3_7

impossible without the use of strategic diagnostics, which allows to assess both the external environment and internal potential. Strategic diagnostics is an important tool for ensuring the sustainable development of aviation enterprises, as it helps to identify prospects, risks and optimal ways to implement the strategy. For aviation enterprises, the procedure for assessing reliability parameters in case of deterioration of the technical condition can be solved by analytical calculations and statistical modeling [1].

The aviation industry plays a key role in ensuring the country's economic stability, regional integration, support for international trade, and therefore needs to adapt to new global challenges. Strategic diagnostics helps analyze market needs, predict changes in demand for transportation services, and find ways to minimize costs. Successful adaptation of aviation enterprises to changing conditions is only possible with a deep understanding of the factors that influence them, such as innovation, energy saving, environmental requirements, digitalization, etc. Information technologies allow automating numerous processes, reducing costs and minimizing human errors. This allows for faster and more accurate management of inventory, transportation routes, and deliveries [2]. In addition, strategic diagnostics helps to increase the competitiveness of enterprises, providing them with the ability to respond to market challenges in a timely manner. This is especially true in the context of economic instability and increasing environmental regulation. By analyzing strengths and weaknesses, as well as identifying threats and opportunities, aviation enterprises can develop effective development strategies.

Thus, the introduction of strategic diagnostics is becoming a prerequisite for maintaining market positions and the long-term success of aviation enterprises. Its importance is growing due to the need to adapt to new technologies, ensure environmental responsibility, and integrate into global transportation networks. In modern conditions, strategic diagnostics forms the basis for making management decisions that determine the prospects for the development of the aviation industry.

It should be noted that in the context of the rapid development of digital technologies and active digitalization of all sectors of the economy, aviation enterprises face the need to revise their approaches to the management system. Automation of business processes has a positive effect on the indicators of the aviation enterprise's activity and increases the need for their assessment in terms of different areas of activity (purchasing, finance, personnel, sales, marketing) [3].

Modern digital technologies provide new opportunities for strategic diagnostics, which allows for more accurate analysis, faster decision-making, and more efficient use of resources. The use of digitalization tools in strategic diagnostics is a prerequisite for achieving competitive advantages in the aviation sector.

Digital tools such as Big Data, Artificial Intelligence (AI) analytics, cloud technologies, and the Internet of Things (IoT) enable the processing of large amounts of information in real time. This allows aviation enterprises to better understand market dynamics, predict changes in demand for services, identify weaknesses in their operations, and develop effective strategies. Digitalization facilitates the integration of various data sources, such as operational, financial, and logistics information, into a single system, which creates the preconditions for a more comprehensive assessment of the external and internal environment of aviation enterprise. In particular, digital platforms allow analyzing risks, tracking new trends, and modeling possible development scenarios. The use of digital tools also significantly speeds up the diagnostic process, allowing for a quick response to market changes or unforeseen challenges. Through business process automation and artificial intelligence, aviation enterprises can reduce the human factor in decision-making, which reduces the likelihood of errors. In addition, digitalization opens up opportunities for more efficient cost and resource management.

Thus, the incorporation of modern digital technologies into the strategic diagnostics system is not only a requirement of the times, but also an important step in ensuring the sustainable development of aviation enterprises.

1.2 Motivation

In order to increase the efficiency of aviation enterprises activity, analyze market needs, predict changes in demand for transport services, and find ways to minimize costs, it is necessary to conducting strategic diagnostics that will ensure the sustainable development of aviation enterprises, identify prospects, risks, and optimal ways to implement the strategy.

1.3 Contribution

The key contribution of this study is developed methodology for conducting strategic diagnostics of aviation enterprises in order to maintain market positions and long-term success, which is growing due to the need to adapt to new technologies, ensure environmental responsibility and integrate into global transport networks. A modeling of the calculation of the probability of an erroneous decision regarding the state of the business process is proposed, which provides the aviation enterprise with a high level of competitive advantages in the market and increases its profitability.

1.4 The Organization of the Paper

The paper contains six sections. The first section describes the introduction. The second section highlights the key findings of the literature review. In the third section, the main features and methods for conducting strategic diagnostics to optimize the activities of aviation enterprises in the context of digitalization are considered. The fourth section presents an example of calculation and modeling. The fifth section is the conclusion. Finally, the sixth section outlines the future scope.

2 Literature Review and Problem Statement

Topical issues of the impact of digital transformation on the development of methods and processes for strategic diagnostics of aviation enterprises remain insufficiently researched. The lack of systematic analysis and clear methodological approaches makes it difficult for aviation enterprises to adapt to the conditions of digital transformation. Key challenges include the integration of Big Data, automation of diagnostic processes, and the use of AI analytics. These technologies require a revision of traditional approaches to assessing the external and internal environment of an enterprise. In addition, there is the issue of ensuring the security of data used for diagnostics and staff training to work with digital tools.

Insufficient theoretical justification of these processes creates risks for aviation enterprises trying to implement digital technologies in their operations. This requires further research aimed at developing new approaches and diagnostic models adapted to the conditions of the digital age. It is important to note that the impact of digital transformations is complex and requires an interdisciplinary approach to study.

Paper [4] shows that digital technologies such as AI and business process automation affect the economy, business and society, as well as enterprises, helping them to adapt their strategies to the conditions of the digital revolution. The authors of the paper [5] analyze how leading companies are using digital technologies to transform business models, increase efficiency, and create new sources of value, emphasizing the importance of strategic management in the digital age. Paper [6] explores the impact of information and communication technologies (ICT) on economic growth in different countries, emphasizing the importance of digitalization for the strategic development of enterprises in a global context. The authors of the paper [7] examine the evolution of the concept of "digital strategy" and its integration into the overall strategic management system, analyzing how digitalization affects the formation of enterprise strategies.

In study [8] identifies the impact of digitalization on the strategic management of enterprise development. Recommendations are formulated to improve the process of strategic management of enterprise development in the context of digitalization in order to ensure sustainable development and adaptation to rapidly changing market conditions. Publication [9] focuses on the issues of strategic management of an enterprise in the digital economy. The authors identify tools for the digital transformation of enterprise business models and propose measures to improve the strategic management of the enterprise, including analyzing the impact of digital technologies on business models, developing a methodology for strategic planning in conditions of instability, planning and assessing the risks of digital transformation. The strategic management of enterprise development in the context of economic digitalization was considered in paper [10], which reveals the strategic aspects of managing Ukrainian enterprises on the basis of digitalization, including the identification of alternative strategies, goals and key areas of digital transformation. Paper [11] substantiates the importance and role of digital business transformation and the need to reorient the relevant management systems to new strategic guidelines. The nature of reasonable digital changes that accompany the economic activities of business entities is outlined. Publication [12] substantiates the main strategic aspects of digitalization in the current conditions of functioning of international companies. The study identified the challenges of digital transformation and analyzed various types of digital business models.

The multidirectionality and ambiguity of the presented approaches requires further in-depth research to create a balanced and sound methodology in the chosen context. Currently, there are a number of unresolved issues related to the development of innovative methods of strategic diagnostics that can meet the challenges of digitalization. This necessitates further research aimed at integrating modern digital tools into the process of strategic analysis and management of aviation enterprises.

The aim of this research is to analyze the theoretical foundations and develop methodological recommendations of strategic diagnostics for optimize the activities of aviation enterprises in the condition of digital transformations.

3 Materials and Methods

In the process of scientific research on the strategic diagnostics of aviation enterprises in the condition of digital transformation, it is advisable to use a combination of quantitative and qualitative methods. Quantitative methods allow to objectively assess the key performance indicators of aviation enterprises, analyze trends and make forecasts based on mathematical modeling and statistical data. At the same time, qualitative methods are aimed at a deeper understanding of the specifics of business processes, taking into account the human factor and subjective factors that influence strategic decision-making. Such an integrated approach provides a comprehensive analysis and increases the efficiency of development strategies in the face of digital challenges.

The application of a systematic approach to the study of strategic diagnostics of aviation enterprises in the condition of digital transformation is necessary to ensure comprehensive analysis and effective management decision-making. A systematic approach allows taking into account the relationship between various elements of the internal environment of aviation enterprise, such as organizational structure, business processes, resources, and technologies. In addition, it promotes a deeper understanding of external factors, including market influences, the competitive environment, regulatory changes, and innovative technologies.

This approach makes it possible to identify weaknesses and development potential of the aviation enterprise, as well as to adapt strategic decisions to the rapid changes caused by digitalization. Given the complexity and multifaceted nature of the digital transformation processes, the systematic approach allows for the creation of a holistic model of interaction between all components of the enterprise aimed at achieving strategic goals, increasing competitiveness and long-term sustainability.

Digital transformations have become one of the key trends in the modern business environment, significantly affecting all aspects of management, including the strategic approach to the development of aviation enterprises [13].

For aviation enterprises operating in a dynamic market, digitalization helps to create integrated logistics systems, automate fleet management processes, and introduce innovative customer services. Safety and regularity of civil aviation aircrafts flights is determined by complex applying of various resources during preparing aircrafts to flight, radio technical flight support (RTFS) by ground-based communications, navigation and surveillance facilities, air traffic control service and aviation security devices usage [14].

A strategic approach in the context of digital transformation requires a revision of traditional management methods, as the need for adaptability, quick decision-making and integration of new technologies is growing.

Digital platforms and tools allow for detailed diagnostics of a company's operations, analysis of internal and external factors affecting its development, and forecasting future trends. Digital transformations create new challenges, including those related to data security, the need to upgrade staff skills, and investments in technological infrastructure [15].

Thus, digital transformations are not only changing the way we do business, but are also pushing aviation enterprises to rethink their management strategies to maximize their adaptability, innovation, and competitiveness.

Aviation enterprise's stronger competitiveness requires ensuring the growth of its sales in products or services. Yet, even rise in sales volumes may produce meaningful results only if it is accompanied by positive outcomes in enterprise's profits and cost of sales [16].

The advantages of using digital technologies in the system of strategic diagnostics of aviation enterprises are the follows:

1. Speed of data collection and processing. Digital technologies provide for the rapid collection of large amounts of information from various sources, such as GPS-systems, vehicle sensors, CRM-systems, and financial databases. This allows you to quickly analyze the current situation and make timely management decisions.
2. Analysis accuracy. The use of modern analytical tools, such as Big Data and Machine Learning (ML) algorithms, improves the accuracy of forecasts and diagnostics. This allows us to identify hidden risks, identify effective areas of activity and determine areas for optimization.
3. Automation of business processes. Digital platforms allow automating routine data analysis processes, reducing errors and increasing productivity.
4. Data integration and visualization. Modern technologies allow combining information from different departments of the aviation enterprises and external sources into a single platform. This provides convenient visualization of indicators in the form of dashboards, graphs or maps, which simplifies strategic decision-making.
5. Improved communication. Digital tools help improve internal and external communication through the integration of cloud services, collaboration platforms, and online data analysis. This ensures better coordination between the aviation enterprise's departments.
6. Adaptation to changes. Thanks to digital technologies, aviation enterprises can quickly respond to changes in the external environment (fluctuations in demand, changes in regulatory conditions, or new technological challenges).
7. Saving resources. The use of digital solutions reduces the cost of paperwork, manual data processing, and other inefficient processes. This helps to optimize costs and increase overall profitability.
8. Support for innovative development. Digital technologies create the basis for the introduction of innovative approaches, such as blockchain for transparency of logistics chains or (IoT) for real-time monitoring of the condition of vehicles.

Thus, the use of digital technologies in the strategic diagnostics of an aviation enterprise allows to increase management efficiency, ensure adaptability to dynamic changes in the market, and create conditions for sustainable development.

The introduction of digital tools contributes to increased competitiveness, customer satisfaction, and profitability, so it is a necessary step to ensure their sustainable development [17].

Digital tools allow for more efficient analysis of the external environment, taking into account regulatory, technological, economic, and social factors. Data visualization through dashboards and interactive graphs makes it easier to present the results of diagnostics to management. Digital tools also facilitate scenario analysis, allowing to simulate different scenarios. As a result of the use of digital technologies, strategic diagnostics becomes more accurate, comprehensive and efficient, providing aviation enterprises with competitive advantages and adaptability to changes in the business environment. Conservation and sustainable management of ecosystems is a necessary condition for ensuring sustainable development. Integrating ecosystem services into business management, decision-making and policy development helps preserve valuable natural resources and ensure long-term economic and social development [18].

The possibilities of expanded use of strategic diagnostics tools in the conditions of digitalization are summarized in Table 1.

Digital platforms and data integration systems make it easier to collect, process and visualize information, which contributes to more effective strategic planning and management. Digital technologies contribute to the formation of the innovative potential of an aviation enterprise and are directly related to the sustainable development of its operation, in particular with regard to such aspects as safety, minimization of negative impact on the environment, and the introduction of modern Industry 4.0 technologies [19]. Due to the rapid development of digital technologies, traditional approaches to strategic diagnostics of aviation enterprises require significant adjustments. Digitalization changes not only the mechanisms for collecting and processing data, but also the very nature of business processes, which requires the adaptation of strategic diagnostics algorithms to new conditions. Adjusting the algorithms allows these tools to be integrated into the strategic diagnostics process, increasing its accuracy, efficiency, and effectiveness. Adapting diagnostic procedures to digital realities will not only improve management decisions but also increase the competitiveness of aviation enterprises.

The algorithm for conducting strategic diagnostics to optimize the activities of aviation enterprises in the condition of digitalization includes key stages that integrate traditional methods of strategic analysis with the latest digital technologies (Fig. 1).

The proposed algorithm for conducting strategic diagnostics of aviation enterprises in the conditions of digitalization ensures the integration of traditional analysis methods with the latest technologies for more efficient management. The collection and processing of large amounts of data in real time using digital platforms allows aviation enterprises to respond quickly to changes in the external and internal environment. The use of analysis tools in combination with Big Data and AI capabilities helps to collect more accurate and timely data to predict future changes [20]. The strategic planning process is becoming more adaptive, and monitoring the implementation of the strategy ensures constant correction based on real data. The use of digital technologies in strategic diagnostics improves the accuracy of decision-making, reduces risks, and ensures the competitiveness of aviation enterprises in the modern digital environment.

Table 1. Possibilities for expanded use of strategic diagnostics tools in the conditions of digitalization.

A tool for strategic diagnostics	Field of application	Possibilities in the conditions of digitalization
PESTLE-analysis	Analysis of the external environment	Use of digital tools to monitor macroeconomic, political, technological and other changes in real time
Analysis of the competitive environment	Study of competitors' market positions	The use of Big Data and AI to analyze the behavior of competitors, identify their strengths and weaknesses
ABC-analysis	Classification of products/services	Automate product and customer segmentation based on Big Data on sales, costs, and profitability
Value chain analysis	Evaluation of business processes	Integration of IoT for real-time process tracking, supply chain optimization through blockchain
GAP-analysis	Identify gaps between plans and reality	Digital dashboards for visualizing current performance and identifying deviations in real time
Benchmarking	Comparison with competitors	Platforms for analyzing industry standards, using analytical reports with access to international databases
Scenario analysis	Forecasting future development options	Use of AI for scenario modeling, risk assessment by specialized software tools
Portfolio analysis	Business/project portfolio management	Digital systems for assessing project performance in real time, automatic calculation of ROI and other indicators
Stakeholder analysis	Manage interaction with key groups	Use CRM-systems to track communications with stakeholders and automate interaction processes
Value stream analysis (VSM)	Optimization of value streams	Use of digital tools to model processes and identify areas for eliminating losses
Risk analysis	Identifying and minimizing risks	Application of software for risk prediction, assessment of their impact through AI and machine learning (ML)
Balanced Scorecard (BSC)	Evaluation of strategic goals	Integration with ERP-systems for real-time monitoring of KPIs, automation of data collection and analysis
Multivariate analysis	Assessment of interdependencies	Using ML to identify key factors that affect the efficiency of aviation enterprise

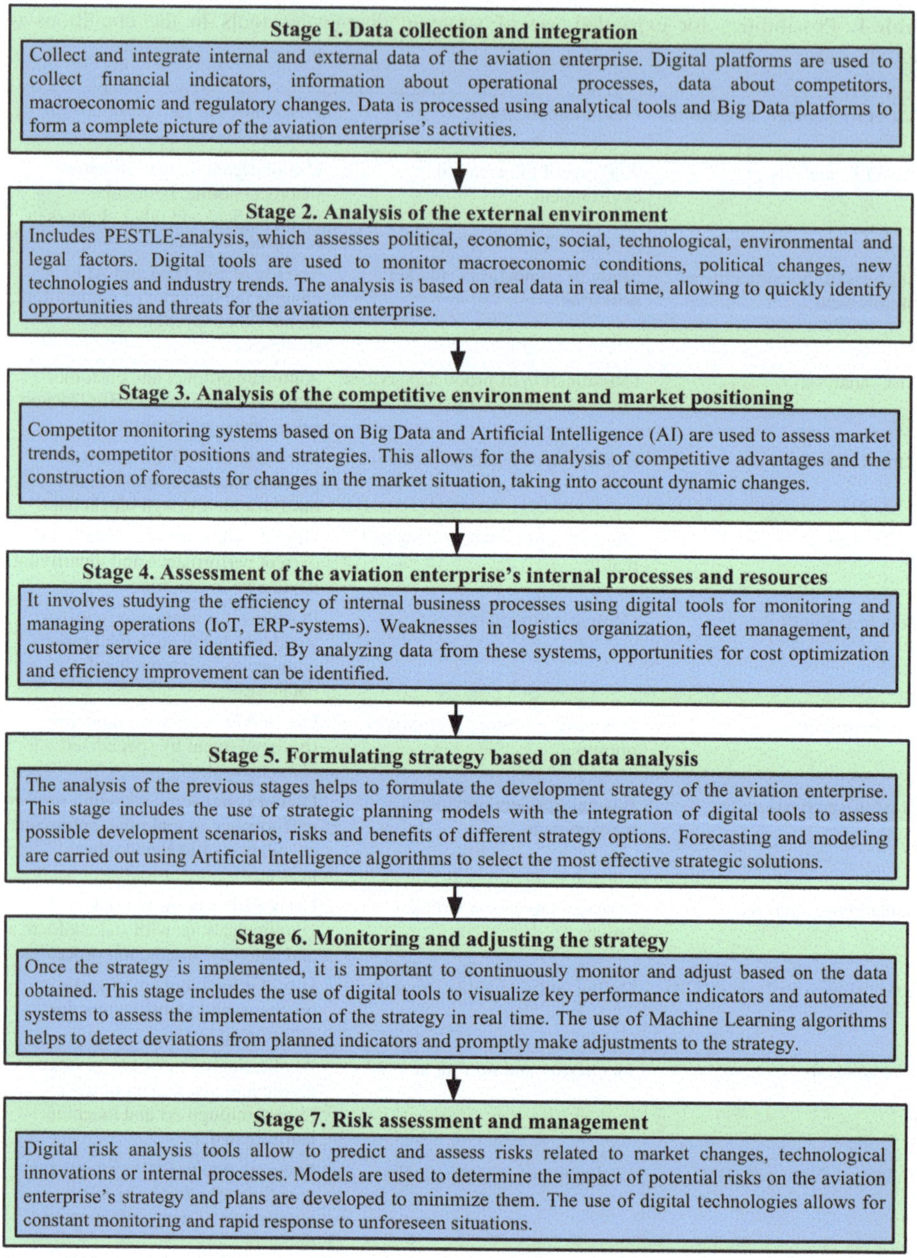

Fig. 1. Algorithm for conducting strategic diagnostics to optimize the activities of aviation enterprises in the conditions of digitalization.

4 Results and Discussions

Managing aviation enterprises in modern economy requires a comprehensive approach to assessing their strategic position. To do this, it is necessary to use a system of indicators that allow to assess both the internal processes of the aviation enterprise and the external environment in which it operates. The use of a set of indicators makes it possible to study in detail various aspects of the aviation enterprise's activities, including financial, operational, market, and innovative ones. In the context of digitalization, the availability of such comprehensive tools allows for operational monitoring of key parameters and real-time strategic diagnostics. At the same time, the use of various indicators improves the accuracy of management decision-making and facilitates timely response to changes in the external environment. Thus, the use of a set of indicators is necessary for in-depth analysis and adjustment of the aviation enterprise's strategy in a dynamic business environment.

The author proposes modeling the probability of an erroneous decision regarding the state of a business process. In order to determine the probability of an erroneous decision, it is necessary to divide the types of errors into the first and second types. Probabilities of an erroneous decision regarding the state of a business process are:

1. An error of the first type (a suitable solution type is recognized as unsuitable) is characterized by a probability:

$$\alpha = \int_{T}^{\infty} \omega(x_{norm}(t)|\tau)dt, \tag{1}$$

where T is the threshold for deciding on the state of the process, $\omega(\cdot)$ is the probability density function, $x_{norm}(t)$ is the trend realization of the controlled business process parameter in the case of its suitable state,τ is the duration of observation of the random process.

2. An error of the second type (an unsuitable type of solution is recognized as suitable) is characterized by a probability:

$$\beta = \int_{-\infty}^{T} \omega\big(x_{fault}(t)|\tau\big)dt, \tag{2}$$

where $x_{fault}(t)$ is the implementation of the trend of the controlled business process parameter in case of its unusable state.

After that, we will simulate the calculation of the error of the first type. The number of modeling procedures is 10,000. The controlled parameter of the business process is measured 1000 times. To evaluate the erroneous decision based on the results of the simulation, we use formula (3):

$$\alpha_{mod} = \frac{n}{N}, \tag{3}$$

where n is the number of exceedances of the threshold T by the random parameter $x_{norm}(t)$ during the measurement period, N is the total number of measurements. In this case, $N = 1000$.

The modeling assumed a normal distribution of the $x_{norm}(t)$ parameter with a mean of 0.8 and a standard deviation of 0.05. The decision threshold was set at 0.91.

An example of a decision-making process is shown in Fig. 2.

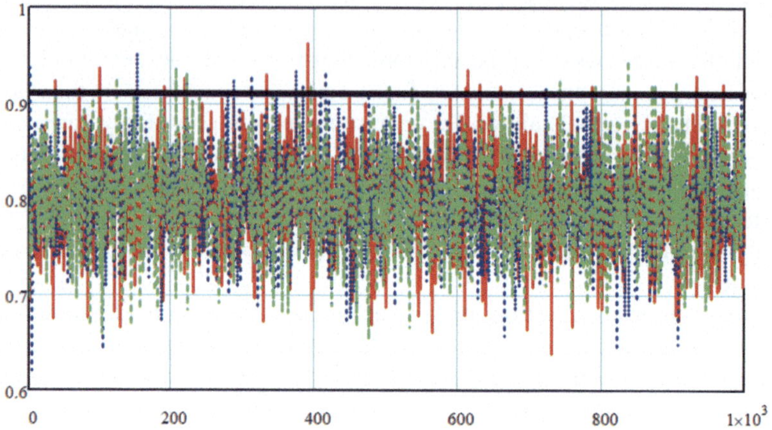

Fig. 2. Examples of implementation of a controlled decision parameter.

Figure 2 shows three examples of the implementation of controlled decision parameter in red, blue, and green colors. The black color indicates the decision threshold. If the controlled parameter crosses the decision threshold, an erroneous decision is recorded.

As a result of repeating the experiment multiple times for 10,000 iterations, a histogram of the probability of an incorrect decision was obtained. This histogram is shown in Fig. 3. Based on the results of the modeling, we calculated the statistical characteristics of the probability of an erroneous decision. The mathematical expectation is 0.014, the standard deviation is 0.003735.

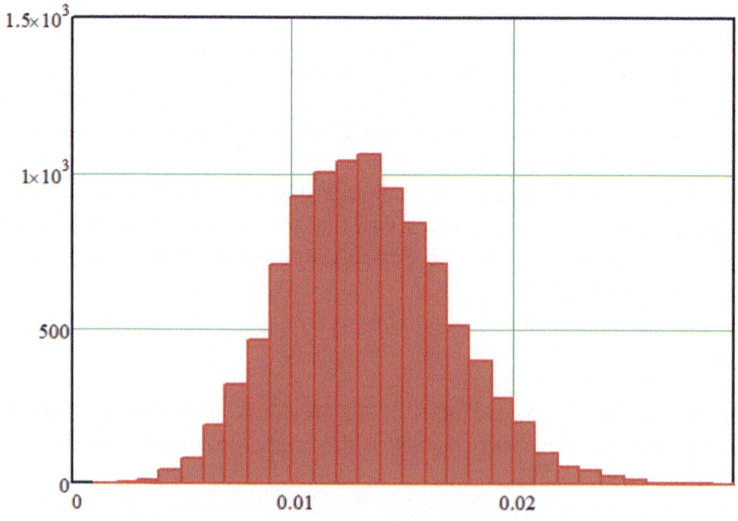

Fig. 3. Histogram of the probability of an incorrect decision.

5 Conclusions

The content of the material presented herein confirms the importance of integrating digital technologies in the process of strategic diagnostics of aviation enterprise in the condition of digital transformations. Digitalization is changing the way data is collected and processed, which opens up new opportunities for optimizing the aviation enterprise's operations. The introduction of the latest digital tools significantly increases the accuracy and efficiency of the processes under study. This allows aviation enterprises to respond more flexibly and quickly to changes in market conditions, reduce risks, and make effective decisions. The use of digital platforms helps to comprehensively assess the company's operations and improve its market position.

Particular attention is paid to the importance of adjusting the algorithms for strategic diagnostics of aviation enterprises in the condition of digital transformations, which allows for improved decision-making accuracy. This makes it possible to quickly assess opportunities and threats to the aviation enterprise, as well as effectively predict changes. The integration of digital technologies can significantly reduce the time spent on data collection and analysis, which is critical for the strategic management of aviation enterprises. A system of comprehensive indicators that includes financial, operational, market and innovation parameters become the basis for strategic diagnostics in the conditions of digitalization. The application of such approaches allows for better control over the strategy implementation in real time. Digital technologies help automate strategic management processes and ensure continuous monitoring of key performance indicators. All of these factors increase the overall efficiency of decision-making and reduce the likelihood of making mistakes. Digitalization is becoming an essential tool for developing strategic diagnostics and improving the efficiency of aviation enterprises in the modern business environment.

6 Future Scope

Future research is related to: (a) development of business models to ensure the sustainable development of the aviation enterprise; (b) development of recommendations for the introduction of innovative technologies in the business processes of aviation enterprise; (c) optimization of the aviation enterprise's operations based on the efficiency of the business process.

References

1. Zaliskyi, M., Solomentsev, O., Kozhokhina, O., Herasymenko, T.: Reliability parameters estimation for radioelectronic equipment in case of change-point. In: Signal Processing Symposium 2017 (SPSympo 2017), pp. 1, IEEE–4, Jachranka Village (2017). https://doi.org/10.1109/SPS.2017.8053676
2. Poberezhna, Z., Petrova, Y., Slimani, K.: Information technologies in logistics processes of enterprises in the aviation industry. CEUR Workshop Proc. **3732**, 90–102 (2024) https://ceur-ws.org/Vol-3732/paper07.pdf

3. Poberezhna, Z.: Formation of a system for optimizing business processes of aviation enterprises based on their automation. In: Ostroumov, I., Zaliskyi, M. (eds.) Proceedings of the 2nd International Workshop on Advances in Civil Aviation Systems Development. ACASD 2024, Lecture Notes in Networks and Systems, vol. 992, pp. 325–338. Springer, Cham (2024). https://doi.org/10.1007/978-3-031-60196-5_24
4. Brynjolfsson, E., McAfee, A.: The Second Machine Age: Work, Progress, and Prosperity in a Time of Brilliant Technologies. W. W. Norton & Company (2014)
5. Westerman, G., Bonnet, D., McAfee, A.: Leading Digital: Turning Technology into Business Transformation. Harvard Business Press (2014)
6. Niebel, T.: ICT and economic growth - comparing developing, emerging and developed countries. World Dev. **104**, 197–211 (2018)
7. Lipsmeier, A., Kuhn, A., Joppen, R., Dumitrescu, R.: Digital strategy: the historical development of the strategic management concept. Sci. Direct. **2020**, 173–178 (2020)
8. Nagorna, I., Koreyba, V.: The impact of digitalization on the strategic management of enterprise development. Effect. Econ. **6**, 1–14 (2024). https://doi.org/10.32702/2307-2105.2024.6.47
9. Hrinko, T., Gviniashvili, T., Kaliberda, M.: Strategic management of the enterprise in the conditions of the digital economy. Econ. Soc. **50**, 1–8 (2023). https://doi.org/10.32782/2524-0072/2023-50-71
10. Tokmakova, I., Shatokhina, D., Melnyk, S.: Strategic management of enterprise development in the context of digitalization of the economy. Bull. Trans. Indus. Econ. **64**, 283–291 (2018)
11. Hrinka, T., Nemchenko, T.: New management strategies for the digital transformation of business in Ukraine. Centr. Ukrainian Sci. J. **9**(42), 49–57 (2023)
12. Melnyk, O., Ruda, M.: Strategic aspects of digital transformation of business. Manag. Entrepren. Ukraine. **2**(12), 196–209 (2024)
13. Strutynska, I.: Definitions of the concept of "digital transformation". Black Sea Econ. Stud. **48**, 91–96 (2024)
14. Solomentsev, O., Zaliskyi, M., Nemyrovets, Y., Asanov, M.: Signal processing in case of radio equipment technical state deterioration. In: Signal Processing Symposium 2015 (SPS 2015), pp. 1–5. IEEE, Debe (2015). https://doi.org/10.1109/SPS.2015.7168312
15. Yakushko, I.: The essence and features of digital transformation. Prob. Prosp. Econ. Manag. **4**(28), 75–82 (2022)
16. Poberezhna, Z.: Comprehensive assessment of the airlines' competitiveness. Econ. Ann. XXI. **167**(9–10), 32–36 (2017). https://doi.org/10.21003/ea.V167-07
17. Smerichevskyi, S., Poberezhna, Z., Kryvovyazyuk, I., Ivanenko, L., Malnov, D.: Formation of principles of a customer-oriented approach by transport enterprises in conditions of sustainable development. E3S Web Conf. **534**(01022), 1–12 (2024). https://doi.org/10.1051/e3sconf/202453401022
18. Smerichevska, S., Poberezhna, Z., Mykhalchenko, O., Shtyk, Y., Pokanevych, Y.: Modeling and evaluation of organizational and economic support for sustainable development of transport enterprises: innovative and ecological aspects. Finan. Credit Act. Prob. Theor. Pract. **4**(51), 218–229 (2023). https://doi.org/10.55643/fcaptp.4.51.2023.4121
19. Smerichevskyi, S., Poberezhna, Z., Mykhalchenko, O., Petrova, Y., Smilyanets, V.: Ensuring the sustainable development of the aviation enterprise in the context of forming innovative potential using digital technologies. CEUR Workshop Proc. **3732**, 174–185 (2024) https://ceur-ws.org/Vol-3732/paper14.pdf
20. Marchenko, V.: The concept and legal support of digital transformation in Ukraine. Legal Sci. Elect. J. **6**, 279–282 (2019)

Interference in Global Navigation Satellite System

Research of the Influence of Retransmission Radio Interference on SBAS Characteristics

Valery Konin[1], Tetiana Maliutenko[1](✉), David Duchet[2], and Natali Caccioppoli[2]

[1] National Aviation University, Liubomyra Huzara Ave., 1, Kyiv 03058, Ukraine
tetiana.maliutenko@npp.kai.edu.ua

[2] EUROCONTROL Innovation Hub, 91222 Bretigny sur Orge, France

Abstract. This research provides experimental data on the effect of radio interference on SBAS characteristics. Radio interference is obtained by retransmission of GPS signals and correction signals from SBAS geostationary satellites. In the navigation receiver configured for SBAS operation mode through microwave power adder receives retransmitted signal and standard signal from two closely located antennas. The messages produced by the receiver are recorded in binary form. In parallel, a second receiver is switched on, which receives SBAS standard signals from a separate antenna. The data are recorded by the receivers and processed by the PEGASUS software system, its own programs and compared. When the receiver was operating in normal mode, navigation characteristics corresponding to an LPV 200 approach were provided. When the receiver was operating in radio interference mode, position deviations and protection levels increased and Stanford plots showed SBAS unavailability at 665 points from 11,000. On separate time intervals, the position deviations are in the horizontal plane 8 m–25 m (95%), in the vertical plane 40 m–25 m (99%).

Keywords: Air Navigation · EGNOS · Pegasus · Position Error · Protection Level

1 Introduction

Satellite Based Augmentation System (SBAS) is designed to improve the accuracy of coordinate determination, provide information on the integrity and availability of GPS signals, and in the future Galileo, GLONASS, BeiDou. Wide Area Augmentation System (WAAS) and European Geostationary Navigation Overlay System (EGNOS) are in active operation. In these systems by monitoring and special processing of navigation satellite signals by ground stations the information is formed in the form of corrections to ephemeris, navigation satellite time, ionospheric corrections, data on satellite integrity and availability. The formed information in the form of messages via geostationary satellite is transmitted to the consumer at the frequency of navigation signal. Under normal operating conditions, the systems meet the requirements for landing on Category I Precision approach (CAT I), Approach with vertical guidance (APV1, APV2), Localizer Performance with Vertical guidance (LPV-200), Non-Precision Approach (NPA) [1, 2]. SBAS technologies are developing and widespread all over the world [3].

I. Ostroumov et al. (Eds.): ACASD 2025, LNNS 1418, pp. 107–121, 2025.
https://doi.org/10.1007/978-3-031-91992-3_8

The availability of EGNOS signals is directly related to the location of ground Ranging Integrity Monitoring Station (RIMS), Navigation Land Earth Stations (NLES), GEO satellites, broadcasting to the consumer corrective information and messages about the integrity and availability of signals. For Ukraine, the availability boundary of EGNOS signals in longitude corresponds to approximately 30° [2]. In order to assess the availability of EGNOS in Ukraine, the National Aviation University conducted experimental research. Reports on data processing, using the PEGASUS software [4] are given in the repository of the National Aviation University [5], also published in [6, 7].

The performed researches have shown potential performance capabilities and prospects of EGNOS application in Ukraine. However, for full-scale use of all functions it is necessary to install several RIMS stations on the territory of Ukraine.

With the widespread implementation of satellite navigation technologies in the world, the threat of unintentional interference and intentional interference with satellite navigation signals is spreading. Unintentional interference is associated with the distribution of radio-electronic devices whose signal spectra and harmonics fall within the spectra of navigation signals. Intentional interference is aimed at deliberately generating electromagnetic oscillations to jam a navigation receiver (jamming) or spoofing satellite signals to force an object to follow an imposed path (spoofing, meaconing). The European Aviation Safety Agency's 2024 update bulletin notes that spoofing is becoming a major issue for flight safety [8]. The bulletin provides recommendations to prevent (mitigate) spoofing, the bulletin provides recommendations for air carriers, ATM/ANS service providers, equipment manufacturers, and government services.

In the scientific and technical literature, much attention is paid to countering spoofing [9, 10]. In publications related to experimental researches on the effect of spoofing on the characteristics of satellite navigation systems, the main focus is on GPS [11, 12]. In this paper, the effect of meaconing type radio spoofing on the performance of SBAS, in particular EGNOS are presented. We assume that the most significant influence will be caused by interference representing retransmitted signals of GPS navigation satellites and geostationary satellites. At the same time, the intensity of retransmitted signals does not differ much from the nominal signals. We assume that EGNOS data are collected synchronously when a navigation receiver receives signals in nominal mode and a second navigation receiver receives nominal retransmitted signals, after that data are processed and analyzed.

2 Research Methods

Experimental study has been grounded on equipment given in Fig. 1. Satellite navigation receivers were configured to receive and process signals in SBAS mode. GPS and GEO satellite signals from antenna 1 were received by OEM 719-1 and the necessary messages were recorded on the computer in the form of log files. Simultaneously from antennas 2 and 3 via Microwave adder GPS and GEO signals were received by OEM 719-2 and the same messages were recorded on the computer for further processing. Since the OEM 719-2 received signals from two antennas, the signals from either antenna can be considered as a retransmitted signal or meaconing.

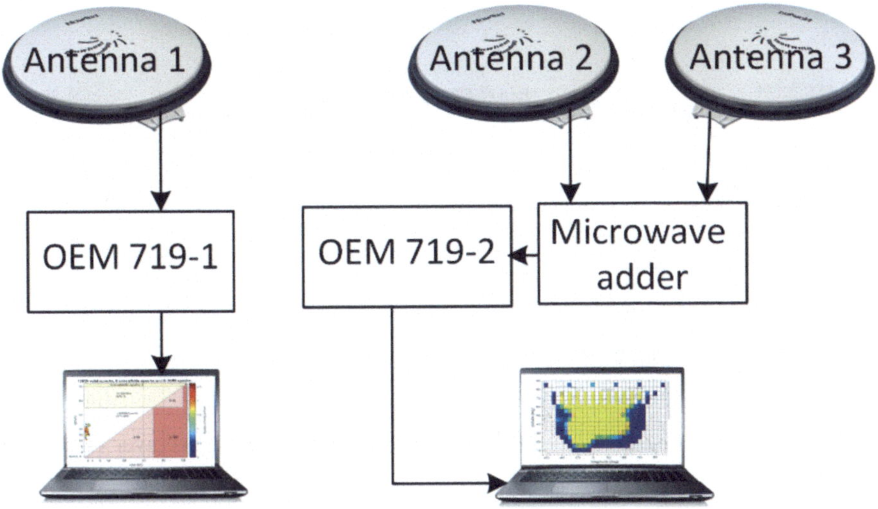

Fig. 1. Block diagram of equipment used in experimental study.

The following messages [13] (log files) were recorded by OEM 719 in binary form for processing:

1. Message 8 is IONUTC GPS, containing ionospheric model parameters (ION) and Coordinated Universal Time (UTC) parameters.
2. Message 73 is ALMANAC, containing an almanac of GPS satellites.
3. Message 43 is RANGEB, containing Pseudorange measurement (m), Carrier phase, in cycles (accumulated Doppler range), Instantaneous carrier Doppler frequency (Hz), Carrier to noise density ratio and other components.
4. Message 42 is BESTPOS, containing Latitude, Longitude, Height above mean sea level and its standard deviations.
5. Message 1043 is SATVIS2, containing Number of satellites, Satellite health, Elevation, Azimuth, Theoretical Doppler of satellite, Apparent Doppler for receiver.
6. Message 973 is RAWSBASFRAME, containing Raw SBAS frame data.
7. Message 41 is RAWEPHEM, containing Satellite PRN number, Ephemeris reference week number, Ephemeris reference time, and data.

The electromagnetic background was formed by the signals of satellite systems and SBAS systems. The research also took into account the state of the ionosphere. The signal spectrum in the researched electromagnetic background and Ionosphere Grid Point are shown in Figs. 2 and 3.

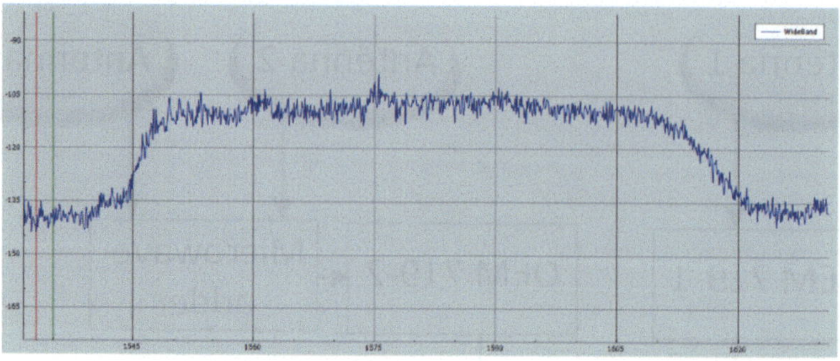

Fig. 2. Signal spectrum in receiver OEM 719.

In Fig. 2 data are obtained from the built-in spectrum analyzer receiver OEM 719.

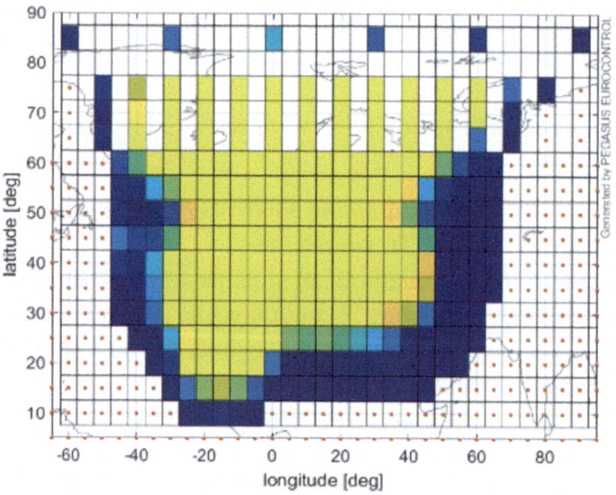

Fig. 3. Ionospheric grid.

In Fig. 3 data are calculated by the software REGASUS Ver. 24_05_RC3 [14].

The window of connection of receiver OEM 719 (interface) to the computer at the moment of start for data recording is shown on the Fig. 4 [13].

Fig. 4. Connection window OEM 719.

Constellation diagram is given in Fig. 5 shows tracks of 7 GPS satellites (represented by green circles), concentric circles correspond to the angles of the place within 0° (horizon level) and 90° (zenith), azimuth is counted hourly from the direction to the north. Signals of GLONASS, Galileo, BeiDou have been fixed, but does not participate in solving the problem of determination of the coordinates when working in SBAS mode. Components of Galileo, BeiDou signals are shown in the red boxes in the second line.

Fig. 5. Constellation diagram.

The Position window represents the current position and the operating mode (Type - SBAS), which confirms that the receiver is receiving correction information from EGNOS geostationary satellites. The Position confidence window is represented the values of the dilution of precision (dilution of precision in coordinates, time and total).

The GPS window is represented information about GPS satellite signals: number of frequencies from each satellite, signal-to-noise ratio, pseudorange, Doppler frequency, residual pseudorange values, and the time of continuous tracking of the satellite signal.

The Interference toolkit window is represented the interference toolbox when out-of-band signals near carrier frequencies arrive at the receiver. Messages of OEM 719 receivers were recorded for 11,200 seconds. The received files were processed using PEGASUS software [14] and sliding window method.

3 Results

3.1 Results in Normal Mode

The normal mode refers to the results of data processing when there is no interference with satellite navigation signals. In the receiver OEM 719 data are received only from the antenna 1, and the receiver is set to SBAS mode. The processing results are shown on the Figs. 6, 7, 8, 9 and 10.

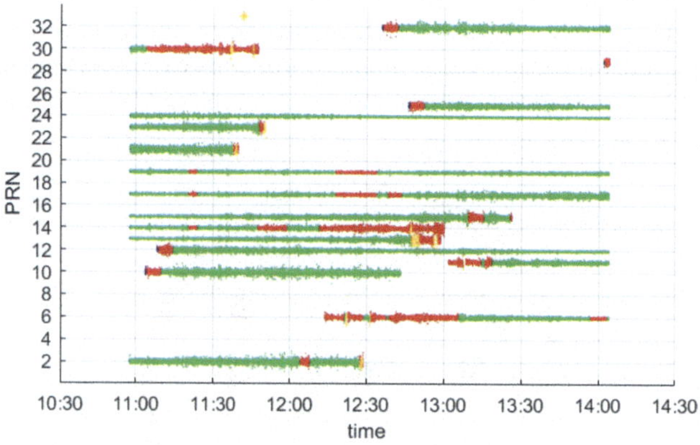

Fig. 6. PRN noise and status.

In Fig. 6 are shown a diagram of visible GPS satellites. The horizontal axis is shown the time, the vertical axis is shown the satellite numbers; the horizontal lines are shown the time during which the satellite signals are received by the receiver.

Deviations (meters) in the horizontal plane of the topocentric coordinate system from the reference point are shown in Fig. 7. Information of vertical deviations could be seen on the Stanford plots and after processing the message BESTPOS. The coordinates of the phase center of antenna 1, previously estimated with geodetic accuracy, are chosen as the reference point.

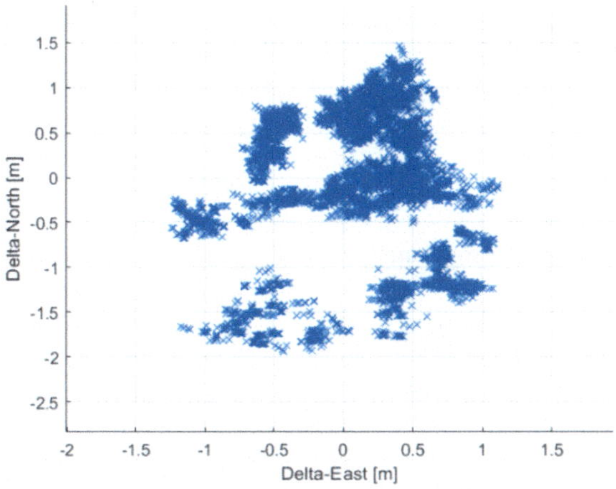

Fig. 7. Horizontation deviation from reference SBAS.

Stanford plots [15] are shown in Fig. 8. Stanford plot provides conclusions about the quality of navigation signals. On the horizontal axes are plotted the Position Error in the horizontal plane (*HPE*) and vertical plane (*VPE*), respectively. The vertical axes show the Protection Levels in the horizontal (*HPL*) and vertical (*VPL*) planes. The Alert Limit thresholds for horizontal (*HAL*) and vertical (*VAL*) planes are indicated by horizontal lines at the levels of 40 m and 35 m, respectively [1]. Each measurement is classified based on region. If the point falls within an uncolored triangle, the system is considered to be available. The following conditions are met Normal Operations:

$$HPE < HPL < HAL, \tag{1}$$

$$VPE < VPL < VAL. \tag{2}$$

If a point falls within any shaded area, the information from the system is considered System Unavailable, Misleading Information (MI), or Hazardously Misleading Information (HMI).

System Unavailable:

$$HPE < HAL < HPL, \tag{3}$$

$$VPE < VAL < VPL. \tag{4}$$

Misleading Information:

$$HPE < HPL < HAL, \tag{5}$$

$$VPE < VPL < VAL. \tag{6}$$

System Unavailable, Misleading Information:

$$HAL < HPL < HPE, \tag{7}$$

$$VAL < VPL < VPE. \tag{8}$$

Hazardously misleading information:

$$HPL < HAL < HPE, \tag{9}$$

$$VPL < VAL < VPE. \tag{10}$$

Data in Fig. 8 shows that system was available for the LPV-200 category at $HAL = 40$ m, $VAL = 35$ m during the 10,839 measurements.

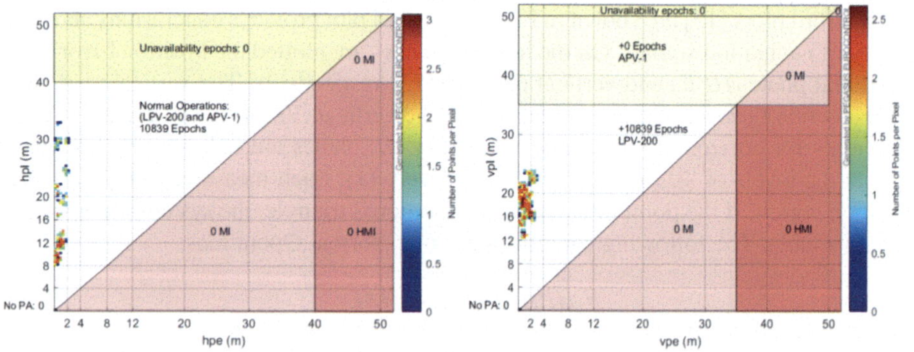

Fig. 8. SBAS Stanford plot for horizontal and vertical planes.

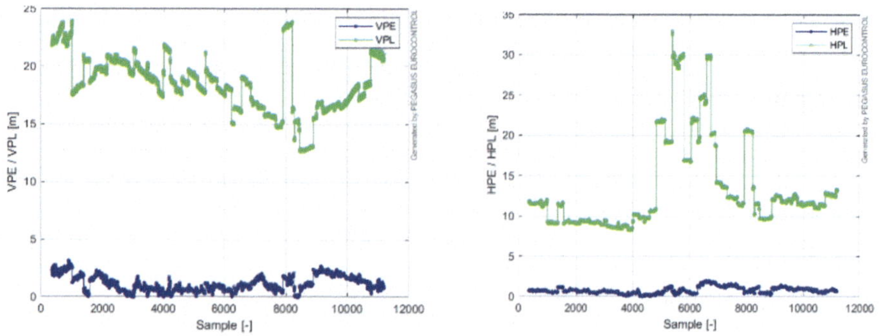

Fig. 9. Time dependencies of VPE, VPL, HPE, HPL without interference.

Values of VPE, VPL, HPE, HPL for the LPV-200 category (Fig. 9) reach $HAL = 40$ m, $VAL = 35$ m.

3.2 Research Results under the Influence of Interference

During the experiment, a significant degradation of the data was obtained when the satellite signals were interfered with. Diagram of visible satellites under the influence of interference on the signals of the GPS system are shown in Fig. 10. On the horizontal axis is plotted the time during which the signals arrive in the receiver OEM 719-2, and on the vertical axis are plotted the numbers of visible satellites. As can be seen from Fig. 10, time of influence of interference on satellite signals has increased.

Deviations (meters) in the horizontal plane of the topocentric coordinate system from the reference point under the influence of interference are shown in Fig. 11. Information about deviations in the vertical plane can be seen in the Stanford plots in Fig. 12. From the plots of Figs. 12 and 13, it could be seen that the system was available at 10075 measurements in the horizontal plane (hpl) and 9454 measurements in the vertical plane (vpl) for the LPV-200 category at $HAL = 40$ m, $VAL = 35$ m.

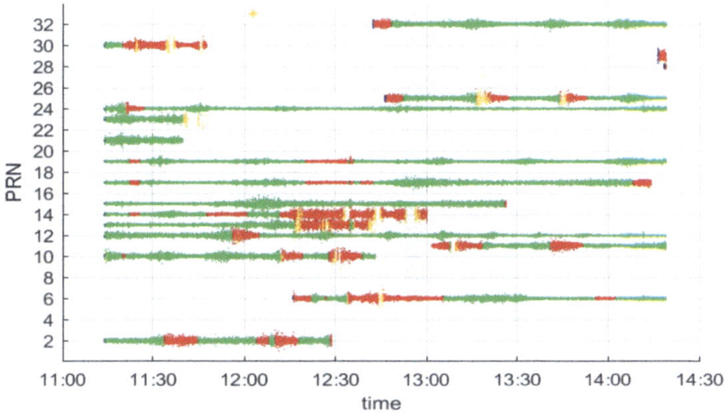

Fig. 10. PRN Noise and Status under influence of interference.

Comparing the results of data processing we obtained that different system availability time for the category LPV-200 in the horizontal and vertical planes. In the absence of interference with GPS signals, the system is available without change at the given levels HAL = 40 m, VAL = 35 m (Fig. 8). On other hand, the presence of interference with satellite signals (Fig. 12) at HAL = 40 m, VAL = 35 m reduces the access time to use GPS navigation signals.

3.3 Position Research (BESTPOS Message)

The next stage of the research consists in processing the BESTPOS message for normal SBAS mode and mode with interferences. BESTPOS message is one of the key messages when solving navigation tasks. The following algorithm was used to process the received message.

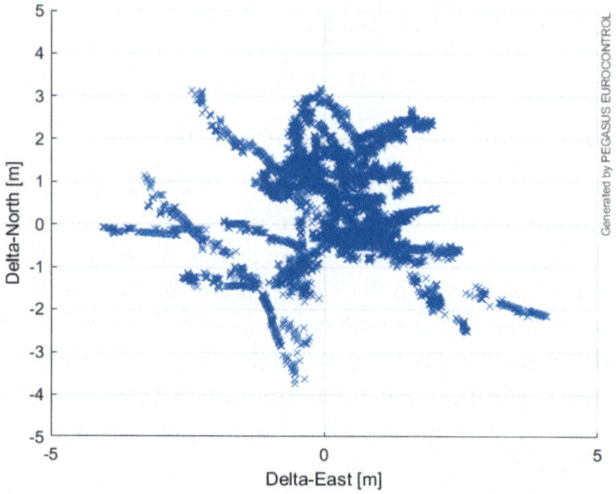

Fig. 11. Horizontation deviation from reference SBAS under influence of interference.

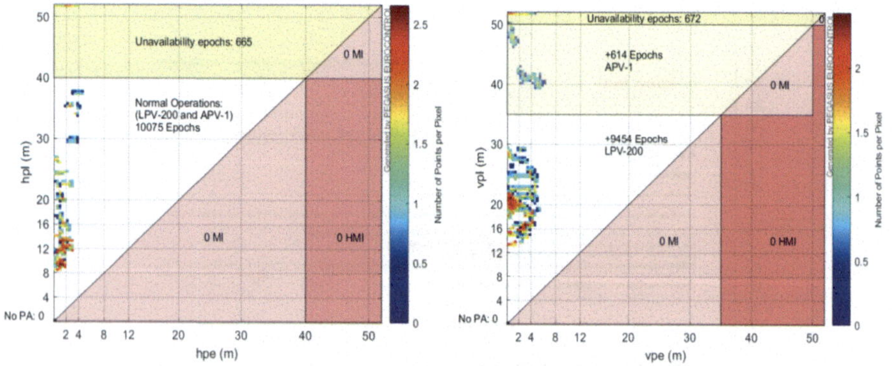

Fig. 12. Stanford_Plot SBAS under influence of interference.

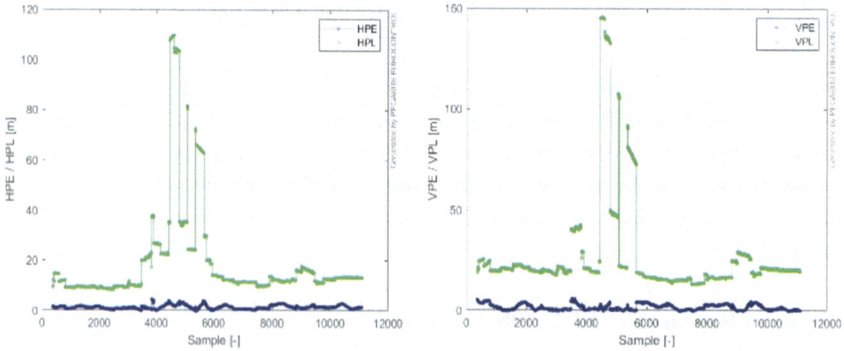

Fig. 13. Time dependences of VPE, VPL, HPE, HPL under interference influence.

BESTPOS data were decoded and latitude (lat), longitude (lon), altitude (h) were determined, which were converted into topocentric coordinate system (East, North, Upper) relative to the reference coordinates of antenna phase centers. (Fig. 1).

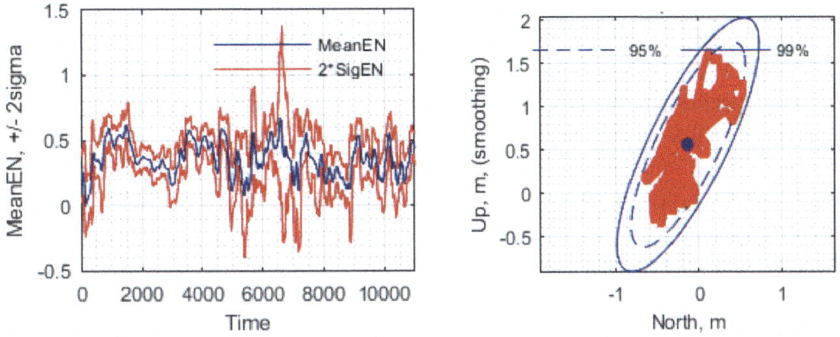

Fig. 14. Data from BESTPOS message for the normal mode: mean values of deviations (MeanEn), variance (2sigma) and ellipse error.

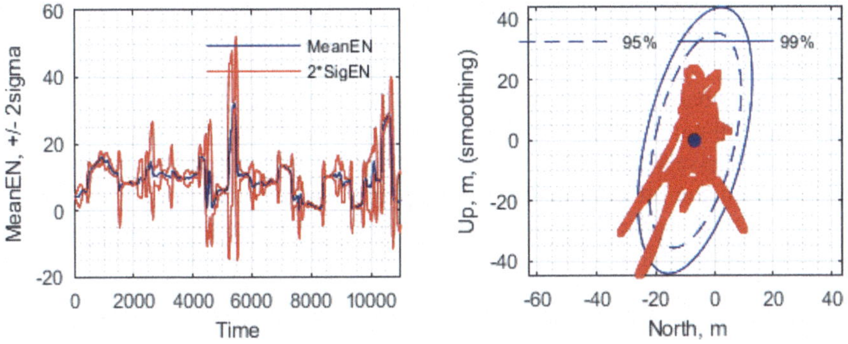

Fig. 15. Data from BESTPOS message for the interfered mode: mean values of deviations (MeanEn), variance (2sigma) and ellipse error.

The obtained coordinates were smoothed using the sliding window method. In 100 samples (slide size) the mean (Mu) standard deviation (std) were determined. The slide was shifted by one sample and the procedure was repeated until the end of the time interval. From the resulting arrays, mean values of deviations in the East-North horizontal plane (MeanEN), variance (sigma), and ellipses of dispersion of deviations (Ellips error) were calculated. The obtained results are shown in Fig. 14 for the normal mode and in Fig. 15 for the interfered mode.

4 Discussions

In these researches the signals of the same GPS satellites were processed both in normal mode and in the mode with interference. When researching the mode with interference, the beginning of GPS signal reception was shifted by about 15 min due to the influence of interference (Figs. 6 and 10). Analyzing the obtained data, signal missing in normal mode were at satellites 32, 30, 13, 6; and in the mode with interference at satellites 32, 30, 25 (2), 14 (4), 13 (2), 11, 10, 6 (the number of signal missing is given in brackets, if there is more than one). Accordingly, in the mode with interference we observe signal missing at more navigation satellites. Also in the interference mode, the change in the signal-to-noise ratio is oscillatory in nature. As a result, the deviations in the horizontal plane for normal mode and mode with interference (Figs. 7 and 11) have the following values: in the direction to the North from −2 m to 1.5 m (mode with interference: −3.9 m to 3.1 m), in the direction to the East from −1.1 m to 1.1 m (mode with interference: −4 m to 4 m). Thus, in the mode with interference the deviations to the north increased two times and deviations to the east four times, which significantly affects the accuracy of navigation data.

The information presented in the Stanford plots (Figs. 8, 12) allows us to conclude that in normal mode the constant availability of EGNOS for the LPV-200 category for the time interval of 10,839 s is provided. On the other hand, in the interfered mode, the system is not available in 665 counts out of 10,740 in the horizontal plane and in 672 counts out of 10,740 in the vertical plane for the given landing category LPV- 200. Analyzing the content of Stanford graphs (Figs. 9 and 13) we can see on which parts of time intervals anomalies take place. For example, it follows from Fig. 13 that the HPL and VPL jumps occur between 4000 and 6000 s, which also affects the system availability.

Also, when analyzing the data from the BESTPOS message, we can conclude that in the mode without interference, the deviations in the horizontal and vertical planes (Fig. 14) are consistent with the data in Figs. 8 and 9. But in the mode with interference there is a systematic offset in the horizontal plane reaching 18 m and deviations of 20 (95%) meters in the vertical plane (Fig. 15). Additional studies are required to research this issue in more detail.

Previously [16–18], it was shown that such phenomena are associated with changes in the signal-to-noise ratio. We have shown the dependences of the signal-to-noise ratio for the normal and interfered modes on Fig. 16.

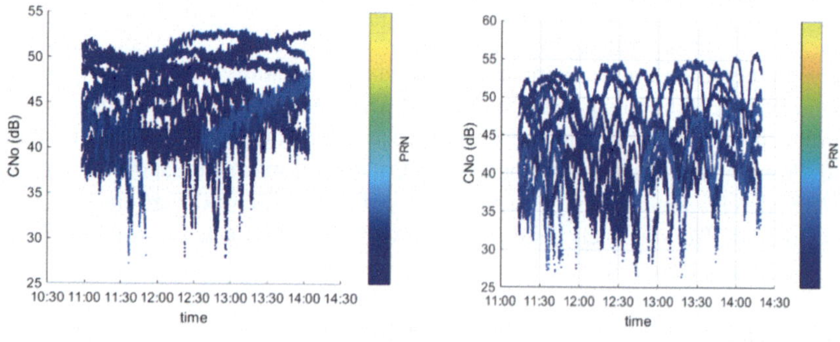

Fig. 16. GPS signal-to-noise ratio in normal mode and in reference mode.

In Table 1 are shown the average positioning errors (Mu) and HPE APV1, HPE LPV, VPE APV1, VPE LPV at 95% intervals and the average values of the protection levels

HPL APV1, HPL LPV, VPL APV1, VPL LPV at 99% intervals, which are obtained by the PEGASUS software package [14]. The data for antenna 1 are presented interference free mode with satellite signals, and the coordinates of antenna 1 are taken as the SBAS reference point. The data for antennas 2 and 3 are presented in the mode of interference influence on the signals, and the coordinates of the phase centers of antennas 2 and 3 are taken as the reference points, respectively. Thus, it follows from the data that in the mode with radio interference position deviations are 1.5–1.9 times greater compared to the normal mode, and protection levels are 1.16–2.0 times.

Table 1. Average values of positioning errors and average values of protection levels.

	Antenna 1	Antenna 2	Antenna 3
Mu/HPE APV1 95%	0.8/1.57	1.44/2.87	1.57/2.81
Mu/HPL APV1 99%	13.4/29.93	14.23/35.63	13.76/35.46
Mu/HPE LPV 95%	0.8/1.57	1.42/2.85	1.56/2.85
Mu/HPL LPV 99%	13.4/29.93	13.39/34.90	13.2/29.66
Mu/VPE APV1 95%	1.16/2.31	1.83/4.43	1.71/3.98
Mu/VPL APV1 99%	18.48/23.64	21.05/48.18	20.49/47.13
Mu/VPE LPV 95%	1.16/2.31	1.78/4.24	1.70/3.99
Mu/VPL LPV 99%	18.48/23.64	19.56/28.97	19.55/29.18

Radio interference of meaconing type was used to research the influence of interference on satellite signals. They were formed by retransmission of signals of navigation satellites located in the line of sight and geostationary satellites transmitting messages with corrective information. Thus, the retransmitted signals and standard satellite signals were fed into the navigation receiver through a microwave adder. The intensity of both types of signals was practically the same. Therefore, it is difficult to distinguish interfering signals from regular signals, even when authentication components are built into regular signals. The fact that radio interference and regular signals are commensurate in intensity is a transition point when it comes to finding ways to detect the presence of radio interference. With radio interference of greater intensity, the navigation receiver will go into radio interference-only mode. To quantify the effect of radio interference on navigation parameters, SBAS signals were processed in normal mode. All data in both modes were recorded simultaneously.

Thus, when operating in normal mode, navigation characteristics corresponding to LPV-200 were provided. On the other hand, in the mode of operation with radio interference, position deviations increased by 1.5–1.9 times, and deviations of defense levels increased by 1.16–2.0 times. At the same time, Stanford charts began to display information about SBAS unavailability in the horizontal plane for 665 seconds and in the vertical plane for 672 seconds. In addition, the protection levels reached 110 m in the horizontal plane and 148 m in the vertical plane (Fig. 13). These are very alarming warnings since the alarm response time is 6 seconds [1, 2].

The processing of the BESTPOS message also confirmed the significant influence of radio interference. In normal mode, when processing the BESTPOS message, the maximum position deviation in the horizontal plane is 1.4 m (95%) and in the vertical plane is 1.6 m (99%) (Fig. 14). Under the influence of interference, the results of BESTPOS processing are showed a position deviation in the horizontal plane greater than 8 m–25 m (95%), and in the vertical plane −40 m–25 m (99%) (Fig. 15).

Finally, analyzing the presented data interference in satellite signals we see reduced of access to satellite navigation systems, as well as an increase in the risks of using navigation information.

5 Conclusions

Radio interference is an increasing threat to SBAS satellite navigation systems. Spoofing and meaconing are the most dangerous as they can replace true position information with false coordinates. Meaconing interference is broadcast on the same frequency as the real signal. This is done in order to confuse the navigation system and users of navigation services, which increases the risks in flight safety.

The conducted researches have shown that the retransmission signal interference significantly affects the navigation performance of SBAS, making the system unavailable for use on LPV-200 category and affecting the offset of protection levels in the horizontal and vertical planes. The research should be continued in the direction of searching for indications of the presence of interference in the navigation signal to improve the safety of navigation service.

References

1. Aeronautical Telecommunications: Annex 10 to the Convention on International Civil Aviation, Volume I (Radio Navigation Aids), 8th edn. ICAO, Montreal (2023)
2. EGNOS Safety of Life (SoL) for Aviation, Issue 3.6. Luxembourg: Publications Office of the European Union (2024). https://doi.org/10.2878/33902
3. EGNOS Homepage. https://egnos.gsc-europa.eu/egnos-system/about-egnos. Last accessed 28 Dec 2024
4. EUROCONTROL: PEGASUS Interface Control Document Manual (2015)
5. EGNOS PEGASUS Reports. National Aviation University, Ukraine (2015–2017). https://er.nau.edu.ua. Last accessed 24 Dec 2024
6. Konin, V., Shyshkov, F.: Extending the Reach of SBAS Some Aspects of EGNOS Performance in Ukraine. https://insidegnss.com/auto/janfeb15-KONIN.pdf. Last accessed 04 Jan 2024
7. Kharchenko, V.P., Konin, V.V., Kondratiuk, V.M., Ilnytska, S.I., Shyshkov, F.O.: Activities of the National Aviation University in "UKRAINE" project of Horizon 2020 to foster EGNSS implementation in aviation sphere in Ukraine. In: Proceedings of the 7th World Congress. Aviation in the XXI- st Century. Safety in Aviation and Space Technologies, Kyiv, Ukraine, pp. 3.6.31–3.6.35 (2016)
8. European Union Aviation Safety Agency: Global Navigation Satellite System Outage and Alterations Leading to Communication / Navigation/Surveillance Degradation, SIB 2022-02R3, Publications EASA Safety Intelligence & Performance Department (2024). https://ad.easa.europa.eu/ad/2022-02R3. Last accessed 28 Dec 2024

9. Morales-Ferre, R., Richter, P., Falletti, E., Lohan, E.S.: A survey on coping with intentional interference in satellite navigation for manned and unmanned aircraft. IEEE Commun Surv Tutor. **22**(1), 249–291 (2020). https://doi.org/10.1109/COMST.2019.2949178
10. Walter, T., Anderson, J., Lo, S.: SBAS Message Schemes to Support Inline Message Authentication. In: 34th International Technical Meeting of the Satellite Division of The Institute of Navigation (ION GNSS+), St. Louis, Missouri, pp. 474–484 (2021). https://doi.org/10.33012/2021.17908
11. Chen, Y., Gao, W., Chen, X., Liu, T., Liu, C., Su, C., Lu, J., Wang, W., Mu, S.: Advances of SBAS authentication technologies. Satell. Navig. **2**, 12 (2021). https://doi.org/10.1186/s43020-021-00043-1
12. Felux, M., Fol, P., Figuet, B., Waltert, M., Olive, X.: Impacts of global navigation satellite system jamming on aviation. Navigation. **71**(3), 657 (2024). https://doi.org/10.33012/navi.657
13. OEM719 Multi-Frequency GNSS Receiver. https://novatel.com/en-gb/products/receivers/gnss-gps-receiver-boards/oem719. Last accessed 01 Feb 2024
14. EUROCONTROL: PEGASUS Interface Control Document Manual (2022)
15. Tossaint, M., Samson, J., Toran, F., Ventura-Traveset, J., Hernandez-Pajares, M., Juan, J.M., Ventura-Traveset, J., Sanz, J., Ramos-Bosch, P.: The Stanford - ESA integrity diagram: a new tool for the user domain SBAS integrity assessment. Navigation. **54**(54), 153–162 (2007). https://doi.org/10.1002/j.2161-4296.2007.tb00401.x
16. Konin, V., Pogurelskiy, O., Prykhodko, I., Maliutenko, T., Zhalilo, A., Yakovchenko, A.: Revealing the criteria for detecting the spoofing and premediated interference of GNSS signals using the experimental simulation model. In: Ostroumov, I., Zaliskyi, M. (eds.) Proceedings of the 2nd International Workshop on Advances in Civil Aviation Systems Development. LNNS, vol. 992, pp. 101–113. Springer Nature, Cham (2024). https://doi.org/10.1007/978-3-031-60196-5_9
17. Sushych, O., Pogurelskiy, O., Konin, V., Kutsenko, O., Prykhodko, I., Maliutenko, T.: Performance analysis of intentional interference on Multi-GNSS receivers. In: Ostroumov, I., Zaliskyi, M. (eds.) Proceedings of the 2nd International Workshop on Advances in Civil Aviation Systems Development. LNNS, vol. 992, pp. 114–124. Springer Nature, Cham (2024). https://doi.org/10.1007/978-3-031-60196-5_10
18. Konin, V., Pogurelskiy, O., Prykhodko, I., Maliutenko, T., Sushych, O., Ishchenko, O.: Multi-GNSS in limited navigation satellite availability. In: Ostroumov, I., Zaliskyi, M. (eds.) Proceedings of the International Workshop on Advances in Civil Aviation Systems Development. LNNS, vol. 736, pp. 126–140. Springer Nature, Cham (2023). https://doi.org/10.1007/978-3-031-38082-2_10

AIM Technology Experimental Testing with Different Types of Antennas

Oleksandr Kutsenko(✉), Volodymyr Kharchenko, and Valeriy Konin

National Aviation University, Liubomyra Huzara Ave., 1, Kyiv 03058, Ukraine
kutsenkooleksandrua@gmail.com

Abstract. Global navigation satellite system usage has led to a significant increase in air traffic. It also led to the emergence of new air traffic participants such as drones and urban air mobility. One of the major weaknesses of the global navigation satellite system is its vulnerability to interference and spoofing. This is particularly dangerous for both manned and unmanned aerial vehicles. One of the methods of mitigating interference is the Advanced Interference Monitoring and Mitigation technology (AIM+). This technology protects against simple narrow-band interference as well as more complex wide-band interference, including jamming and spoofing. This paper presents the results of experimental testing of this technology. We investigated the accuracy and availability of the navigation solution when the navigation receiver is operating under the influence of interference. Three types of navigation antennas were used in this research. The research results are presented in text and graphical form. The conclusions regarding the use of the studied technology were made.

Keywords: GNSS · Position-Navigation-Timing · Interference · Jamming · Spoofing · Septentrio · AIM+ · Accuracy · Availability

1 Introduction

Global Navigation Satellite System (GNSS) technology has been used in civil aviation for quite some time, which has led to a significant increase in air traffic. According to the European Union Agency for the Space Programme report [1] this continued growth has increased dependency on GNSS and ensured a continuous revision and assessment of the performance needed to support traditional and emerging users, such as drones and urban air mobility. Now, GNSS is essential in the aviation domain, a leading and indispensable asset, supporting all aspects of aviation: Performance Based Navigation, flight planning, operations, airspace design, and service provision to airlines, pilots, airports and air navigation service providers. Unmanned aviation has experienced significant growth in recent years. GNSS is key for positioning and enabling integration with manned aviation; safe, efficient, and conflict-free operations rely on GNSS as the primary position navigation and time (PNT) source.

I. Ostroumov et al. (Eds.): ACASD 2025, LNNS 1418, pp. 122–132, 2025.
https://doi.org/10.1007/978-3-031-91992-3_9

1.1 Motivation

One of the disadvantages of GNSS is the low power level of radio signals from navigation satellites that are received by the navigation receiver. This can cause interference when a low signal is overpowered by a stronger signal operating at the same frequency. The source of interference can be antennas of radio communication and data transmission systems, powerful electric motors, power lines, electric welding, etc. We should also mention commercially available, inexpensive jammers to prevent the detection of vehicle location. Interference is hazardous when flying at low altitudes in urban areas. For Ukraine, an additional source of interference is the powerful jammers activated during air raid alarms.

So far, Controlled Reception Pattern Antenna (CRPA) technology has been considered an effective way to prevent interference. By changing the shape of the GNSS receiving antenna's radiation pattern in real time, this technology suppresses the reception of signals from the direction of the interference. This technology is described in more detail in the paper [2]. However, this technology has disadvantages, such as high product price, high energy consumption, and increased weight and dimensions.

An alternative to CRPA is the technology Advanced Interference Monitoring and Mitigation (AIM+) presented by Septentrio company (Belgium) [3], the top performing GNSS receivers developer.

1.2 Contribution

The purpose of the work presented in this article was to test the AIM+ technology in experimental conditions. Testing aimed to determine the accuracy and availability of the navigation solution obtained under the influence of interference in the L1 and L2 GNSS bands. At the same time, navigation antennas from three manufacturers were used simultaneously during the tests. This allowed us to provide recommendations for the selection of components for the optimal navigation system.

1.3 Organization

The second section of the article provides a detailed description of the Advanced Interference Monitoring and Mitigation technology and reviews the works of its authors. The third section presents the experimental research itself. It describes the composition of the experimental equipment, the course of the experiment, and the results of the research.

2 AIM+ Technology Description

AIM+ technology detects and neutralizes interference resulting in faster set-up, reduced downtime, and secure operation [3]. AIM+ protects against simple narrow-band interference as well as more complex wide-band interference, including jamming and spoofing. Septentrio receiver's web interface allows to analyze interference with the spectral plot, to be able to determine the type of interference and its possible source.

The Septentrio navigation receiver's web interface allows you to configure three Notch filters and enable or disable Wideband Interference Mitigation.

A notch filter is a well-known in digital signal processing filter that allows you to suppress a narrow band in the signal spectrum. Finding information about the notch filter in open sources is relatively easy. For example, reference [4] contains a description of the filter with an example of its software implementation. This filter can be represented as an expression:

$$y_n = \frac{\alpha x_n + 2(\alpha - 8)x_{n-1} + \alpha x_{n-2} - 2(\alpha - 8)y_{n-1} - (\alpha - \beta)y_{n-2}}{\alpha + \beta}, \tag{1}$$

where x_n, x_{n-1}, x_{n-2} are current and two previous counts of the filter input signal; y_n, y_{n-1}, y_{n-2} are current and two previous counts of the filter output signal; α and β are filter coefficients.

Filter coefficients α and β are defined by the expressions:

$$\alpha = 4 + (w_{0d}T)^2, \tag{2}$$

$$\beta = 2w_w T, \tag{3}$$

$$w_{0d} = \frac{2}{T}\tan\left(\frac{w_0 T}{2}\right), \tag{4}$$

where w_w is filter's notch width [rad/sec]; w_{0d} is the pre-warp center frequency of the filter's notch; w_0 is the center frequency of the filter's notch [rad/sec]; T is the sampling period.

The web interface of the receiver allows the user to set parameters w_0 and w_w, automatic selection is also possible. In the automatic selection mode, the selected parameters are not displayed on the web interface. The Wideband Interference Mitigation parameters are also not presented in the web interface in any way. This approach is quite understandable, since, as declared in [3], the AIM+ technology is patented. Using a patent search in open sources allowed us to obtain patents: "Device and method for multipath mitigation" [5] and "System and method for detecting spoofing of GNSS signals" [6]. The first patent describes a device and method for eliminating pseudo-range errors in a satellite navigation receiver channel that occur as a result of multipath. The method allows rejecting the re-reflected signals which arrive with a time delay relative to the original ones. This is achieved by improving the scheme for tracking the delay of the received signal from the satellite relative to its receiver's internal replica. The second patent describes a system and method for detecting spoofing signals based on monitoring the relative amplitude and phase values for pairs of received signals from left-hand and right-hand circularly polarized antennas. The authors claim that spoofing, unlike signals from navigation satellites, keeps the clusters of the mentioned parameters unchanged for a long time. This makes it possible to detect spoofing. It is also possible to eliminate spoofing, and where this is not possible, to exclude the signal of a spoofed satellite from the navigation solution.

Papers [7, 8] in which had been described the GNSS signal post-correlation method along with the premise of having the number of exploitable observables and measurements that are possible to monitor and cross-correlate by applying machine learning algorithms with the purpose of detecting the presence of the false signals.

Identification of excessive values of GNSS signal monitoring could be done with Pearson's correlation:

$$r_{x,y} = \frac{\sum_{i=1}^{N}(x_i - \overline{x})(y_i - \overline{y})}{\sqrt{\sum_{i=1}^{N}(x_i - \overline{x})^2}\sqrt{\sum_{i=1}^{N}(y_i - \overline{y})^2}}, \quad (5)$$

where N is number of samples; x_i and y_i are individual samples of correlated parameters; $\overline{x}$and $\overline{y}$ are mean values of correlated parameters.

As GNSS signal parameters, the authors used: lock time, signal noise ratio, pseudorange, carrier Doppler frequency, full carrier phase, multipath correction, code variance, carrier variance, carrier multipath correction, receiver clock bias, receiver clock drift, and spoofing indication. Only those parameters that had a statistically significant correlation with the indication of whether the GNSS signal was manipulated or not were considered for the following, supervised machine learning C-Support Vector Machines (C-SVM) based approach.

The paper presents experimental studies' results for different training and test data sets. Both simulated and real-world spoofing signals were used for testing.

Even though it is unclear whether the methods described above were used in the AIM+ technology, the information obtained is sufficient to perform field tests of the technology.

3 Experimental Testing

The AIM+ technology experimental testing aimed to determine the accuracy and availability of the navigation solution obtained under the influence of interference in the L1 and L2 GNSS bands.

3.1 Experimental Equipment

During the experimental testing, the equipment shown in Fig. 1 was used.

The experiment's core was the mosaic-X5 multi-constellation navigation receiver manufactured by Septentrio [9]. The manufacturer packaged the receiver in a land grid array (LGA) surface mount module, which was mounted on a print circuit board (PCB). The board was made without extra shielding from electromagnetic emissions. Receiver basic performances are presented in Table 1.

In total, three same GNSS receivers were used. Each of the navigation receivers was connected to a personal computer via a USB interface to control and data collection. The receiver's web interface was used as a software application. The receivers were powered via personal computer USB port 5 volt. Each of the receivers received signals from the same navigation satellites via an antenna. Navigation antennas from three manufacturers were used, namely: AJ977XF by Calian (Canada) [10], HA20 by Swegeo (Sweden) [11], and ANN-MB by Ublox (Switzerland) [12]. The performances of the antennas are presented in Table 2.

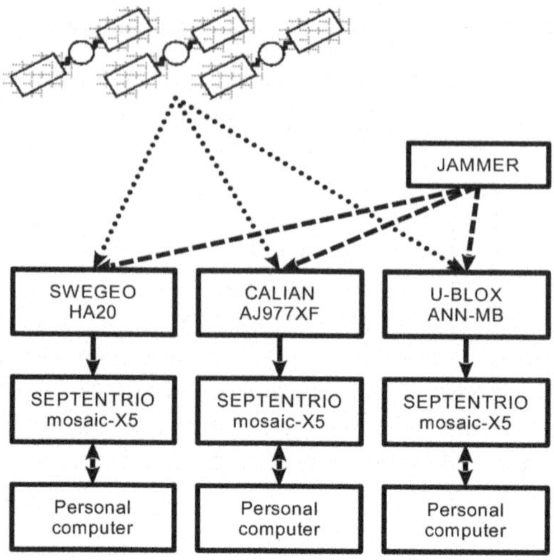

Fig. 1. Hardware used during experimental testing.

Table 1. GNSS receiver performances.

Performance	Value
Supported navigation systems and signal types	GPS: L1, L2, L5; GLONASS: L1, L2, L3; Beidou: B1, B2, B3; Galileo: E1, E5, E6; QZSS: L1, L2, L5; SBAS: Egnos, WAAS, GAGAN, MSAS
Positioning modes accuracy. Horizontal/vertical.	Standalone: 1.2 m/1.9 m SBAS: 0.6 m/0.8 m DGNSS: 0.4 m/0.7 m RTK: 0.6 cm/1 cm
Velocity accuracy	3 cm/s
Size	31 × 31 × 4 mm
Weight	6.8 g
Power consumption	0.6 W

A low-cost, low-power cramp jammer was used as a source of interference. The signal spectrum emitted by the jammer measured on the spectrum analyzer is shown in Fig. 2. Marker set to L1 band center frequency.

At a noise level of −45 dBm, the jammer emits signals with a power of up to −20 dBm. This power is enough to suppress the GNSS signal at a distance of up to 5 m. The jumper was powered by a built-in battery.

Table 2. GNSS antennas performances.

Performance	Calian AJ977XF	Swegeo HA20	Ublox ANN-MB
Antenna type	Helix	Helix	Patch
Frequency band	GPS: L1, L2, L5 GLONASS: G1, G2, G3 Beidou: B1, B2 Galileo: E1, E5	GPS: L1, L2, L5 GLONASS: G1, G2, G3 Beidou: B1, B2, B3 Galileo: E1, E5, E6	GPS: L1, L2 Beidou: B2 Galileo: E5
LNA gain	40 dB typ.	35 dB typ.	28 dB typ.
LNA noise	2.5 dB typ.	1.5 dB typ.	2.8 dB typ.
Size	90.0 mm (dia.) × 180 mm (h.)	27.5 mm (dia.) × 55.6 mm (h.)	82.0 × 60.0 × 22.5 mm
Weight	245 g	18 gram	173 g

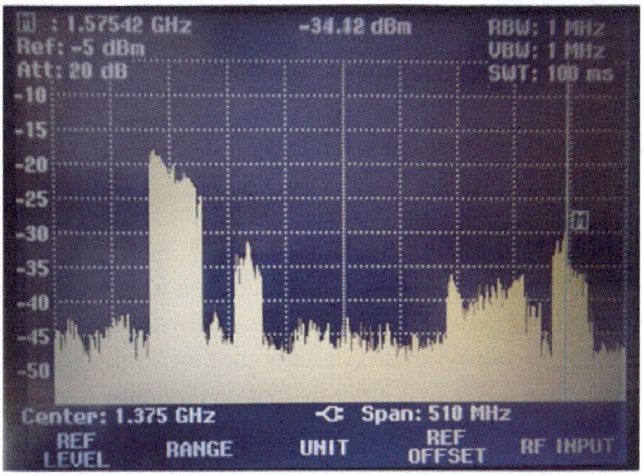

Fig. 2. Jammer spectrum.

3.2 Performing the Experiment

Experimental testing was performed at a special test site located in Ukraine to prevent any negative impact on navigation receivers not involved in the experiment. The experimental equipment at the test site is shown in Fig. 3.

Laptops were placed on the table. Above them, GNSS antennas were placed on some elevation. The jammer was on the ground, near the table. Such an arrangement made it possible to simulate the typical arrival angle of an interference for aircraft or UAV.

Fig. 3. Experimental equipment on test site.

The navigation receivers had been set to receive data with a 1-second update period. Initially, the receivers operated without interference for 5 min. Then the jammer was switched on. After that, the data was recorded for another 5 min. Due to the interference, the receiver with the Ublox ANN-MB antenna lost its navigation solution. The receivers with the other two antennas were able to keep the navigation solution. Thus, two sets of data were obtained, each consisting of latitude, longitude, and altitude WGS-84. The data sets also contained values of horizontal (HDOP) and vertical (VDOP) geometric dilution of precision factors. These factors allow us to estimate the signal availability from navigation satellites during the experimental testing period.

3.3 Experimental Data Processing

The methodology presented in [13] can be used to estimate accuracy and availability, but the following methodology is used in this paper.

The coordinates obtained in each dataset were converted to the local Cartesian coordinate system NED. The center of this coordinate system is the average value of the coordinates in the datasets. The axes of the coordinate system are directed to the north, east, and down, respectively. As an intermediate step, coordinates were obtained in the Earth-Centered Earth Fixed (ECEF) Cartesian coordinate system. For this purpose, the NavPy library for the Python programming language was used [14]. As a result, the deviation of the coordinates in the dataset from the average value in meters on each world side was obtained. Accuracy criteria for the horizontal (h_{acc}) and vertical (v_{acc}) planes could be obtained as follows:

$$h_{acc} = \sqrt{N^2 + E^2}, \tag{6}$$

$$v_{acc} = |D|, \tag{7}$$

where N, E end D are coordinates of Cartesian coordinate system NED.

Values (6) and (7) take only positive values and allow comparing navigation errors. Figure 4 shows the histogram of horizontal accuracy criteria for the HA20 and AJ977XF antennas. Data for the ANN-MB antenna is absent since no navigation solution was found under the influence of the interference.

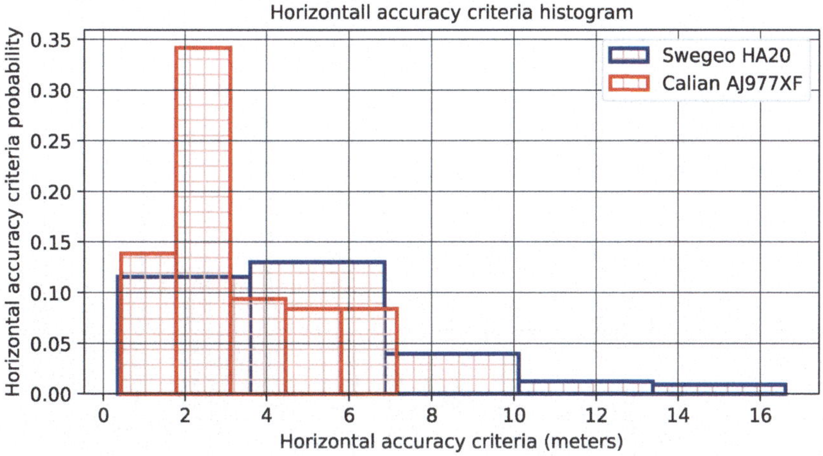

Fig. 4. Histogram of horizontal accuracy criteria.

The probability distribution can be seen from the figure. The horizontal accuracy criteria during the experiment reached a maximum value of 16.63 m for the Swegeo HA20 antenna and 7.15 m for the Calian AJ977XF antenna. The average values of the criteria were 4.93 and 3.12 m, respectively.

Figure 5 shows the histogram of vertical accuracy criteria. Data for the ANN-MB antenna is absent since no navigation solution was found under the influence of the interference. The probability distribution can be seen from the figure. The vertical accuracy criteria during the experiment reached a maximum value of 16.29 m for the Swegeo HA20 antenna and 12.92 m for the Calian AJ977XF antenna. The averaged values of the criteria were 8.78 and 4.69 m, respectively.

The time dependence of the horizontal dilution of the precision factor is shown in Fig. 6.

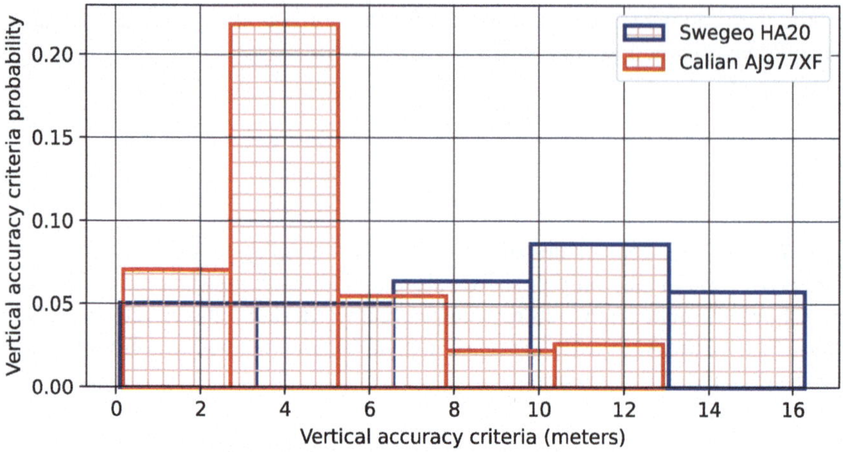

Fig. 5. Histogram of vertical accuracy criteria.

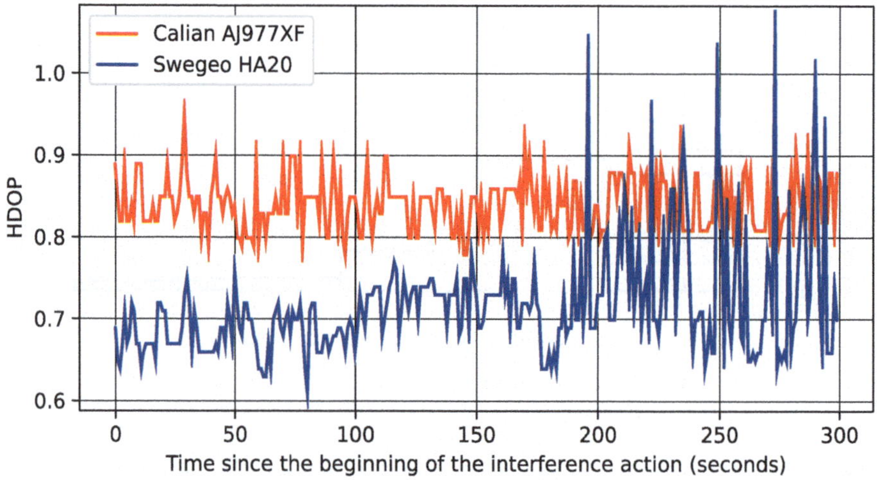

Fig. 6. Horizontal availability.

Data for the ANN-MB antenna is absent since no navigation solution was found under the influence of the interference. The starting point of time is the beginning of the interference influence. Data for the ANN-MB antenna is absent since there was no navigation solution under the interference's influence. The HDOP factor for the Swegeo HA20 and Calian AJ977XF antennas reached a maximum value of 1.06/0.95, a minimum value of 0.61/0.78, and an average value of 0.72/0.84, respectively.

Figure 7 shows the time dependence of the vertical dilution of the precision factor. The starting point of time is the beginning of the interference influence. Data for the ANN-MB antenna is absent since there was no navigation solution under the interference's influence.

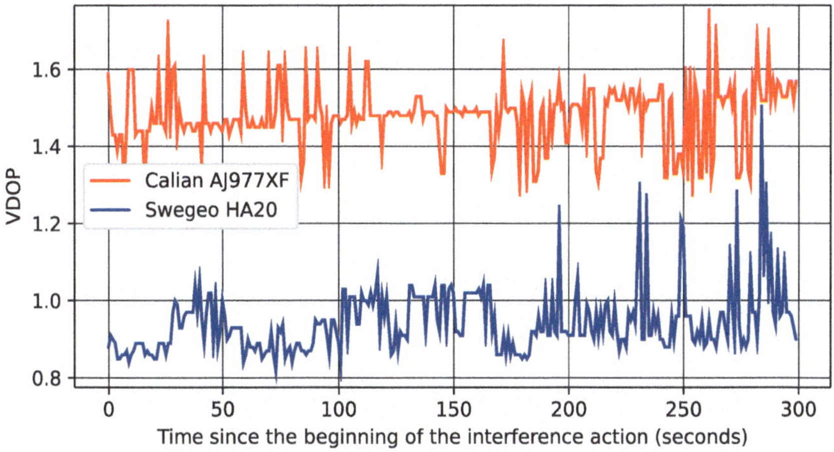

Fig. 7. Vertical availability.

The VDOP factor for the Swegeo HA20 and Calian AJ977XF antennas reached a maximum value of 1.47/1.72, a minimum value of 0.82/1.31, and an average value was 0.94/1.48, respectively.

4 Conclusions

The following conclusions can be drawn from the results of the experimental tests:

- The AIM+ technology avoids the effects of relatively low-power GNSS signal interference. This technology can be especially useful for unmanned aerial vehicles with small dimensions and low costs.
- The AIM+ technology does not work with planar antennas. Helical antennas must be used.
- When using Calian AJ977XF-like antennas, the accuracy of coordinate determination increases approximately by twofold. Therefore, it is recommended to use antennas of this type if weight, size, and cost restrictions allow it.
- When using Calian AJ977XF like antennas, there is a slight decrease in the availability of the navigation solution. However, this does not hurt the accuracy of position determination.

In the future, research on other navigation receivers types containing AIM+ technology and an examination of its compatibility with classic CRPA technology are planned.

References

1. EUSPA EO and GNSS Market Report. https://data.europa.eu/doi/10.2878/73092. Last accessed 03 Dec 2024
2. Averyanova, Y., Kutsenko, O., Konin, V.: Method of GPS, GLONASS, Galileo, and Beidou systems spoofing suppression. Telecommun. Radio Eng. **80**(7), 51–64 (2021). https://doi.org/10.1615/TelecomRadEng.2021037162

3. Septentrio Homepage. https://www.septentrio.com/en. Last accessed 03 Dec 2024
4. Phil's Lab Youtube channel. https://youtu.be/ysS4bIXFAsU. Last accessed 03 Dec 2024
5. De Wilde, W., Sleewaegen, J. M.: Device and method for multipath mitigation (EP Patent No. 2,034,617). https://worldwide.espacenet.com/publicationDetails/biblio?FT=D&date= 20090311&DB=&locale=en_EP&CC=EP&NR=2034617A1&KC=A1&ND=1. Last accessed 03 Dec 2024
6. De Wilde, W., Sleewaegen J.M.: System and method for detecting spoofing of GNSS signals (U.S. Patent No. 11,714,199). https://worldwide.espacenet.com/patent/search?q=US11714199. Last accessed 03 Dec 2024
7. Semanjski, S., Semanjski, I., De Wilde, W., Muls, A.: Use of supervised machine learning for GNSS signal spoofing detection with validation on real-world meaconing and spoofing data—part I. Sensors. **20**(4), 1171 (2020). https://doi.org/10.3390/s20041171
8. Semanjski, S., Semanjski, I., De Wilde, W., Gautama, S.: Use of supervised machine learning for GNSS signal spoofing detection with validation on real-world meaconing and spoofing data—part II. Sensors. **20**(7), 1806 (2020). https://doi.org/10.3390/s20071806
9. Mosaic-X5, Compact multi-constellation GNSS receiver, product datasheet. https://www.septentrio.com/en/products/gnss-receivers/gnss-receiver-modules/mosaic-x5#resources. Last accessed 03 Dec 2024
10. Calian. AJ977XF, Anti-Jam Multi-Constellation Triple-Band Antenna, product datasheet. https://sites.calian.com/app/uploads/sites/8/2024/06/Calian%C2%AE-AJ977XF-Datasheet.pdf?_gl=1*yhn971*_gcl_au*OTcyNjEzMjQwLjE3MzUyMTQzMzI. Last accessed 03 Dec 2024
11. Swegeo HA20 product web page. https://www.swegeo.com/product/swegeo-ha20. Last accessed 03 Dec 2024
12. U-blox AG. ANN-MB series. Multi-band, high precision GNSS antennas, product datasheet. https://content.u-blox.com/sites/default/files/documents/ANN-MB_DataSheet_UBX-18049862.pdf. Last accessed 03 Dec 2024
13. Sushych, O., Pogurelskiy, O., Konin, V., Kutsenko, O., Prykhodko, I., Maliutenko, T.: Performance analysis of intentional interference on multi-GNSS receivers. In: Ostroumov, I., Zaliskyi, M. (eds.) Proceedings of the 2nd International Workshop on Advances in Civil Aviation Systems Development. ACASD 2024. Lecture Notes in Networks and Systems, vol. 992, pp. 114–124. Springer, Cham (2024). https://doi.org/10.1007/978-3-031-60196-5_10
14. NavPy's documentation web page. https://navpy.readthedocs.io/en/latest. Last accessed 03 Dec 2024

Impact or Russian-Ukrainian war on Global Air Transportation System and Recovery of Ukrainian Civil Aviation

Area Analysis of DME/DME Positioning Method in Ukrainian Airspace

Ivan Ostroumov(✉) and Nataliia Kuzmenko

National Aviation University, Liubomyra Huzara Ave., 1, Kyiv 03058, Ukraine
ostroumov@ukr.net

Abstract. Airplane position measurements by data from the network of navigation aids have been highly useful during recent years due to a wide jamming of global positioning system in the eastern side of Europe. Distance Measuring Equipment (DME) is one of the reliable sensors that could be used for civil airplane position measurements. Requirements of area navigation RNAV 1 could be guaranteed by the DME/DME positioning algorithm in case of long interruption of a primary positioning system. In this paper, an area analysis approach is proposed for DME network configuration estimation. Configuration of the Ukrainian ground network is studied in three scenarios: initial, reduced functionality, and suspended to analyze network performance in correspondence with RNAV 1 requirements. Area analysis is performed by parameters of the number of ground beacons and DME/DME pairs available to use, mean standard deviation error of positioning by effective pair, and estimation area in which requirements of area navigation specification are guaranteed. Results of the study show that the area of RNAV 1 in Ukrainian airspace is reduced from 95.3% in the initial scenario to 87.7% in the reduced functionality scenario, and reaches 79.8% in a suspended scenario.

Keywords: Air Navigation · CNS/ATM · DME/DME · Navigation Aids · Performance · Positioning · Aerospace · Aviation

1 Introduction

Air transportation services are developed on the idea of the safety and security of civil aviation. Ensuring the safety of air traffic requires reducing mid-air collisions and minimizing airplane deviations from preplanned trajectories. Aviation uses a minimum of separation between airspace users to ensure the safety perils for each airplane [1, 2]. Separation minimums are estimated based on error actions in airplane maintaining at a particular track. The main part of this error is error of positioning sensor [3].

A receiver of the Global Positioning System (GPS) is a primary navigation sensor on board of most modern airplanes. During the last century quarter, GPS had many improvements which increased positioning performance significantly. Nowadays multi-constellation positioning algorithm integrated into most receivers could provide position measurements with an accuracy of less than 10 m [4, 5]. Moreover, the wide implementation of ground-based and satellite-based augmentation systems made precise positioning

I. Ostroumov et al. (Eds.): ACASD 2025, LNNS 1418, pp. 135–147, 2025.
https://doi.org/10.1007/978-3-031-91992-3_10

in performance less than 1 m, which satisfy requirements for landing operations [6, 7]. Also, high levels of reliability, small size, weight, and global coverage of operation made GPS is number one navigation system for air traffic support services.

Short interruptions in GPS could be easy recovered by Attitude Heading and Reference System (AHRS) or by simple Dead Reckoning Methods (DRM) application in Flight Management System (FMS). Modern AHRS includes a strapdown inertial reference approach, which is grounded on the measurement of airplane orientation angles by gyroscopes and accelerations by local reference frame to recover airplane position [8]. The sequential process of coordinates measurement in AHRS stimulates accumulation of errors [9, 10]. The required precision level significantly limits AHRS usage [11, 12]. DRM uses airplane airspeed and heading angle measurements with given wind speed and wind direction angle to predict airplane position as a sequential process with increasing error at each step, also.

In the case of long-term GPS lock algorithms of positioning by pair of navigation aids are used [13, 14]. There are DME/DME, VOR/VOR, and VOR/DME position calculation methods used in common FMS configuration. In this case, FMS uses measurements of distances or angles to ground navigation beacons: Distance measuring equipment (DME) and VHF omnidirectional range (VOR) [15]. The performance of the positioning approach by navigation beacons depends on the geometry of airplane location and the available configuration of ground beacons.

Several studies use linear grid to evaluate navigation aids network performance at each point of airspace [16, 17]. This approach is grounded on the estimation of navigation system error of positioning at each node of the grid with further data visualization as a heat plot. Unfortunately, these approaches could work only in a horizontal plane. Realization for 3D requires multiple runs for different input altitudes, which significantly increases overall computation time.

One of the perspective approaches used for network performance analysis could be based on a global spatial indexing system that provides a partition of airspace with a set of elementary cells [18, 19]. Hierarchical spatial indexing systems with hexagonal, rectangular, or triangular cells are mostly used in different geospatial analyses.

Airspace could be represented through a set of rectangular boxes. The value of performance could be estimated for a box reference point and is considered constant for a particular box geometry.

Also, there are studies of network optimization based on generic algorithms that follow a sequential process of nodes relocation or adding a new one to fill the gaps in overall network performance [20].

War in Ukraine and militaries actions in countries of Middle East Region caused strong interference action in neighboring airspaces [21, 22]. Jamming and spoofing of GPS equipment of civil airplanes are well documented by different surveillance systems. Also, the problem of GPS interference in Eastern Europe is highlighted in several articles [23, 24]. Wide areas of interference made methods of airplane positioning by navigation aids highly important in civil aviation.

In the paper, a ground navigation aids network configuration of Ukraine is analyzed in different historical periods. The ongoing war in Ukraine caused losses in territories of the country since 2014. This process coincides with navigation aids network reduction and reconfiguration processes. The paper provides network analysis based on graph theory and computer simulation of positioning accuracy provided for airspace users. Also, an area analysis of the required level of performance is proposed to get network productivity to meet area navigation specification minimum levels.

2 Navigation Aids

There are three basic types of navigation aids: DME(TACAN), VOR, and Non-directional beacons (NDB). Each navigation aid type includes ground beacons and specific on-board equipment to measure particular navigation parameter. Modern roadmaps of air navigation services development consider only DME and VOR based on the high accuracy of parameter measurements.

Any navigation aid is primary navigation equipment that is used to maintain an airplane position within a predefined flight path. The precise position of ground beacons is shared with airspace users in electronic Aeronautic Information Publications (eAIP). Also, position of a ground beacon is chosen based on flight-routes network configuration to make direct flight route track angle readings.

DME provides a continuous slant range measuring from a particular ground beacon to the airplane. Each ground beacon uses radio-frequency channel access to perform range measuring [25]. Simultaneously measured ranges from two ground beacons to an airplane are used for exact airplane positioning by the DME/DME positioning method.

VOR ground beacon transmits a specific signal which includes the phase-coded value of the airplane relative location to the VOR ground station. The receiver of VOR signals receives a navigation signal at a particular ground beacon channel and measures the angle of airplane orientation. FMS could recalculate many other angles required for airplane piloting from angle provided by VOR.

Co-located installation of DME and VOR provides full navigation capability in range and angle measuring related to point-of-ground beacon installation.

Ground beacons of navigation aids form a network. Each air navigation service provider (ANSP) designs and deploys a particular network configuration based on certain criteria of optimality. The minimum positioning error is the common objective function in the network configuration design.

The performance of the ground beacon network is estimated for each point (cell) of analyzed airspace with the help of the following parameters:

- Amount of ground beacons available to use;
- Amount of ground beacons pairs;
- Accuracy of positioning by pairs of ground beacons based on a method used;
- Binary value of area navigation specification which are guaranteed.

Also, parameters that characterize system reliability could be estimated based on the number of available pairs [26].

All listed above parameters are distributed continuously across airspace area. To analyze these parameters distribution possible airspace is represented as a matrix of cells which forms an airspace model. Commonly a cartesian grid of rectangles is generated based on a particular approach of node fragmentation: uniformly, centripetal, or any other. A global spatial indexing system (GSIS) is also could be used [27, 28]. There are different software libraries which support GSIS and could be used in various programming languages available for practical application. GSIS could use different cell geometry in the airspace model. Rectangular or hexagonal are the most commonly used cell configurations. GSIS provides reliable geo-data addressing and processing including airspace representation [28, 29].

The center of a cell is usually used as a reference point for network parameters estimation. However, a mean value for three inside points is also used quite often.

The number of available ground beacons is estimated based on the service area of each beacon. ANSP informs airspace users common service volume of each beacon in AIP. These areas are areas of guaranteed service that could be used in FMS to identify effective pairs of ground beacons. The intersection of service areas of different ground beacons is analyzed to get the number of available navigation aids at each cell [30].

Accuracy of positioning is estimated by the root mean square error (RMSE) of positioning based on the method of navigation. In the case of DME/DME positioning RMSE is estimated based on the following model [3]:

$$\mathrm{RMSE}^2_{DME/DME} = \left(\mathrm{RMSE}^2_{DME\ A} + 2\mathrm{RMSE}^2_{sis} + \mathrm{RMSE}^2_{DME\ B}\right)\sin(\rho)^{-2}, \quad (1)$$

where $\mathrm{RMSE}_{DME\ A}$ and $\mathrm{RMSE}_{DME\ B}$ are root mean square error of distance measuring by DME A and DME B; RMSE_{sis} is a root mean square error of signal propagation in space; ρ is an inclusion angle between directions to ground beacons A and B.

Error RMSE_{DME} is estimated as 0.125% from the distance to the ground beacon, but bigger than 157.42 m. Error RMSE_{sis} is limited by a constant value of 92.6 m. The inclusion angle is calculated from the cosine theorem and the known position of ground beacons DME A and DME B in the local Cartesian reference frame (as a reference point a position of one of DMEs is used).

Most continental airspace should be equipped with RNAV 1 navigation service quality. It means that positioning methods have to ensure airplane positioning with a total error not higher than 1852 m [15]. This error peril considers overall airplane maintaining error including flight technical error. Statistical analysis of different piloting modes shows that flight technical error for manual mode can not reach 926 m and for common autopilot mode 463 m.

The total error budget gives that the RMSE of positioning with the DME/DME method should be less than 801.93 m for 95% of the confidence band or 534.62 for 99.7% of the confidence band. According to a particular confidence band the area of correspondence to $\mathrm{RMSE}_{DME/DME}$ could be identified. Due to the known size of each cell, an area could be calculated for each level of $\mathrm{RMSE}_{DME/DME}$. Area function could be identified for each of the network parameters to make an overall network performance.

3 Ukrainian Navigation Aids Network

Initially, the Ukrainian navigation aids network was designed to support the conventional navigation method. Since 2000, a process of network optimization to follow the requirements of performance-based navigation began [15]. Since 2014, the configuration of the navigation aids network has been changed several times due to ongoing war in the source and Eastern sides of Ukraine. In our study, we evaluate network performance in different scenarios to do the following:

- Measure the impact of war in Ukraine on navigation aids network configuration;
- Estimate airspace that is fully compatible with RNAV 1 requirements;
- Identify areas within Ukrainian airspace with low performance.

Because of the ongoing war air navigation services are not provided in Ukrainian airspace. Therefore, we use navigation aids network configuration valid on February 24, 2022, which is in suspended state now. To study network configuration evolution we use two network scenarios: initial (valid on October 17, 2013) and reduced functionality (valid on August 18, 2016).

Navigation aids network configuration of the initial scenario includes 9 VOR/DMEs and 7 stand-alone DMEs (also 32 NDB). A scenario of reduced network configuration, caused by lost territories in the South and Eastern parts of Ukraine, includes only 8 VOR/DMEs and 6 stand-alone DMEs (also 15 NDB). The suspended scenario includes 7 VOR/DMEs and 9 stand-alone DMEs (but only 4 NDB) [31]. Comparison of node data identifies the core network set that has been used. Core VOR/DME services include BRP, IVF, KHR, LIV, ODA, and SLV ground beacons. Core DMEs include 6 stand-alone ground beacons: BAH, KSN, KVR, STB, VIN, and YHT. Each scenario has its extra beacons:

- Initial scenario: DNP, DON, SMF, and MRP;
- Reduced functionality scenario: DNP and KVH;
- Suspended scenario: RVN, KVH, TER, and UZH.

Core navigation aids list in different scenarios is given in Fig. 1. These network modifications are caused by war and the ongoing process of network optimization.

Ukrainian national network of navigation aids is shown in Fig. 2. Configuration of the network is organized to provide the highest position performance in the geometrical center of the network and cover flight information regions: Kyiv (UKBV), Lviv (UKLV), Odesa (UKOV), Dnipro (UKDV), and Simferopol (UKFV).

Also, services of neighboring networks of navigational aids could be used in Ukrainian airspace. Navigational aids of neighboring countries increase the performance of positioning methods used in Ukraine. Thus, during Ukrainian network performance estimation, we use national networks of neighboring countries: Moldova (1 DME), Romania (24 DMEs), Hungary (9 DMEs), Poland (42 DMEs), Slovak Republic (12 DMEs), Turkish, Belorussia and Russian Federation.

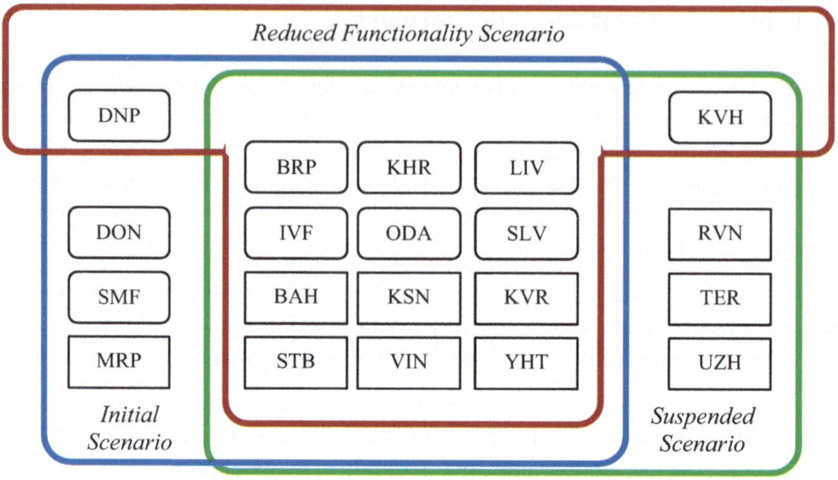

Fig. 1. Network configuration evolution through the scenarios.

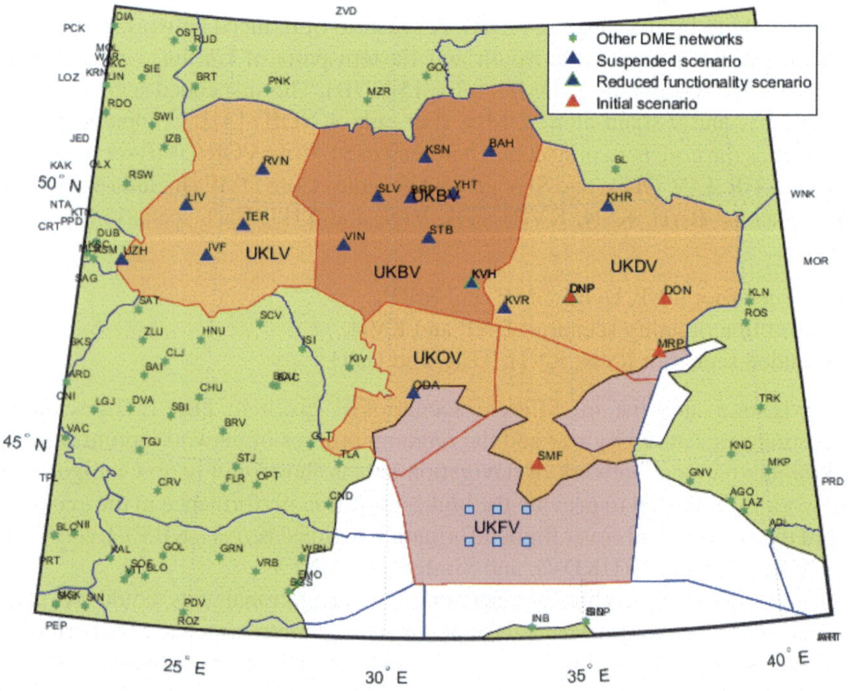

Fig. 2. Geometry of Ukrainian navigation network configuration to support DME/DME positioning method.

4 Performance Analysis

Navigational aids performance is analyzed with the help of the following parameters: amount of ground beacons available to use, Amount of ground beacons pairs formed by available DMEs, accuracy of positioning by DME/DME pairs by (1), and estimation area in which requirements of area navigation specification is guaranteed.

The multi-cylindrical service volume of each DME is used to estimate availability. Parameters of maximum range, altitude levels, and possible closed sectors are used in the DME model. Each DME model is fixed in reference point at a particular altitude. In availability estimation, the rectangular elementary cell with size 2.8 × 1.34 km is used (that gives uncertainty area in 3.76 km^2). Ukrainian airspace has been segregated into 230,485 cells. Results of availability analysis and a corresponding number of pairs are shown in Figs. 3, 4 and 5 for different scenarios of network configuration. Results of NSE with correspondence of RNAV requirements are shown in Figs. 6, 7 and 8.

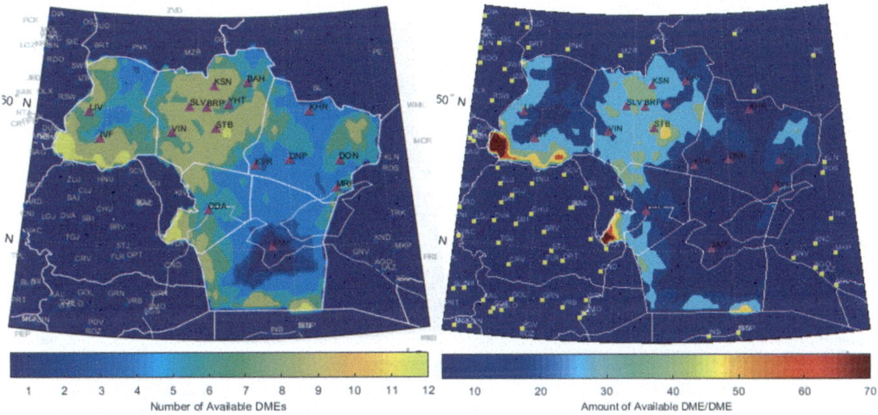

Fig. 3. Amount of navigation aids and DME/DME pairs available for the initial scenario.

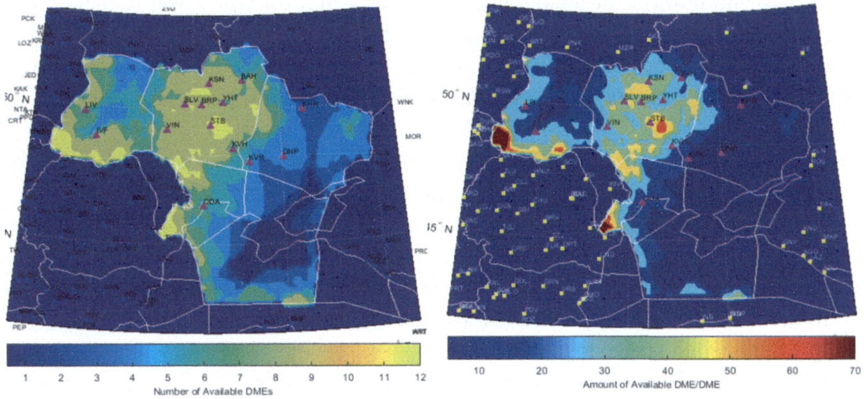

Fig. 4. Amount of navigation aids and pairs available for reduced functionality scenario.

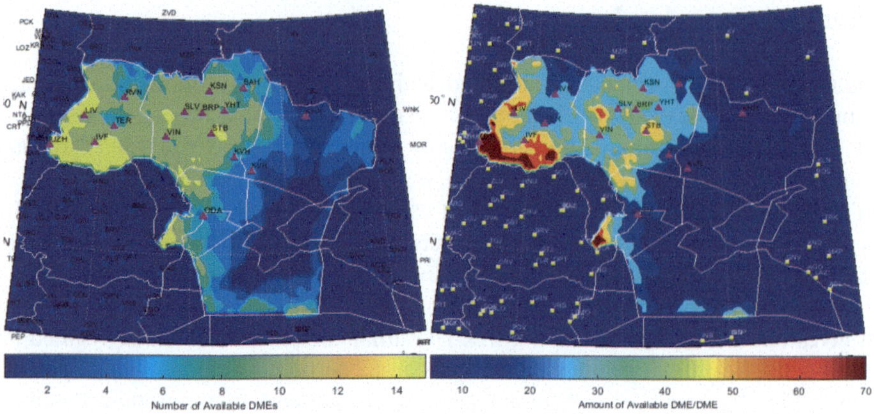

Fig. 5. Amount of navigation aids and DME/DME pairs available for suspended scenario.

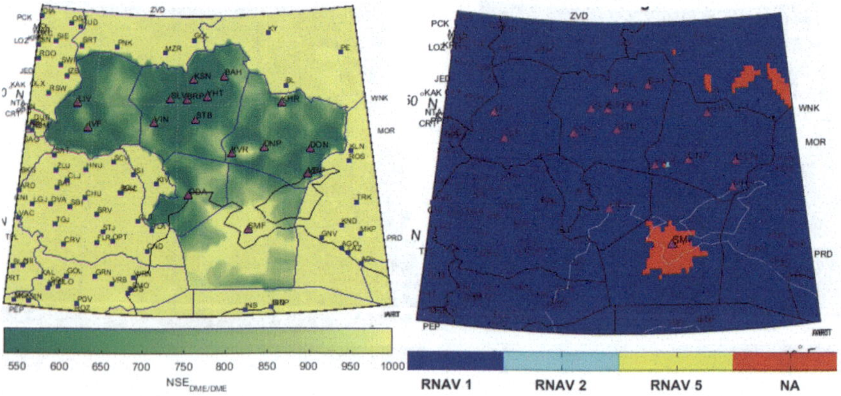

Fig. 6. NSE and area of RNAV criteria correspondence for the initial scenario.

Results of the number of available DMEs show that for the initial scenario, it was an area of low service in the central part of UKFV, only. Reduced functionality and suspended scenarios show a significant increase in the size of the low service area to UKDV. At the same time, network configuration in the central part of Ukraine has been improved.

RMSE of available pairs estimated for each center of the elementary cell has been used in Navigation System Error (NSE) estimation based on a 95% confidence level. NSE with the lowest value is used as an accuracy of the DME/DME positioning method which could be provided within a particular airspace cell. Results of the correspondence level of RNAV 1 are given in Figs. 6, 7 and 8. The area of RNAV 1 is reduced from 95.3% in the initial scenario to 87.7% in the reduced functionality scenario and reaches 79.8% in a suspended scenario.

Finally, results of area analysis are given in Figs. 9, 10 and 11. Results of available nodes analysis for DME (Fig. 9) show that the area with simultaneously available 4 DMEs

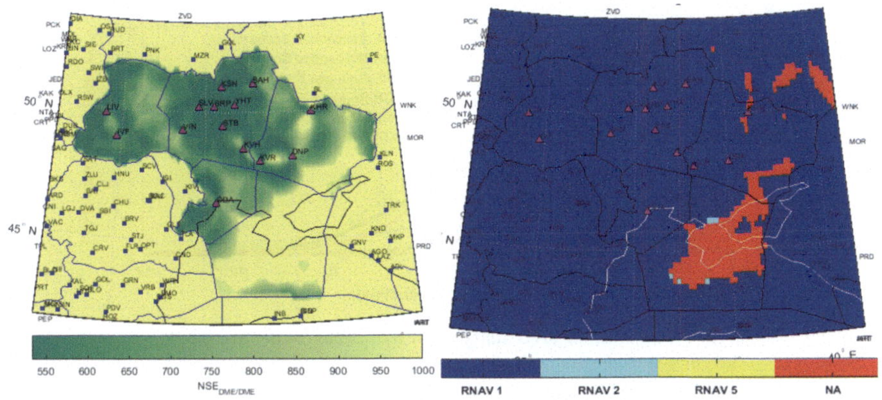

Fig. 7. NSE and area of RNAV criteria correspondence for reduced functionality scenario.

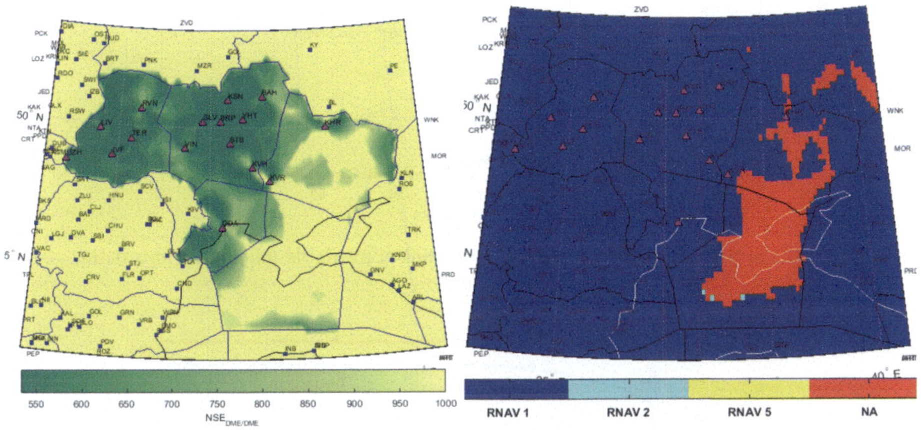

Fig. 8. NSE and area of RNAV criteria correspondence for the suspended scenario.

is reduced from 90% in the initial scenario to 70% in suspended. However, availability area of 9 DMEs has been increased from 9.4% to 21%. These indicate about reduction in the number of available pairs up to 15 and a significantly increased area for the available above number of pairs for the suspended scenario (Fig. 10).

Results of a comparison study of DME/DME and VOR/DME positioning capability of Ukrainian network show that accuracy of VOR/DME is significantly less in comparison to DME/DME. VOR/DME meets the requirements of RNAV 1 only in a circular area close to the ground beacon.

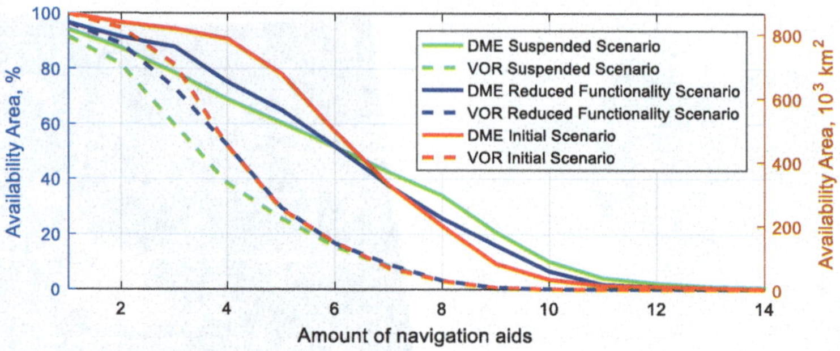

Fig. 9. Results of area analysis of available VOR and DME.

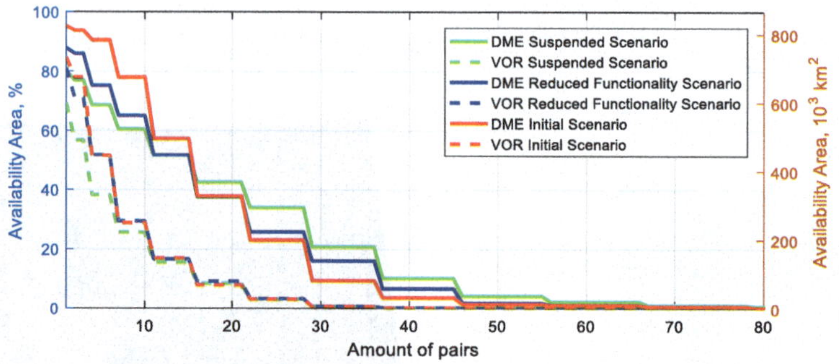

Fig. 10. Results of area analysis in the amount of available navigation aids pairs.

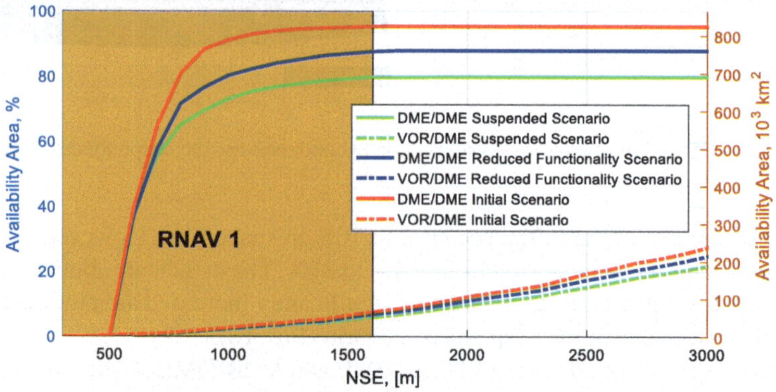

Fig. 11. Results of NSE distribution with area.

5 Conclusions

During the last years importance of alternatives to GPS positioning has increased rapidly in civil aviation. That is caused by an increasing number of aviation events with GPS interference. The DME/DME positioning method is one of the possible alternative methods of position calculation by data provided by the navigation aids network. Proposed in the paper area analysis considers airspace as a matrix of rectangular cells. In numerical validation, a Ukrainian navigation network performance has been evaluated in three different scenarios caused by the ongoing war. The result of analysis indicates that the area with simultaneously available 4 DMEs is reduced from 90% in the initial scenario to 70% in suspended. However, availability area of 9 DMEs has been increased from 9.4% to 21%. Cumulated functions of parameters that have been analyzed help to study the network development process and identify gaps in service which should be guaranteed at the level of RNAV 1 requirements.

A proposed approach could be used in vary of applications including DME, VOR, and NDB network deep study. Also, obtained results could be interesting for the Ukrainian Air Navigation Service Provider to develop a national strategy for air navigation services improvement.

Acknowledgment. This project has received funding through the EURIZON project, which is funded by the European Union under grant agreement No.871072. (Project EU #3035 EURIZON "Research and development of Ukrainian ground network of navigational aids for increasing the safety of civil aviation").

References

1. Qing, C.A., Ang, H.J., Sameer, A.L.: Collision risk assessment of reduced aircraft separation minima in procedural airspace using advanced communication and navigation. Chin. J. Aeronaut. **36**(4), 315–337 (2023). https://doi.org/10.1016/j.cja.2022.11.016
2. Kim, J., Nam, G., Min, D., Kim, N.M., Lee, J.: Determining minimum separation boundary for UAM tactical deconfliction. In: Proceedings of the ION 2024 Pacific PNT Meeting, pp. 502–513. ION, Honolulu (2024). https://doi.org/10.33012/2024.19601
3. Kuzmenko, N.S., Ostroumov, I.V., Marais, K.: An accuracy and availability estimation of aircraft positioning by navigational aids. In: 5th International Conference on Methods and Systems of Navigation and Motion Control (MSNMC), pp. 36–40. IEEE, Kiev (2018). https://doi.org/10.1109/MSNMC.2018.8576276
4. Wang, X., Zhao, Y., Lin, G., Zhou, J.: Navigation and positioning with multi-constellation LEO satellite collaboration signals. In: China Conference on Command and Control, pp. 213–224. Springer Nature, Singapore (2024). https://doi.org/10.1007/978-981-97-7774-7_20
5. Tao, X., Liu, W., Wang, Y., Li, L., Zhu, F., Zhang, X.: Smartphone RTK positioning with multi-frequency and multi-constellation raw observations: GPS L1/L5, Galileo E1/E5a, BDS B1I/B1C/B2a. J. Geod. **97**(5), 43 (2023). https://doi.org/10.1007/s00190-023-01731-3
6. Ludwig, T., Dautermann, T., Korn, B.: Enabling LPV for GLS equipped aircraft using a SBAS to GBAS converter. In: 36th International Technical Meeting of the Satellite Division of the Institute of Navigation (ION GNSS+ 2023), pp. 1747–1749. ION, Denver, CO (2023). https://doi.org/10.33012/2023.19270
7. Rajwant Singh, B.K., Mohd Ali, A., Bahari, S.A., Abdullah, M.: A review on GBAS: ionospheric effects at low-latitude regions. In: International Conference on Space Science and Communication, pp. 337–346. Springer Nature, Singapore (2023). https://doi.org/10.1007/978-981-97-0142-1_33
8. He, Y., Li, J., Liu, J.: Research on GNSS INS & GNSS/INS integrated navigation method for autonomous vehicles: a survey. IEEE Access. **11**, 79033–79055 (2023). https://doi.org/10.1109/ACCESS.2023.3299290

9. Ostroumov, I.V., Kuzmenko, N.S.: An area navigation (RNAV) system performance monitoring and alerting. In: First International Conference on System Analysis & Intelligent Computing (SAIC), pp. 1–4. IEEE, Kyiv (2018). https://doi.org/10.1109/SAIC.2018.8516750
10. Ivashchuk, O., et al.: A configuration analysis of Ukrainian flight routes network. In: In: 16th International Conference on the Experience of Designing and Application of CAD Systems, pp. 6–10. IEEE, Lviv (2021). https://doi.org/10.1109/CADSM52681.2021.9385263
11. Zaliskyi, M., Solomentsev, O., Larin, V., Averyanova, Y.: Model building for diagnostic variables during aviation equipment maintenance. In: In: 17th International Conference on Computer Sciences and Information Technologies (CSIT), pp. 160–164. IEEE, Lviv (2022). https://doi.org/10.1109/CSIT56902.2022.10000556
12. Solomentsev, O., Zaliskyi, M., Zuiev, O.: Estimation of quality parameters in the radio flight support operational system. Aviation. **20**(3), 123–128 (2016). https://doi.org/10.3846/16487788.2016.1227541
13. Kuzmenko, N.S., Ostroumov, I.V.: Performance analysis of positioning system by navigational aids in three dimensional space. In: First International Conference on System Analysis & Intelligent Computing (SAIC), pp. 1–4. IEEE, Kyiv (2018). https://doi.org/10.1109/SAIC.2018.8516790
14. Ostroumov, I.V., Marais, K., Kuzmenko, N.S.: Aircraft positioning using multiple distance measurements and spline prediction. Aviation. **26**(1), 1–10 (2022). https://doi.org/10.3846/aviation.2022.16589
15. ICAO: Performance-Based Navigation (PBN) Manual. Doc 9613, 4th edn. ICAO (2013)
16. Kim, E.: Investigation of APNT optimized DME/DME network using current state-of-the-art DMEs: Ground station network, accuracy, and capacity. In: Proceedings of the 2012 IEEE/ION Position, Location and Navigation Symposium, pp. 146–157. ION, Myrtle Beach, SC (2012). https://doi.org/10.1109/PLANS.2012.6236876
17. Ostroumov, I.V., Kharchenko, V.P., Kuzmenko, N.S.: An airspace analysis according to area navigation requirements. Aviation. **23**(2), 36–42 (2019). https://doi.org/10.3846/aviation.2019.10302
18. Elayam, M.M., Kerhoas, G., Cursay, V.L., Ray, C., Menard, A.: On the interest of hexagonal abstraction of maritime information. In: OCEANS 2022, Hampton Roads, pp. 1–6. IEEE (2022)
19. Bousquin, J.: Discrete global grid systems as scalable geospatial frameworks for characterizing coastal environments. Environ. Model Softw. **146**, 105210 (2021). https://doi.org/10.1016/j.envsoft.2021.105210
20. Topkova, T., Pleninger, S., Hospodka, J., Kraus, J.: Evaluation of DME network capability using combination of rule-based model and gradient boosting regression. J. Air Transp. Manag. **119**, 102637 (2024). https://doi.org/10.1016/j.jairtraman.2024.102637
21. Ostroumov, I., Ivashchuk, O., Kuzmenko, N.: Preliminary estimation of war impact in Ukraine on the global air transportation. In: 12th International Conference on Advanced Computer Information Technologies (ACIT), pp. 281–284. IEEE, Ruzomberok (2022). https://doi.org/10.1109/ACIT54803.2022.9913092
22. Ostroumov, I., Ivannikova, V., Zaliskyi, M., Kuzmenko, N.: Impact analysis of Russian-Ukrainian war on airspace. J. Air Transp. Manag. **124**, 102742 (2025). https://doi.org/10.1016/j.jairtraman.2025.102742
23. Novella, G., Macabiau, C., Garcia-Pena, A., Martineau, A., Ladoux, P., Estival, P.: From ICAO GNSS interference mask to jamming protection area for safe civil aviation operation. In: Proceedings of the 34th International Technical Meeting of the Satellite Division of The Institute of Navigation (ION GNSS+ 2021), pp. 834–854. ION, St. Louis, MI (2021). https://doi.org/10.33012/2021.17936

24. Obiols-Bernaus, E., Tortajada-Ropero, L., Creus-Blanch, A., Gonzalez-Novell, A., Fabra, F., Seco-Granados, G.: GNSS interference monitoring and detection (GIMAD) system. Eng. Proc. **54**(1), 25 (2023). https://doi.org/10.3390/ENC2023-15460
25. Aeronautical Telecomunications. Volume I Radio Navigation Aids. Annex 10 to the Convention on International Civil Aviation. ICAO (2006)
26. Zaliskyi, M., Solomentsev, O., Kozhokhina, O., Herasymenko, T.: Reliability parameters estimation for radioelectronic equipment in case of change-point. In: Signal Processing Symposium 2017 (SPSympo 2017), pp. 1–4. IEEE, Jachranka Village (2017). https://doi.org/10.1109/SPS.2017.8053676
27. Diallo, F.P., Preda, M.A., Mustata, A.I., Mocanu, B.C.: Using spatial indexing systems to optimize emergency response routes. In: 24th International Conference on Control Systems and Computer Science (CSCS), pp. 544–550. IEEE (2023)
28. Aini, A.N., Dewantari, O.A., Mandala, D.P., Bisri, M.B.: An enhanced earthquake risk analysis using H3 spatial indexing. IOP Conf. Series Earth Environ. Sci. **1245**(1), 012014 (2023)
29. Sakan, D., Zuskin, S., Rudan, I.: Spatial database creation applicable for global path planning in integrated navigational system. In: Proceedings of the International Association of Maritime Universities Conference, pp. 335–342. Massachusetts Maritime Academy, Boston, MA (2024)
30. Ostroumov, I.V., Kuzmenko, N.S., Marais, K.: Optimal pair of navigational aids selection. In: 5th International Conference on Methods and Systems of Navigation and Motion Control (MSNMC), pp. 32–35. IEEE, Kiev (2018). https://doi.org/10.1109/MSNMC.2018.8576293
31. AIP Ukraine: Ukrainian air navigation service provider (2024)

Mental Health of University Students in Wartime

Elvira Luzik(✉), Valentyna Semychenko, Nataliia Ladohubets, Anzhelika Kokarieva, and Lesia Konoplianyk

National Aviation University, Liubomyra Huzara Ave., 1, Kyiv 03058, Ukraine
elvira.luzik@npp.kai.edu.ua

Abstract. The paper deals with the issue of preserving students' mental health in a crisis society, demonstrating its multifaceted nature and the necessity for in-depth research. It highlights the importance of assessing students' mental health in wartime and engaging them in the structured system of resources to sustain their mental health during social crises. These mental health concerns are related to the broader transformations in Ukrainian higher education during the post-pandemic stage and ongoing armed conflict. This period has been marked by contradictions, uncertainties, and insufficient conceptual development of methodological foundations and mechanisms for professional training. Today higher education struggles to meet the demands of the labor market, which increasingly requires specialists capable of solving complex professional tasks. Contributing factors include the decline in educational quality due to the widespread transition to online learning, reduced academic and psychological motivation, superficial knowledge acquisition, a lack of interpersonal interactions, and restricted opportunities for social and emotional development. A theoretical model of mental health has been proposed based on a system-synergetic approach. This model conceptualizes mental health as a system including six interconnected levels. The study also provides a set of methodological tools for diagnosing and proposes mathematical models to analyze stress and mental health dynamics.

Keywords: Health · Mental Health · Theoretical Model of Mental Health · Students · Higher Education Institution (HEI) · Global Social Crisis · Wartime

1 Introduction

1.1 Introduction to the Problem

Rapid changes in today's dynamic conditions demand that young people attain not only a high level of professionalism but also maintain stable and robust health. In addition to enriching and updating their knowledge and developing their skills, they must also develop the ability to recognize and regulate their emotional and mental states. During periods of dynamic social transformations, the need for developing human resources of society inevitably increases. The solution to these challenges involves developing specialists capable of rapid adaptation while preserving their mental health. This, in

I. Ostroumov et al. (Eds.): ACASD 2025, LNNS 1418, pp. 148–165, 2025.
https://doi.org/10.1007/978-3-031-91992-3_11

turn, will foster the value-emotional and worldview development of future specialists, enhance their reflective skills, support the holistic shaping of their personalities, and strengthen their resilience against the destructive impacts of crisis conditions. In this context, the global academic community, including researchers in Ukraine, is actively engaged in research aimed at developing a holistic, systemic understanding of the goals, objectives, and means of preserving students' mental health in a crisis society.

Recent statistics demonstrate the significant decline in students' mental health during the ongoing war [1], revealing increased rates of anxiety, depression, and other psychological disorders [2]. Several studies and reports provide detailed insights into the impact of this crisis:

1. General data on students' mental health during the war in Ukraine:
   - according to a report by the Ukrainian Institute for Social Research, approximately 70% of Ukrainian students experience heightened anxiety and stress because of military operations, with 40% reporting that stress affects their ability to study and concentrate [3];
   - a survey conducted by the Ukrainian Association of Psychologists reveals that 25% of students exhibit symptoms similar to post-traumatic stress disorder (PTSD), including intrusive memories, fears, and sleep disturbances.
2. Data from the United Nations Children's Fund (UNICEF) and the World Health Organization (WHO):
   - UNICEF reports that one in three adolescents and young adults (aged 15–24) in Ukraine experiences symptoms of depression or anxiety caused by the war, with 57% of young people noting a decline in their overall mental health [4];
   - the WHO indicates that the risk of mental disorders in conflict-affected areas has increased to 22%, nearly double the rate observed during peacetime [1].
3. The impact of war on students' sleep and physical health:
   - nearly 50% of the surveyed students report sleep disturbances, losing the ability to rest adequately due to stressful circumstances, complicating the recovery and affecting their mental and physical health;
   - more than 60% of students experience chronic fatigue, emotional exhaustion, and symptoms of burnout due to constant stress [5].
4. Academic burnout and reduced academic performance:
   - about 45% of students report difficulties with concentration and reduced motivation for studying, which negatively affects academic performance;
   - more than 30% of students find it difficult to cope with their academic load due to psycho-emotional stress and difficulties with adaptation to the online learning format [5].
5. Need for psychological support: about 50% of students report a need for psychological support and consultations, however, only half of them have consulted specialists due to a lack of resources or fear of seeking help [5].

The theoretical analysis of the issue reveals that the concept of mental health remains insufficiently substantiated, and specific research procedures for data collection have not been defined. Researchers outline various components of mental health: physical, psychophysical, psychosomatic, mental, psychological, social, spiritual, etc. The integration of these components ensures the level of mental health as an individual's perception of their quality of life. New priorities in interpreting health-related issues in science and practice have been identified: previously societal attention was primarily focused on physical health preservation but today increasing emphasis is placed on health-preserving tasks such as overcoming traumatic effects of external environments, managing extreme situations without psychological losses, and preserving the potential for development and existence in society (psychological, social and spiritual health). The importance of assessing students' health status in wartime and involving them in the system of resources for preserving mental health in the context of global crisis changes in society is emphasized.

1.2 Motivation

The motivation for this study arises from the need to analyze the concept of health, especially mental health, in the context of contemporary challenges. Traditional approaches in the past were largely focused on preserving physical health, while the mental aspects of well-being have gained significant attention recently. The ongoing global crises, including the COVID-19 pandemic and the military conflict in Ukraine, have intensified the importance of understanding and supporting mental health as a dynamic, multi-layered construct.

This paper seeks to fill gaps in the existing investigations, where definitions of mental health remain fragmented, and related terms lack clear distinctions and evidence-based foundations. Moreover, the current socio-political and economic realities require a rethinking of how to motivate students to maintain resilience, cope with stress, anxiety, and depression, and develop their potential.

By investigating these critical issues, this paper aims to contribute to developing strategies for preserving mental health, enhancing motivation, and fostering adaptability in an unpredictable world.

1.3 Contribution

Extensive research has been conducted on health-preserving disciplines, related to the professional training of future medical specialists and learning by every individual specific technique to improve their physical health, following a valeological approach. However, the scope of this issue extends beyond physical well-being to encompass the preservation of mental health. For university students, this requires determining the appropriate educational content, methodologies, and practical activities that integrate diagnostic and corrective programs into the learning process.

The present research aims to cope with these challenges by providing psychological assistance to students, focusing on the assessment and monitoring of mental health dynamics during wartime. This objective aligns with the common definition of health as a state of complete physical, mental, and social well-being, which emphasizes the importance of individual responsibility in maintaining their health. Understanding health as a value that determines the adoption of an appropriate behavioral strategy – commonly referred to as a healthy lifestyle – enables us to develop a theoretical model of a student's mental health. This model includes six levels: physical, psycho-emotional, mental, psychological, social, and spiritual, all structured within specific algorithms related to open dynamic and synergetic systems.

A theoretical analysis of existing studies in the field of valeology has revealed the following findings:

- human health can be conceptualized across distinct but interdependent levels of functioning: biological, psychological, social, spiritual, and mental;
- mental health, as an outcome of an individual's experience of specific life circumstances, inevitably depends not only on external life events but also on the results of their integration;
- when considering health at any level, it is essential to take into account both objective and subjective factors, i.e., the state of health is determined by objective circumstances (genetic factor, material living conditions, the state of society, the influence of various systems in which an individual is involved), and the personal characteristics of the individual, such as their attitude to health as a value, their ability for self-regulation and self-development, and a belief in the ability to improve the situation.

The following procedures are required for organizing our research:

- to define key components of human health at each level and the results of their integration;
- to develop a robust methodological base;
- to determine mathematical methods for processing empirical data;
- to identify trends in the development of student's mental health (invariant and variable trends);
- to detect key destructive tendencies and propose ways to overcome them;
- to develop, implement and test the means and tools to overcome destructive tendencies.

By integrating these steps, this study will contribute to the theoretical and practical understanding of mental health preservation among university students in the context of wartime challenges.

1.4 The Organization of the Paper

The paper contains seven sections. The introduction points out the significance of students' mental health in a crisis society, particularly during wartime, by analyzing its impact on emotional state, academic performance, and physical well-being. It also emphasizes the necessity of developing systematic approaches to its preservation and support. The second section presents key findings from the literature review. The third section focuses on identifying the conditions necessary for ensuring the sustainability of the educational process during global crises and proposes conceptual approaches and methodological recommendations. The fourth section analyzes the transformations of higher education in Ukraine in the context of the pandemic, military operations, and social distancing, assessing their impact on the quality of education, academic motivation, and socio-emotional development of university students. This section also integrates theoretical research on maintaining students' mental health which will be the basis of further experimental confirmations. The fifth section provides recommendations for assessing and enhancing personal mental health. The sixth section presents the conclusions of the study, while the seventh section outlines the future scope.

2 Literature Review and Problem Statement

In the context of ongoing global crises in the modern world, the higher education system of Ukraine faces multifaceted challenges related to the transformation of the educational process, changes in the conceptual foundations of its organization, and the need to adapt students' individual educational trajectories to contemporary realities. These factors, among others, require researchers and educators in higher education institutions (HEIs) to recognize the substantial decrease in motivation among young people and society as a whole to obtain high-quality higher education.

One of the key factors contributing to the insufficient readiness of Ukrainian and international higher education to meet the new challenges is the diminishing desire for "success and comfort in life" among future specialists. This decline is particularly evident among students engaged in online or blended learning within a crisis society. Such a shift in motivation negatively influences not only academic engagement, but also the depth of interdisciplinary knowledge and skills, effective communication, and social and emotional development of future specialists. Simultaneously, stress, anxiety, and depression have become increasingly prevalent phenomena in modern society, largely as a result of the long-term destructive effects of ongoing social crises. For instance, Chen and Lucock reported high levels of depression and anxiety among students in the United Kingdom [6]; Worsley et al. noted significant mental health difficulties, including depression, anxiety, academic stress, and expectation stress, among university students [7]; Hotez et al. studied the mental health challenges faced by college students during COVID-19 pandemic, using both quantitative and qualitative methods to reveal notable increases in depression and anxiety, particularly among female and older students, while highlighting academic, professional, and relational stressors and uncertainty [8]. However, these phenomena are especially noticeable in Ukraine, where the ongoing war has exacerbated the situation, as confirmed by studies conducted by Ukrainian researchers [9–11]. For instance, Osokina et al. in their investigation of the psychological impact of war on Ukrainian adolescents, reported a prevalence of post-traumatic stress disorder (PTSD), anxiety, and depression, with significantly higher risks for mental health issues observed among adolescents exposed to war-related trauma and stress [12]. These conditions undermine the emotional and psychological well-being of individuals, resulting in a drastic decline in their mental health, which is responsible for the quality of life and the ability to function effectively in society.

Thus, the challenges of a protracted social crisis as a result of the global COVID-19 pandemic and Russia's military aggression against Ukraine have forced a globalized society to reconsider on a planetary scale approaches to international cooperation in all spheres of life (international partnership, production, provision of economic, social and educational services, etc.) as well as the means and methods of communication, which are the main regulator of establishing partnership interaction and is among the most significant means for organizing and improving the efficiency of the educational process in Ukrainian HEIs.

Simultaneously, these challenges arise not only from conceptual and organizational transformations within society but also from the presence of several destructive processes:

- the limited success in implementing market-based models of functioning;
- insufficient motivation among contemporary students to learn;
- a lack of scientific grounding in combining distance learning and in-person (face-to-face) learning;
- declining student enrollments in Ukrainian HEIs and outflow of a significant number of young people who obtain education abroad;
- an increase in educational costs over income due to rising tuition fees and increasing unemployment in society as a result of the forced evacuation of the population from the frontline regions of Ukraine.

These and related processes not only impede the creation of favorable conditions for students to access and benefit from educational services but also pose significant challenges to maintaining their mental health. While the issue of preserving human health is not new, it provides an opportunity to clarify and systematize the relevant theoretical provisions [13] and effective intervention strategies [14], as well as to identify modern trends in addressing these concerns. Currently, such issues are primarily considered by medicine and psychiatry, with their insights integrated into the specialized interdisciplinary domain of valeology [15]. The rapid pace of change in modern conditions requires that students maintain stable physical and mental health, alongside the understanding that health represents the most valuable asset in human life and cannot be neglected. Health is defined by the World Health Organization (WHO) as "a state of complete physical, mental, and social well-being and not merely the absence of disease or infirmity" [1]. Consequently, health serves as a resource and an objective for development, as well as a cornerstone of prosperity, confirming the principle of a people-centered state policy in Ukraine's social and economic spheres.

A recent review by Alegría et al. highlights significant advancements in understanding the role of social determinants of health (SDH) in shaping mental health outcomes. The authors synthesize research conducted over the past 5 years, emphasizing the expanded frameworks and theories in SDH that now encompass a broader spectrum of social conditions – from immigration-related trauma to psychosocial and community resources affecting mental health and well-being [16]. The review also underscores the heightened risk of psychiatric and mental disorders associated with systemic oppression, such as racism and marginalization. Moreover, the COVID-19 pandemic has exposed the inequitable impact of social determinants on health outcomes. Addressing mental health disorders requires studies examining factors across multiple levels of influence, extending beyond the individual level. Such studies prioritize understanding how social and structural factors, including structural racism and discrimination, affect mental health outcomes at individual, community, and population levels. Furthermore, interactions between these determinants and levels of influence must be studied to understand the complexities of their interconnected impacts on the mental health of individuals, families, communities, and populations.

The program for preserving the mental health of modern students in Ukraine is largely based on the research conducted by the teaching staff of the Department of Pedagogy and Psychology of Professional Education at the State University "Kyiv Aviation Institute" (KAI). For instance, Luzik et al. consider the challenges caused by the current state of higher education and society as a whole [17]. Through the analysis of Ukrainian and global experience in organizing distance learning, the organizational and methodological problems arising from long-term social crises are identified and their psychological consequences are determined. A survey conducted among teaching staff, administrative personnel and students on the further prospects of the national education system in HEIs has revealed key problems and identified barriers to educational communication while ensuring high-quality professional training. Based on the collected empirical data, the challenges encountered by the teaching staff in Ukraine during periods of crisis have been identified.

Studies examining the issue of uncertainty, which characterizes the current state of society explain tolerance for uncertainty as a fundamental attribute of personal adaptability and a factor influencing behavioral strategies in unpredictable circumstances [15, 18, 19]. Through the analysis of the roles of contemporary educators, the researchers emphasize the necessity of adopting a synergistic approach to teaching as a self-organizing system capable of flexibly responding to social challenges. One of these challenges is the necessity of preserving the mental health of both teaching staff and students [17, 20].

Proskurka et al. present and analyze empirical data directly or indirectly related to the mental health challenges of students majoring in practical psychology and engineering. Their research examines components of mental health, such as optimism and students' adaptability to new conditions [10]. Trends in the personal and professional development of future practical psychologists are revealed, including the need to overcome emotional burnout of this category of specialists as a critical strategy for maintaining their mental health [20]. Proskurka et al. emphasize the role of mobile applications as a tool to support mental health of university students during wartime [21]. Key indicators of mental health are determined, including constructive and destructive forms of psychological protection in challenging life circumstances [22], features of students' self-perception [21], coping strategies used when resolving conflicts [14], and the role of emotional intelligence in interpersonal interactions [23]. Luzik et al. advocate for examining the higher education system through a synergistic approach, emphasizing the potential for organizing the educational process as an autonomous system [17, 23]. In addition, the values and behavioral strategies of individuals in wartime are analyzed with a focus on military personnel as an illustrative example [22, 24]. Empirical data also reveal the impact of emotional changes on the psychological health of teaching staff, highlighting a decline in mental well-being under today's conditions. Based on the findings of the study, the effectiveness of art therapy as a preventive measure against emotional burnout during the war has been confirmed [20].

Semychenko investigates the need to analyze human relations with the world from the perspectives of systemic and synergistic approaches. She analyzes the essence of these approaches, focusing on how they align with an individual's acceptance of the logic of interaction with stable and dynamic systems. The method of analyzing system-synergetic modes of organizing human interaction with the surrounding world is proposed based on the principle of polarity [25], opening new possibilities to design relevant research methods. A psychological perspective is applied to the system-synergetic strategies of organizing human activity, encompassing the individual's perception of the world (personal worldview) and the tactics for solving current life problems. These strategies are based either on adapting to the requirements of stable systems or overcoming uncertainty in unstable systems. A methodology for identifying system-synergetic strategies of human interaction with the world has been developed and is currently undergoing validation [13].

The theoretical analysis of the experience in studying and practically solving mental health problems in the world has made it possible to generalize the assessment of the situation in Ukraine.

1. The problem of human mental health is known to have been considered for a long time in the context of motivating its preservation in a stable society. The main focus of the analysis was aimed at the fact that an individual can increase their potential by adequately responding to life challenges [26]. However, today, people live in a world where the conditions of existence contain direct threats to their lives. This fact determines the relevance of research aimed at identifying the actual state of various types of mental health disorders and identifying appropriate prevention techniques to correct unpredictable negative manifestations [27].
2. The analysis of the scientific papers on health issues has shown that, in the past, researchers predominantly focused on preserving physical health [16, 23]. However, today, in conditions of long-term social isolation and ongoing military actions in Ukraine, mental health – the ability of a person to overcome difficult and unpredictable life circumstances – has come to the forefront [22, 28].
3. Recent scientific literature lacks a clear definition of the structure of mental health. Numerous related terms are used in the syllabi for relevant academic disciplines, such as: "physical health", "psycho-emotional health", "emotional health", "somatic health", "mental health", "psychological health", "social health", "spiritual health", "cognitive health" without providing evidence-based distinctions or overlaps among these concepts [13, 15]. Furthermore, the terms such as "healthy personality", "health potential", "social aspects of health", "creative aspects of health", "spiritual potential of health" and others are often used without sufficient theoretical and practical substantiation [19, 25]. Based on the identified issues, there is a need to develop a theoretical model of mental health as a multi-level system construct with a structure comprising several levels. The development of such a model should include various system components, define criteria and indicators, and, on this basis, use diagnostic tools that allow asessing the state of students' mental health [29–31].
4. The process of preserving and restoring the mental health system requires an appropriate approach, as mental health is considered in our research as a result of the interaction between two forms of personal activity: external (psychological assistance provided by specialists through consultations and training) and internal (the individual's efforts

and aspirations to enhance their psychological and auto-psychological competencies) [18].

3 Materials and Methods

The main idea and key points of the study emphasize that transformations and social challenges in higher education in Ukraine, caused by the global social crisis (quarantine restrictions, combat zones, lockdowns, relocation, and forced social distancing) create a threat to the mental health of the population. These challenges also affect the higher education system, as students represent a social category that significantly influences the further intellectual, economic, social, and demographic development of society.

The main provisions of the study require the development of a new concept of organizing higher education as a whole and formulating prospective directions for its further development; the implementation of effective and high-quality functioning of higher education in Ukraine and organizing high-quality educational process in Ukrainian HEIs [32, 33]; the readiness of the teaching staff and students to teach and study in the altered working and learning conditions in the post-pandemic era, wartime or other natural or man-made disasters; overcoming psychological, social, economic, communicative, methodological and cultural barriers and providing appropriate balanced administrative and academic support aimed at activating their adaptive capabilities and overcoming these barriers.

The development of methodological recommendations is crucial for ensuring psychological, methodological, organizational, information and communication support in implementing the positive experience of Ukrainian and foreign HEIs in organizing the educational process during periods of COVID-19 pandemic, wartime and social distancing.

The basic components of the theoretical model of mental health have been identified by using general scientific methods (analysis, comparison, synthesis, measurement, abstraction, and formalization). This model represents an open system built by using Fibonacci numbers, a hierarchical structure of which by reflecting a medical perspective of considering the state of health (the presence of physical defects or frequent illnesses) will indicate worsening of physical health, reduced need for movement and habitual forms of physical activity, unexplained pains in the body, and harmful habits (excessive smoking, alcohol or drug use).

At the psychophysical level, health reflects the state of the body and its ability to function at a sufficient level of activity. The criteria of impaired physiological health include a significant decline in well-being and overall activity, the predominance of negative moods, sleep disturbances (excessive sleep or insomnia), eating disorders (overeating or loss of appetite), apathy, and indifference to the surrounding world.

At the psycho-emotional level, health reflects the state of psychosomatics. The indicators of mental health disturbances include heightened anxiety, fears, phobias, and a predominance of depressive states.

The mental level of health, which reflects the nature of the relationship between normal psychological functioning and pathologies, has the following indicators of mental health disorders: panic attacks during danger, symptoms of PTSD, feelings of helplessness or hopelessness, inability to perform daily tasks (such as self-care, caring for family, work, or study), destructive addictions, the presence of clearly expressed, uncompensated character accentuations [24].

The psychological level of health reflects the absence of personality deformations. The indicators of disturbances include high levels of maladaptation, complications of responding in situations of uncertainty, and negative self-perception.

The social level of health, which characterizes interpersonal relationships, has the following indicators of disturbances: a lack of empathy, underdeveloped emotional intelligence, reduced social optimism, and inability to behave constructively in conflict situations [34].

At the spiritual level, health encompasses an individual's orientation toward higher values. The indicators of spiritual health disturbances include a lack of life prospects, an inability to comprehend the meaning of life, significant change in value orientations, reduced life optimism, and loss of interest in learning and integration into society.

Thus, mental health, as the result of integrating the aforementioned levels, can be understood as an emergence phenomenon – a new quality arising from the system's integration at a definite level. The indicators of mental health disturbances are as follows: an inability to experience happiness and life satisfaction, failure to realize creative skills, and a lack of psychological well-being.

4 Results and Discussions

Theoretical and empirical studies of the transformations in the structure of higher education in Ukraine, which are marked by uncertainty and caused by the pandemic, military invasion, lockdowns, forced social distancing, and temporary displacement, have revealed insufficient preparedness for higher education to face contemporary challenges. The main negative consequences of these disruptions include a decline in the quality of education, a loss of academic motivation, reduced depth of knowledge, limited personal interactions with teachers and peers, and restricted opportunities for social and emotional learning.

In response to these challenges, researchers in Ukraine and worldwide are conducting studies to develop a comprehensive vision for creating system-synergetic resources to preserve students' mental health in a crisis society. Furthermore, efforts are being made to develop a holistic approach to further preserving students' mental health as a foundation for shaping higher education in the near future.

An analysis of the positive and negative experiences of transforming higher education in Ukraine during crises – such as the pandemic, military operations, etc. – has contributed to forecasting further transformations and development of higher education in Ukraine. Specifically, this includes directing the educational process in HEIs towards integrated interdisciplinary courses designed to solve complex tasks of the future specialty; and the development of twenty-first-century skills, the "4Cs" (critical thinking, communication, collaboration, and creativity) that will contribute to the realization of the creative abilities of the student community, the development of critical thinking and basic communication skills in teamwork.

As a part of this research, a set of scientific and methodological recommendations has been developed to organize educational processes in HEIs under conditions of uncertainty and forced social distancing that may arise in the future. A complex of methodological recommendations has been proposed to ensure methodological, information and communication, psychological, and communicative support to incorporate the positive practices of both Ukrainian and foreign HEIs in preserving students' mental health in the post-pandemic and post-war recovery periods in Ukraine. Furthermore, methodological approaches have been refined to better prepare research and teaching staff in HEIs for professional activities in conditions of unpredictability, uncertainty, and new challenges of modern society.

While investigating, the key definitions such as "transformation of higher education", "transformation of the educational process", "open educational space", "challenges of higher education", "strategy for higher education development", and "prospects for transformations in higher education" have been clarified within the context of their use by the Ukrainian, European and American scientific communities. Additionally, critical aspects of the concept of higher education functioning and development in the post-pandemic and post-war Ukraine have been elaborated.

Expected results and implementation procedure:

- deepening the theoretical foundation of psychology concerning the issues of mental health development;
- expanding the methodological base for diagnosing mental health states;
- assessing the state of students' mental health in wartime and during distance learning through comprehensive empirical research;
- developing and implementing a program to preserve and improve students' mental health in a crisis environment;
- developing strategies to improve students' psychological and auto-psychological competencies related to mental health preservation during wartime [35].

The study uses the methodology of a system-synergetic approach to conceptualize mental health as a systemic formation containing six levels: physical, psycho-emotional, mental, psychological, social, and spiritual.

Based on the theoretical research, a methodological basis has been determined for diagnosing individual characteristics and group indicators of mental health development among students during military aggression. The results of this theoretical determination of students' mental health will help specify destructive trends in mental health impairment.

The programs aimed at improving students' mental health in crises are expected to be developed and implemented as practical outcomes of this research.

5 Recommendations

The study has resulted in formulating recommendations. As previously noted, the WHO defines mental health as a state of happiness and well-being in which individuals realize their creative abilities, cope effectively with life stresses, work productively, and contribute to social life [1]. However, achieving this state is an integrative outcome of an individual's interaction with their external environment, taking into account the psychological consequences that arise in this process. This interaction may occur spontaneously or be controlled either externally (e.g. individual's involvement in a preventive program or overcoming destructive tendencies), or internally (e.g. where a person is motivated and technologically equipped with the means to preserve their health and undertakes appropriate procedures for self-knowledge and self-development).

The state of mental health depends on many factors, including social, environmental, and economic. A mentally healthy person can recognize and manage emotions, function harmoniously, communicate and build relationships, develop and learn, positively evaluate, accept, and love themselves and others, adapt well to new life circumstances, and make firm decisions. These guidelines should be added to the list of general skills that determine the interaction between teachers and students.

During the development of the research program, a set of methods was determined that can be applied to diagnose students' health, which can also be adapted for other population groups in the future.

The following tools can serve as methodological guidelines for evaluating various aspects of health:

1. The evaluation of physical health is facilitated by tools such as the Methodology for Self-Assessment of Physical Health developed by Stepanov and the Test for Self-Assessment of Health Status by Voitenko. These tests assess key aspects, including sleep, nutrition, work-rest balance, physical activity, bad habits, and the individual's perception of their health.
2. For psychophysical health, Stepanov's Self-Assessment Test for Psychophysical Health and the WAM Test (well-being, activity, mood) are applied. Criteria for self-assessment include mobility, speed, functional tempo (activity), strength, overall health, fatigue, and predominance of positive or negative emotions. Wessman-Rick's method of self-assessment of emotional state evaluates the mental state on the scales such as calmness-anxiety, energy-fatigue, elation-depression, and confidence-helplessness.
3. Psychosomatic health is measured by using the Differential Diagnostics of Differential States Questionnaire, which identifies the levels of apathy, hypothermia, dysphoria, confusion, anxiety, and fear; the Nervous-Psychological Tension Questionnaire and Self-Assessment Questionnaire evaluating anxiety, frustration, aggressiveness, and rigidity (adapted from Eysenck's Personality Test).
4. Mental health is diagnosed using the Mini-Mult Questionnaire, which contains scales for hypochondria, depression, hysteria, psychopathy, paranoia, psychasthenia, schizoid personality disorder, and hypomania; Leonhard-Shmishek's diagnostic tools for studying character accentuations across scales for hyperthymia, anxiety, dysthymia, pedantry, excitability, stuckness, demonstrativeness, emotionality, and cyclothymia.
5. Psychological health is assessed by using Gilbukh's Personality Maturity Questionnaire which measures achievement motivation, self-attitude, sense of civic duty, life attitude, and psychological closure; Shtepa's Personality Maturity Questionnaire which examines responsibility, decentration, depth of experiences, philosophy of

life, tolerance, autonomy, contact, self-acceptance, creativity, and synergy; Pantileev-Stolin's Self-Attitude Test, which includes such scales as: closure, self-confidence, self-direction, self-attitude, self-worth, self-acceptance, internal proneness to conflict, and self-accusation;
6. Social health is assessed using a methodology for diagnosing empathy including the Emotional Intelligence Questionnaire (by Liusin) which checks the ability to understand and manage other people's emotions, understand and manage one's own emotions, and control expression; the Social Optimism Questionnaire which examines attitude to social relations at the mega (world-person), macro (society-individual), and meso (interpersonal relations) levels. Holland's Emotional Intelligence Test measures social awareness, emotion management, self-motivation, empathy, and recognizing other people's emotions.
7. Spiritual health is measured through the Test of Meaningful Life Orientations, which evaluates life goals, life process or interest in the emotional intensity of life, life effectiveness or satisfaction with self-realization, locus of control over self and life; Rokeach's Personal Values Survey, which categorizes values by blocks (terminal values, concrete values, abstract values, values of professional self-determination, values of personal life; instrumental values: ethical values, values of communication, values in activity, individualistic values, values of conformity, values of self-affirmation).
8. Holistic mental health is examined using the Attitude to Health Questionnaire, which includes cognitive, emotional, behavioral, and value-orientation scales; the Questionnaire of Personality Psychological Resources, which assesses the individual's ability for self-actualization; the Oxford Happiness Questionnaire (by Meyer), which is based on the following indicators: a person's experience well-being satisfaction, self-realization potential, and health preservation efforts.

These guidelines provide a robust foundation for identifying methodological approaches to health diagnostics. Even though this list is not full, it outlines the general principles for selecting appropriate instruments for the study. Thus, the defined program of work can create initial, methodological, theoretical, and organizational conditions for wider coverage of the problem of mental health preservation in a crisis society and in the future.

Mathematical modeling of mental health, as a promising field, can help systematize knowledge and facilitate decision-making in psychological support. Taking into account the complexity of the human psyche – psychological processes are multidimensional and difficult to formalize, the precise model parameters are difficult to determine due to the lack of complete data, individual features of the model may be general, but less effective for specific cases [30, 31].

Hans Selye's mathematical model, considered in our study, represents an attempt to formalize the concept of the adaptation syndrome for quantitative analysis of stress processes, based on three main phases of stress (anxiety, resistance, exhaustion) and the use of variables to describe physiological processes. Such a nonlinear model has been applied to the dynamic system of mental health, as it allows modeling the dynamics of the body's response to stressful influences, predicting the time to exhaustion, and assessing the effectiveness of stress management methods.

The following variable models can be used to formalize the model: $S(t)$ is the stress level in the body at time t; $R(t)$ is body reserves (resources for adaptation); $H(t)$ is the homeostasis level (the body's ability to maintain a stable state); $E(t)$ is energy resources expended for adaptation; $I(t)$ is the intensity of external influence.

To solve dynamic equations:

1. Stress level (S):

$$\frac{dS(t)}{dt} = \alpha I(t) - \beta R(t)S(t), \tag{1}$$

where α is the coefficient of external influence, β is the body's ability to reduce stress through reserves.

2. Body reserves (R):

$$\frac{dR(t)}{dt} = -\gamma S(t) - \delta R(t), \tag{2}$$

where γ is the expenditure of reserves to combat stress, δ is a natural decline in reserves.

3. Homeostasis (H):

$$\frac{dH(t)}{dt} = \varepsilon S(t) + \zeta, \tag{3}$$

where ε is the negative impact of stress on homeostasis, ζ is the restoration of homeostasis through rest or other mechanisms.

4. Energy resources (E):

$$\frac{dE(t)}{dt} = -\eta S(t) + \mu R(t), \tag{4}$$

where η is energy expenditure for adaptation, μ is energy replenishment through recovery.

The interpretation of the phases is presented below:

- Anxiety phase: $I(t)$ increases sharply, $S(t)$ rises, $R(t)$ and $H(t)$ begin to decrease.
- Resistance phase: $S(t)$ stabilizes at a certain level, $R(t)$ gradually decreases, $E(t)$ is spent to support adaptation.
- Exhaustion phase: $R(t)$ is almost used, i.e. $R(t) \to 0$; $H(t)$ and $E(t)$ decrease significantly; $S(t)$ becomes uncontrollable.

Our future study aims to compare the results of studies obtained using Hans Selye's nonlinear model with models of post-traumatic stress disorder (PTSD) that take into account the level of traumatic impact and the recovery rate through resilience

$$P(t) = P_0 e^{-\lambda t}, \tag{5}$$

where $P(t)$ represents PTSD symptoms over time, P_0 is the initial level, and λ is the recovery rate.

6 Conclusions

The study highlights the transformations and social challenges in the structure of the higher education system in Ukraine resulting from contemporary crises, including the pandemic, military aggression, and forced social distancing. These challenges have revealed gaps in the system's preparedness to cope with them, resulting in decreased educational quality, reduced academic motivation, limited interpersonal interactions, and restricted opportunities for social and emotional learning. Such consequences create a real threat to students' mental health, which is a critical factor, taking into account that this social group greatly influences the intellectual, economic, social, and demographic development of society.

The investigation into the issue of identifying the system-synergetic resources for preserving students' mental health has led to the development of the theoretical model of mental health. This model includes six levels: physical, psycho-emotional, mental, psychological, social, and spiritual, structured in proportions corresponding to open, dynamic synergetic systems. The adoption of this model serves as a foundation for identifying, developing, and preserving an individual's mental health as a value that results in enabling the acceptance of a corresponding behavioral strategy commonly referred to as a psychologically and ecologically healthy lifestyle. The study also presents a methodology for diagnosing mental health indicators and offers recommendations for developing programs to reduce the negative psychological impacts of crises on students. Additionally, the potential of mathematical modeling in stress management offers a promising direction for future research.

7 Future Scope

The investigation into the issue of determining the system-synergetic resources for preserving students' mental health creates opportunities for solving a number of topical issues that have not yet found their final coverage in this study. This involves solving the following tasks:

1. The methodological basis for the study of mental health as a complex, multi-level system (content, structure, criteria, indicators, levels of development) will be improved, which involves not only the use of existing diagnostic methods but also the development and validation of new methods that meet contemporary requirements.
2. Based on the proposed theoretical model and the established methodological base, an empirical study will be conducted to identify the state of development of each component and an integrated indicator of mental health, which will contribute to identifying the relationships between the individual components of health and mental health, thereby determining compensatory resources; and detecting the most vulnerable components of students' health, which will determine the main directions of psychotherapeutic interventions.
3. The special programs for working with students will be developed based on system-synergetic principles and their organic combination. Furthermore, based on the general theory of systems, it is planned to develop and test a program for group and individual work with students in the form of psychologist-led training sessions. Based

on the theory of self-managed systems, a program will be implemented to develop students' auto-psychological competence and to master self-monitoring techniques based on methods and self-regulation tools.

References

1. World Health Organization. https://www.who.int/data/gho/data/major-themes/health-and-well-being. Last accessed 06 Jan 2025
2. Kostruba, N., Fishchuk, O.: Mental Health in Times of Crisis. Vezha-Druk, Lutsk (2023) (In Ukrainian)
3. Ukrainian Institute for Social Research. https://www.uisr.org.ua/en/. Last accessed 06 Jan 2025
4. UNICEF: Ukraine. https://www.unicef.org/ukraine. Last accessed 06 Jan 2025
5. UNICEF: Situation Analysis of Children in Ukraine: 2024. UNICEF for Every Child. https://www.unicef.org/ukraine/media/49206/file/UNICEF_SitAn_2024_ENG.pdf
6. Chen, T., Lucock, M.: The mental health of university students during the COVID-19 pandemic: an online survey in the UK. PLoS One. **17**, e0262562 (2022). https://doi.org/10.1371/journal.pone.0262562
7. Worsley, J., Dryburgh, A., McIntyre, J.C., Corcoran, R.: Academic and non-academic predictors of common mental health difficulties among university students during the COVID-19 pandemic. Front. Public Health. **17**, 1441176 (2024). https://doi.org/10.3389/fpubh.2024.1441176
8. Hotez, E., Gragnani, C.M., Fernandes, P., Rosenau, K.A., Wang, K., Chopra, A., et al.: A mixed methods investigation of college student mental health during the first year of the COVID-19 pandemic. J. Am. Coll. Heal. **72**(5), 1632–1639 (2024). https://doi.org/10.1080/07448481.2022.2089842
9. Hozak, S.V., Yelizarova, O.T., Stankevych, T.V., Chorna, V.V., Tovkun, L.P., Lebedynets, N.V., et al.: Mental health and lifestyles of students during the war. Rep. Vinnytsia National Med. Univ. **27**(4), 628–634 (2023). https://doi.org/10.31393/reports-vnmedical-2023-27(4)-17
10. Proskurka, N., Demchenko, N., Kokarieva, A.: Mobile applications as a tool for supporting higher education students' mental health during the war. In: Proceeding of the XII International Scientific and Practical Conference "Current Problems of Higher Professional Education", pp. 45–49. NAU, Kyiv (2024). (In Ukrainian)
11. Kurapov, A., Pavlenko, V., Drozdov, A., Bezliudna, V., Reznik, A., Isralowitz, R.: Toward an understanding of the Russian-Ukrainian war impact on university students and personnel. J. Loss Trauma. **28**(2), 167–174 (2022). https://doi.org/10.1080/15325024.2022.2084838
12. Osokina, O., Silwal, S., Bohdanova, T., Hodes, M., Sourander, A., Skokauskas, N.: Impact of the Russian invasion on mental health of adolescents in Ukraine. J. Am. Acad. Child Adolesc. Psychiatry. **62**(3), 335–343 (2023)
13. Semychenko, V., Artiushyna, K.: The uncertainty problem in the theory and practice of higher school. Proc. National Aviat. Univ. Pedag. Psychol. **16**, 141–152 (2019) (In Ukrainian)
14. Bondar, L.V., Proskurka, N.M., Boichenko, M.M.: Coping strategies of adult gamers. Actual Prob. Psychol. **11**(21), 38–52 (2020) (In Ukrainian)
15. Luzik, E., Semychenko, V., Artiushyna, K.: Uncertainty in methodological and methodical perspectives of analysis: subjective and objective dimensions. Proc. National Aviat. Univ. Pedag. Psychol. **24**, 85–99 (2024). https://doi.org/10.18372/2411-264X.24.19253. (In Ukrainian)

16. Alegría, M., Alvarez, K., Cheng, M., Falgas-Bague, I.: Recent advances on social determinants of mental health: Looking fast forward. Am. J. Psychiatry. **180**(7), 473–482 (2023). https://doi.org/10.1176/appi.ajp.20230371
17. Luzik, E.V., Ladohubets, N.V., Demchenko, N.I., Kokarieva, A.M., Proskurka, N.M., Polishchuk, O.S.: Stress resistance as a factor in the development of tolerance to uncertainty in the professional development of future aviation specialists. In: Proceeding of The Tenth World Congress "Aviation in the XXI-st century – Safety in Aviation and Space Technology". NAU, Kyiv (2022) https://conference.nau.edu.ua/index.php/Congress/Congress2022/paper/viewFile/8755/7160. Last accessed 06 Jan 2025 (In Ukrainian)
18. Luzik, E.V., Demchenko, N.I., Melnyk, N.I., Semychenko, V.A., Proskurka, N.M.: Influence of tolerance to uncertainty on the personal and professional development of pilots during aviation specialists' training. INCAS Bull. **13**(Special Issue), 143–157 (2021). https://doi.org/10.13111/2066-8201.2021.13.S.14
19. Luzik, E., Melnyk, N., Ladohubets, N., Polishchuk, O., Lariontseva, A.: Model for assessing the formation of special (professional) competences of practical psychologists in the educational and information environment of higher education establishments under conditions of dynamic uncertainty. Multidiscip. Sci. J. **5**, 2023ss0517 (2023)
20. Datsun, O., Demchenko, N.: Features of the prevention of emotional burnout of teachers by art therapy in the conditions of war. Proc. National Aviat. Univ. Pedag. Psychol. **21**, 108–114 (2022) (In Ukrainian)
21. Proskurka, N., Synchuk, V., Boichenko, M.: Manifestations of psychological protection of creative personality of technical specialization students. Proc. National Aviat. Univ. Pedag. Psychol. **14**(133–138) (2019). https://doi.org/10.18372/2411-264X.14.13734. (In Ukrainian)
22. Olefirenko, Y., Proskurka, N.: Value orientations and behavioral strategies of military personnel in war conditions. Proc. National Aviat. Univ. Pedag. Psychol. **23**, 127–136 (2023). https://doi.org/10.18372/2411-264x.23.18183. (In Ukrainian)
23. Luzik, E., Ladohubets, N., Komisarenko, N., Mikheyeva, T.: Cognitive-professional development strategies as the basis of formation of success trajectory of subjects of the educational process. In: EDULEARN Proceedings, pp. 3695–3705. IATED, Palma (2022). https://doi.org/10.21125/edulearn.2022.0902
24. Luzik, E., Shvydchenko, V., Hrybun, N.: Post-traumatic stress disorder in military personnel during military operations. Proc. National Aviat. Univ. Pedag. Psychol. **25**, 1–12 (2024). https://doi.org/10.18372/2411-264X.25.19358. (In Ukrainian)
25. Semychenko, V.: Differentiation of theoretical systems of depth psychology based on the principle of polarity. In: Psychology of Consciousness: Theory and Practice of Scientific Research, pp. 14–18. Liha-Press, Lviv (2023) (In Ukrainian)
26. Luzik, E., Kokarieva, A., Hurska, O., Veselska, O., Konivitska, T.: Information technology for the formation and implementation of the personal trajectory of a future specialist's success in the information and educational environment. CEUR Workshop Proc. **2654**, 556–570 (2019) https://ceur-ws.org/Vol-2654/paper44.pdf
27. Slimmen, S., Timmermans, O., Mikolajczak-Degrauwe, K., Oenema, A.: How stress-related factors affect mental well-being of university students: a cross-sectional study to explore the associations between stressors, perceived stress, and mental wellbeing. PLoS One. **17**(11), e0275925 (2022). https://doi.org/10.1371/journal.pone.0275925
28. Martsenkovskyi, D., Shevlin, M., Ben-Ezra, M., Bondjers, K., Fox, R., Karatzias, T., et al.: Mental health in Ukraine in 2023. Eur. Psychiatry. **67**(1), e27 (2024). https://doi.org/10.1192/j.eurpsy.2024.12
29. Caponigro, M., Fornasier, M., Piccoli, B., Trelat, E.: Sparse stabilization and control of alignment models. Math. Model Method Appl. Sci. **25**(03), 521–564 (2015). https://doi.org/10.1142/S0218202515400059

30. Tapiero, E.A.: The mathematical model that describes the behavior (2024). https://www.linkedin.com/pulse/el-modelo-matem%C3%A1tico-descifra-comportamiento-el%C3%ADas-azulay-wytkf\. Last accessed 06 Jan 2025 (In Italian)
31. Cox, C.: Mathematical modelling tackles mental health issues. The University of Western Australia (2024). https://www.uwa.edu.au/news/article/2024/july/mathematical-modelling-tackles-mental-health-issues. Last accessed 06 Jan 2025
32. Luzik, E., Kokarieva, A., Khomenko-Semenova, L.: Organizational principles of implementing the model of future specialist personal and professional development in educational and information environment of a technical university. Proc. National Aviat. Univ. Pedag. Psychol. **16**(1), 86–96 (2020). https://doi.org/10.18372/2411-264x.16.14683
33. Tkachuk, V., Yechkalo, Y., Semerikov, S., Kislova, M., Hladyr, Y.: Using mobile ICT for online learning during COVID-19 lockdown. In: Bollin, A., Ermolayev, V., Mayr, H.C., Nikitchenko, M., Spivakovsky, A., Tkachuk, M., Yakovyna, V., Zholtkevych, G. (eds.) Information and Communication Technologies in Education, Research and Industrial Applications, Proceedings of ICTERI, vol. 1308, pp. 46–67. Springer, Cham (2020). https://doi.org/10.1007/978-3-030-77592-6_3
34. Carod-Artal, F.J.: Social determinants of mental health. In: Bährer-Kohler, S., Carod-Artal, F. (eds.) Global Mental Health, pp. 34–46. Springer, Cham (2017). https://doi.org/10.1007/978-3-319-59123-0_4
35. Trofimov, A., Drobot, O., Kokarieva, A., Lovochkina, A., Kozytska, I.: The influence of management style and emotional intelligence on the formation of employees' commitment and loyalty. Human. Soc. Sci. Rev. **5**, 393–404 (2019). https://doi.org/10.18510/hssr.2019.7544

Military Impact on Civil Aviation: Case Study of AHY8243 Flight Based on ADS-B Data Processing

Ivan Ostroumov[1](✉), Denys Kolesnichenko[1], Daria Tkachuk[1], and Olena Kyzymchuk[2]

[1] National Aviation University, Liubomyra Huzara Ave., 1, Kyiv 03058, Ukraine
ostroumov@ukr.net

[2] ITM Technische Universitat Dresden, Dresden 01069, Germany

Abstract. Services of global air transportation are grounded on criteria safety and security of civil aviation. Unfortunately, during the last decade, there are several catastrophes occurred caused by direct military action in civil aviation. Flight MH17 was shot in Ukrainian airspace by a surface-to-air missile on 17 July 2014. Passenger flight PS 752 was classified as an intruder fighter by the air defense system and shot by a surface-to-air missile on 8 January 2020 in Iranian airspace. Both cases happened in open airspace for civil aviation with normally provided air-navigation services. During 2024 the airspace of Eastern Europe is under the wide GNSS jamming action which significantly affects civil aviation. In the paper, we provide deep trajectory data processing of the J28243 (AHY8243) flight, which crashed on December 25, 2024, whith data transferred by Automatic Dependent Surveillance-Broadcast (ADS-B) technology. The algorithm of the invert dead reckoning method is used for the recovery of lost data as a sequential process. Results of the analysis show clear, well-documented cases of direct GNSS jamming and spoofing action of civil passenger airplane which resulted in the crash. Results of the analysis show problems in civil-military coordination which reduce the safety of international civil aviation.

Keywords: CNS/ATM · air navigation · air transportation · ADS-B · trajectory data analysis · J28243 · aviation safety

1 Introduction

Global air transportation operates under the Convention on International Civil Aviation issued by the International Civil Aviation Organization (ICAO), which has been supported by 193 members [1, 2]. According to numerous international regulations services of air transportation are organized based on ensuring the safety and security of civil aviation. Operation of global civil aviation grounds on ICAO normative documents which provide strict rules for airspace usage.

I. Ostroumov et al. (Eds.): ACASD 2025, LNNS 1418, pp. 166–186, 2025.
https://doi.org/10.1007/978-3-031-91992-3_12

Special attention is given to Civil-military coordination to legislate a process related to military actions and ensuring the safety of civil aviation [3, 4]. Militaries have high priority in any airspace manipulation including segregation for particular actions. However, any military operations have to be based on the regulative documents of ICAO.

During the last decade, there have a several issues of military actions on the direct side of civil aviation that raised a serious problem with civil-military coordination.

Flight MH17 of Malaysia Airlines was shot down in Ukrainian airspace by a surface-to-air missile on 17 July 2014. Direct military action in a civil passenger airplane caused 298 people killed [5].

Flight PS 752 of Ukraine International Airlines was classified as an intruder fighter by the air defense system and shot down by a surface-to-air missile on 8 January 2020 in Iranian airspace near Tehran International Airport. All 176 persons on board including passengers and crew members were killed [6].

Flight J28243 of Azerbaijan Airlines crashed near Aktau Airport on 25 December 2024. J28243 was affected by Russian militaries including jamming and spoofing during the most flight trajectory. In the catastrophe, 38 people were killed and 29 had serious injuries [7].

These three catastrophes of intentional direct military action in civil passenger airplanes highlight the need to improve legislative documents, especially in civil-military coordination for the prevention of similar cases in the future.

Nowadays civil aviation is held by serious airspace limits caused by the War in Ukraine. Airspaces of Ukraine (since Feb 24, 2022) and Syria (since Dec 2024) are closed. Airspaces of the Russian Federation, Belarus, and some countries in the Middle East region are limited in use due to the high risk of military action. Involved airspace is large enough to limit direct flight connections from Asia to North America and European destinations [8, 9].

An interference of the primary positioning system is another serious problem of civil aviation. Numerous studies indicate a tension of the Global Navigation Satellite System (GNSS) for high-power interference [10, 11]. War in Ukraine and military conflicts in Mid-East countries caused militaries to use high-power jamming equipment in the involved airspaces. Interference of GNSS navigation signals in on-board equipment of airplanes has been indicated many times in the Eastern Europe region [12, 13]. Jamming of GNSS causes reduced positioning performance or can lead to GNSS lock [14, 15]. In case the primary positioning system is unavailable, inertial navigation equipment of navigation by navigation aids could be used as a stand-by system [16, 17]. The effect on a civil airplane from high power interference could be easily detected with the help of data processing data transferred down-link under Automatic Dependent Surveillance-Broadcast (ADS-B) technology [18, 19].

In this paper, we provide a comprehensive analysis of trajectory data of the J28243 flight based on ADS-B data sets to highlight military action on civil passenger airplane which resulted in catastrophe. The trajectory of this flight is under high interest due to clear jamming, spoofing, and probably physical action on a nominal flight of a civil airplane. Also, we propose a methodology of airplane data recovery based on available data sets of parameters transmitted by ADS-B messages.

The structure of the paper in the following second section describes the data that are used in trajectory analysis and methods applied for trajectory data interpolation recovery. Section 3 provides data analysis of the J28243 flight from Heydar Aliyev International Airport to Grozny Airport crashed on December 25, 2024, near Aktau Airport. Section 4 provides a comparative study of the J28243 crash with other catastrophes caused by direct military action on MH17 and PS752 flights.

2 Materials and Methods

Surveillance in civil aviation is provided by radars (primary and secondary surveillance radars) and ADS-B. ADS-B technology was globally integrated in recent years during the last decade. Primary and secondary surveillance radars are widely used by Air Traffic Management services for active measuring of airspace users' position. ADS-B is grounded on measuring trajectory data and parameters of flight onboard and transferring it by wireless digital communication channel on 1090 MHz. Heavy airplanes use an airplane transponder of mode 1090ES to support ADS-B data transmission [20]. The top and Bottom antennas of airplane transponders made an omnidirectional area of data transmission within the maximum range of the radio-communication link. The common maximum range of ADS-b radio communication line is limited by 300–400 km. ADS-B communication channel uses open-access data formats, that make it possible to receive and decode the data by anyone including low-cost software-defined radios (SDR). Open access data format and low-cost user equipment ($<$120$) stimulated the development of commercial networks of SDR for collecting the data from wide areas [21, 22].

2.1 Data Transferred by ADS-B

ADS-B data transmission grounds on 112-bit long messages which is issued asynchronicall mode. Each airplane transponder begins transmission based on its internal timestamp settings. It may lead to overlap or interference of messages in one data channel (1090 MHz). Damaged messages could not be recovered and will be lost for both transmissions. Absents of synchronization in the ADS-B data channel is a common problem in congested airspace that causes multiple gaps in trajectory data. Parameters of onboard equipment transferred by ADS-B need to apply specific algorithms of data processing [23, 24].

Each ADS-B mode 1090ES message has a common structure which includes: format specification (5 bits), transponder capability (3 bits), unique ICAO registration code of airplane (24 bits), message (56 bits), and parity control (24 bits) [25, 26]. All trajectory data is transferred in the field of Message which is only 56 bits long. To make it possible to transfer all required data in 56 bits of message field ADS-B uses different types of messages like airplane identification, surface or airborne positions, velocities, data and airplane status. Each transferred parameter is coded according to a specific coding algorithm to minimize the number of data for transferring. For example, airplane coordinates are coded Compact Position Report (CPR), altitude and velocities use a scale-based coding.

Vary of data messages in ADS-B made a set of messages transmitted within a one-second data transmission cycle. The repetition frequency of data transfer varies based on settings and particular mode of operation. Common data transfer required at least 6–8 messages per second. ADS-B uses 1 μs bit modulation thus each data message required 112 μs accompanied by 8 μs preamble (total 120 μs).

Common trajectory data transmitted by ADS-B includes: airplane position (latitude and longitude) measured by the on-board positioning system, barometric altitude, Height measured by GNSS in relation to WGS84 ellipsoidal surface, vertical rate (GNSS or barometrical), ground speed, airspeed (indicated or true), magnetic heading. In addition, ADS-B data includes navigation accuracy, integrity, and uncertainty category for different parameters. These categories specify the precision level of data transferred.

Nowadays GNSS in nominal state provides the highest level of precision in airplane position measurement [27]. Therefore, data transferred by ADS-B is a high performance which in most cases is more precise than any radar measurements.

2.2 Recovery of Missing Data

Lack of ground coverage area, interference of ADS-B messages, or jamming of GNSS navigation signals may be a case when some ADS-D data a present but another one is missing due to unsuccessful data transmission. In this case, algorithms of data recovery could be initiated based on a math model or relation between available and lost data.

Recovery Gaps in Trajectory Data. Short gaps in airplane trajectory could be recovered by tracking filters or by linear regression model. An α-β-γ-η tracking filter or Kalman filter is used mostly to minimize error action in trajectory data. Increasing the frequency of data or recovering gaps could be easily done by polynomial function or linear regression model with non-linear B-splines basic function (S):

$$S_{i,n}(t) = \frac{t - a_i}{a_{i+n-1} - a_i} S_{i,n}(t) + \frac{a_{i+n} - t}{a_{i+n} - a_{i+1}} S_{i+1,n-1}(t), \tag{1}$$

where a is a node point; i is a basis function number; n is a basis function order.

The location of nodes is selected based on an input data set and centripetal method. Input data is used to estimate a control point location (C) for the spline function:

$$C = \left(S^T S\right)^{-1} S^T X, \tag{2}$$

where X is a matrix of input trajectory data usually transformed to a Cartesian reference frame accompanied by timestamps.

$$X = \begin{bmatrix} x_1 \; y_1 \; t_1 \\ x_2 \; y_2 \; t_2 \\ \vdots \;\; \vdots \;\; \vdots \\ x_m \; y_m \; t_m \end{bmatrix}; C = \begin{bmatrix} c_1 & t_1 \\ c_2 & t_2 \\ \cdots & \cdots \\ c_k & t_k \end{bmatrix}; C = \begin{bmatrix} C_{1,n}(t_1) & \cdots & C_{k,n}(t_1) \\ \vdots & & \vdots \\ C_{1,n}(t_m) & \cdots & C_{k,n}(t_m) \end{bmatrix}, \tag{3}$$

where m is the size of an input data set; k is the number of selected control points.

Recovered data for the required set of timestamps (R) are calculated as follows:

$$R = SC. \tag{4}$$

Recovered by (4) coordinates will have identical to X matrix structure. Most recovery processes should be performed in a local Cartesian reference frame and final coordinates need to be converted to a geodesic reference frame (LLA).

Estimation of Coordinates from Given True Heading Angle and Ground Speed. Coordinates of airplane could be recovered from known ground speed and heading angle based on the Dead Reckoning Method. DRM also works in the Cartesian reference frame with the assumption of the previous airplane position in the reference point of a frame. DRM is grounded on input true heading angle (H) and airplane ground speed (V). The following navigation equations are used in North-East horizontal frame orientation based on inverse reference point:

if $0 < H \leq 90°$, then

$$\begin{aligned} x_i &= -V_i(t_i - t_{i-1})\cos(H_i) \\ y_i &= -V_i(t_i - t_{i-1})\sin(H_i) \end{aligned} \tag{5}$$

if $90 < H \leq 180°$, then

$$\begin{aligned} x_i &= -V_i(t_i - t_{i-1})\cos(H_i - 90) \\ y_i &= V_i(t_i - t_{i-1})\sin(H_i - 90) \end{aligned} \tag{6}$$

if $180 < H \leq 270°$, then

$$\begin{aligned} x_i &= V_i(t_i - t_{i-1})\cos(H_i - 180) \\ y_i &= -V_i(t_i - t_{i-1})\sin(H_i - 180) \end{aligned} \tag{7}$$

if $270 < H \leq 360°$, then

$$\begin{aligned} x_i &= V_i(t_i - t_{i-1})\cos(H_i) \\ y_i &= -V_i(t_i - t_{i-1})\sin(H_i)' \end{aligned} \tag{8}$$

where x and y are coordinates of airplane position in North-East horizontal frame; V is airplane ground speed; H is a true heading angle; t_i and t_{i-1} are time at different iterations.

Calculated by (5)–(8) airplane coordinates in the Cartesian reference frame should be converted from NED to geographic LLA reference frame. The true heading could be recalculated from a magnetic one minus magnetic variation (declination) angle which could be obtained from a world magnetic model for a particular region and date. Ground speed (GS) could be recovered from true airspeed (TAS) based on known airplane orientation and weather data. A wind triangle equation is used to recover ground speed based on wind speed (W) and wind direction angle (WD):

$$GS = \sqrt{TAS^2 + W^2 - 2WTAS\cos(180 - (WD - TH))}. \tag{9}$$

3 Flight Data Analysis

As an example of flight data analysis let's consider data of the J28243 flight that crashed on December 25, 2024.

3.1 Description of Flight

Flight AHY8243 (J28243) served by Azerbaijan Airlines with a flight connection from Heydar Aliyev International Airport (UBBB, Baku, Azerbaijan) to Grozny Airport (URMG, Grozny, Russian Federation) on December 25, 2024. The scheduled takeoff time was 4:10 UTC (Actual takeoff time is 3:56 UTC) and scheduled landing time was 5:10 UTC. The airplane crashed near Aktau Airport (UATE, Aktau, Kazakhstan) at 6:28:04 UTC. Registration number of airplane 4K-AZ65, operated by Embraer E190-100IGW. The airplane has been in use since 19.07.2013. The airplane was equipped with two engines CF34-10E5. The flight was served by 5 crew members including 2 pilots and 3 flight attendants. The flight was served for 62 passengers. In the crash, 38 persons were killed and 29 got serious injuries due to the segregation of the tail part of airplane caused by hitting the ground.

Flight was performed in 3 flight information regions UBBA (Baku FIR, Azerbaijan), URRV (Rostov-na-Donu FIR, Russian Federation), and UATT (Aktobe FIR, Kazakhstan). The engines started and airplane moved to the holding point of Runway 34 via Echo taxiway. At 03:56 airplane took off from Baku airport and followed the SID (EKRAM 1D) then climbed to cruise FL300 and followed the flight plan with the following waypoints: Nobva, Bumar, Agato, Osdon, MKL (Makhachkala DVOR), Dubki, Remka, and arrive to URMG with STAR scheme REMKA1R based on NDB WK.

In the destination airport, there was a valid NOTAM "G0446/24 NOTAMR G0414/24" (starts from 29.11.2024 until 30.12.2024) informing about possible disruption of GNSS [28].

The ground network of SDR receivers of FlightRadar24 received 6229 data from messages transmitted by ADS-B. However, only 4269 of them included airplane position. There are two trajectory datasets available:

1. From departure airport to near MKL DVOR the dataset included 3153 data (only 3041 included coordinates). The data sequence covered the period from 3:50:00 UTC to 4:25:22 UTC;
2. Near Aktau Airport dataset included 1079 data (only 1045 included coordinates). The data sequence covered the period from 6:07:50 UTC to 6:28:04 UTC.

Due to wire area jamming of GNSS navigation signals in the airspace of the Russian Federation, airplane position was absent from 4:25:22 UTC to 6:07:50 UTC. During that time, pilots have to use an inertial navigation system. However, due to sequentially increased errors, they have to provide positioning by navigation aids [29, 30].

Two survivors Aidan Rahimli (flight attendant) and Subkhon Rakhimov (passenger) in the video informed that the airplane arrived in SID of URMG and made an unsuccessful approach due to fog and low visibility. This information is approved by valid weather data issued in the METAR message [31]:

URMG 250430Z 26001MPS 3600 BR OVC007 03/02 Q1025 R26/290250 NOSIG RMK MT OBSC OBST OBSC QFE754/1005

Decoding of METAR message gives presented mist (BR), wind 1 m/s directed 260°, overcast at 700 ft, and low horizontal visibility 3600 m. These data approve information given by Aidan and Subkhon about low visibility and mist.

AIP data of URMG [32] include only two types of STARs: RNAV and using NDB. The most probable case is airplane used the NDR approach due to GNSS jamming and unavailability of RNAV. Also, based on previous flights, it used runway RWY26 for landing. A decision altitude for the NDB approach is 731 ft which is higher than informing in METAR overcast in 700 ft. Thus, it makes sense that pilots at 731 ft did not have visual contact with the runway and had two missed approaches. RNAV STAR has a minimum decision height at 381 ft for RNAV with LNAV, 270 ft for RNAV with LNAV/VNAV, and 200 ft for RNAV with GLS.

Also, both Aidan and Subkhon informed that the airplane after the missed approach went to turn and after following missed approach scheme got approach the second time. The second approach also was missed and the airplane followed for the third time. During the moving on missed approach scheme at the second time, both Aidan and Subkhon heard an explosion outside of the airplane. The airplane body was damaged with small holes from shrapnel. Also, both Aidan and Subkhon informed about persons who had injuries from elements coming outside the airplane. Multiple holes in the passenger cabin had been filmed on mobile phones including documented damaged of the left wing. This information is approved by photos of the tail and stabilizer which have multiple holes from the active part of the missile at the left side airplane only. Also, they informed about sudden airplane climbing after the missile attack.

3.2 Trajectory Data Analysis

Unfortunately, information from the Flight Data Recorder (FDR) system is not available for open access, and after the investigation process by authorized bodies, these data will not be provided for analysis for free. In data analysis, we use a full sequence of data transmitted by ADS-B including all parameters [7]. We consider two trajectory datasets separately. Also, the available dataset includes three previous flights of AHY8243 on December 22, 18 and 15, 2024. Trajectories of three previous flights and the first data set between UBBB and URMG are shown in Fig. 1. All three data sets follow identical flight paths in URRV airspace: BUMAR-MKL-REMKA-WK.

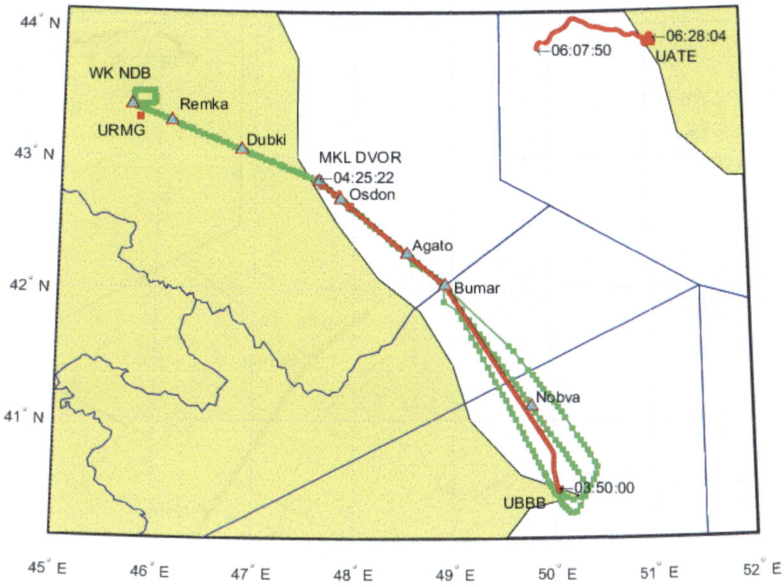

Fig. 1. Trajectory data of AHY8243 on Dec 25, and three previous flights.

Trajectory data from 04:25:22 UTC to 06:07:50UTC is not available. However, there are several series of airspeed and heading data fixed in the period between 05:13:51 UTC to 06:07:50 UTC. Making an assumption that the airplane follows the same trajectory as it was used during previous flights a missing data could be recovered up to NDP WK. Three data sets at this period were mixed based on time synchronization by spline functions. Trajectory data up to WK has been shown in Fig. 2. Trajectory modeling made it possible to estimate the time of arriving in WK at 4:44:00 UTC. The results of waypoints data estimation are specified in Table 1.

Arriving at WK airplane moved to STAR based on NDB. The trajectory of airplane movement by STAR and missed approach scheme have been recovered by flight simulator X-Plane and airplane model. Simulation shows that from arriving in WK it requires 5:15 to move by STAR and reach the decision point. Also, getting one missed approach requires 7:30 mean.

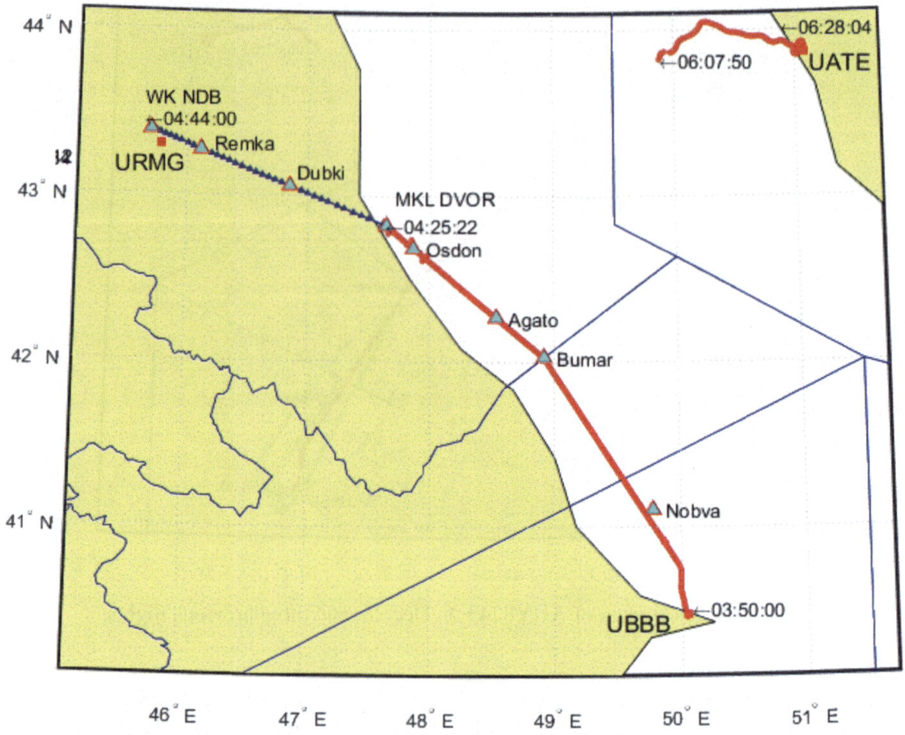

Fig. 2. Result of airplane trajectory recovery for AHY8243 on Dec 25.

Table 1. Calculation of flight path parameters.

Waypoint	Coordinates	Distance from the previous waypoint, [km]	Time of Arrival, UTC, hh:mm:ss	Time to Fly, TOE, mm:ss
Nobra	40°06 28 N 49°46 52E	74.83	04:04:14 Actual	14:14
Bumar	42°01 42 N 48°55 18E	124.96	04:14:15 Actual	10:01
Agato	42°15 59 N 48°32 19E	41.55	04:17:33 Actual	03:18
Osdon	42°41 05 N 47°52 01E	72.02	04:23:19 Actual	05:46
MKL (DVOR)	42°49 17 N 47°38 44E	23.48	04:25:12 Actual	01:53

(*continued*)

Table 1. *(continued)*

Waypoint	Coordinates	Distance from the previous waypoint, [km]	Time of Arrival, UTC, hh:mm:ss	Time to Fly, TOE, mm:ss
Dubki	43°03 33 N 46°50 53E	70.21	04:32:21 Estimated	07:09
Remka	43°16 11 N 46°07 01E	63.91	04:39:42 Estimated	07:21
WK (NDB)	43°23 22 N 45°43 42E	36.33	04:44:00 Estimated	04:18

All other data available by the ADS-B data set is shown in Figs. 3, 4 and 5 include:

- Track angle (is a True heading measured by an on-board GNSS sensor);
- Heading (is a magnetic heading (MH) measured by attitude heading and reference system);
- Barometric altitude;
- Altitude measured by GNSS;
- Vertical speed;
- Ground speed (measured by GNSS);
- True Air Speed (measured by air data system);
- Indicated Air Speeds (measured by air data system);
- Also, a few messages include roll angles and outside temperature.

Recovery of airplane trajectory from 05:13:51 UTC to 06:07:50 UTC could be done by the dead reckoning method based on parameters of True Heading (TH) and ground speed (GS) calculated on the opposite side from the available airplane position at 6:07:50 UTC. Unfortunately, both TH and GS are unavailable for a given period.

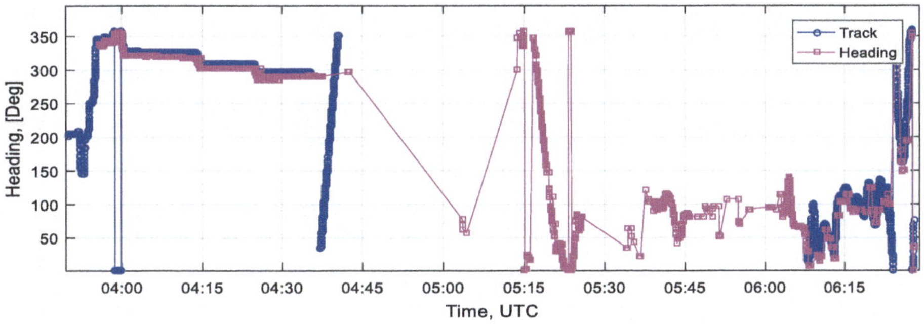

Fig. 3. Track and heading angles for AHY8243 on Dec 25.

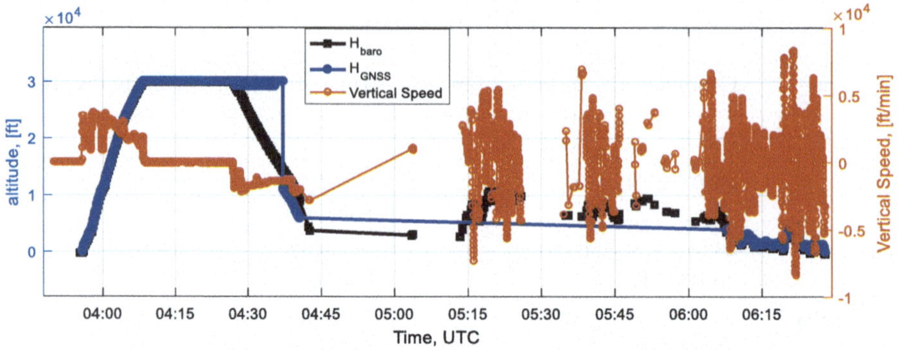

Fig. 4. Altitude data and vertical speed for AHY8243 on Dec 25.

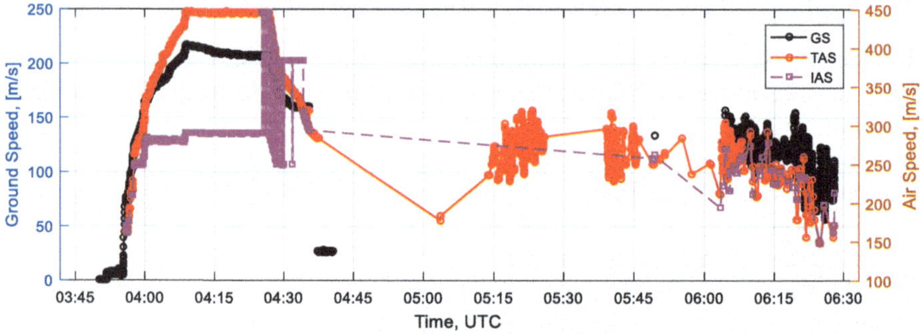

Fig. 5. Ground speed, True and Indicated Air Speeds for AHY8243 on Dec 25.

TH could be recalculated from MH taking into account magnetic variation in the region. A global world magnetic model is used to estimate magnetic variation. The magnetic declination angle for altitude 2100 m based on the WMM2020 model which is valid on epoch 2020–2025 gives 7.295°. GS could be recalculated with wind triangle Eq. (8) based on wind speed and direction. We use a historical weather database archive to access wind data for the required time [33].

Recovered both TH and GS use different timestamps and are not synchronized by time. We use spline data interpolation for both parameters to recalculate TH and GS for each second of the flight. Both recovered parameters TH and TAS for each second since 05:13:51 UTC are given in Fig. 6. Results of trajectory recovery with DRM based on TH and GS as the invert sequential process are given in Fig. 7 as Scenario 1.

Analysis of input data for Track and TAS gives their cleared portion of data:

1. From 5:13:50 to 5:25:44 UTC include circle;
2. From 5:38:10 to 5:45:57 UTC;
3. From 6:03:04 to 6:07:50 UTC final part.

Configuration of recovered trajectory at these series has a high level of confidence however a reference point is not known.

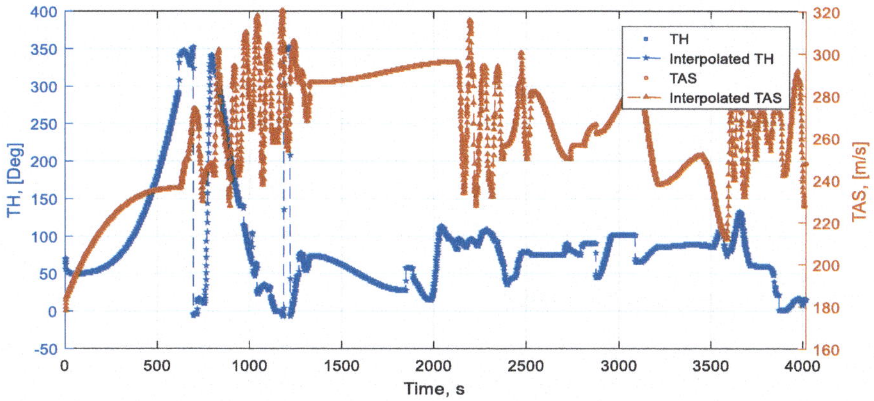

Fig. 6. Interpolated TH and GS for AHY8243 since 05:13:51 UTC.

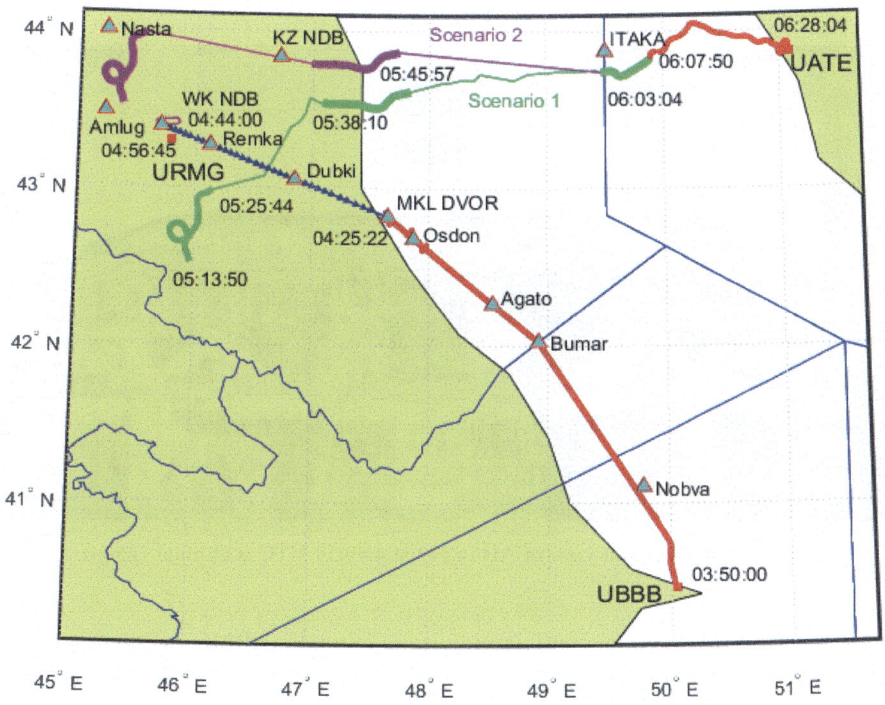

Fig. 7. Recovered trajectory for AHY8243.

Analysis of heading, airspeed, and vertical speed series shows that since 5:13:50 UTC, excessive parameters variation possibly caused by problems with airplane guidance especially with the elevator and rudder has begun.

Due to possible problems with the elevator, a method of airplane altitude guidance with speed was probably used. It is also approved by a good correlation of barometrical altitude, air speed, and vertical speed.

Taking into account the configuration of the flight route chart during trajectory data processing, one of the possible scenarios was a flight through the Caspian Sea using KZ NDB via R227 to ITAKA waypoint (Scenario 2). Results of the Scenario 2 simulation is shown in Fig. 7.

3.3 Simulation in Flight Simulator

Flight simulator X-Plane v 11.55 with airplane model "SSG - E-Jets 195LR Evolution" has been used to simulate particular scenarios. As an example, simulation of weather conditions (based on METAR data) in URMG at the decision point shows that pilots cannot have a chance to have visual contact with runway RW26 (Fig. 8) at altitude of 1250 ft MSL.

Fig. 8. Simulation of cockpit view of AHY8243 at 4:49:17 UTC at altitude 1250 ft. MSL.

Also, results of landing configuration simulation in URMG at given weather conditions is shown in Fig. 9.

3.4 Jamming and Spoofing Analysis

Available datasets indicate cleared jamming action on civil airplane. GNSS data was not available from 04:25:22 UTC to 6:07:50 UTC. Thus, the total time of airplane jamming was 1:42:28. Also, this is approved by the Navigation Integrity Category (NIC) (Fig. 10). Overall analysis of GNSS interference presented on Dec 25, 2024, is given in Fig. 11.

Jamming of GNSS equipment of civil airplane has a dramatic influence on this flight. If GNSS works, then RNAV STAR could be available for an approach with low levels

Fig. 9. Simulation of AHY8243 landing configuration in URMG.

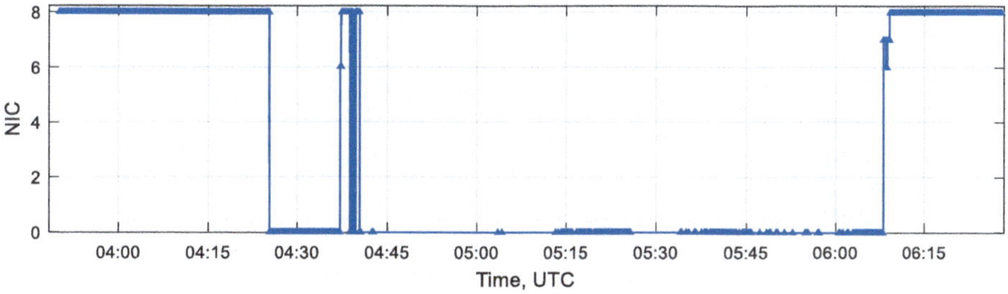

Fig. 10. Navigation Integrity Level for AHY8243.

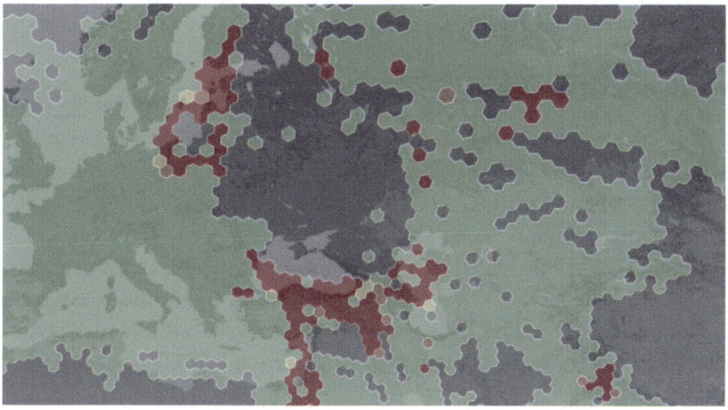

Fig. 11. GNSS interference map on Dec 25, 2024.

of visibility, and airplanes with particular weather conditions have a high chance of successfully landing in URMG.

Another dangerous action is the spoofing of GNSS signals. Spoofing replaced the navigation signals of GNSS with a wrong one. The airplane under spoofing attack could identify its position, track angle, ground speed, and GNSS altitude. However, this data is wrong.

During the AHY8243 flight, spoofing action was identified from 4:37:12 to 4:40:34 UTC (the duration of the spoofing attack on the civil airplane was 3 min 22 s). Spoofing equipment used a simulation of airplane movement by circular trajectory with a radius of about 1 km with reference point 43°10′29″N 44°17′16″E. Total amount of spoofed data is 183. A spoofed trajectory is given in Fig. 12.

Fig. 12. Spoofed airplane trajectory for AHY8243.

4 Comparison with Other Cases

4.1 Catastrophe of MH17 Flight

On 17 July 2014, flight MH17 operated by Malaysia Airlines en-route from Amsterdam to Kuala Lumpur served by B 777-200ER, was shot down by a Russian missile over the territory of Donbas within Ukrainian airspace at FL330. All 298 people on board, including 15 crew members, were killed.

On the night of 16–17 July 2014, a BUK anti-aircraft military missile system was delivered to Ukraine from the Kursk region, belonging to the 53rd Brigade based in Kursk (Russian Federation). The BUK was delivered to the village of Pervomaiske and was subsequently placed in a field in an area from which it was possible to fire at targets at high altitudes.

At 13:20 UTC, it detected an air target at an altitude above 10 km, civil airplane was mistaken for a military aircraft and, due to the guidance, the military crew launched a 9M38 missile from the BUK anti-aircraft system. A few minutes later, the missile exploded near the cockpit and hit the aircraft. The front of the aircraft was destroyed by the missile strike. The central part of the fuselage and cockpit were destroyed instantly and debris began to fall to the ground (Fig. 13). Most of the wreckage fell near the village of Hrabove. After the tragedy, the BUK launch pad was transported back to the Russian Federation, which is confirmed by satellite data.

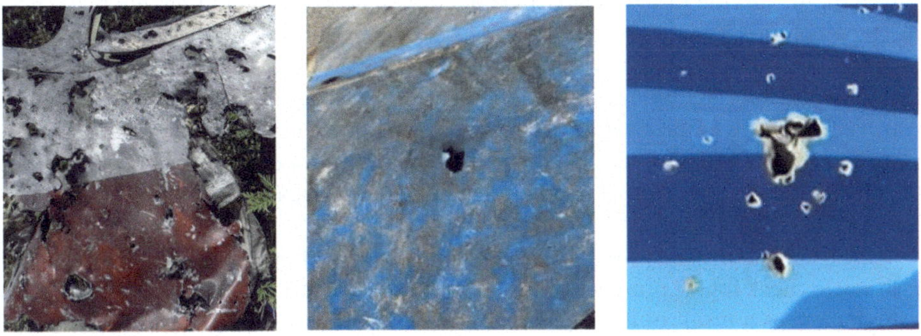

a) MH17 *b) PS752* *c) AHY8243*

Fig. 13. Holes caused by the active part of missiles. *(Taken from videos available on YouTube)*

The international investigation team analyzed photos and videos showing the movement of the BUK system from the Russian Federation to Ukraine. The Security Service of Ukraine provided recordings of phone conversations between the militants discussing the delivery of this complex. Local residents also confirmed the presence of the complex. The experts also examined the wreckage of the aircraft and the remains of the missile, which allowed them to identify a 9M38 missile belonging to the BUK complex.

The investigation nevertheless confirmed and established that it was the use of 9M38 missile weapons from the BUK anti-aircraft missile system. Russian Federation did not approve the results of the international investigation team and proposed an alternative investigation by Almaz-antey.

4.2 Catastrophe of PS752 Flight

Flight PS752 of Ukrainian International Airlines served by B737-800, departed from Imam Khomeini (OIIE, Iran) to Boryspil (UKBB, Ukraine) international airport, on January 8, 2020, and was shot down by two Iranian air defense missiles.

At 02:25 UTC, the ground controller issued a take-off and departure clearance for PS752 from the parking area, which was read out by the pilot. At 02:42 UTC, the aircraft took off from IKA's right-hand Runway 29 and headed west to land in Mehrabad. The pilot contacted the approach controller, announcing IKA radar procedure 1A as the SID procedure. The Mehrabad controller then determined and approved a climb clearance of FL.260. The controller instructed the pilot to turn right at 6000 ft and continue on a direct course to PAROT.

The aircraft reached an altitude of 8100 ft and a speed of 500 km/h. At 02:44:56 UTC PS752 was shot down by two Tamuz missiles (a type of Tor-M1 missile) from a surface-to-air missile system launched by Iranian militaries. The 167 passengers and 9 crew have been killed.

Evidence indicates severe damage to the lower nose section of the aircraft, including the lower half of the cockpit. As for the upper half found at the scene, the cockpit windows remained in place. Also, a large number of holes were found on the fuselage of the aircraft, varying in size, shape and direction of rupture, and falling into two types. One type comprises smaller holes with dense smoke emitting towards the outside of the fuselage. The other type, with a larger number of holes, contains only torn holes (Fig. 13). Two aircraft engines were found at the crash site, which had been completely destroyed. Damage to the rotating parts indicates that the engines were operating at the time of the incident. However, there was no evidence of an explosion or severe fire in the engines, and it is likely that most of the damage was caused by deformation caused by the impact on the ground.

Holes were found in the passenger seat cushions, so they were collected from the crash site to check for the presence of explosive ordnance that could have been introduced through them as a result of the explosion. Initial detection of foreign object debris was carried out using X-ray scanners, and the samples were submitted to the air accidents investigation branch. After cutting the pads, the metal parts were extracted and sent to a metallurgical laboratory for evaluation and comparison with rocket fragment alloy.

The official cause of the catastrophe was an anticipated error in the thread identification system which misidentified a civil airplane as an intruder UAV.

4.3 Comparison of AHY8243 Crash with Other Catastrophes

There are too many facts that prove military missiles hit of AHY8243 flight. The most valuable ones are the following:

4. Witnesses. Two survival Aidan and Subkhon informed about a loud explosion outside the plane during the movement of an airplane on the standard scheme of missed approach. Also, they told about persons that have injuries from elements coming outside the airplane.
5. Photos and videos of passengers after missiles explosion. Many survival passengers shared videos from onboard after the missiles explosion. Multiple holes in the passenger cabin have been recorded on mobile phones including documented damage of the left wing. Also, videos indicate normal engine operation with no fire, smog, or other damages. In addition, all videos show oxygen masks operated which admits a pressurization problem with the passenger cabin.
6. Photos and videos of the tail side of an airplane with multiple holes from possible active parts of the missiles action (Fig. 13).
7. Airplane trajectory data since 05:13:51 UTC illustrates problems with airplane stabilization in space which could possibly caused by problems with the elevator and rudder.

Based on videos and photos of tail part of airplane, left elevator has multiple damages including holes from possible missiles. Also, using videos it could be clearly identified damages of the left elevator on the side of the actuators installation (Fig. 14). Destroyed actuators could made the elevator non-operated. The vertical stabilizer from the left side has also multiple holes (Fig. 13). At the same time, there are no damages in the left elevator. Position of the left and right elevator is different. The left elevator is deviated up, right one is in a horizontal position. Both elevators work synchronically and should be in identical positions. Also, during the landing elevators should be deviated down as seen in Fig. 9.

Fig. 14. Holes caused by the active part of missiles on the left elevator. *(Taken from videos available in YouTube)*

All facts documented above, point that the AHY8243 flight was attacked by military missiles on December 25 at timeframe 4:56:45–5:13:50 UTC during movement by standard scheme of missed approach of URMG or in airport vicinity. Also, the airplane previously made two missed approach attempts. Therefore, Russian militaries who operate missile launch system, made at least two contacts with the airplane previously.

5 Conclusions

Data transferred from the onboard side by ADS-B technology is a key technology for real-time airplane data documentation with a high precision level. Open access to ADS-B trajectory data made the investigation of any airplane crash open and available for everyone with no need for flight data recorder data. Parameters transferred by ADS-B have the same values as fixed in FDR, however with a low frequency.

Results of the AHY8243 crash study show that bad weather conditions were at perilous for landing by NDB STAR. However, if GNSS had been available, the airplane would have successfully landed by RNAV scheme with no problems. The on-board receiver of GNSS was jammed during 01:42:28. Also, on-board equipment was spoofed for 3 min 22 s. On-board equipment can not identify spoofing data and uses it as the correct signal (approved by information from the Navigation integrity category shown in Fig. 8). Missile attack of a civil airplane damaged the elevator and caused serious problems with altitude and heading control of the airplane.

The increased number of catastrophes caused by military action during the last decade requires improving normative documents to avoid particular events and reduce the risk in future air transportation systems.

References

1. ICAO: Convention on International Civil Aviation, Doc 7300 ICAO, Chicago (1944)
2. Ostroumov, I.V., Kharchenko, V.P., Kuzmenko, N.S.: An airspace analysis according to area navigation requirements. Aviation. **23**(2), 36–42 (2019). https://doi.org/10.3846/aviation.2019.10302
3. ICAO: Manual on Civil-Military Cooperation in Air Traffic Management. Doc 10088 (2021)
4. Markiewicz, T.M.: Methodology for the performance evaluation of airspace use by military aviation. Safe. Defen. **2**, 9–17 (2022). https://doi.org/10.37105/sd.180
5. JIT: Findings of the JIT MH17 investigation into the crew members of the Buk TELAR and those responsible in the chain of command. Report MH17 (2023). https://www.prosecutionservice.nl/documents/publications/mh17/map/2023/report-mh17. Last accessed 2024/01/02
6. Flight PS752 Accident Investigation. Final Report. Aircraft accident Investigation board of the Islamic republic of Iran (2021)
7. Petchenik, I.: Azerbaijan Airlines E190 crashes near Aktau (2024). https://www.flightradar24.com/blog/azerbaijan-airlines-e190-crashes-near-aktau. Last accessed 2025/01/02
8. Ostroumov, I., Ivashchuk, O., Kuzmenko, N.: Preliminary estimation of war impact in Ukraine on the global air transportation. In: 12th International Conference on Advanced Computer Information Technologies (ACIT), pp. 281–284. IEEE, Ruzomberok (2022). https://doi.org/10.1109/ACIT54803.2022.9913092
9. Ostroumov, I., Ivannikova, V., Zaliskyi, M., Kuzmenko, N.: Impact analysis of Russian-Ukrainian war on airspace. J. Air Transp. Manag. **124**, 102742 (2025). https://doi.org/10.1016/j.jairtraman.2025.102742
10. Rados, K., Brkic, M., Begusic, D.: Recent advances on jamming and spoofing detection in GNSS. Sensors. **24**(13), 4210 (2024). https://doi.org/10.3390/s24134210
11. Curran, J.T., Bavaro, M., Closas, P., Navarro, M.: On the threat of systematic jamming of GNSS. In: 29th International Technical Meeting of the Satellite Division of the Institute of Navigation (ION GNSS), pp. 313–321. ION, Portland, OR (2016). https://doi.org/10.33012/2016.14672
12. Felux, M., Figuet, B., Waltert, M., Fol, P., Strohmeier, M., Olive, X.: Analysis of GNSS disruptions in European Airspace. In: Proceedings of the 2023 International Technical Meeting of The Institute of Navigation, pp. 315–326. ION, Long Beach, CA (2023). https://doi.org/10.33012/2023.18626
13. Figuet, B., Waltert, M., Felux, M., Olive, X.: GNSS jamming and its effect on air traffic in Eastern Europe. Eng. Proc. **28**(1), 12 (2022). https://doi.org/10.3390/engproc2022028012
14. Osechas, O., Fohlmeister, F., Dautermann, T., Felux, M.: Impact of GNSS-band radio interference on operational avionics. J. Inst. Navig. **69**(2), 516 (2022). https://doi.org/10.33012/navi.516

15. Nasser, H., Berz, G., Gomez, M., la Fuente, A.D., Fidalgo, J., Li, W., Troller, M.: GNSS interference detection and geolocalization for aviation applications. In: Proceedings of the 35th International Technical Meeting of the Satellite Division of The Institute of Navigation (ION GNSS+), pp. 192–216. ION, Denver, CO (2022). https://doi.org/10.33012/2022.18358
16. Kuzmenko, N.S., Ostroumov, I.V.: Performance analysis of positioning system by navigational aids in three dimensional space. In: First International Conference on System Analysis & Intelligent Computing (SAIC), pp. 1–4. IEEE, Kyiv (2018). https://doi.org/10.1109/SAIC.2018.8516790
17. Ostroumov, I.V., Kuzmenko, N.S.: An area navigation (RNAV) system performance monitoring and alerting. In: First International Conference on System Analysis & Intelligent Computing (SAIC), pp. 1–4. IEEE, Kyiv (2018). https://doi.org/10.1109/SAIC.2018.8516750
18. Li, J., Wang, H., Dan, Z., Xu, J., Wang, Z., Zhu, Y.: Civil Aviation GNSS Interference Detection and Location Based on Genetic Algorithm Using ADS-B Data. In: Proceedings of the 36th International Technical Meeting of the Satellite Division of The Institute of Navigation (ION GNSS+), pp. 4168–4182. ION, Denver, CO (2023). https://doi.org/10.33012/2023.19392
19. Kazmierczak, J., Joseph, A., Cook, G.: Aviation GNSS interference analysis based on ADS-B out data. In: Proceedings of the 34th International Technical Meeting of the Satellite Division of The Institute of Navigation (ION GNSS), pp. 1108–1121. ION, St. Louis, MI (2021). https://doi.org/10.33012/2021.18039
20. Holubnychyi, O., Zaliskyi, M., Sushchenko, O., Solomentsev, O., Averyanova, Y.: Self-organization technique with a norm transformation based filtering for sustainable infocommunications within CNS/ATM systems. In: Ostroumov, I., Zaliskyi, M. (eds.) 2nd International Workshop on Advances in Civil Aviation Systems Development (ACASD 2024), LNNS, vol. 992, pp. 262–278. Springer, Cham (2024). https://doi.org/10.1007/978-3-031-60196-5_20
21. Ferrara, A.: ADS-B in context: A Comprehensive Threat Analysis (2023), https://www.politesi.polimi.it/handle/10589/222642. Last accessed 2025/01/02
22. Habibi, M., Amrhar, A., Gagne, J.M., Landry, R.J.: NextGen ADS-B software-defined reception with enhanced techniques. Appl. Sci. **13**(1), 493 (2022). https://doi.org/10.3390/app13010493
23. Solomentsev, O., Zaliskyi, M., Zuiev, O.: Radioelectronic equipment availability factor models. In: Signal Processing Symposium 2013 (SPS 2013), pp. 1–4. IEEE, Serock (2013). https://doi.org/10.1109/SPS.2013.6623616
24. Zaliskyi, M., Solomentsev, O.: Method of sequential estimation of statistical distribution parameters in control systems design. In: 3rd International Conference Methods and Systems of Navigation and Motion Control (MSNMC), pp. 135–138. IEEE, Kyiv (2014). https://doi.org/10.1109/MSNMC.2014.6979752
25. Junzi, S.: The 1090 Megahertz Riddle: A Guide to Decoding Mode S and ADS-B Signals (2021). doi:https://doi.org/10.34641/mg.11
26. ICAO: Technical Provisions for Mode S Services and Extended Squitter, Doc 9871, 1st edn. ICAO, Monreal (2008)
27. Ostroumov, I., Kuzmenko, N.: Statistical analysis and flight route extraction from automatic dependent surveillance-broadcast data. In: Integrated Communication, Navigation and Surveillance Conference (ICNS), pp. i–ix. IEEE, Dulles, VA (2022). https://doi.org/10.1109/ICNS54818.2022.9771515
28. Defense Internet NOTAM Service DINS. FAA, https://notams.aim.faa.gov last accessed 2025/01/02
29. Ivashchuk, O., et al.: A configuration analysis of Ukrainian flight routes network. In: 16th International Conference on the Experience of Designing and Application of CAD Systems, pp. 6–10. IEEE, Lviv (2021). https://doi.org/10.1109/CADSM52681.2021.9385263

30. Averyanova, Y., Larin, V., Kuzmenko, N., Ostroumov, I., Zaliskyi, M., Solomentsev, O., et al.: Turbulence detection and classification algorithm using data from AWR. In: IEEE 2nd Ukrainian Microwave Week (UkrMW), pp. 518–522. IEEE, Lviv (2022). https://doi.org/10.1109/UkrMW58013.2022.10037172
31. Aviation Weather Center. NOAA, https://aviationweather.gov last accessed 2025/01/02
32. AIP Russia, Federal Air Transport Agency (2024)
33. Historical weather data collection, https://openweathermap.org/api last accessed 2025/01/02

Aircraft Noise

Reference Aircraft Noise Model for the New Supersonic Airplane Design with Corrected Noise Propagation and Engine Installation Effect Contributions

Oleksandr Zaporozhets[1](✉), Vitalii Makarenko[2], Vadim Tokarev[2], and Kateryna Kazhan[2]

[1] Łukasiewicz Research Network – Institute of Aviation, Aleja Krakowska 110/114, Warsaw 02-256, Poland
oleksandr.zaporozhets@ilot.lukasiewicz.gov.pl

[2] National Aviation University, Liubomyra Huzara Ave., 1, Kyiv 03058, Ukraine

Abstract. EU SENECA project considers two basic design configurations for new European supersonic business jets with two engines and two airliners—the quad-engine and triple-engine airframe integrated platforms. Besides the necessary exploration and optimization for flight performances aircraft noise and emission are the subject of investigation. In assessing noise levels at ICAO certification points the sound propagation and engine installation effects contribute to sufficient value. Semi-empirical and ray acoustic modeling show that their contribution is determined by sound reflection effects from the terrain surface along the propagation path to the microphones at all noise certification points for all acoustic sources in consideration, reflection from the wing surface for the fan inlet noise under the flight path and twin-engine effect of the jet noise for the sideline certification point. The twin-engine effect for the triple-engine airplane designs may reach 3 EPNdB, and for the quad-engine designs—5 EPNdB of the jet noise reduction. The shielding effect by the wing may contribute to fan noise reduction and further increase overall noise, but only if the engine nacelle is installed over the wing surface.

Keywords: Aircraft · Reference Model · Noise · Fan · Jet · Installation · Shielding · Reflection

1 Introduction

Multidisciplinary design analysis and optimization plays a key role in the conceptual design studies of novel aircraft configurations [1]. In order to design within the EU project "Landing and Take-Off (LTO) noiSe and EmissioNs of supErsoniC Aircraft" (EU SENECA) the baseline supersonic transport aircraft (SST) configurations were used for PAX airliners of Mach number 1.8 (Ma1.8) and Mach 2.2 airframe-engine integrated platforms (Fig. 1). For their exploration and optimization several steps have been followed and different tools have been implemented [2]. A purpose

I. Ostroumov et al. (Eds.): ACASD 2025, LNNS 1418, pp. 189–210, 2025.
https://doi.org/10.1007/978-3-031-91992-3_13

of the current research is to define during the design process the possible contributions of sound propagation and engine installation effects in certification noise level at the designing stages "Fuselage-Wing-Engines" configuration and mostly for the stage of aircraft/engine configuration.

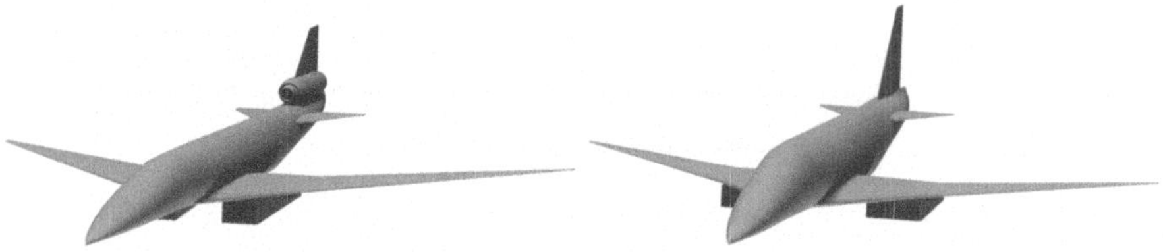

a) *b)*

Fig. 1. Perspective view of the supersonic airliner platform: a) tri-jet configuration; b) quad-jet configuration.

There are three basic components of aircraft noise (AN) phenomenon should be considered in assessment of the sound level at point of noise control: AN generation in source, sound propagation and installation effects [3]. Every of them are the complex phenomena including the number of contributions [4]. Overall AN generation in source includes the contribution of the dominant acoustic sources produced by engines (jet, fan, turbine and combustion chamber acoustic sources) and airframe (wing, landing gear, flaps and slats acoustic sources) in flight—their contribution to Effective Perceived Noise Level (*EPNL*) calculations at aircraft noise control points defined by ICAO requirements [5].The real aircraft designs provide several obstacles for the sound rays linking the engine and airframe noise sources with the point of noise control, their shielding, reflection and transmission of these rays depend on how the sources are located or installed in aircraft design—producing the AN installation effect which is a character for every aircraft design. Sound propagation effects usually include sound wave divergence from the source, air absorption and refraction, reflection and shielding by ground surfaces.

The ICAO requirements [5] are simplified a little bit in consideration of the propagation effects because of justification of the atmosphere as International Standard Atmosphere (ISA) with zero wind and temperature at sea level of 25 °C decreasing with altitude at an ISA rate. ICAO requirements for AN calculation [6] simplify them even more considering the atmosphere conditions without any refraction effect for the sound waves. If the task of the modelling is to define the *EPNL* values by ICAO requirements and to compare them with the ICAO noise limits the accuracy of calculations will be a subject of interest because of their importance in final decision about the modelled aircraft configuration.

2 Requirements to AN Assessment Model

The AN calculation tool NoiTra was designed for noise abatement procedures (NAP) assessment in various flight conditions including the defined by ICAO [5] and described in general in [7–9]. In particular, the predicted by NoiTra and measured noise levels

EPNL for aircraft being in operation that time were compared showing the necessary accuracy for their further usage—all their calculated values were found inside the measured intervals $EPNL_{meas} \pm \sigma$, where both the *EPNL* and σ were defined in certification flight noise tests [3]. Later improvements necessary for comparison of the calculated results with noise monitoring data [10] provided the quite high accuracy for spectral and overall *SPL* assessment ($\sigma_{\Sigma} = \pm 1..2$ dB) of the aircraft in any possible flight event but only in case if all the necessary for calculation conditions were available [10]. The combined standard uncertainty of the AN measurements by unattended monitoring systems with confidence level of 95% is ~1.3 dB. It can be raised up on 0.5 dB when simple detection techniques are used and can be reduced to approximately 0.1 dB when radar tracking is used to enhance the identification task, the calculation uncertainty is looking the same as by current measurement technique [9].

The gap between the calculated and measured values (uncertainty) was covered and further reduced by inclusion of the transfer function defined by solving the identification task to derive the basic overall acoustic model of an aircraft [3, 8]. In this previous approach the transfer function covered the installation effect. Further NoiTra improvements with BELTASS and TRANASS modules [3] and inclusion of INSTALL module (described elsewhere [9, 10]) the transfer function became so small that the acoustic models of some aircraft types do not use the transfer function at all.

Big advantage of the NoiTra calculation tool is its unification with IsoBella model used for the calculation of airport noise exposure scenario in accordance with [6]. In case of the correct implementation of the input data the results by NoiTra and IsoBella are of very close accuracy—as it was shown even by some measurement surveys made in Ukrainian airports [11]. The results of measurements are required for proving aircraft noise contour calculations, especially they are used for noise zoning around the airports. The ADS-B turners used during the measurements for aircraft flight path supervising [12] and further usage of their data in comparison assessment of modelled and measured sound levels. Current NoiTra tool includes the module NoiRad to calculate the Noise-Power-Distance (NPD) data for any type of the aircraft under consideration due to requirements of the ICAO [6] to be used further by IsoBella for airport noise scenario assessment. Together with flight path data from ADS-B the newly or re-calculated NPD data provides closer *SEL* or *EPNL* or any other sound level/index calculations to measured ones in survey.

The presented currently technical discussions [13, 14] concerning the definition of Landing and Take-off (LTO) AN certification limits for novel SST still are showing the same value of accuracy of around ±2.0 EPNdB on the cumulative *EPNL*. The findings were based on flight noise single-event data [15] and the same approach for quantification of the uncertainty of AN assessment as used by studies for NoiTra modelling tool.

Aircraft Design Space (ADS) was developed to realize complex analysis and multidisciplinary optimization (MDO) during the aircraft design, maintenance and operation of the environmental factors including the AN assessment by means of NoiTra tool. The ADS module determines flight parameters (Fig. 2) by solving aircraft flight dynamics eqs. [3, 16] using convenient interface usually dividing flight path on several segments. Flight segments refer to specific portions of the aircraft flight, each with its own unique set of operation procedures. Currently, the ISA conditions are considered but with availability of the necessary data for other meteorological conditions calculations are possible [16]. Because of the highest priority of fuel burn inside the set of environmental factors considered in MDO of new aircraft design a system of aircraft flight motion [3, 7, 9] is widened by inclusion an equation aircraft flight mass change due fuel consumption [16], which is necessary for correct analysis of the aircraft flight profiles due to their TLAR.

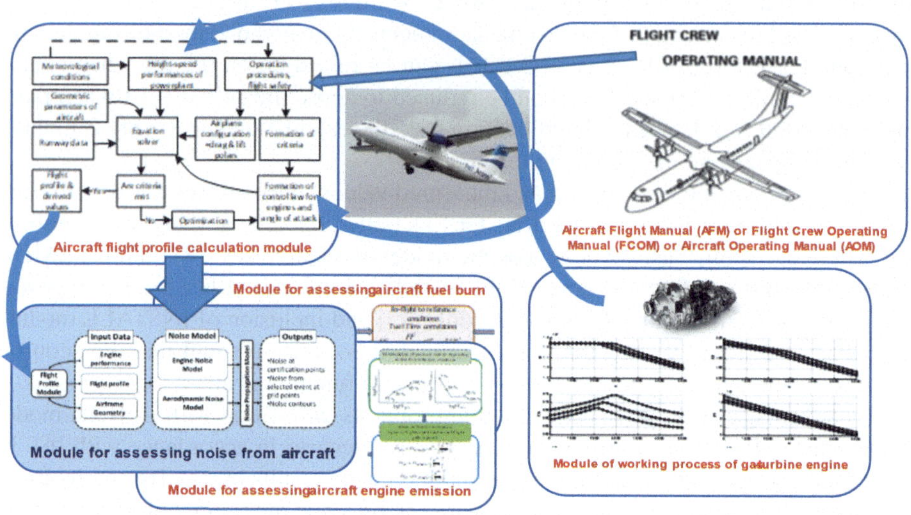

Fig. 2. Modular approach to evaluation of fuel burn, engine emission and noise metrics.

This ADS platform is used for the analysis of the reference aircraft also. For every flight segments their specific input, output and limitation data are predefined [16]. Optimization method is included defining flight control parameters within flight safety constraints listed in the flight crew operation manual of the reference aircraft. For the trajectory segments close to noise control points there are following parameters estimated for noise level assessment: distance and thrust for estimating radiated sound power, and pitch of the flight, which is necessary for correct directivity calculation of sound propagation.

3 Reference Aircraft Model

Reference aircraft model is used as a base platform for designing new aircraft—baseline model. It is used also for the comparison with new baseline aircraft performances in operation and maintenance to show the achieved advantages and attraction among others in fleet. For better understanding of the comparison between the baseline and reference models the clear difference should be defined between the specific acoustic sources in their aircraft/engine design.

Correct implementation of the fan and jet noise models which are different from the low bypass engines or even turbojets is necessary in distinguishing the better accuracy results during the aircraft LTO noise modelling. For these purposes under the umbrella of SENECA project several benchmarking comparisons were done by partners in assessment of jet and fan noise contributions in two principal steps. At first step, there were four test cases selected from the NASA coaxial jet modelling research [17] and considered in benchmarking the calculation tools involved in SENECA for supersonic aircraft LTO noise assessment. Last Case 4 [17] included even the efficient NRT for the jet noise reduction [18]—external plug insertion in jet stream which was further investigated in NLR wind tunnel additionally and also calculated by SENECA partners. The predicted spectra show a good agreement with the corresponding published [18] and measured results [19] with relatively small variations between the implementations from different SENECA partners—all the tools were considered as validated, which is an important criterion for the envisaged prediction of the noise certification levels with and without the NRT. For the NoiTra purposes the Case 5 was introduced to consider jets exhausting by circular nozzles [20] which are the subject for turbojets and bypass engines with mixed exhaust nozzles.

Similar approach has been used for calculations of the fan noise—second dominant acoustic source of latest SST technology. Current NoiTra fan model is fully correspondent with Heidmann methodology [21] improved for high bypass ratio engines [22]. Predictions were made for representative engine for which detailed noise measurements were available—the NASA Energy Efficient Engine (E3) program: bypass ratio 7.7, exhaust type—mixed, fan pressure ratio 1.4, core jet velocity 270 m/s. Calculated spectra by Heidmann and Heidmann/Kantos enhanced models are shown in Fig. 3 in comparison with measured spectrum for the E3 engine. Enhanced model (Fig. 3) is closer to measured spectrum on few decibels inside the frequency bands where the fan noise contribution is dominant for the high bypass engine noise. The benchmark test of fan models involved in SENECA project is shown in Fig. 4 presenting the similar value of deviation between the fan spectra as it was achieved in the jet noise benchmark test [17].

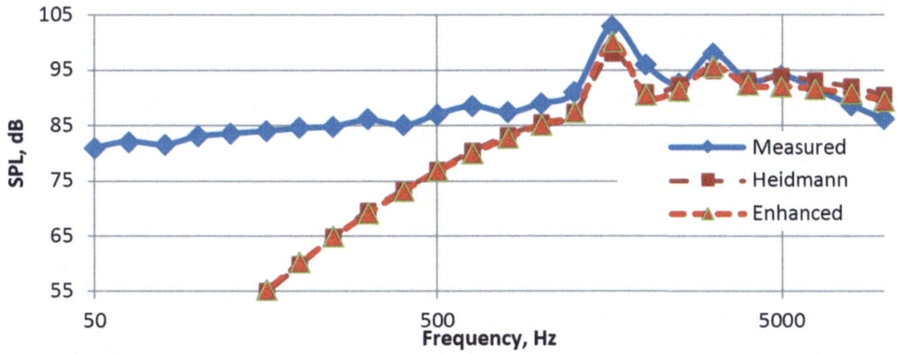

Fig. 3. Comparison of the calculated spectra by Heidmann and Heidmann/Kantos fan noise models with measured spectrum for the E3 engine at angle 40 to inlet axis.

The old and new SST designs have a relatively poor low-speed performance compared to similarly sized subsonic aircraft. The Concorde was not equipped with high-lift devices and its elevons were used for trim and control only, although they could be used

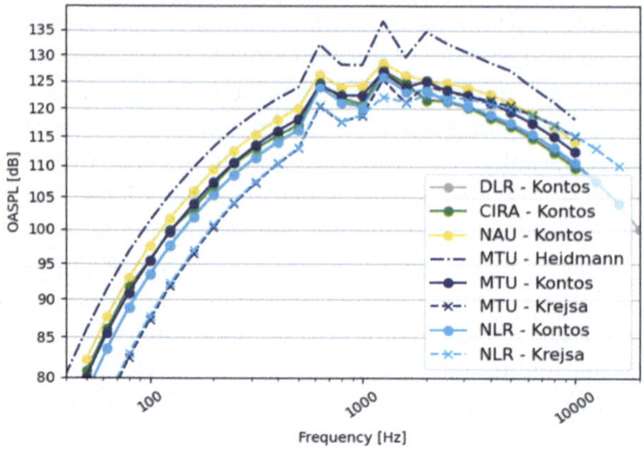

Fig. 4. Benchmark comparison between the fan spectra calculated by different SENECA tools.

to adjust the necessary aerodynamic properties of the wing during take-off and landing. Also, the delta wing noise is different from the usual wing noise of the similarly sized subsonic aircraft. NoiTra includes the complex model [23] to calculate the noise generated by wing, flaps and landing gear, taking in consideration the large dominance of engine noise over the airframe acoustic sources.

Complex model [23] considered the flight noise tests of the F-106B aircraft (flyovers with Mach number of 0.4 at altitude ~90 m [24]) to assess the specific delta wing noise spectral and directivity patterns which were extracted from overall aircraft noise by excluding the engine static noise at the same engine operation mode as it was used in flight tests (Fig. 5). Wing noise spectra calculated with (red lines) and without (blue lines) ground effect produced by sound reflections from the ground surface are shown in comparison with measured flight noise in good correspondence within the low/middle frequency bands where the wing *SPL* maximum is observed. The high-frequency portion was dominated by compressor noise from the main turbojet at flight-idle thrust. Several benchmark tests were defined from the data presented in the report [23] for the reliability of airframe noise calculations for the clean wing, flaps and landing gear noise sources to prove the correct usage of their component modelling algorithms within the FRAME module of the NoiTra tool.

For sound wave propagation effects the ground effect contribution is essential [3, 25]—it is confirmed by numerous aircraft noise measurements like shown contribution in Fig. 5. Because the outdoor surfaces are locally reacting on incident sound rays the acoustical properties of the reflected rays are modified in comparison with direct ray (reaching the microphone without the reflection from the ground surface) properties in accordance with acoustic impedance of the surface covering (represented completely by normal surface impedance Z, which is reciprocal normalised admittance β) [3, 25]. A widely used model [26] for the acoustical properties of outdoor surfaces involves a single parameter, the effective flow resistivity σ_e, to characterize the ground surface. According to this single-parameter model, the propagation constant normalized characteristic impedance Z is defined as recommended by the standard [27] and scientific knowledge [25] of the effect under consideration. A standard [27] provides a calculation test which

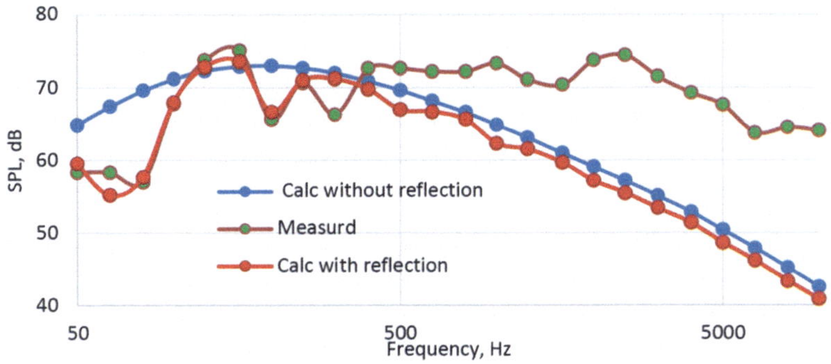

Fig. 5. Comparison of measured F-106B aircraft flight noise spectra with calculated by NoiTra tool for conditions with (red line) and without (blue line) ground reflections.

may be used as a benchmark test for comparison of the models involved in calculation of the ground effect in scenario under consideration.

Calculations of Concorde noise were performed to verify the improved noise source models included in NoiTra by comparison with measured data in US airports [28, 29, 30]. The measured noise spectra indicated that the Concorde generated more energy in the low frequency spectral bands than do long-range subsonic jet aircraft. During take-off/climbing and approach the noise abatement procedures were used. Engine power throttling is usual element of both, but during the climbing flight stage the throttling was used with various programmed power lapse rate (PLR) (Figs. 6, 7 and 8). At approach the effect of Concorde decelerating along the glideslope was assessed in comparison with stabilized flight. It should be accompanied with engine throttling, and it was found to be similar with noise exposure from long-range subsonic aircraft arrivals. The weather conditions during the Concorde noise measurements were fixed at every specific sites—they were different from the ICAO standard requirements for AN certification.

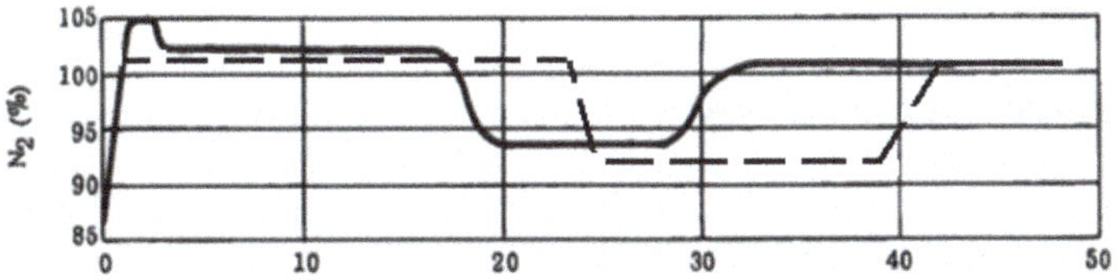

Fig. 6. Comparison between calculated from ANP-database and the observed in operation departure profiles when engine PLR defined by rotation frequency of the high-pressure compressor shaft N_2.

Maximum 1/3-octave band sound pressure level (*SPL*) spectrum data for close departure (~7 km from the brake release point) and distant sites (~11 km from the brake release point) are shown in Fig. 9, and their details in Table 1. Figure 9 displays the measured spectra for stabilized and decelerating Concorde approaches (a) and close location departure in comparison with calculated by NoiTra. For both flight stages the jet is a dominant source of noise defining the noise exposure close to 100% of the overall noise value.

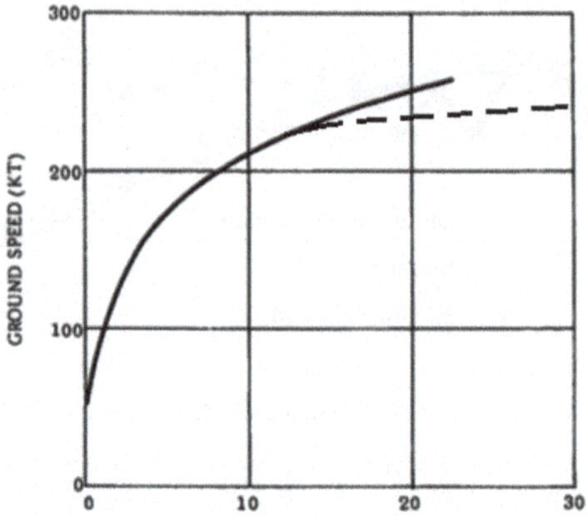

Fig. 7. Comparison between calculated from ANP-database and the observed in operation departure profiles when flight velocity defined as ground speed.

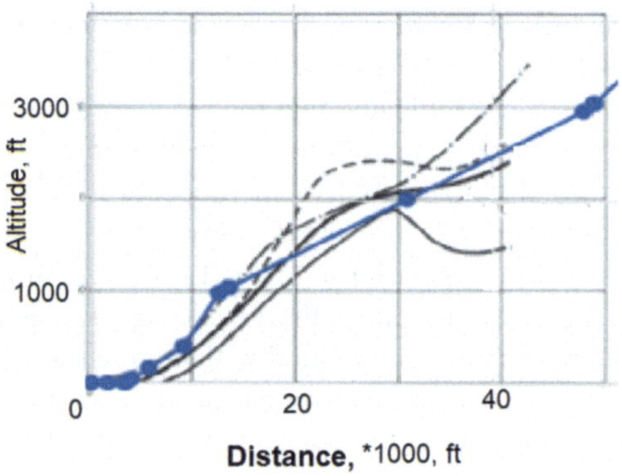

Fig. 8. Comparison between calculated from ANP-database and the observed in operation departure profiles when flight profile defined by altitude via distance dependence.

3.1 A Concorde Approach Noise

The approach flight is simpler for noise calculation due to the strictly defined glide slope (−3°) and stabilized flight parameters. Moreover, the INM database includes this approach flight path for the Concorde, and it was used for input flight performance data.

The input data for Olympus engine operational performance were calculated due these flight parameters and put in the NoiTra tool. Also, the Concorde airframe geometry (for the fuselage, wing and landing gear) was provided for airframe noise assessment. For the distance 1 nautical mile (1852 m) before the runway end the altitude if the flight path over the point of noise control No 3 H = 114 m, the stable flight with velocity 85 m/s is provided by engine thrust ~4000 kg per each engine. For these flight and engine operational performances the noise was calculated with $EPNL = 117.54$ EPNdB, which is much close to 116.6 EPNdB fixed by flight testing/measurements in airport (Table 2). For the dominant noise sources of the Concorde alone the values of $EPNL$ [EPNdB] are the following: $EPNL_{jet} = 116.53$; $EPNL_{fan} = 109.22$; $EPNL_{airframe}$

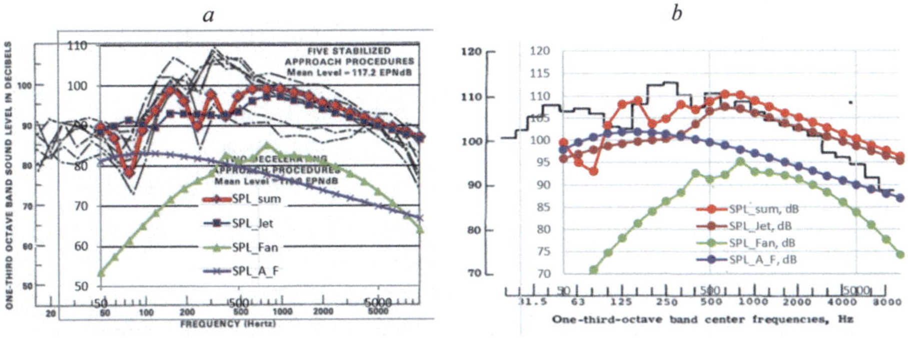

Fig. 9. The Concorde approach (a) and departure (b) noise spectra at time of maximum perceived noise level: black line - measured *SPL* (jet, fan, airframe acoustic source contributions are displayed by their spectra without ground effect, it is included in overall noise spectra).

Table 1. Comparison between Concorde and Boeing-707 acoustical performances at departure closein and distant sites

Site	Airplane type	*OASPL*, dB	*EPNL*, dB	*SPL* (20 Hz), dB	*SPL* (<125 Hz), dB	*SPL*3333333 (<500 Hz), dB
Close	Concorde	119.5	121.7	100.5	113.8	119.2
Close	Boeing-707	104.0	110.1	77.5	97.9	101.8
Distant	Concorde	98.5	101.2	74.8	97.4	98.5
Distant	Boeing-707	87.6	92.6	61.3	87.1	87.8

= 100.09; $EPNL_{\text{turbine}} = 76.55$; $EPNL_{\text{com chamber}} = 88.68$. The dominance of the jet noise is confirmed by the values of the spectral indexes of agreement, which definitions are fully explained in [3]: if $d_2 = 1$ the resulting prediction model is reliable, accurate and consistent with the observed spectral data. It was also proved that for evaluation of the exposure indexes/levels like *EPNL* and *SEL* the indexes of agreement must be highest in time of the $PNL_{\max}$ and $L_{A\max}$ accordingly.

From the Table 2 ground effect (GE) contribution to sound levels is close to 3.4 dB—much bigger than contributions from other acoustic sources character for the Concorde but not dominant. If the jet noise will be included in the Concorde noise alone the *EPNL* at approach will be on 1 EPNdB less only. Its index of agreement d_2 at point of $PNL_{\max}$ will be on 4% less accordingly. If in this case (jet noise alone) the GE will be excluded the *EPNL* will be on 3.5 dB and d_2 17% less accordingly. This analysis emphasizes the importance of the GE contribution in AN exposure assessment.

For the Concorde noise calculations the conditions of normal uncompacted (forest floor, pasture field) and compacted ground (compacted lawns, park area) surface with flow resistivity $\sigma = 200$–500 kPa s m^{-2} were considered predominantly because the conditions of surface covering in September are more appropriate with these values. To look

Table 2. Assessment of the contribution of the main acoustic sources in overall noise from Concorde at approach.

Acoustic sources	*EPNL*, EPNdB	PNL_{max}, PNdB	*OASPL*, dB	Index of agreement d_1	Index of agreement d_2
All sources included	117.54	121.95	109.81	0.9863	0.8306
Jet, fan & Airframe	117.53	121.93	109.79	0.9862	0.8290
Jet & fan	117.36	121.82	109.63	0.9860	0.8200
Jet with airframe	116.76	121.51	109.35	0.9857	0.8099
Jet with GE	116.53	121.40	107.03	0.9854	0.7995
Jet without GE	114.06	118.45	104.10	0.9792	0.6634

on any other possibilities of GE influence on noise exposure assessment—the extreme cases of compacted dense ground surface like for parking lot ($\sigma = 2000$ kPa s m^{-2}) and uncompacted loose ground ($\sigma = 80$ kPa s m^{-2}) were assessed (Fig. 10) also. GE is dependent from surface covering and from the incidence angles for sound waves—direct to microphone and reflected from the surface. Because in calculations the flight path was considered from ANP data base which is different from the flight paths realized in measurements these angles are different from calculation case and may define an additional inaccuracy of calculation results. The difference of the calculated *EPNL* for this set of flow resistivities is about 2 dB, *EPNL* = 116.42…118.23 EPNdB between the extreme values of resistivity in consideration $\sigma = 80$–2500 kPa s m^{-2}.

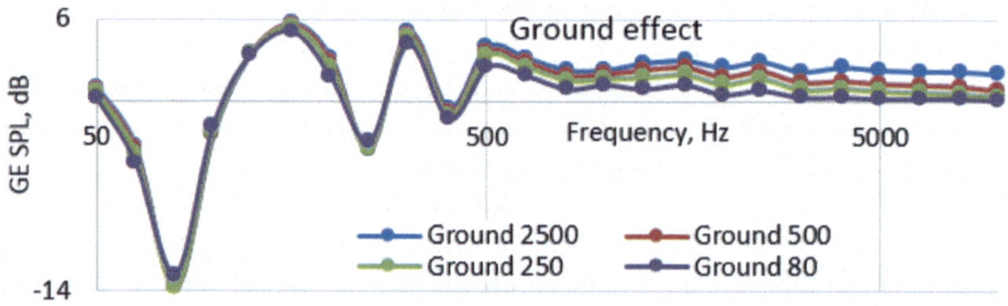

Fig. 10. Ground effect in dependence with surface flow resistivity σ during Concorde flight noise measurements in US airports.

Also the location of the aircraft at point of the PNL_{max} was analysed to show how the source location (Fig. 11) and angle of sound wave incidence may influence the GE taking in mind that flight deceleration was used during the approach for reducing the noise level at point of AN control. The location of the aircraft at point of the PNL_{max} should be different from the stable flight along the glideslope and to show such influence of aircraft location two points before and after t_{max} (when the PNL_{max} is produced) were used for the analysis. The interference dimples in GE spectra in Fig. 11 are less for the aircraft locations aside the t_{max} and shifted to higher frequency bands after and lower frequency bands before t_{max}. The calculated Concorde spectra (Fig. 12) at these points are identical with measured for at approach (Fig. 9a) confirming the chosen methodology. The calculated spectra for Concorde are compared with engine jet noise (red line in Fig. 12) for which the installation effect is absent.

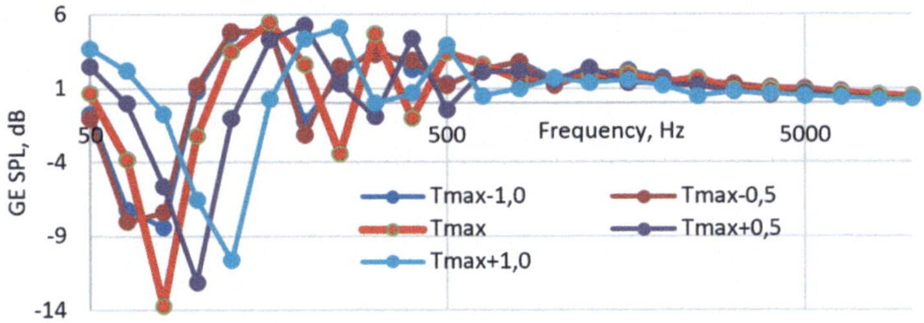

Fig. 11. GE effect in dependence with aircraft location on glideslope from the point of PNL_{max} during Concorde approach noise measurements.

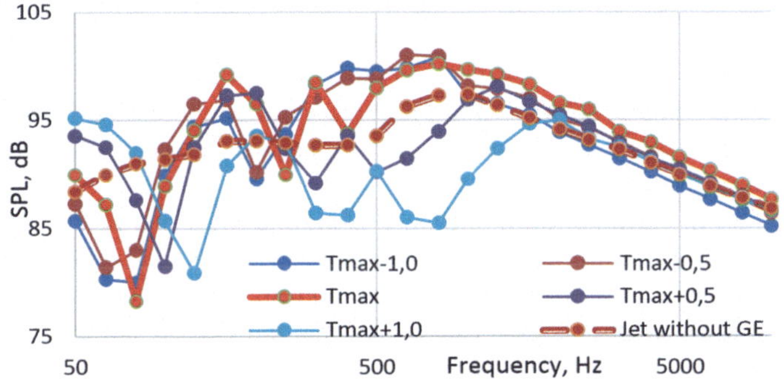

Fig. 12. The Concorde approach spectra due to their GE effects shown in Fig. 10 in dependence surface flow resistivity σ: red line—calculated spectra for t_{max}; calculated *SPL* are compared with jet noise spectrum in dash line.

New SST baseline models are designed and investigated without a base model for all of them, but a Concorde aircraft is proposed to be considered as a reference model for baseline supersonic airliners. Reasons for that were the similarity of their TLARs, including their dimensions and weights, number of engines in power plants, mission profiles, cruise Mach numbers, etc.

The Concorde was not certified for noise due to US FAR or ICAO Annex 16 requirements [5], but its take-off/climbing and approach/landing noise levels were accurately assessed during the environmental impact assessment studies made in Dallas airport in USA and Heathrow airport in UK [29–31]. Noise exposure, flight path and complaints data gathered at both airports during the first year of scheduled Concorde operations are presented in the reports and some comparisons made with noise levels from other aircraft in operation that time.

3.2 A Concorde Departure Noise

During the departure two points of AN control used by ICAO standard for noise certification purposes: sideline (450 m aside the flight/runway axis) point at the end of take-off of the aircraft along the runway (ICAO AN point No 1) and a point on flight/runway axis under the climbing flight stage on distance 6.5 km aircraft run start (ICAO AN point No 2). Due to current ICAO requirements operational mode of the engines is maximum at point No 1 and with PLR providing the climbing gradient 4% (and possibility of one engine inoperative) for the aircraft flight height above runway level 210 (aeroplanes with four engines or more) and speed not less $V_2 + 19$ km/hours [5].

Standard operating procedures and configurations were normally used during the flights in operation. If the requirements exist to manage the noise exposure less in comparison with normal the noise abatement procedures (NAP) are recommended for operation. The normal departure NAP for the Concorde involves climbing at full throttle to 230 m (750 ft) altitude, at which point two engines are throttled back out of reheat. One second later, reheat power on the other two engines is cut. Two seconds after that, slow throttling is initiated to reach cutback power after 5 s. Thus, a total power reducing period of 8 s is scheduled. Secondly, the nozzle area control schedule for Flight 80 [28] was set to give a lower nozzle area than the production aircraft. The production nozzle area will give noise levels about 1.5 PNdB quieter. Most of the measurements of Concorde departure noise like shown in Fig. 9b are done with usage of the PLR during the climbing (Fig. 6) and their flight performance are very close to ICAO requirements for testing flight over the point No 2 [5].

Ground effect in certification point No 2 is similar with one in point No 3 during the approach (Figs. 9, 10). In ICAO point No 1 the GE effect is higher due to sideline distance from flight axis 450 m for current requirements and 650 m in time of the Concorde operation (Chapter 2 [5]). Because of that the grazing angle of the direct and reflecting sound waves are very small providing the bigger GE values. For the point No 1 so called twin-jet effect contributes to noise exposure (Fig. 13). For both departure points the reflection of the engine inlet noise (defined by compressor or fan acoustic sources) by wing surface may contribute to noise exposure at microphone location. The main influence on the effect is provided by engine nacelle location in airplane design: under the wing, over the wing, at tailplane, or combination of these main three cases (Fig. 13).

The most desirable result of the installation effect is a shielding of sound ray and expected sound level reduction at the point of noise control accordingly. The sound reflections from wing and fuselage may enhance the overall noise exposure (Fig. 14). Also sound wave transmission effect may contribute and being included in noise installation effect—for example in case of jet noise radiation and propagation to microphone located at sideline point of noise control (Fig. 13).

The jet noise shielding (sound refraction during its transmission through jet close to microphone location, Fig. 13), because of the high flow temperature and speed with respect to the immediate surroundings, acts as a partial barrier between the source and

Fig. 13. Jet-by-jet shielding—twin-jet effect—at noise control point at aircraft take-off (ICAO sideline point No 1).

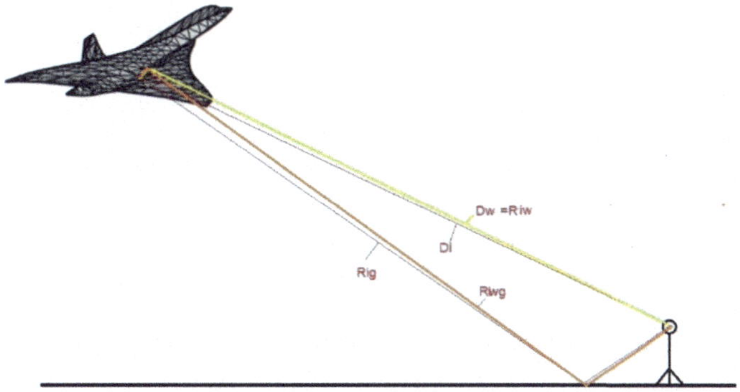

Fig. 14. Geometry of the numerous reflection effects of the sound rays from the aircraft engine intake in their propagation along the ground surface to microphone.

the receiver. This effect is known as a twin-jet shielding: the jet noise reduction achieved by shielding one jet exhaust flow with a second, closely spaced, identical jet [32]. Due to design specifications of the new SST (Fig. 1a, b) this effect may contribute to noise level assessment at sideline point No 1 [5].

The developed analytical model for the twin jet effect consists of the sound field emitted from a stationary, discrete frequency point source, which impinges on a cylinder of locally parallel flow as shown in Fig. 13. The flow temperature and velocity profiles are assumed uniform in the jet at any location downstream of the nozzle. To calculate the refraction effect, the engine jet can be treated as a homogeneous moving layer [3]. The transmission of sound through the transmitting jet is controlled by the flow and temperature discontinuity. Under certain conditions, these discontinuities partially or totally reflect (shield) sound since the transmitting jet has a different impedance compared to the surrounding ambient air.

Finally, the sound can be diffracted by the discontinuity at the jet interface or scattered by the turbulent flow regime of the shielding jet. The difference between the *SPL* measured by an observer in the presence and in the absence of the fluid shield defines the noise reduction due to the phenomena of reflection, refraction, diffraction and scattering. This twin jet effect (ΔSPL) is defined as the *SPL* for the twin jets (SPL_{Twin}) minus the

SPL of a single jet at the same conditions (SPL_{Single}) plus 3 dB as following:

$$\Delta SPL = SPL_{\text{Twin}} - \left(SPL_{\text{Single}} + 3\right). \quad (1)$$

Acoustic experiments were performed on twin round and rectangular jets to investigate the impact of nozzle spacing, jet temperature and Mach number, flight speed, and nozzle aspect ratio [32]. The twin round jet data from this test were used to create an empirical model that accounts for twin jet variations over frequency, observer angles, jet spacing, jet static temperature ratio, and flight Mach number beyond the simplistic doubling of the sound levels:

$$\Delta EPNL = EPNL_{\text{Twin}} - EPNL_{\text{Single}} + 3EPNdB. \quad (2)$$

where $\Delta EPNL$ is changing inside the data interval [2., −2.5] with repeatability of ±0.3 *EPNdB*—95% confidence interval.

The formula (2) is like formula (1) but only to show that the model [32] was proposed for equivalent and effective noise levels also.

The noise calculations for the Concorde at departure (ICAO point No 2) with and without the reflection by wing of fan noise on engine inlet is shown in Table 3 and Fig. 15. Departure flight over the control point No 2 (similar for arrival/approach flight over the point No 3) with microphone height 1.2 m over the ground surface with flow resistivity for the covering character for ICAO conditions $\sigma = 300$ Pa·s/m^2. Reflection from the wing surface was calculated for the $\sigma = 20{,}000$ Pa·s/m^2. The theoretically expected ground/wing reflection effect for the compressor tonal noise is assessed ~8 dBA by the semi-empirical model implemented in NoiTra calculation tool.

Table 3. Concorde noise level *EPNL* at departure point No 2.

Noise source	Without wing reflection		With wing reflection		Difference	
	ΔPNLT	*ΔEPNL*	*ΔPNLT*	*ΔEPNL*	*ΔPNLT*	*ΔEPNL*
Aircraft overall	124.1	119.5	124.2	119.7	0.15	0.15
Jet	120.9	116.1	120.9	116.1	0	0
Fan	112.0	107.4	113.4	108.8	1.4	1.43
Turbine	86.45	76.09	86.45	76.09	0	0
Combustion	94.79	88.36	94.79	88.36	0	0
Airframe	103.36	99.73	103.36	99.73	0	0

These calculated data show the difference between the compressor inlet noise contributions with and without its reflection by wing surface of 1.4 dB. This engine inlet noise exposure difference provides the quite small increase of the overall Concorde noise level on 0.15 dB—because of the dominant contribution of the jet noise in AN noise exposure of the Concorde.

Sound level calculation at the microphone location provides the following ground and installation effects (Fig. 16) for the calculation scheme in Fig. 14 and used in calculations for *PNL* in Fig. 15:

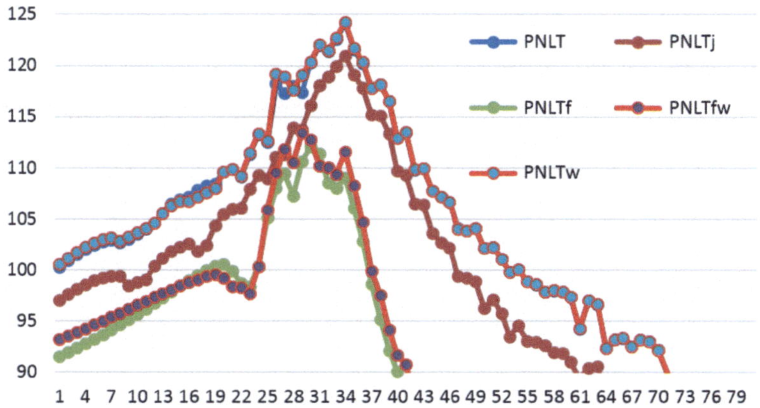

Fig. 15. Concorde noise at ICAO departure point No 2 including engine inlet noise with (PNLTfw—red line) and without (PNLTf—green line) reflection by wing surface: PNLT—overall aircraft noise PNLT (blue line), PNLTm—overall aircraft noise PNLT including the reflection by wing surface of the compressor inlet noise (red line); PNLTj—jet noise PNLT.

1. the direct ray *Di* and the reflected ray *Rig* form an interference model of the composition of direct and reflected waves from the ground surface—clean ground effect DelGr is shown in Fig. 16a;
2. the direct beam Di and the reflected beam *Riw (Dw)* form an interference model of the composition of direct and reflected waves from the surface of the wing—clean wing reflection effect DelWin, Fig. 16b;
3. the effect of summing up the results of the interference of reflected waves from the ground surface and from the wing *Riw (Dw)*—probably there is no interference between them, the waves simply add up (i.e. 3 components of sound waves)—DelWGS, Fig. 16c;
4. the effect of summing the results of the interference of reflected waves from the ground surface, from the wing and from the wing and then from the ground surface (i.e. 3 components of sound waves)—DelGWG, Fig. 16d;
5. comparison of all the effects between each other, Fig. 16e.

A simple theoretical model for surface reflection and edge diffraction, using ray theory, was introduced and compared with experimental results [3, 31]. For acoustically hard surfaces such as paved covering of the ground surface or of the aircraft wing and/or fuselage the interference of direct and reflected rays (sound reflection effect) generally results in a higher sound level in comparison with single direct sound wave D_i at the point of noise control due to the contributions of reflected rays from the airframe R_{iw}, which is direct sound ray D_w from the airplane to the microphone, and ground surface reflected rays from the source R_{ig} and the double reflected ray from the wing and the ground surface, as shown in Fig. 14.

Theoretically the combined reflection of the sound wave from the wing or fuselage and ground surfaces R_{wg} may be expected in case of the engine installation under the wing and ground effects. Numerical Ray Acoustics results for the same calculation scheme for the Concorde are shown in Figs. 17 and 18. It was assessed theoretically

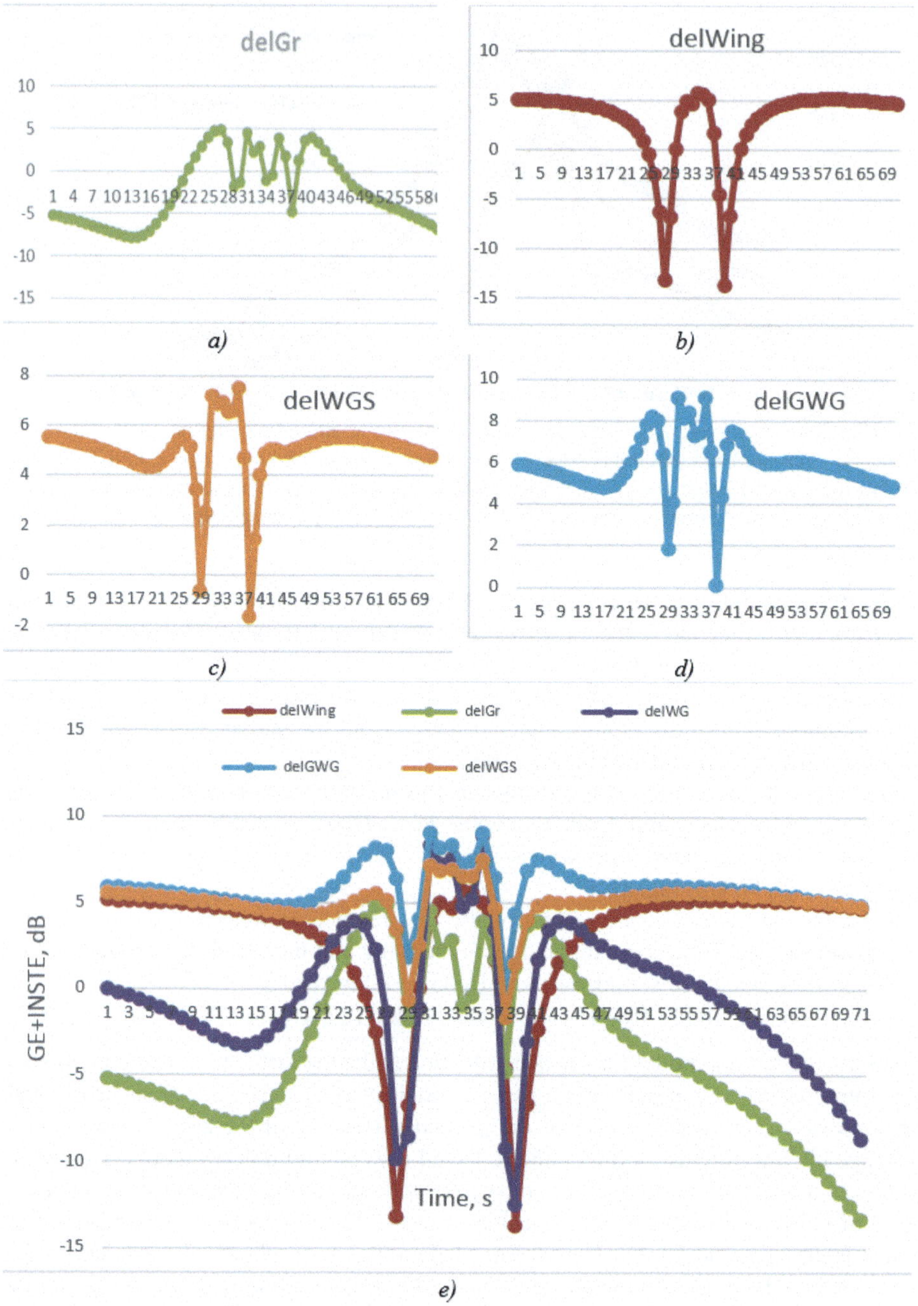

Fig. 16. Comparison of all the sound reflection effects—from the ground surface (GE) and wing (INSTE) surface—contributing to the noise receiver point shown in Fig. 13.

for the acoustic point (in a centre of engine inlet) under the wing surface and evidently for real nacelle entrance design is absent or quite small in comparison with calculated. For the Concorde due to the strict model used in calculations even the effect of sound shielding was found behind the aircraft flight provided for the sound waves radiated from the engine inlet (negative *SPL* values in Fig. 17b and blue coloured area in Fig. 18 depicting the sound level reduction).

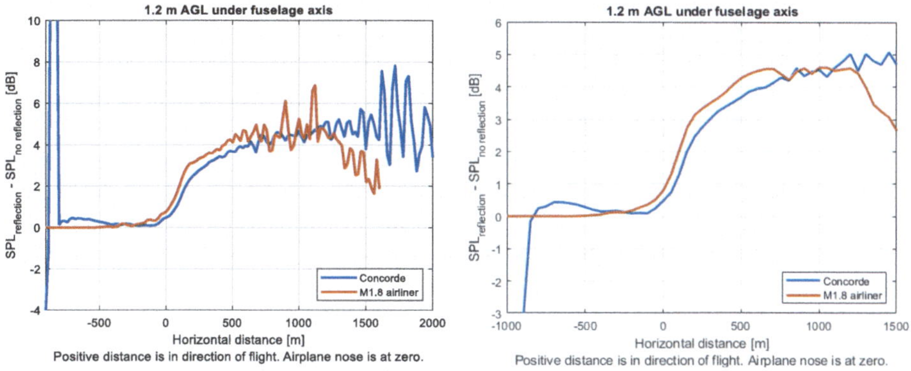

a) *b)*

Fig. 17. Difference between *SPL* for case with reflection from airplane wing (**a**) and case without reflection from airplane wing (**b**) on projection of fuselage axis at 1.2 m surface above ground level. Engine compressor noise is radiating at 1 kHz from the engine/nacelle inlet.

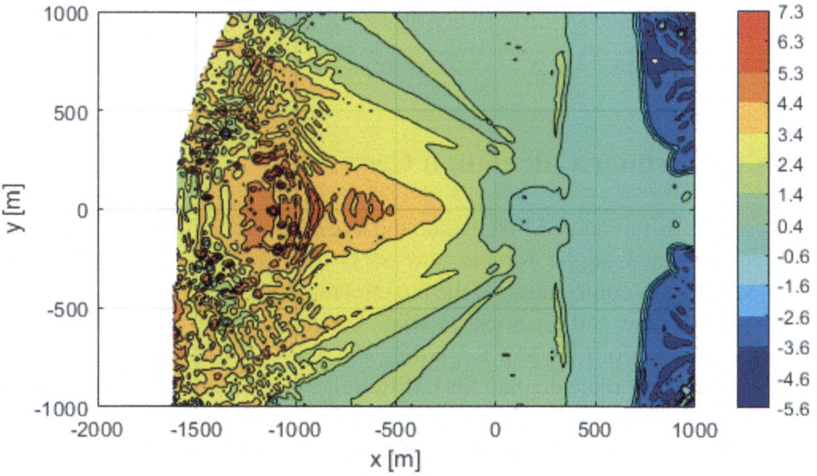

Fig. 18. Ray Acoustics simulation of the flight geometry scheme shown in Fig. 14 but with opposite flight direction: flight axis is along the centre line (y = 0 m), orange and red areas display the contribution of wing reflected waves of about 4–7 dB similar with values on centre line shown in Fig. 16a; blue area displays the shielding effect by wing edge of the engine inlet noise.

Only ground reflection effect in Fig. 17a is equal by value to the ground effect by semi-empirical model (delGr). The combined reflection effect—from the ground and the aeroplane wing—is about 7.3 dB and equal to the combined reflection effect of about 8 dB shown in Fig. 16d (delGWG). Large curve oscillations in Fig. 17a are caused by decrease in accuracy of Ray Acoustics simulation. This decrease in accuracy is caused by limited number of rays interacting with elements of the evaluated wing surface (Fig. 18).

These simplified semi-empirical and ray acoustic approaches show that engine inlet noise—compressor for jet engines or fan inlet noise for the by-pass engine—may be increased by wing reflections and contribute more in overall noise level at point of noise control.

Concorde were the first SST in aviation sector. The Concorde was investigated for noise exposure in detail in London/Heathrow [30] and Washington/Dallas [28, 29] airports. The measured values for the Concorde noise in US airports in accordance with Chapter 1 of the FAR-36 for the comparison with calculated values are shown in Table 4. The calculated data are quite close to the measured data in usual flight operation in conditions quite similar with US Federal Acquisition Regulation (FAR) aircraft noise certification requirements (also as with ICAO [5]).

Table 4. Concorde calculated via measured noise levels at ICAO/FAR points.

Aircraft	*EPNL*, EPNdB		
	Take-off	Departure	Arrival
Concorde measured	113.1	119.3	116.6
Concorde calculated	117.7 112.7–114.7 (including the twin jet affect)	119.5 (max thrust) 115 (cut-off thrust)	116.3

4 Mach 1.8 Airliner Calculation Case

The values of the differences ΔPNLTmax and Δ*EPNL* in the Table 3 defined for the Concorde are expected bigger for the new SST—Ma1.8 airliner (with 4 engines in design)—due to bigger contribution of the fan inlet noise in comparison with compressor noise in Concorde case and jet noise in Ma1.8 airliner to the overall sound level at the point of noise control [33, 34]. Regarding the Ma1.8 climbing noise at the ICAO certification point No 2 the calculate level difference (with sound reflection by aircraft wing minus without sound reflection by aircraft wing) when airliner is flying along the baseline take-off profile is shown in Fig. 19. The nacelles were removed from calculation and the sound reflection by them was not taken into account. The fan inlet noise sources were placed at the coordinates of the entrance to the nacelles. The specific flow resistance of 300 kPa·s/m^2 (grass) with a mirror reflection without surface roughness was used for the ground surface. Maximal calculated difference in Fig. 19 between *SPL*s with and without sound reflection from airplane wing for the airliner in climbing flight was found ~5 dB—quite similar with calculated effect at approach noise point No 2 in Fig. 17b for ray acoustics model and Fig. 16a for the semi-empirical model. This sound reflection effect should be expected for the fan inlet noise only. Its contribution to overall aircraft noise is expected on ~0.2 EPNdB rise for the Concorde at both certification points and a little bit higher ~0.5 EPNdB for the Ma 1.8 airliner.

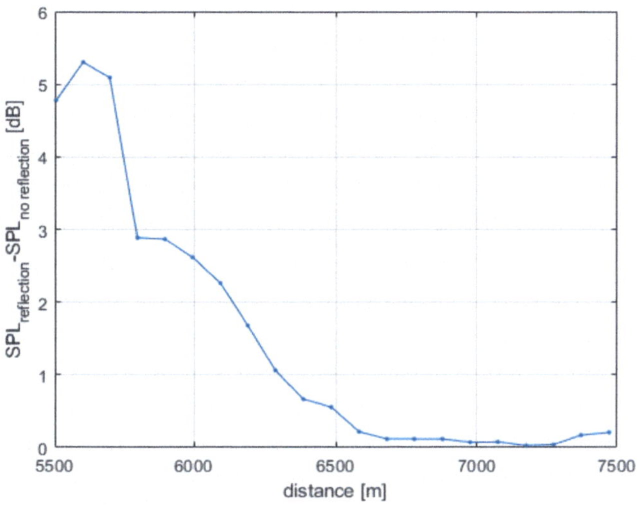

Fig. 19. Difference between *SPL* for case with sound reflection from airplane wing and case without reflection from airplane wing on projection of fuselage axis at 1.2 m surface above ground level. Engine fans are radiating at 1 kHz from the engine/nacelle inlet.

Table 5. Ma 1.8 airliner calculated noise levels by NoiTra at ICAO certification points.

Status of the data	Noise Level, EPNdB		
	Take-off No1	Climbing No2	Approach No3
Certification requirements	99	100	102.57
Minimum calculated	105.43	102.49	102.47
Maximum calculated	107.00	105.55	103.23
Acoustic liners for the fan noise			
100% eff	91.7	90.5	94.97
50% eff			95.34
Engine installation effect included	88.3 (twin-jet effect)	91.0 (fan inlet noise reflected by win)	95.5/96.0 (fan inlet noise reflected by win)

In accordance with the preliminary data of the EU SENECA project the expected values of *EPNL* at ICAO certification points for Ma 1.8 quad-engine airliner with and without noise reduction treatments are shown in Table 5. Also, the noise data for approach flight data set were calculated was calculated with contribution of the landing gear noise (the landing gear geometry was defined similar with Concorde) and without it.

5 Conclusions

An aircraft noise model of Concorde was designed and proved as a reference model for new supersonic aeroplanes in design. For new aircraft designing the noise certification levels should be assessed and compared with ICAO limits [5]. Correct models for the dominant acoustic sources must be proved in calculation scheme but the noise installation effect and sound propagation effects contribute essentially.

Engine installation and ground effects were analysed from the point of view of their contribution to the certification noise levels of the reference and new supersonic aircraft designs, which were the subject of investigation in the EU project SENECA. The most important components of the overall installation effects are the following: shielding effect by wing if the engine acoustic sources are located (for example the engine is installed over the wing) over the wing surface; reflection effect by wing if the engine acoustic sources are located under the wing surface; shielding effect by fuselage if the engine acoustic sources are located over the fuselage surface for the points of noise control under the flight path or in tail of the fuselage if the point of noise control is a sideline point; reflection effect by wing if the engine acoustic sources are installed under the fuselage surface; twin jet effect of noise reduction at the sideline point of noise control.

The platforms for the supersonic airliners with quad engine design still use the installation of the engines under the wing—their noise installation effect will be determined by reflection effect from the wing surface for the fan inlet noise. Preliminary assessment of the installation effect provides the possible increase of the calculated noise level *EPNL* at points No 2 and 3 up to 3 EPNdB. For the sideline point No1 the twin jet effect is possible to contribute with expected efficiency of 3–5 EPNdB reduction.

Acknowledgments. Acknowledgement to EC Horizon SENECA project № 101006742 "((LTO) noiSe and EmissioNs of supErsoniC Aircraft)".

References

1. Munoz, C., Bonavolonta, G., Lawson, C., Riaz, A.: Conceptual design of a next generation supersonic airliner for low noise and emissions. AIAA. **0215** (2023). https://doi.org/10.2514/6.2023-0215.c1
2. Jaron, R., Viola, N.: The future of civil supersonic transport in Europe: the SENECA and MORE&LESS projects. In: ICAS-2024 Congress Proceedings, Paper 820, Florence, Italy, pp. 1–10 (2024)
3. Zaporozhets, O., Tokarev, V., Attenborough, K.: Aircraft noise: assessment, prediction and control. Glyph International, Taylor & Francis (2011)

4. ICAO: In: 8 (ed.) Annex 16 to the Convention on International Civil Aviation. Environmental protection. Vol. I — Aircraft noise, ICAO, Montreal. ICAO (2018)
5. ICAO: Recommended Method for Computing Noise Contours Around Airports, ICAO Doc 9911. ICAO, 2 (2018)
6. Tokarev, V., Zaporozhets, O., Straholes, V.: Noise decreasing during the aircraft operating [in Russian]. Technika, Kyiv (1990)
7. Zaporozhets, O., Tokarev, V.: Aircraft noise modelling for environmental assessment around airports. Appl. Acoust. **55**(2), 99–127 (1998). https://doi.org/10.1016/S0003-682X(97)00101-1
8. Zaporozhets, O., Tokarev, V.: Predicted flight procedures for minimum noise impact. Appl. Acoust. **55**(2), 129–143 (1998). https://doi.org/10.1016/S0003-682X(97)00108-4
9. Zaporozhets, O., Jagniatinskis, A.: Improving single flyover noise prediction for subsonic aircraft. In: INTER-NOISE and NOISE-CON Congress and Conference Proceedings, Washington, USA, pp. 2949–3943 (2021). https://doi.org/10.3397/IN-2021-2352
10. Zaporozhets, O., Levchenko, L.: Accuracy of noise-power-distance definition on results of single aircraft noise event calculation. Aerospace. **8**, 121 (1998). https://doi.org/10.3390/aerospace8050121
11. Zaporozhets, O., Gulevets, V., Karpenko, S., Kazhan, K., Konovalova, O., Paraschanov, V.: Aircraft noise measurements in Ukrainian airports. In: Karakoc, T.H., Usanmaz, O., Rajamani, R., Oktal, H., Dalkiran, A., Ercan, A.H. (eds.) Advances in Electric Aviation. Sustainable Aviation. Springer. https://doi.org/10.1007/978-3-031-32639-4_21
12. Kazhan, K., Synylo, K., Zaporozhets, O., Karpenko, S.: ADS-B data usage for aircraft noise and air quality modelling and measurement during specific stages of LTO cycle. Int. J. Sustain. Aviat. **9**(1), 1–25 (2022). https://doi.org/10.1504/IJSA.2023.127489
13. Bertsch, L., Loble, F., Noding, M.: Considerations toward uncertainty quantification of single event noise predictions for novel SST concept aircraft. In: AIAA SCITECH 2025 Forum Proceedings, 1797, Orlando, USA (2025). https://doi.org/10.2514/6.2025-1797
14. Akatsuka, J., Ishii, T.: System Noise Assessment and Uncertainty Analysis of a Conceptual Supersonic Aircraft. Aerospace. **9**(4), 212 (2022). https://doi.org/10.2514/6.2020-0265
15. Schaffer, B., Pluess, S., Thomann, G.: Estimating the model-specific uncertainty of aircraft noise calculations. Appl. Acoust. **84**, 58–72 (2014). https://doi.org/10.1016/j.apacoust.2014.01.009
16. Makarenko, V., Tokarev, V., Kazhan, K., Synylo, K., Yakushenko, O., Zaporozhets, O.: Analysis of fuel consumption and noise of a regional turbo-propeller aircraft with a hybrid power plant. Int. J. Sustain. Aviat. **10**(4), 315–349 (2024). https://doi.org/10.1504/IJSA.2024.142562
17. Jaron, R., Grabert, M., Having, R., Meulen, M., Huet, M., LeGriffon, I., Petrosino, F., Barbarino, M., Lefarth, K., Zaporozhets, O.: Jet Noise Prediction Benchmark for Landing and Takeoff Noise of Supersonic Aircraft. In: InterNoise-2023 Proceedings, Chiba, Japan, pp. 6994–7510 (2023). https://doi.org/10.3397/IN_2023_1064
18. Stone, R., Zola, C., Clark, B.: An improved model for conventional and inverted-velocity profile co-annular jet noise. In: 37th Aerospace Sciences Meeting and Exhibit, 78, Reston, Virigina (1999). https://doi.org/10.2514/6.1999-78
19. Having, R., Van der Meulen, M., Huet, M., Jaron, R., Petrosino, F., Barbarino, M., Zaporozhets, O.: Dual-stream jet noise test with internal mixer design variations for LTO noise of supersonic aircraft. In: ICAS 2024 proceedings, Florence, Italy. Paper 64, pp. 1–19 (2024)
20. Stone, J. R.: An improved prediction method for the noise generated in flight by circular jets NASA-TM-81470 (1980). https://ntrs.nasa.gov/citations/19800013561, last accessed 2025/01/25
21. Heidmann, M. F.: Interim prediction method for fan and compressor source noise. NASA-TM-X-71763 (1975). https://ntrs.nasa.gov/citations/19750017876, last accessed 2025/01/25

22. Kontos, K. B., Janardan, B. A., Gliebe, P. R.: Improved NASA-ANOPP Noise Prediction Computer Code for Advanced Subsonic Propulsion Systems. NASA-CR-195480 (1996). https://ntrs.nasa.gov/citations/19960048499, last accessed 2025/01/25
23. Fink, M.R.: Airframe noise prediction method. US Federal Aviation Administration, FAA-RD-77-29, GOVDOC: TD 4.509:77–29 (1977). https://apps.dtic.mil/sti/tr/pdf/ADA039664.pdf, last accessed 2025/01/25.
24. Burley, R. R.: Preliminary Measurement of the Airframe Noise From an F-l06B Delta Wing Aircraft at Low Flyover Speeds. NASA TM X-71527 (1974). https://ntrs.nasa.gov/citations/19740011555, last accessed 2025/01/25
25. Attenborough, K.: Sound propagation in the atmosphere. In: Rossing, T. (ed.) Springer Handbook of Acoustics. Springer Handbooks. Springer, New York (2007). https://doi.org/10.1007/978-0-387-30425-0_4
26. Delany, M.E., Bazley, E.N.: Acoustical properties of fibrous absorbent materials. Appl. Acoust. **3**(2), 105–116 (1970). https://doi.org/10.1016/0003-682X(70)90031-9
27. Ground surfaces: Determination of the acoustic impedance. NT ACOU 104, 1999-11. (1999). https://www.nordtest.info/wp/1999/11/01/617, last accessed 2025/01/25
28. Chien, C.F., Soroka, W.W.: Sound Propagation Along an Impedance Plane. J. Sound Vibr. **43**(1), 9–20 (1975). https://doi.org/10.1016/0022-460X(75)90200-X
29. Tanner, C.S.: Noise measurement of Concorde 02 approach and takeoff at Dallas-Ft.Worth and Dallas International Airports. EPA 550/9-74-13, Washington, USA. (1974). https://nepis.epa.gov/Exe/ZyNET.exe/9101VL7E.TXT, last accessed 2025/01/25
30. Final environmental impact statement. Noise regulation and type certification alternatives for civil supersonic aircraft. 1978-06-ST-FEIS, US DoT, FAA, Washington, USA (1978). https://files.hawaii.gov/dbedt/erp/EA_EIS_Archive/1978-06-ST-FEIS-Noise-Reg-and-Type-Certification-Alternatives-for-Civil-Supersonic-Aircraft.pdf, last accessed 2025/01/25.
31. CAA Paper 77007: Noise Data from the First Year of Scheduled Concorde Operations at Heathrow Airport – London. London, UK (1977). https://www.caa.co.uk/our-work/publications/documents/content/caa-paper-77007/, last accessed 2025/01/25.
32. Lieber, L., Brown, D.: Wing Reflection Code. ANOPP Noise prediction for small engines. NASA/CR-2000-210630 (2000). https://ntrs.nasa.gov/api/citations/20010045806/downloads/20010045806.pdf, last accessed 2025/01/25.
33. Bozak, R.: The impact of subsonic twin jets on airport noise. In: 18th Aeroacoustics Conference, AIAA–2012–2299, Colorado Springs, USA (2012) https://ntrs.nasa.gov/api/citations/20120013785/downloads/20120013785.pdf, last accessed 2025/01/25
34. Grabert, M., Jaron, R., Plohr, M., Zenkner, S., Richter, C., Stohrer, C., Munoz, C.-V., van der Meulen, M.-J., Having, R., Huet, M., Petrosino, F., Barbarino, M., Fischer, F., Zaporozhets, O.: Landing and takeoff noise prediction of conceptual supersonic aircraft. In: ICAS-2024 Congress Proceedings, Florence, Italy (2024)

Local Air Quality Assessment for SENECA Aircraft

Kateryna Synylo[1(✉)], Andrii Krupko[1], and Oleksandr Zaporozhets[2]

[1] National Aviation University, Liubomyra Huzara Ave., 1, Kyiv 03058, Ukraine
synyka@gmail.com

[2] Institute of Aviation, Aleja Krakowska 110/114, 02-256 Warsaw, Poland

Abstract. SENECA (LTO noiSe and EmissioNs of supErsoniC Aircraft) project aims to provide a basis for decision making of environmentally friendly supersonic aircraft design taken into account the detailed modelling of the emission indexes and the local/global environmental impacts. Emission indices of NOx have been calculated for supersonic airliner M1.8 throughout the supersonic mission profile for the 2000 and 4000 nmi ranges. The EINOx calculation results correlate well between the two methods considered (BFFM2, P3T3) for the take-off and climb subsonic conditions. The difference between BFFM2 and P3T3 was found in the climb and cruise supersonic conditions. The observed difference is due to the fact that BFFM2 (the version used) does not include correction factors for pressure (kp) and temperature (kt) as a function of Mach number, which is quite important for the supersonic case. The calculation results of hourly concentration for airport fleet with contribution of supersonic airline M1.8 (through large aircraft group) demonstrates the similar levels of NOx and twice less of CO in comparison with baseline airport fleet. (Concorde is as reference aircraft for large aircraft group.)

Keywords: Emission Index · Hourly Concentration · Area of Influence · Local Air Quality

1 Introduction

Following the retirement of Concorde in 2003, development of the supersonic aircraft is crucial task for several manufacturers. Key environmental challenges to address include the use of significantly more fuel on a per passenger kilometre basis compared to subsonic commercial aircraft [1], and noise, specifically the impact of the sonic boom generated when flying at supersonic speed.

Previously the data basis regarding emissions and noise of supersonic aircraft is limited to the Concorde era, and commercial as well as research prototypes and studies, e.g. the outcome of the EU project FP6 HISAC6, and the supersonic flight program of NASA.

SENECA aims at developing deepened understanding and detailed modelling for the emissions, the LTO noise, and the global environmental impact of supersonic aircraft. Building on this, the development of beyond state-of-the-art technologies of TRL 6 to further reduce the environmental impact of supersonic aviation.

I. Ostroumov et al. (Eds.): ACASD 2025, LNNS 1418, pp. 211–228, 2025.
https://doi.org/10.1007/978-3-031-91992-3_14

SENECA aims to develop four different supersonic aircraft platforms including the airframes (WP2) and the engines (WP3) ranging from supersonic business jets designed for cruise Mach numbers of Ma = 1.4 and Ma = 1.6 up to large airliners designed for 100 passengers and cruise Mach numbers of Ma = 1.8 and Ma = 2.2. The development includes a Multi-Disciplinary Design Optimization (MDO) targeting firstly to meet at least the current noise regulations for subsonic aircraft during take-off and landing and secondly to reduce the emission levels, Fig. 1.

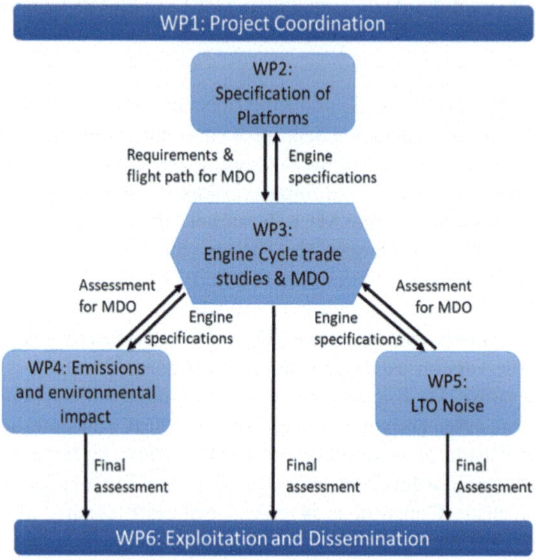

Fig. 1. SENECA project structure.

SENECA project includes the holistic studies of the environmental impact of supersonic aircraft, taking into account emissions and noise in the vicinity of airports and the global environmental impact to support the decision making of the European and international regulatory and certification processes during the CAEP/13 cycle.

2 Methodology of Aircraft Emission and Air Pollutants Dispersion Calculation

The main purpose of PolEmiCa (own software) is to provide dispersion (pollution) and inventory (emission) calculations for aircraft engine emissions during the LTO cycle of aircraft movement within the airport area. It also includes aircraft emissions from take-off procedures, Auxiliary Power Unit (APU) and Ground Support Equipment (GSE), Fig. 2 [2, 3].

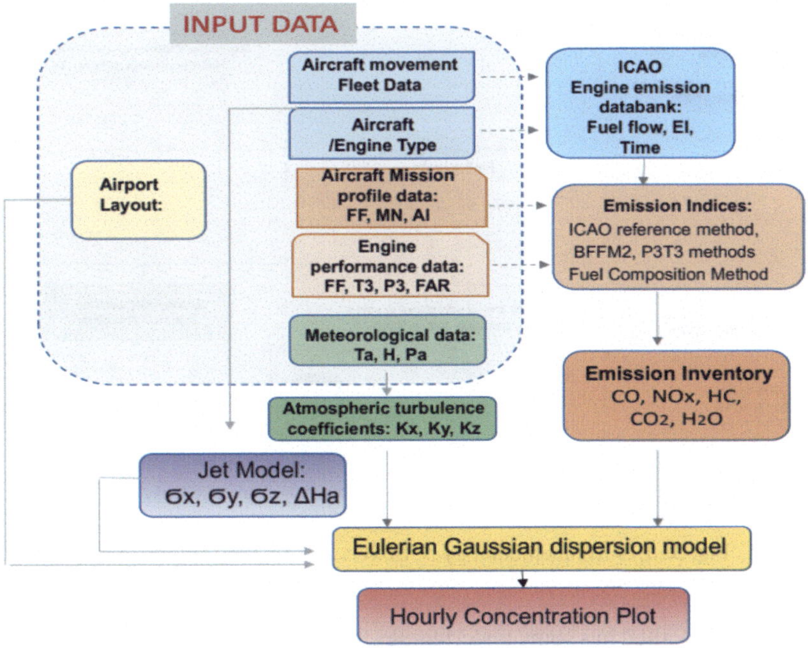

Fig. 2. PolEmiCa model - emission inventory and LAQ tasks (aircraft contribution).

PolEmiCa uses the BFFM2 (Boeing Fuel Flow Method Version 2) and P3T3 methods to predict engine emissions over a flight path based on flight conditions (flight profile data), engine performance data (fuel flow, fuel-air ratio, P3, T3) and meteorological conditions (temperature, humidity, atmospheric pressure) [4, 5].

2.1 The Boeing Fuel Flow Method (BFFM2)

The BFFM2 method uses ICAO (or similar) certification type emission databases to estimate emission indices (EI) of CO (carbon monoxide), THC (hydrocarbons) and NOx (nitrogen oxides) [3, 6]. The key components of the BFFM2 method are semi-empirical techniques to correct for atmospheric effects and the development of a relationship between the emission index and fuel flow, Fig. 3.

The BFFM2 method is used by several stakeholders not only to evaluate emissions during cruise, but also to estimate the total number of pollutants emitted by an aircraft over a reference trajectory, and to support the generation of 3D emission inventories and the consequent assessment of the impact of aviation on climate change.

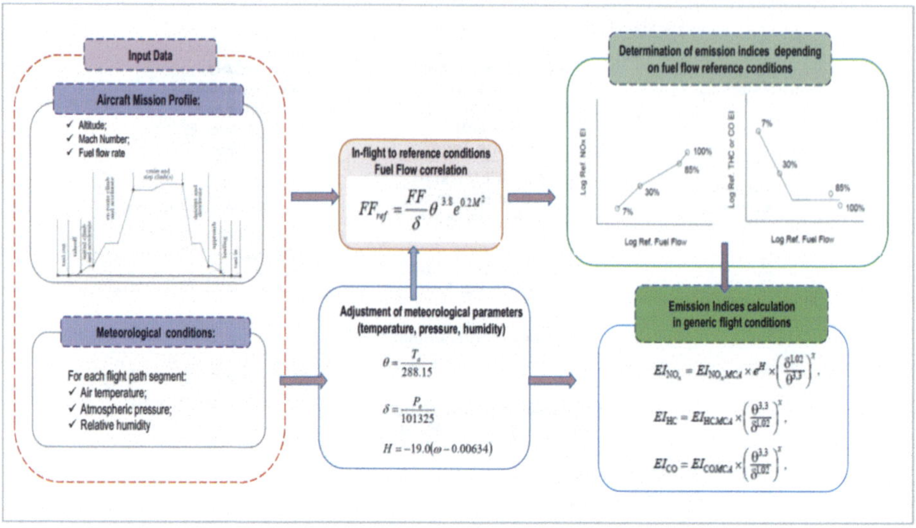

Fig. 3. BFFM2 methodology of EINOx, EICO, EIHC prediction.

The BFFM2 method has some uncertainties that need to be taken into account as part of the clarification of the predicted emission indices. Although there are no problems with NOx prediction, as reported by Kim and Rachami [7], Wood et al. [8], the calculation of HC and CO can be quite challenging. For example, for throttle settings less than 7% (for low fuel flow values), recalculation of EIHC, EICO for reference conditions can result in unrealistically high values. According to Lee [9], an ICAO study found agreement between direct measurements and estimated fuel flow "with a standard deviation of 6% and a maximum error of 13%".

The BFFM2 method has certain limitations for supersonic aircraft because the process between the free flow and the combustor inlet cannot be considered isentropic in the case of subsonic aircraft. Fusaro and others [10] focused on updating the BFFM2 algorithm to take into account the variation of the pressure (k_p) and temperature (k_t) correction factors as a function of Mach number (in the range Mach 1-Mach 3).

This improvement of BFFM2 allows to complement the conceptual design of future supersonic aircraft with the estimation of pollutant and greenhouse gas emissions.

2.2 P3T3 Method

This method [11] is based on the knowledge of the engine parameters (i.e. the combustor inlet pressure (P3) and temperature (T3)), hence the name P3T3. In addition, the fuel/air ratio at both reference and operating conditions must be known. According to Wit et al. [12], the P3T3 method is considered to be the most accurate method for calculating NOx emissions.

The methodology of EINOx calculations is shown in Fig. 4, but is also applied to EICO and EITHC with different parameters m, n.

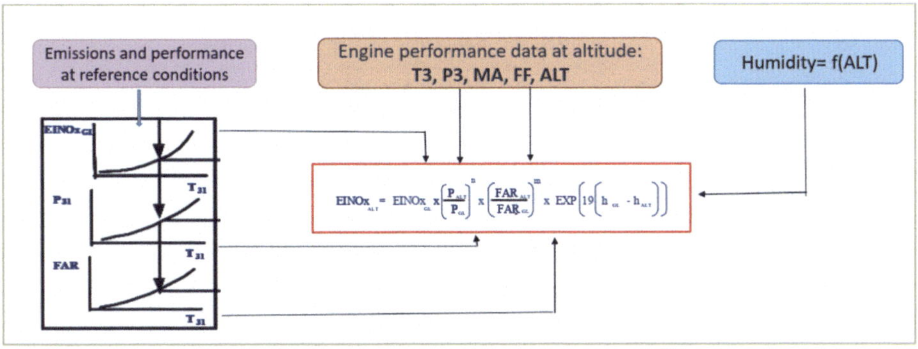

Fig. 4. P3T3 Altitude NOx prediction methodology.

Similar to the Boeing fuel flow method, three graphs are plotted. The NOx emission index, the combustor inlet pressure (P3) and the fuel/air ratio at reference conditions are plotted against the combustor inlet temperature (T3).

The relationships provided by these plots are used to determine the reference EINOx, reference combustor inlet pressure and reference FAR corresponding to the inlet temperature at flight conditions. The reference EINOx is then corrected for the difference in combustor inlet pressure and FAR between the ground reference conditions (REF) and the flight condition (FLT).

The values for the pressure and FAR exponents, a and b, are determined by parametric testing in either a combustor test rig or an engine altitude facility. In addition to the P3 and FAR corrections, a humidity correction is also required as indicated by the exponential term.

The P3T3 method is currently limited to jet engines. Turboprops are not covered by the method. The accuracy of NOx estimates using recommended pressure exponents has been estimated to be $\pm 11\%$ when P3 and T3 at reference conditions are known. The accuracy of CO and HC estimates has not been established [4, 11].

2.3 Dispersion Model

Basic expression for definition of maximum instantaneous value of concentration in grid point is a solution of turbulent diffusion equation according to Eulerian approach for moving point source with preliminary transport and dilution of contaminants by an exhaust gases jet and wing vortices [12, 13].

The equation of the PolEmiCa model for the definition of an instantaneous concentration from a moving source (from a single exhaust event) with preliminary transport on a distance $\boldsymbol{X}_{\mathrm{A}}$ and rise on an altitude Δh_A and dilution $\sigma_{x,\,y,\,z}$ of pollutants by the jet takes a following form (1):

$$c(\mathrm{x, y, z, t}) = \frac{\mathrm{Q}\ \exp\left[-\frac{(\mathrm{x}-\mathrm{x}')^2}{2\sigma_x^2+4\mathrm{K}_x t} - \frac{(\mathrm{y}-\mathrm{y}')^2}{2\sigma_y^2+4\mathrm{K}_y \mathrm{t}}\right]}{\left\{8\,\pi^3\left[\sigma_x^2 + 2\mathrm{K}_x t\right]\left[\sigma_y^2 + 2\mathrm{K}_y t\right]\right\}^{1/2}}$$

$$\times \frac{\exp\left[\frac{(z-z'-H)^2}{2\sigma_z^2+4K_z t}\right]+\exp\left[\frac{z+z'+H}{2\sigma_z^2+4K_z t}\right]}{\left[\sigma_z^2+2K_z t\right]^{1/2}}. \quad (1)$$

The aircraft is considered as a moving emission source, thus current co-ordinates (x', y', z') of the emission source in movement during time t' are defined as:

$$x' = x_0 + u_{PL}t' + 0.5a_{PL}t'^2 + u_w(t+t'), \quad (2)$$

$$y' = y_0 + v_{PL}t' + 0.5b_{PL}t'^2, \quad (3)$$

$$z' = z_0 + w_{PL}t' + 0.5c_{PL}t'^2, \quad (4)$$

where (x_0, y_0, z_0) are initial coordinates of the source; (u_{PL}, v_{PL}, w_{PL}) are velocity vector components of the emission source; (a_{PL}, b_{PL}, c_{PL}) are acceleration vector components of the emission source; K_x, K_y, K_z are coefficients of atmospheric turbulence [13].

As with any dispersion model, the initial properties of a plume are important to model its rise and location. Such plume or jet parameters, as rise height Δh_A due to buoyancy effect, horizontal σ_y^2 and vertical σ_z^2 dispersion parameters are needed as input to dispersion modeling of aircraft sources.

The jet transport model evaluates basic mechanisms of contaminants transportation and dilution by jet of exhausted gases from aircraft engine and provides basic parameters of the jet for further dispersion analysis. The process of contaminant transport by exhaust gases jet is described by the semi-empirical theory of turbulent jets. Buoyancy of a jet is caused by action of Archimedes forces due to excess of temperature of jet gases above air temperature (Fig. 1). The Archimedes number (5) is used for the estimation of the plume rise height (6) [12, 13]:

$$Ar_0 = gD_0(Q_T - 1)/U_0^2, \quad (5)$$

$$\Delta h_A = 0.013 Ar_0 \overline{X_A^3} R_0, \quad (6)$$

where parameter $Q_T = T_0/T_H$ for engines currently in operation changes within the limits of 1.15-2; $\overline{X}_A$ is the longitudinal coordinate of jet axis in relation to radius of engine exhaust nozzle, $R_0 = D_0/2$.

The complex model PolEmiCa has been sufficiently improved in subject of jet transport model by using CFD package (FLUENT 6.3) [13] to investigate the physics and characteristics of ground vortices, which are generated between the ground surface and aircraft engine nozzle, to assess the ground surface impact on the jet structure, parameters and properties of jet development.

Complex model PolEmiCa calculates the concentration filed inside airport area taking into account intensity of flights of airplanes, a loading factor of different taxiways and runways, and other operational circumstances.

3 Modelling of EIs for Airliner (Ma = 1.8)

Emission indices of NOx (EINOx) have been calculated for supersonic airliner (Ma = 1.8) using BFFM2 and P3T3 methods.

Cranfield University (WP2) was developed the mission trajectory profiles of the SENECA M1.8 airline for four scenarios: two with a 4000 nm range and two with a 2000 nm range, covering both supersonic and subsonic cruise conditions.

DLR (WP3) produced the finalized cycle design of the SENECA M1.8 supersonic airliner engine. An engine deck has been calculated for this optimized engine design, covering all potential flight conditions of the aircraft, and provided to the project partners of WP2, WP4 and WP5.

Engine performance data (T3, P3 and fuel flow) and EINOx at ICAO points were used to estimate the reference EINOx, the reference combustor inlet pressure corresponding to the combustor inlet temperature (Figs. 5 and 6) and the reference fuel flow (Fig. 7) in the flight condition using the appropriate P3T3 and BFFM2 methods.

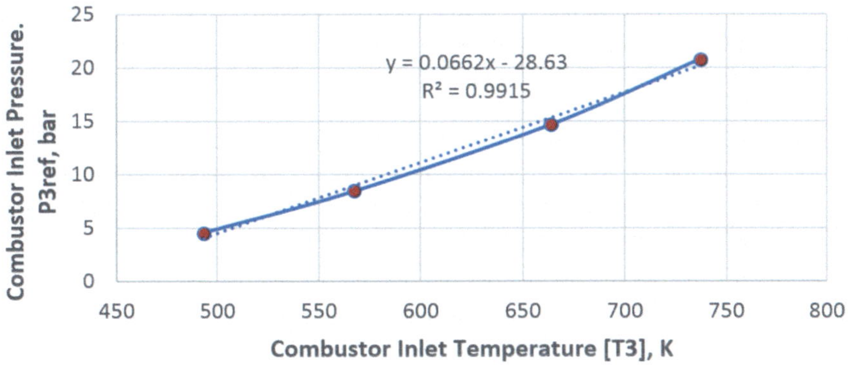

Fig. 5. Estimation of reference P3 and EINOx.

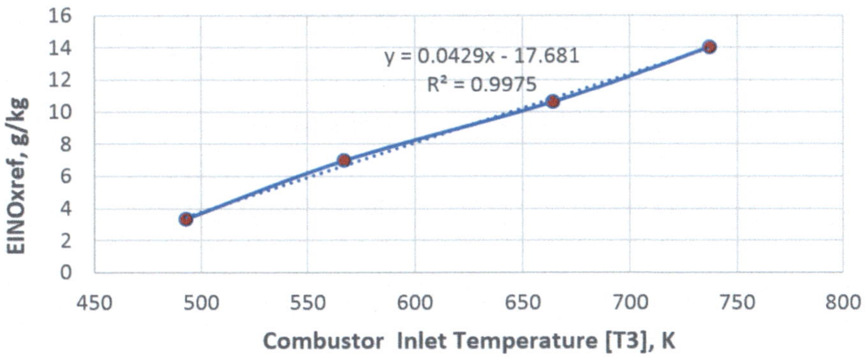

Fig. 6. Estimation of reference P3 as a function of T3 for LTO cycle (P3T3).

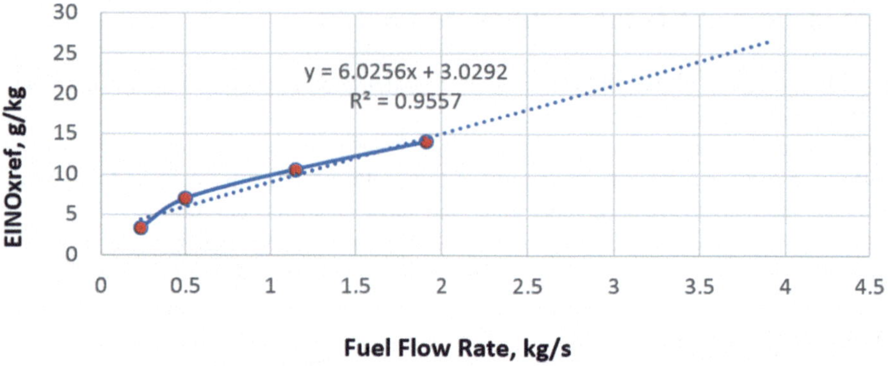

Fig. 7. Estimation of reference EINOx as a function of fuel flow rate for LTO cycle (BFFM2).

The EINOx were calculated using the BFFM2 method and the P3T3 method (module of the PolEmiCa complex model) [2, 14] for each of the four mission profiles mentioned. The distribution of meteorological parameters (temperature, pressure) was estimated according to the ICAO Standard Atmosphere Manual [15]. The certification data in the ICAO database are corrected to a humidity of 0.00634 g of water per kg of dry air (60% relative humidity, ISA day).

Figures 8 and 9 respectively show the dependence of the modelled EINOx on distance throughout the subsonic mission profile for the 2000 nmi range and the subsonic mission profile for the 4000 nmi range. The EINOx calculation results correlate well between the BFFM2 and P3T3 methods throughout the flight profiles considered.

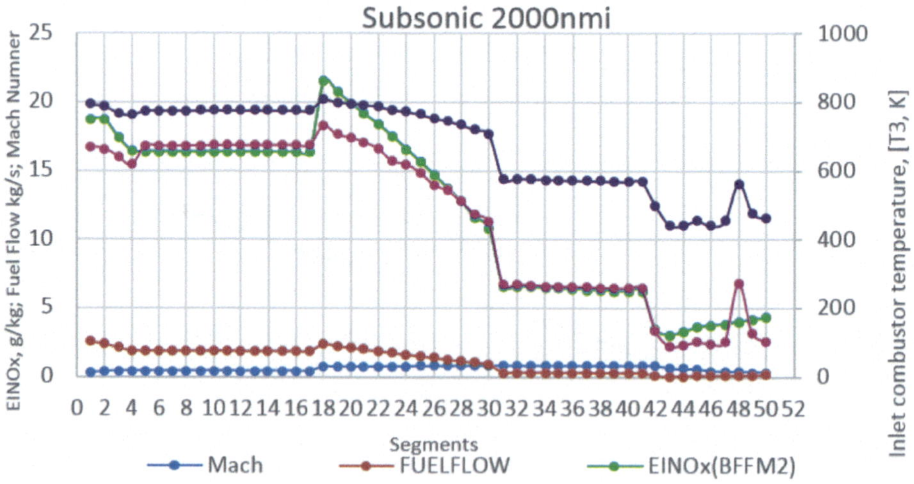

Fig. 8. Distribution of EINOx throughout the subsonic mission profile for the 2000 nmi range.

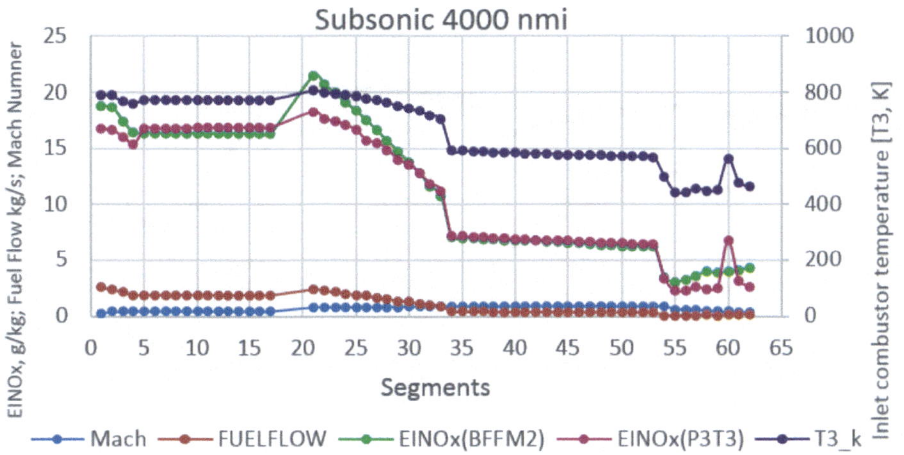

Fig. 9. Distribution of EINOx throughout the subsonic mission profile for the 4000 nmi range.

Accordingly, Figs. 10 and 11 show the calculated EINOx for each segment of the total flight profile for the supersonic mission – 4000 nmi and for the supersonic mission – 2000 nmi.

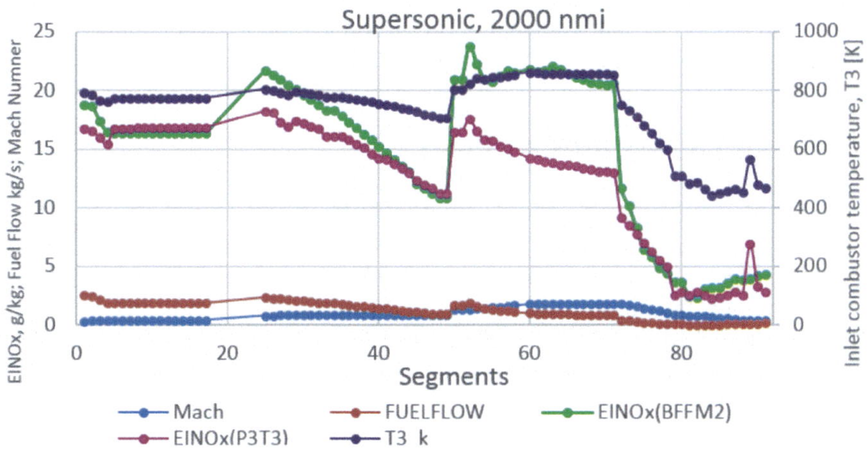

Fig. 10. Distribution of EINOx throughout the supersonic mission profile for the 2000 nmi range.

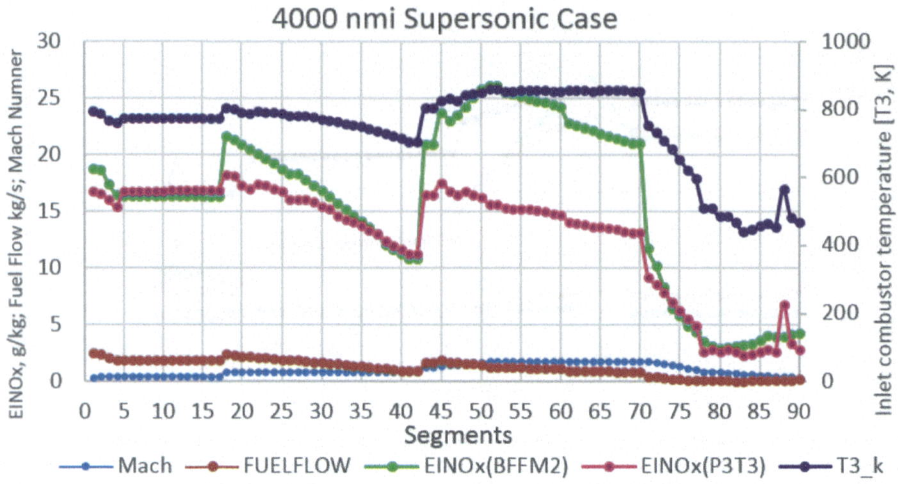

Fig. 11. Distribution of EINOx throughout the supersonic mission profile for the 4000 nmi range.

The EINOx calculation results correlate well between the two methods considered (BFFM2, P3T3) for the take-off and climb subsonic conditions. The difference between BFFM2 and P3T3 was found in the climb and cruise supersonic conditions. The observed difference is due to the fact that BFFM2 (the version used) does not include correction factors for pressure (kp) and temperature (kt) as a function of Mach number, which is quite important for the supersonic case [6].

4 Local Air Quality for CAEPport

An artificial regional airport has been set up (CAEPport) for local air quality studies. It is used by Modelling and Database Group of CAEP to verify the abilities of Local Air Quality models (including PolEmiCa) to handle the various emission groups at an airport in emission and dispersion calculations (Table 1).

The fleet structure of CAEPport includes seven aircraft categories, the Concorde and supersonic airliner M1.8 involves through the contribution by large category, Fig. 12.

Table 1. Annual aircraft movements by category.

No	Category	Arrivals	Departures	Movements	Fractions, %
1	Large	3177	3179	6356	7.20
2	Medium	712	713	1425	1.61
3	Small	24,604	24,109	48,713	55.11
4	Regional	5571	5536	11,107	12.57
5	Business Jet	113	103	216	0.24
6	Turboprop	10,102	9891	19,993	22.62
7	Pistons	290	290	580	0.66
	Total:	44,569	43,821	88,390	100%

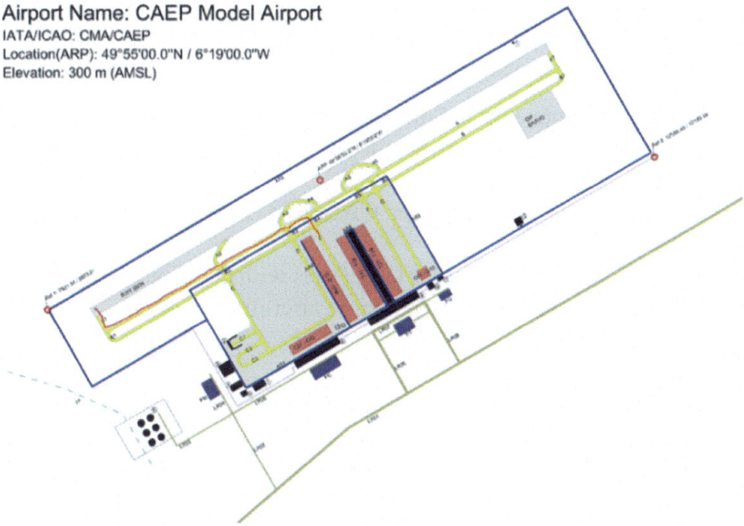

Fig. 12. CAEPport Layout.

To demonstrate the contribution of SeNeCa aircraft to local air pollution, the PolEmiCa GIS based model was used (Fig. 13). The GIS capacity as a test bench simplifies the process of defining the various airport elements (runway, taxiway, terminal, roadways and etc.) and allows the spatial distribution of emissions to be visualized.

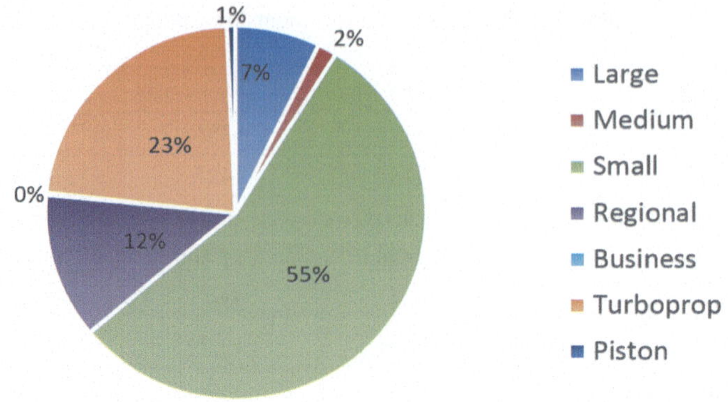

Fig. 13. Aircraft fleet structure for CAEPport.

Based on the annual aircraft movements analysis, including the determination the maximum number operations per peak-hour (25 operations), three scenarios were set for LAQ studies for supersonic aircraft (Table 2):

1. Baseline Scenario 1. Conventional aircraft fleet for CAEPport, which is used also, as reference case within EFACA project to assess the effect of hydrogen and hybrid electric aircraft;
2. Baseline Scenario 2. Aircraft fleet for CAEPport with contribution of Concorde through large aircraft category;
3. Predicted Scenario 3. Aircraft fleet for CAEPport with contribution of the supersonic airliner M1.8 (SeNeCa aircraft) through large aircraft category.

Table 2. Aircraft movements by Category for Peak-Hour (LAQ-scenario).

Aircraft category	Number operations	%	Baseline Scenario 1	Baseline Scenario 2	Predicted Scenario 3
			RA/Eng	RA/Eng	RA/Eng
Large	2	0.08	A343/CFM56-5C2	Concorde/Olymp593	M1.8
Regional	3	0.12	E170/AE3007A1	E170/AE3007A1	E170/AE3007A1
Turboprop	5	0.20	AT76/PW127A	AT76/PW127A	AT76/PW127A
Small	13	0.52	A320/CFM56-5A1	A320/CFM56-5A1	A320/CFM56-5A1
Medium	1	0.04	A332/CF6-80E1A2	A332/CF6-80E1A2	A332/CF6-0E1A2
Business	1	0.04	C550/JT15D-4	C550/JT15D-4	C550/JT15D-4
Total	25	1.00			

For LAQ tasks the hourly concentration of CO and NOx for considered peak hour (Table 2) under both scenarios were modelled for unfavourable meteorological conditions (neutral stability – Monin Obukhov Scale = 99999 m; roughness surface, z0 =

0.2 m; wind velocity u(10 m) = 3 m/s; air temperature T(10 m) = 10 °C; atmospheric pressure P (10 m) = 101325 Pa). Calculation results of hourly concentrations of CO and NOx for baseline and predicted scenarios are presented correspondingly in Figs. 14, 15, 16, 17, 18 and 19.

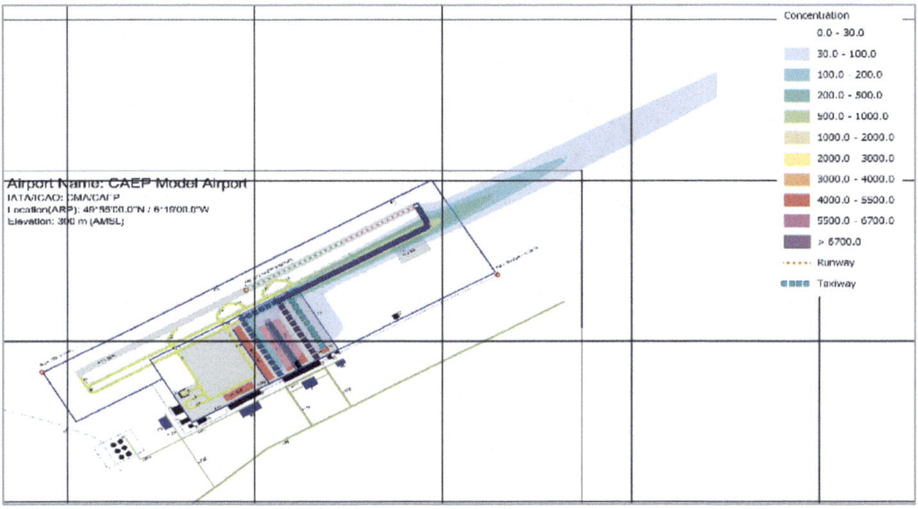

Fig. 14. Hourly concentration of CO [μg/m^3] for peak hour (25 operations) in baseline scenario 1 (maximum value concentration CO = 1763 μg/m^3).

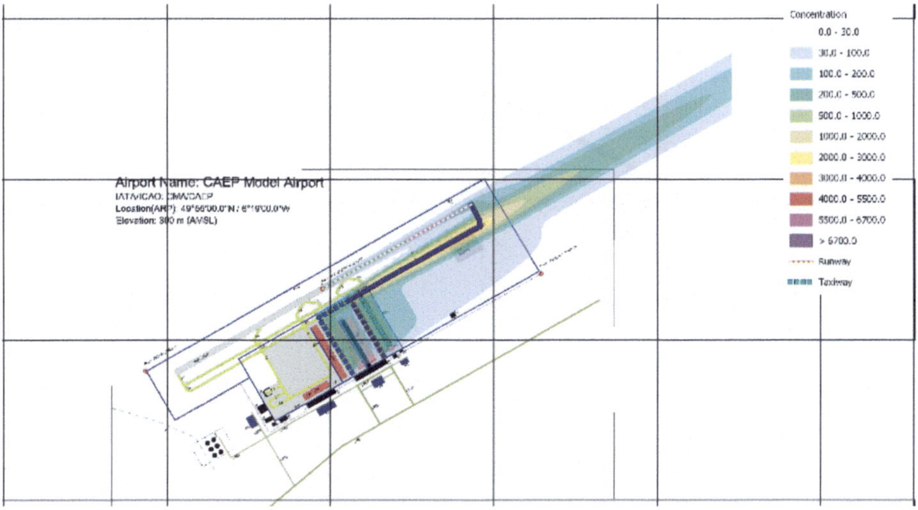

Fig. 15. Hourly concentration of CO [μg/m^3] for peak hour (25 operations) in baseline scenario 2 (maximum value concentration CO = 5811 μg/m^3).

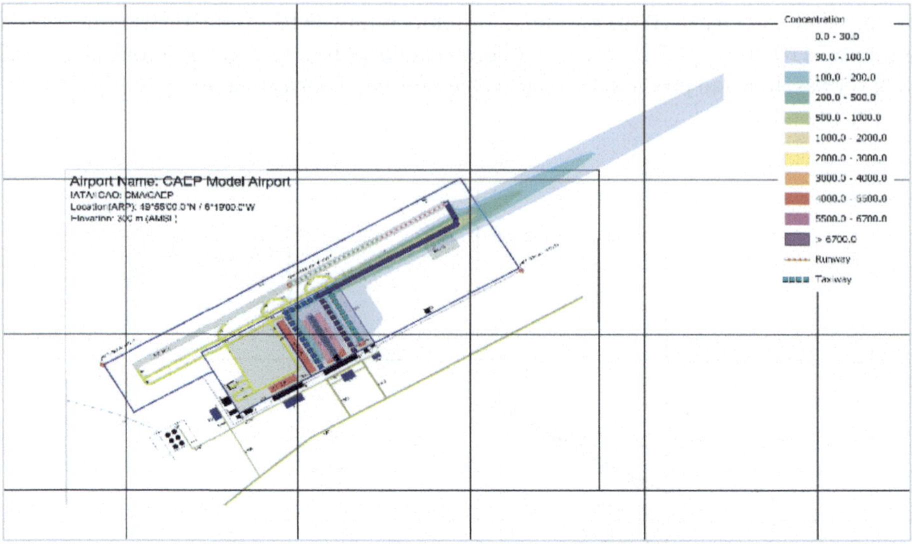

Fig. 16. Hourly concentration of CO [μg/m^3] for peak hour (25 operations) in predicted scenario (maximum value concentration CO = 1569 μg/m^3).

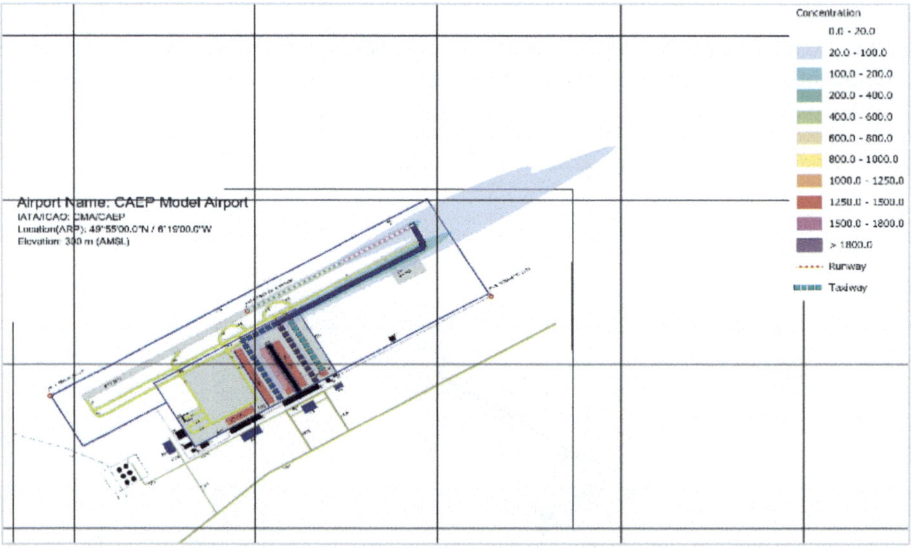

Fig. 17. Hourly concentration of NOx [μg/m^3] for peak hour (25 operations) in baseline scenario 1 (maximum value concentration NOx = 501 μg/m^3).

Figures 14, 15 and 16 demonstrates, that CO pollution levels are same for baseline scenario 1 (conventional fleet, A343 is reference aircraft for large category) and predicted scenario (updated fleet in part of large category – supersonic airliner M 1.8), but in 3 times higher in case of contribution of Concord (baseline scenario 2).

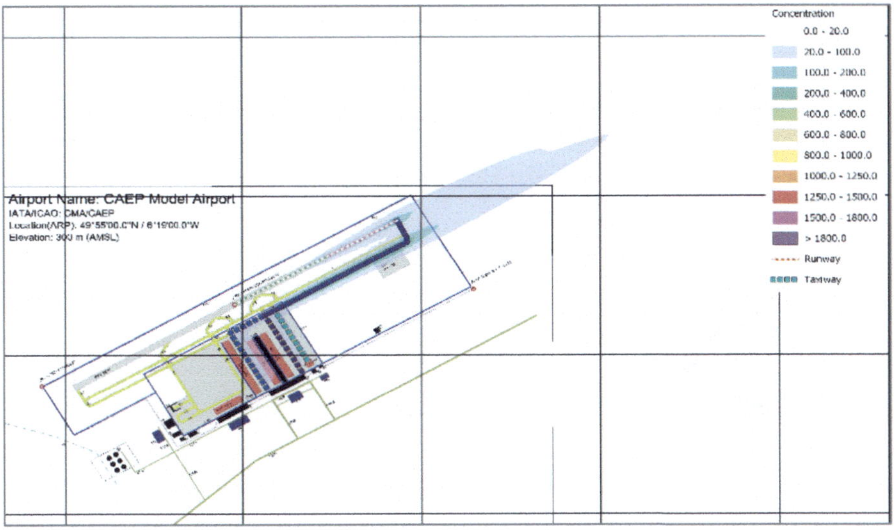

Fig. 18. Hourly concentration of NOx [μg/m^3] for peak hour (25 operations) in baseline scenario 2 (maximum value concentration NOx = 537 μg/m^3).

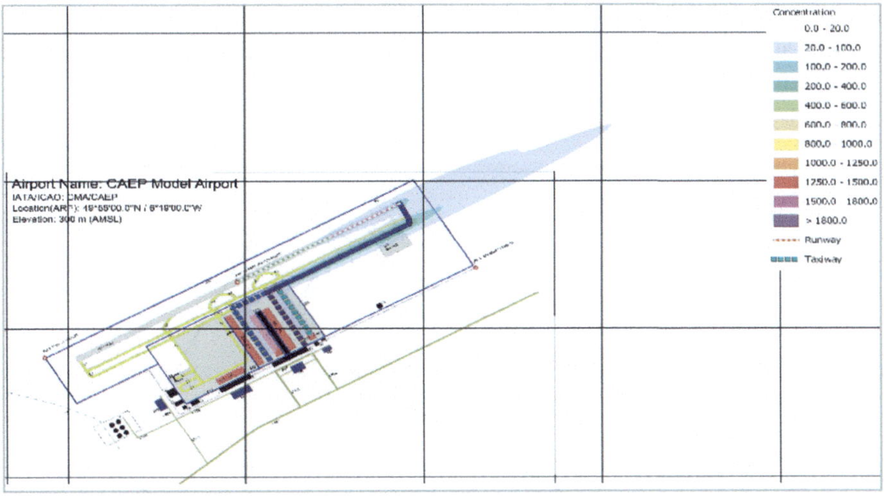

Fig. 19. Hourly concentration of NOx [μg/m^3] for peak hour (25 operations) in predicted scenario 3 (maximum value concentration NOx = 581 μg/m^3).

Figures 17, 18 and 19 demonstrates, that NOx pollution levels are same for all three considered scenarios.

Figure 20 illustrates the estimated area of influence for CO predicted by PolEmiCa for correspondingly for baseline scenario 1 (conventional fleet for CAEPport, A343 – reference aircraft for large aircraft group), baseline scenario 2 (contribution of Concorde) and predicted scenario (contribution of M1.8 airliner).

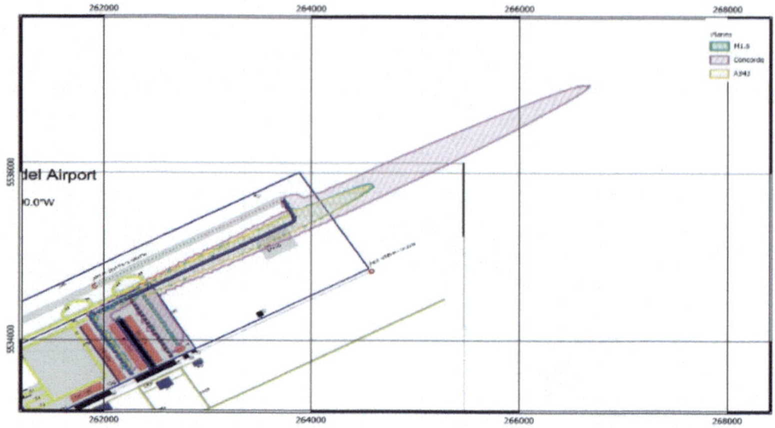

Fig. 20. Area influence of fleet for peak hour (25 operations) for considered three scenarios. Hourly concentration range of CO: 200 μg/m^3 [0,01 MPC] ÷ maximum concentration.

Figure 20 demonstrates the low level of CO pollution for the considered scenarios (25 operations per peak-hour). So, the area of influence for CO in case of M1.8 airliner contribution (SpCO = 0,21 km^2) is smaller than those predicted ones for Concord (SbCO = 0,36 km^2). However, the area of influence for NOx predicted in case of M1.8 supersonic airliner (SpNOx = 0,31 km^2) is similar to modelled for Concord (SpNOx = 0,33 km^2) (Fig. 21).

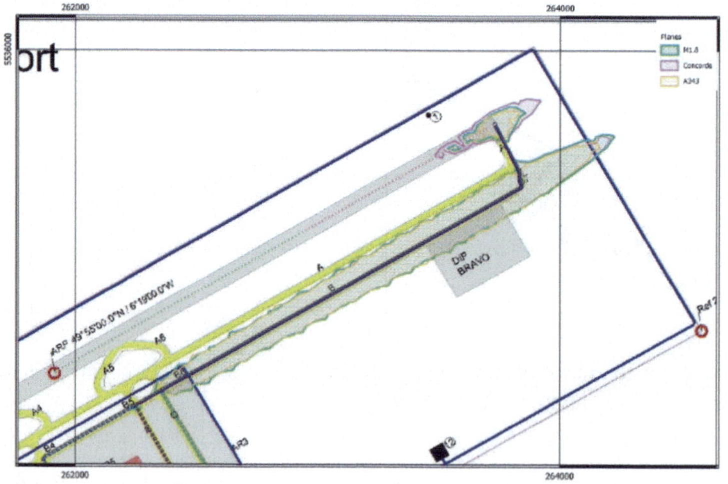

Fig. 21. Area influence of fleet for peak hour (25 operations) for considered three scenarios. Hourly concentration range of NOx: 100 μg/m^3 [0,5 MPC] ÷ maximum concentration.

The calculation results of hourly concentration for airport fleet with contribution of supersonic airline M 1.8 (through large aircraft group) demonstrates the similar levels of NOx and twice less of CO in comparison with baseline airport fleet. (Concorde is as reference aircraft for large aircraft group.)

5 Conclusions

The EINOx were calculated using the BFFM2 method and the P3T3 method (module of the PolEmiCa complex model) throughout supersonic mission profile for the 2000 nmi and 4000 nmi range. The EINOx calculation results correlate well between the two methods considered for the take-off and climb subsonic conditions. The difference between $BFFM_2$ and P_3T_3 was found in the climb and cruise supersonic conditions. The observed difference is due to the fact that BFFM2 (the version used) does not include correction factors for pressure (kp) and temperature (kt) as a function of Mach number, which is quite important for the supersonic case.

The modelled CO pollution levels are same for baseline scenario 1 (conventional fleet, A343 is reference aircraft for large category) and predicted scenario (updated fleet in part of large category – supersonic airliner M 1.8), but in 3 times higher in case of contribution of Concord (baseline scenario 2).

The considered cases the areas of influence for baseline scenario2 (contribution of Concorde through large aircraft group) and predicted scenario (supersonic airliner M1.8) are restricted to areas near the aircraft pathway, well within the airport boundaries.

Future tasks will be focused on the assessment air pollution levels with for updated fleet via contribution M 1.6 business jet (Seneca aircraft).

Acknowledgement. Acknowledgement to SENECA project № 101006742 "((LTO) noiSe and EmissioNs ofsupErsoniC Aircraft)" funded by Horizon 2020.

References

1. Environmental Limits on Supersonic Aircraft in 2035. ICCT (2022)
2. Zaporozhets, O., Synylo, K.: New and improved LAQ models for assessment of aircraft engine emissions and air pollution in and around airports. In: On Board a Sustainable Future: ICAO Environmental Report 2016, International Civil Aviation Organization 999, pp. 82–84. University Montreal, Montreal (2016)
3. Synylo, K., Kurtenbach, R., Zaporozhets, O., Wiesen, P.: Modelling of aircraft emissions in the airport area. In: 2nd ECATS Conference Making Aviation Environmentally Sustainable, pp. 67–71 (2016)
4. Duchene, N., Synylo, K.: P_3T_3 NO_x model of turbofan engine. In: The International Symposium on Sustainable Aviation, Rome, p. 103 (2018)
5. Duchene, N., Synylo, K.: NOx emission model of turbofan engine. Int. J. Sustain. Aviation. **1**, 72–84 (2014)
6. SAE Committee A-21: SAE AIR 5715. Aircraft Emissions (2009)
7. Kim, B., Rachami, J.: Aircraft emissions modelling under low power conditions. In: 101st Annual Conference and Exhibition of the A&WMA's, Portland, OR, USA, pp. 24–27 (2008)

8. Wood, E., Herndon, S., Miake-Lye, R., Nelson, D., Seeley, M.: Aircraft and airport-related hazardous air pollutants: research needs and analysis. Aircraft Cooperative Research Program, ACRP Report 7 (2008)
9. Lee, J.J.: Modelling aviation's global emissions, uncertainty analysis and applications to policy. Ph.D. thesis, Massachusetts Institute of Technology (2005)
10. Fusaro, R., Viola, N., Galassini, D.: Sustainable supersonic fuel flow method: an evolution of the Boeing fuel flow method for supersonic aircraft using sustainable aviation fuels. Aerospace. **8**, 331 (2021)
11. Madden, P., Park, K.: Methodology for predicting NOx emissions at altitude conditions from ground level engine emissions and performance test information. Rolls-Royce Technical Report DNS 90713, Issue 1 (2003)
12. Zaporozhets, O., Synylo, K.: Estimation of buoyancy effect and penetration length of jet from aircraft engine by large eddy simulation method. Int. J. Sustain. Aviation. **2**(1), 34–51 (2016)
13. Synylo, K., Zaporozhets, O., Stiller, J., Frohlich, J.: Improvement of airport local air quality modeling. J. Aircraft. **54**(5), 1750–1759 (2017). https://doi.org/10.2514/1.C033803
14. Wit, R.C.N., Boon, B.H., Velzen, A., Cames, M., Deuber, O., Lee, D.S.: Giving wings to emission trading: inclusion of aviation under the European Emission Trading System (ETS)—design and impacts. Report for the European Commission, DG Environment, No. ENV.C.2/ETU/2004 (2005)
15. Standard Atmosphere Manual, ICAO (1993)

Communication Systems for Civil Aviation Systems

Antennas Combination for Ground and Airborne Segments of UAV Communication Systems

Olga Shcherbyna(✉), Oleksandr Zadorozhnyi, and Fedir Katushonok

National Aviation University, Liubomyra Huzara Ave., 1, Kyiv 03058, Ukraine
olha.shcherbyna@npp.kai.edu.ua

Abstract. The unmanned aerial vehicle complex is a rather complex structure, which includes not only the aircraft itself, but also various devices for exchanging information signals, united into a single system. Antennas occupy one of the most important places in this system. The performance indicators of the entire complex directly depend on the quality characteristics of antenna devices. Antennas used for onboard and ground segments of communication systems have a number of requirements for design and parameters. These requirements directly depend on the specifics of the use and mission of the aircraft. There is a fairly large list of antennas of various designs that can be used for communication systems with drones. But often during the development process the problem arises that it is impossible to use a complex antenna system due to certain operational limitations. Therefore, it is advisable to use relatively simple antenna designs. Such antennas are relatively easy to calculate and manufacture for any frequency range as needed. This paper discusses the basic principles of construction, modeling and manufacturing features of simple antenna structures that can be used in the ground and airborne segments of communication systems with unmanned aerial vehicles. Particular attention was paid to such types of antennas as dipole, discone antenna, collinear antenna, log-periodic antenna.

Keywords: Unmanned Aerial Vehicle · Dipole Antenna · Discone Antenna · Antenna Array · Collinear Antenna · Log-periodic Antenna

1 Introduction

1.1 Introduction to the Problem

Unmanned aerial vehicles (UAVs) and individual segments of their designs are being improved continuously and quite quickly. This is happening both in the defense sector, which is the main area of development today, and in the civilian sector. As is known, UAVs are oriented towards work in border missions, for humanitarian aid, search and rescue operations, scientific research, meteorology, fire monitoring, agriculture, infrastructure inspection, police surveillance, cargo delivery, transmission of communication signals, etc. There is no general classification system for UAVs by design yet. They can differ in weight and size, ranging from devices of several centimeters to aircraft with a wingspan of up to ten meters.

I. Ostroumov et al. (Eds.): ACASD 2025, LNNS 1418, pp. 231–250, 2025.
https://doi.org/10.1007/978-3-031-91992-3_15

The UAV complex comprises not only the aircraft itself but also a sophisticated system that communicates with the ground control station. This requires a large number of electronic devices for recording, receiving and transmitting data, as well as avionics functions. Antennas are one of the most important electronic devices in any UAV system.

Antennas used for the onboard and ground segments of UAV communication systems have several design and parameter requirements. These requirements directly depend on the specifics of the aircraft's use and mission [1, 2]. Let's try to group these requirements:

- radiation pattern (RP) is sometimes the most critical factor when choosing an antenna; basically, for the onboard communication segment, it makes sense to choose an omnidirectional RP, since the UAV occupies different positions in space during movement; for the ground segment of the communication system, different RP options are possible depending on the specific task;
- half power beamwidth (HPBW) is crucial in determining the coverage area of communication systems; when selecting omnidirectional antenna, HPBW is specified in the vertical plane; in the case of directional antenna, it is specified in two mutually perpendicular planes (according to the symmetry axis of the beam pattern);
- the gain directly depends on the previous two points of the list; the more directivity (lower HPBW) the selected antenna has, the greater the gain can be obtained and, accordingly, the greater the range of reliable communication;
- matching with the feeder is estimated by the value of the voltage standing wave ratio (VSWR); good matching allows avoiding significant power losses in the feeders of the communication system, which is inextricably linked with the radiation power of the antenna and, in accordance with the range; VSWR should be less than 1.5, or at least less than 2;
- compactness, robustness, simplicity and lightness of design; these requirements are valid for both on-board and ground antenna segments of UAV communications, since many tasks are associated with the need for rapid folding or deployment of the ground segment.

It can be concluded that all the listed antenna parameters are inseparably linked with each other. Changing one will lead to changing the other. That is, depending on the task, it is necessary to find such an antenna design that would allow to optimally combine all these characteristics.

Typically, both onboard and ground segments of the UAV communication system quite often include simple or modified linearly polarized dipole antennas, simple circularly polarized antennas (clover antenna or Pagoda antenna), or patch antennas [3]. For example, the article [4] presents broadband dipole blade antenna of the UAV onboard segment for frequencies from 20 MHz to 1200 MHz with the gain of no more than 4 dBi. The authors of the work [5] propose 3-blade and 4-blade clover-type antennas for the frequency of 2.45 GHz, containing a built-in reflector, which allows increasing the gain to 6.38 dBi. The advantages of such antennas include wide angular coverage, which is important characteristic when the UAV location is constantly changing.

However, when using omnidirectional dipole antennas, we cannot expect the gain greater than 2 dBi, which limits communication over long distances. Also, omnidirectional antennas are more vulnerable to electronic warfare (EW). To solve this problem, it is advisable to use either directional antennas or antenna arrays (AA) on the ground control segment [6]. For example, article [7] proposes a simple and compact Yagi-Uda antenna on a Rogers RO4003 substrate for use in multi-band radars (1.9, 2.5, and 3.5 GHz with a gain of 6.29, 4.63, and 6.77 dBi, respectively).

A combination of spiral elements of different lengths allows the implementation of spiral antennas for several ranges. The work [8] describes the design of the twelve-element slot helical antenna with right-hand circular polarization (RHCP) for frequencies of 1.227 and 1.575 GHz with the gain of 4.44 and 3.87 dBi, respectively. The principles of constructing wire and microstrip four-element helical antennas are given in [9, 10].

The article [11] compares the parameters of wideband microstrip log-periodic antennas (LPA) in the frequency band of 2–5 GHz using different dielectric substrates. As a result of modeling and experimental study of three samples, the gain of more than 6 dBi was obtained over the entire frequency band. As the result of research [12], LPA design was obtained for the frequency band of 1.4–12 GHz with the gain of 4.51 dBi.

The paper [13] presents the design of the 12-element slotted AA, which can provide HPBW in the meridional plane of about 7 degrees with the gain of 17 dBi. The paper [14] is devoted to development and study of the 10-element coaxial collinear AA for mobile satellite communication at the frequency of 4.2 GHz. This antenna provides a gain of about 4.5 dBi in the 20 and 240 degree directions, which is suitable for satellite communications. The paper [15] developed the 2×2 microstrip antenna design for the frequency of 3.8 GHz, having an HPBW of 33 degrees and the gain of 13.2 dBi. The compact two-element vertically polarized antenna array design for portable UAV remote control is considered in work [16]. This AA operates at the frequency of 2.4 GHz and gives the gain of 9.14 dBi at HPBW of 60 degrees in the E-plane and 65 degrees in the H-plane. The study [17] presents the design and implementation of four linear AA consisting of microstrip rectangular antennas. The linear AA operate in the frequency range from 2.62 to 2.69 GHz and provide cosecant square-law RP in the E-plane and HPBW of about 90 degrees in the H-plane with maximum gain of 11.1 dB to 12.2 dB in the operating frequency band.

In the studies above, the antennas are considered for different segments of radio electronics and telecommunications, but such designs can be easily adapted for UAV communication systems.

Often, researchers also propose special modified antenna designs for UAVs taking into account their features. In [18], the beam-reconfigurable antenna is proposed for the 5–5.5 GHz band with wide coverage angle by applying the beam combining technology of several antennas with different radiation patterns. The proposed system consists of circular patch antenna, low-profile printed meander monopole antenna and coplanar waveguide. The maximum gain, overall efficiency and HPBW of such an antenna system are approximately 5.8 dBi, 76%, 145 in the elevation plane and 360 in the azimuth plane, respectively. The paper [19] proposes the compact flat panel antenna with optimized radiation patterns for small UAVs. It operates in the 2.11–2.58 GHz and 5.06–7.5 GHz bands and is suitable for mounting on the outer side of the non-metallic wing of small UAVs.

1.2 Motivation

There is an extensive list of antennas of various designs on the market that can be used for UAV communication systems. But, quite often in the development process there is a problem of the lack of freely available antennas for non-standard frequency range or the impossibility of using complex antenna system due to certain operational limitations. Therefore, it is advisable to use relatively simple antenna designs that are well described in classical literature. Such antennas are relatively easy to calculate and manufacture for any frequency range as needed.

Theoretically, any type of antenna can be used for both ground and airborne communication segments. But in practice, everything depends on the needs, convenience and possibility of mounting the antenna on different types of housing. Let's consider several types of simple antennas of classic design without reference to the installation segment. For modeling, the Altair Feko 2024.0.1 software environment for modeling microwave electromagnetic structures was used. To measure the VSWR, the LiteVNA 64 and Rigexpert AA-1400 antenna analyzers were used.

1.3 Contribution

This paper contributes to the development of the theory and practice of antenna devices. The practical value is due to the fact that the article shows the features of the design and manufacture of simple antennas that can be used for both ground and airborne segments of the UAV communication system.

1.4 The Organization of the Paper

The paper is organized as follows. The first section describes the main types of antennas that can be used in the ground and airborne segments of UAV communications and analyzes the main problems in this area. The second section presents the principles of construction, modeling and experimental study of such antennas for UAV communication systems as a dipole, discone antenna, collinear antennas, log-periodic antenna, as well as some features of their manufacture. The last chapter presents conclusions and directions for future research.

2 Materials and Methods

2.1 Symmetrical Dipole Antennas

The symmetrical dipole is very widely used in the ranges from decameter to centimeter waves. Its extensive practical use is due to both the simplicity of the antenna itself and the simplicity of its engineering calculations [20–23]. Its importance for antenna technology is great: the symmetrical dipole is used as an independent antenna and as an element of antenna arrays.

The symmetrical dipole is two metal rods of the same shape, placed on one axis, between which high-frequency signal generator is connected (Fig. 1). The rods are

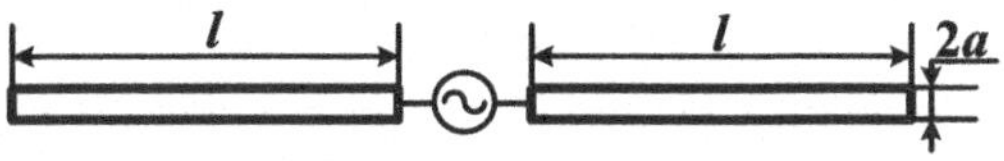

Fig. 1. Symmetrical dipole.

characterized by radius a, length l and are called dipole arms. Usually the inequality holds: $2a \ll 2l$.

The distribution of the dipole field in space is determined by the relative length of the dipole l/λ. For a half-wave dipole ($l = 0.25\lambda$), widely used in practice, the normalized directivity characteristic is determined by the formula

$$F(\theta) = cos(0.5\pi cos\theta)/sin\theta.$$

The HPBW of the half-wave dipole is $2\theta_{0.5} = 80°$. As the dipole relative length increases, the main lobe narrows, and side lobes appear and grow. An even greater increase in the dipole length leads to the fact that the radiation intensity in the direction perpendicular to the axis will be less than the radiation intensity in the direction of the side lobe maxima. And with the length of $l = \lambda$, there is no radiation in the azimuthal plane.

Approximate expression of the input impedance of a symmetrical dipole:

$$Z_A = \frac{R_\Sigma}{\left(R_\Sigma^2/W_A^2\right) + sin^2kl} - i\frac{W_A}{2}\frac{sin2kl}{\left(R_\Sigma^2/W_A^2\right) + sin^2kl} \approx \frac{R_\Sigma}{sin^2kl} - iW_A ctgkl, \quad (1)$$

where R_Σ is radiation impedance, $W_A = 120(ln(l/a) - 1)$ is dipole wave impedance, a is dipole radius.

Based on the nature of the change in input impedance of a symmetrical dipole, it can be concluded that the symmetrical dipole behaves as oscillatory circuit. At $l = 0.25\lambda$, as follows from formula (1), the input impedance is active $R_A = R_\Sigma = 73.1$ Ohm. As can be seen from formula (1), the input impedance depends significantly on the dipole wave impedance. The lower the wave impedance value, the narrower the limits of the active and reactive components of the input impedance when changing the wavelength. This property is used in the band antennas design, which have the following requirements:

- in the operating frequency range, the maximum radiation direction must remain unchanged;
- in the operating frequency range, the antenna input impedance must not change or its changes must be limited in such a way that the matching of the antenna to the feeder is satisfactory without adjustment ($VSWR < 1.5...2$).

As was said above, the maximum radiation direction of symmetrical dipole is perpendicular to the dipole axis when the wavelength changes within $l/\lambda \leq 0.7$. But the dipole input impedance changes very strongly. With the increase in the wire diameter, the linear capacitance increases, resulting in decrease in the wave impedance. Therefore, by making the dipole arms in the form of large-diameter metal cylinders, it is possible to significantly reduce the wave impedance of the antenna and reduce the fluctuations in the input impedance when changing the frequency.

In this work, three samples of symmetric dipoles with different modifications of the arms were modeled, manufactured and investigated.

Symmetrical Dipole with Extended Radius of Arms The first sample of symmetrical half-wave dipole was made according to the classical principle. The arms were made of brass cylinders with a diameter of 2.5 cm (Fig. 2). A greater increase in diameter was unacceptable due to the limitation of installation on the UAV body. The main characteristics of this design are shown in Figs. 3, 4 and 5.

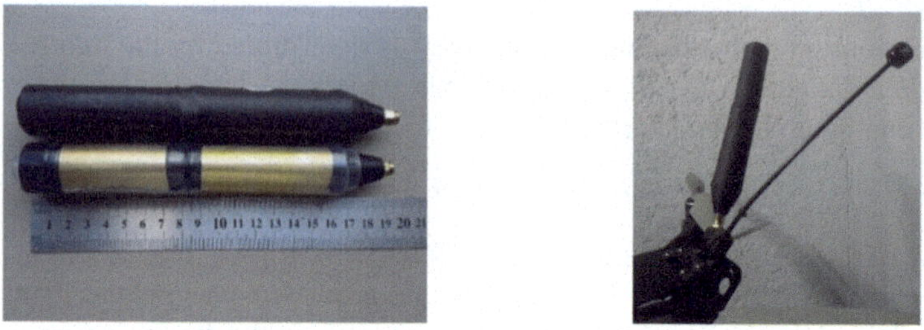

Fig. 2. Symmetrical dipole antenna with extended arm radius and the example of placement on the UAV body.

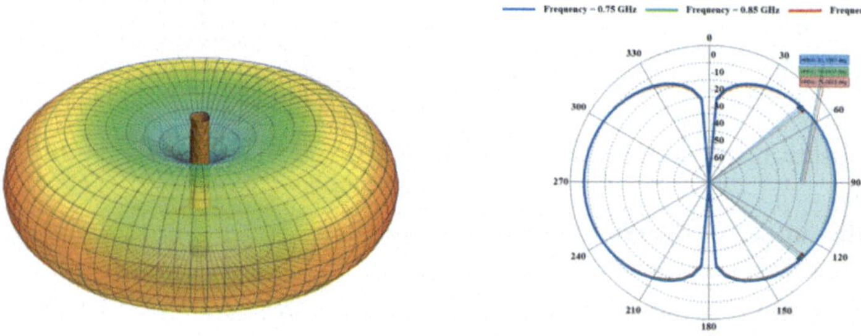

a *b*

Fig. 3. Radiation patterns of the dipole antenna: *a* – 3D; *b* – in the *E*-plane.

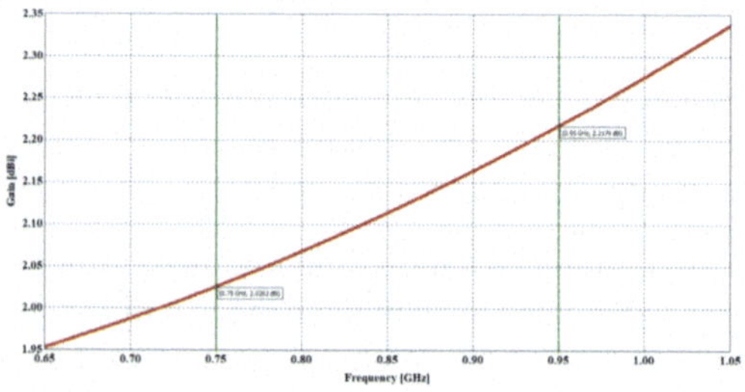

Fig. 4. Dependences of the gain of the dipole antenna on the frequency.

Flat Dipole Antenna 500–950 MHz. When it is necessary to reduce one of the transverse dimensions of the dipole antenna and simplify the manufacture of the dipole arms,

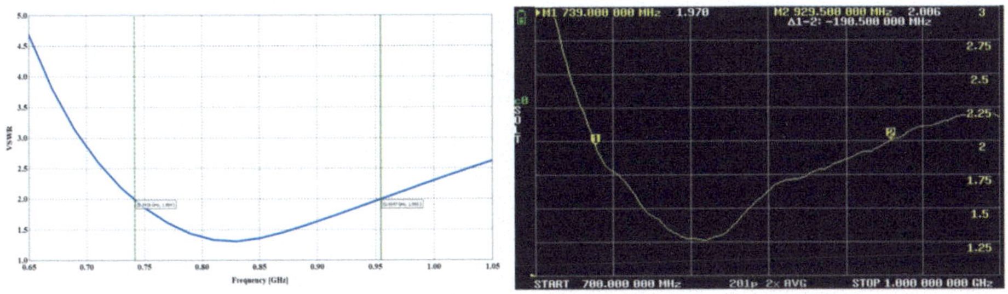

a *b*

Fig. 5. Dependences of the VSWR of the dipole antenna on the frequency: *a* – modeling; *b* – experiment.

they can be made flat. But for the similar frequency band (in the previous example it is 150 MHz) the dipole width can increase significantly. This can be partially compensated by different arm configurations.

If it is possible to install wider flat dipole design on the UAV body (Fig. 6), it is possible to significantly expand the operating frequency band for communication. In some cases, this is necessary for countering EW. The main radiation and matching characteristics of such flat dipole are shown in Figs. 7, 8 and 9.

Fig. 6. Flat dipole antenna 500–950 MHz and the example of placement on the UAV body.

Analyzing the graphs of the RP (Fig. 7c) and the dependence of the gain on operating frequency (Fig. 8), one can note the only drawback of such expansion of the flat dipole. In the *H*-plane, with increasing operating frequency, the gain begins to decrease in the direction of the dipole plate edges. This disadvantage can be compensated by using two dipoles simultaneously in the system (similar to an AA), as shown in Fig. 6.

2.2 Discone Antennas

In cases where it is necessary to provide omnidirectional radiation in wide frequency range, a discone antenna (DCA) is used (Fig. 10).

The disk with diameter 2*R* is connected to the coaxial cable central conductor. The cone, the generatrix of which has the length l, is connected at the top to the coaxial cable braid. The cone angle affects the bandwidth and the input impedance of the antenna. To

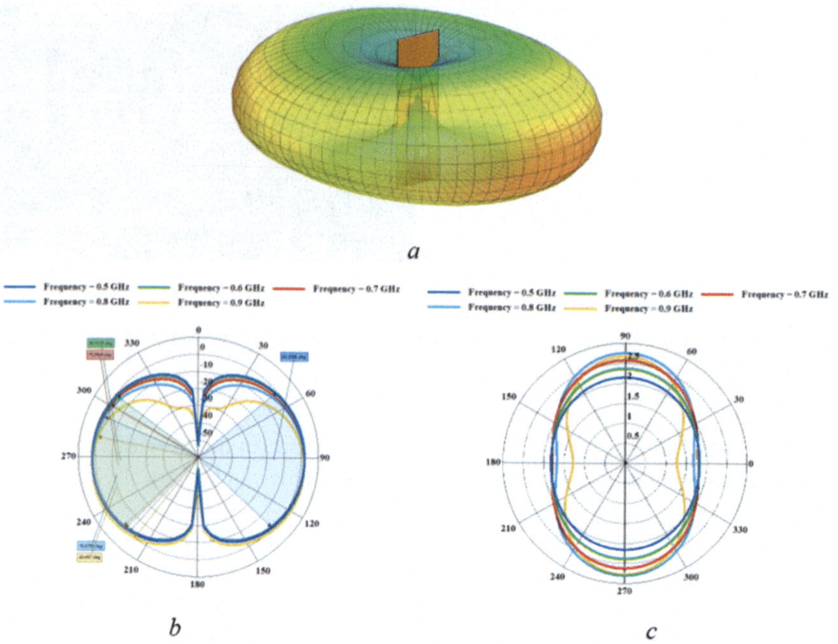

Fig. 7. Radiation patterns of the flat dipole 500–950 MHz: *a* – 3D; *b* – in the *E*-plane; *c* – in the *H*-plane.

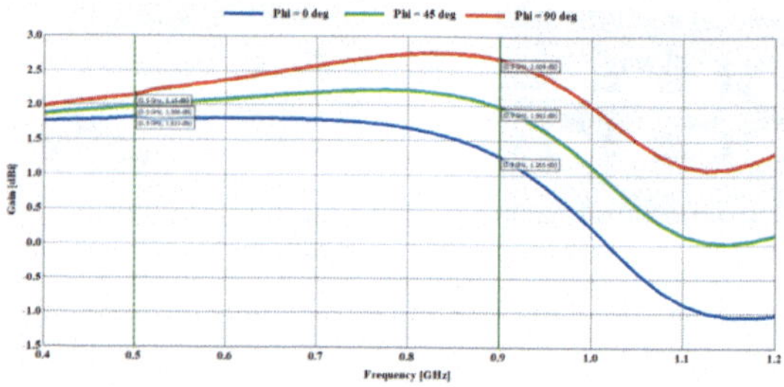

Fig. 8. Dependences of the gain of the flat dipole 500–950 MHz on the frequency.

match the antenna with standard feeder, the angle ψ is taken within the range of $15 - 30°$. As the angle ψ increases, the bandwidth increases due to the increase in the upper limit frequency. The coaxial diameter d narrows the operating frequency band when increased. The lowest operating frequency at which the VSWR is 2 is called the cutoff frequency, since further decrease in frequency leads to sharp increase in the VSWR. The longest wavelength corresponding to the cutoff frequency is called the limiting wavelength and is determined as $\lambda_{lim} \approx 3.6\, l$. The distance between the disk and the cone top s also affects the operating frequency band. As the size s decreases, the frequency band expands toward higher frequencies. The relation between the antenna's design dimensions are as follows:

– the generatrix length is determined through the limiting wavelength: $l = 0.28\lambda_{lim}$;

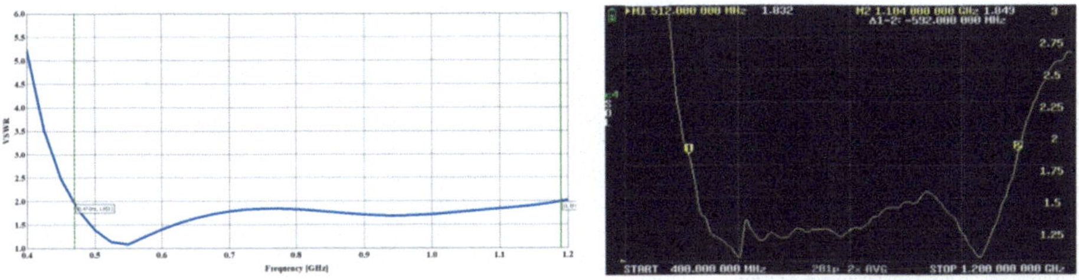

a *b*

Fig. 9. Dependences of the VSWR of the flat dipole 500–950 MHz on the frequency: *a* – modeling; *b* – experiment.

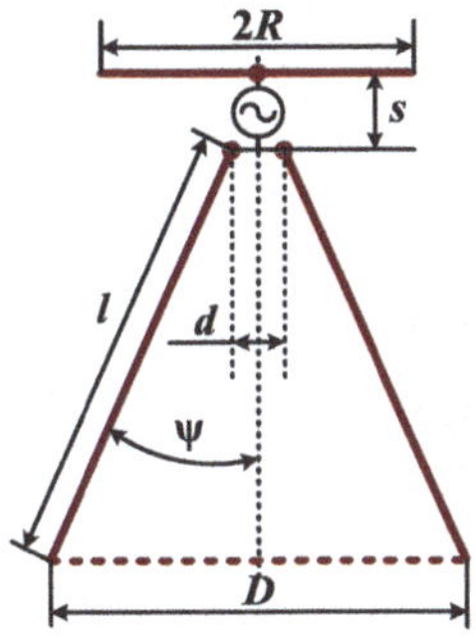

Fig. 10. Structure of the discone antenna.

- the diameter of the cone base depends on the generatrix length, the angle ψ and the coaxial diameter at the cone top d: $D = 2(lsin\psi + 0.5d)$;
- the optimal value of the distance between the disk and the cone top is determined as $s = 0.3d$;
- the diameter of the disk is chosen equal to $2R = 0.7D$.

If the coaxial cable is used, the characteristic impedance of which is 50 Ohms, then the VSWR will be no more than 2, that is, the antenna input impedance in the operating frequency band matches well with the feeder characteristic impedance. The overlap coefficient of the DCA range can reach values of 4 and more.

In this work, two samples of the DCA for two ranges were modeled, manufactured and studied: 1.6–2.6 GHz (Fig. 11) and 300–600 MHz (Fig. 15). Brass with the thickness of 0.2 mm was chosen as the manufacturing material. The main characteristics of the first sample of the DCA are presented in Figs. 12, 13, 14 and 15, the second sample of the DCA – in Figs. 16, 17 and 18.

As can be seen, the DCA for the frequency range of 1.6–2.6 GHz, due to its dimensions, is ideal as omnidirectional broadband antenna for the onboard segment of UAV communication systems. The second sample (300–600 MHz), due to its relatively large dimensions, can be used if there is a need for omnidirectional reception/transmission of communication signals in wide frequency range on the ground segment. If necessary, the frequency band can be further expanded by increasing the cone angle.

Fig. 11. The DCA 1.6–2.6 GHz and an example of its possible placement on the UAV body.

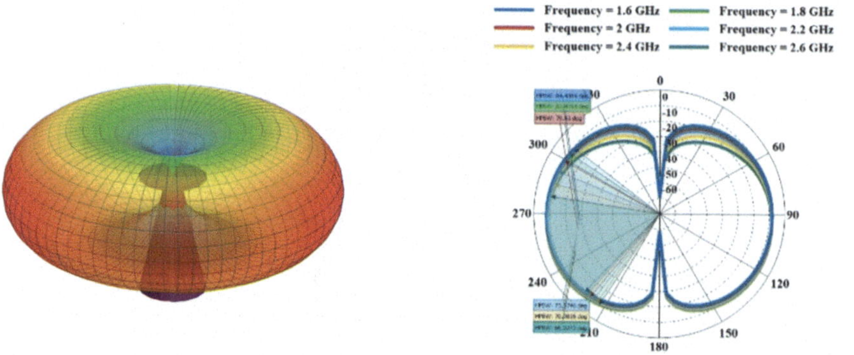

a *b*

Fig. 12. Radiation patterns of the DCA 1.6–2.6 GHz: *a* – 3D; *b* – in the *E*-plane.

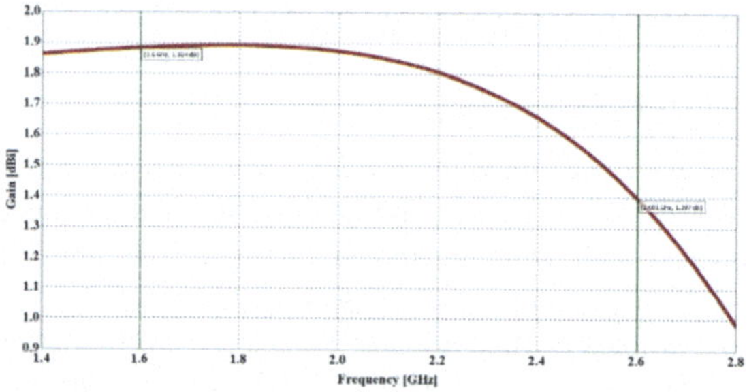

Fig. 13. Dependences of the gain of the DCA 1.6–2.6 GHz on the frequency.

2.3 Collinear Antennas

The antenna designs described above, despite the wide operating frequency band, have one common drawback. This is the low gain (about 2 dBi). There are cases when it is necessary to increase the range of the system without increasing the radiation power. Antenna arrays are used for this. Theoretically, it is possible to build AAs using the

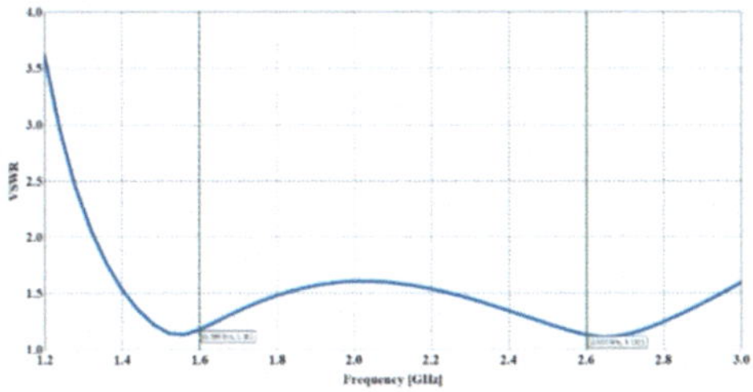
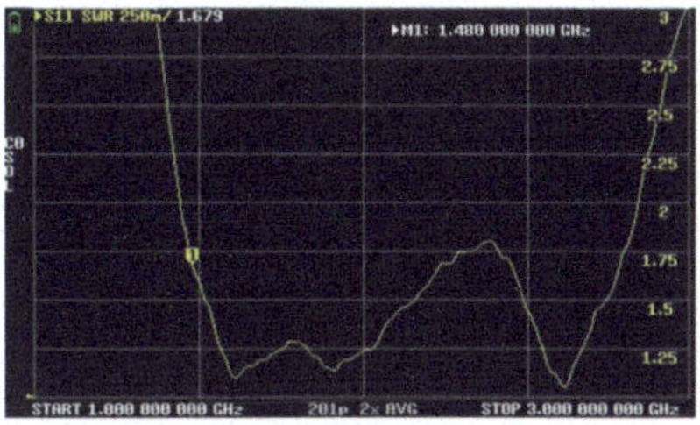

a *b*

Fig. 14. Dependences of the VSWR of the DCA 1.6–2.6 GHz on the frequency: *a* – modeling; *b* – experiment.

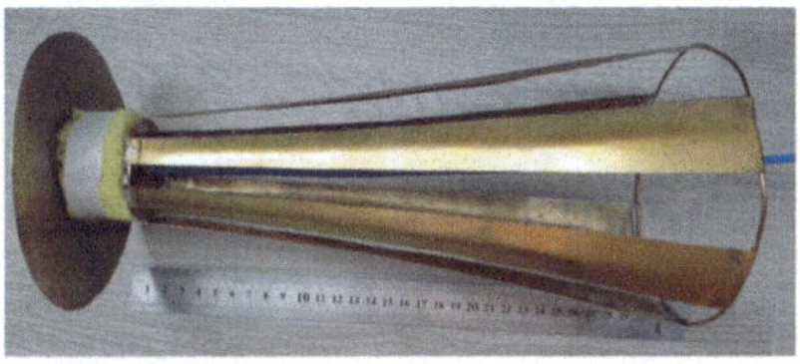

Fig. 15. The DCA 300–600 MHz.

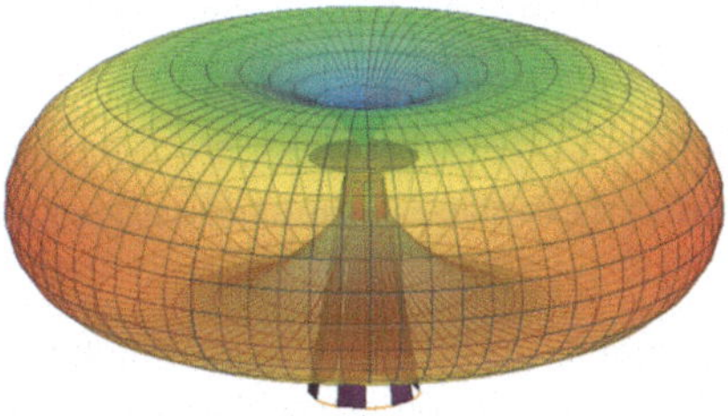
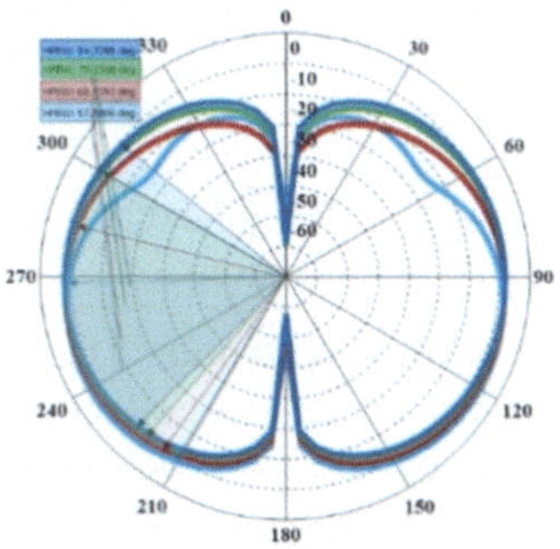

a *b*

Fig. 16. Radiation patterns of the DCA 300–600 MHz: *a* – 3D; *b* – in the *E*-plane.

broadband elements described above. But this will lead to significant increase in size, installation inconvenience and complexity of use in mobile systems. Therefore, there is a need to manufacture other AA designs.

Collinear antennas are widely used, although they do not have such wide operating frequency band as symmetrical and asymmetrical dipoles with modified arm design. But their gain is higher than 5 dBi, depending on the elements number. Collinear antennas of series type belong to the class of linear in-phase AA, which are made of symmetrical dipoles of different lengths and designs. Let's consider two types of collinear antennas,

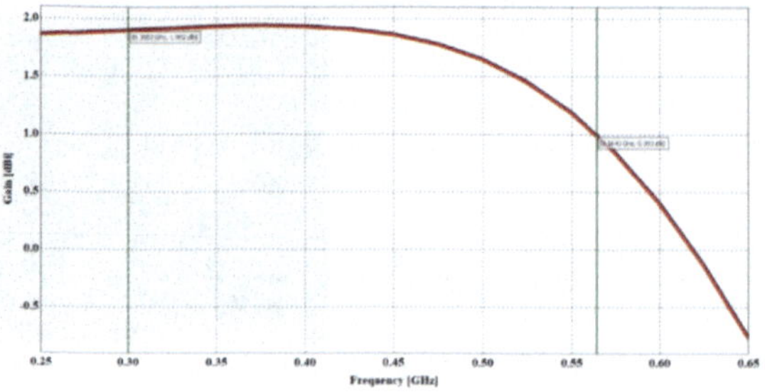

Fig. 17. Dependences of the gain of the DCA 300–600 MHz on the frequency.

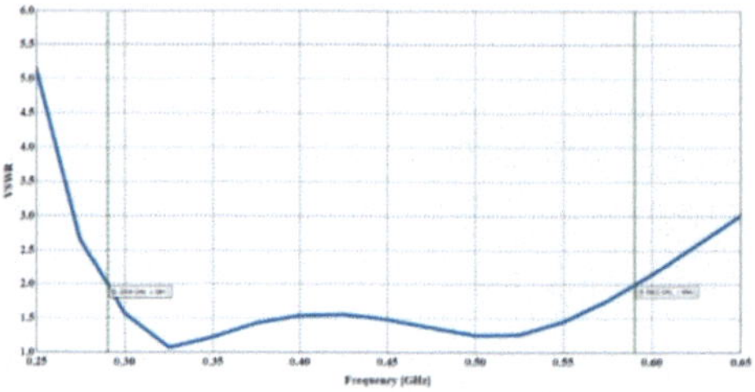

a

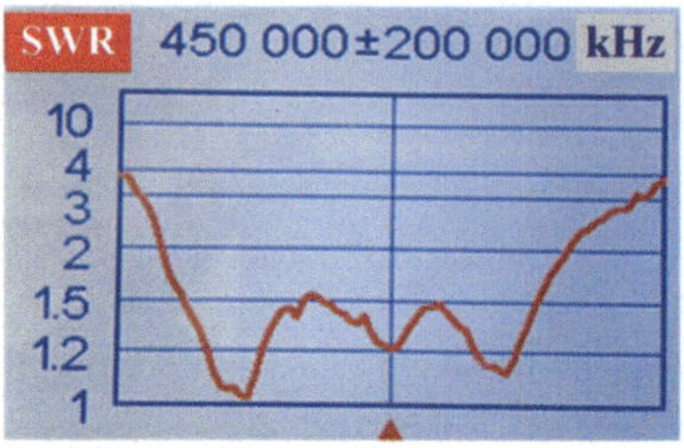

b

Fig. 18. Dependences of the VSWR of the DCA 300–600 MHz on the frequency: *a* – modeling; *b* – experiment.

which, by their design and parameters, have found their place in the on-board segment of UAV communications.

Wire Collinear Antenna. The wire collinear antenna (Fig. 19a) consists of metal rods, 0.5λ long, and inductive coils connecting the rods. With the appropriate choice of generator frequency, half the wavelength is laid on the rods and the rods radiate electromagnetic waves in the same way as half-wave symmetrical dipoles. The inductive coils absorb half the standing wavelength, which ensures in-phase excitation of the dipoles. The wire collinear antenna (Fig. 19, *b*) was investigated for the frequency range of 2.02–2.11 GHz. The main characteristics are presented in Figs. 20, 21, and 22.

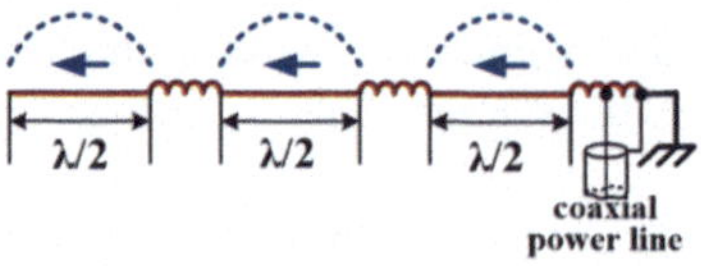

a

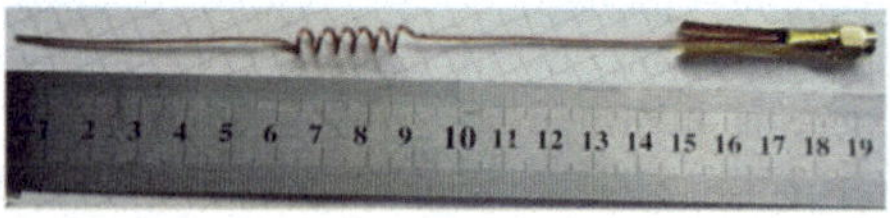

b

Fig. 19. The wire collinear antenna.

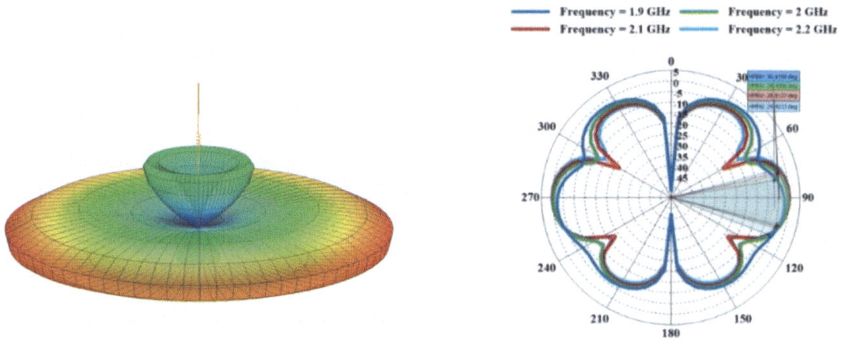

a *b*

Fig. 20. Radiation patterns of the wire collinear antenna: *a* – 3D; *b* – in the *E*-plane.

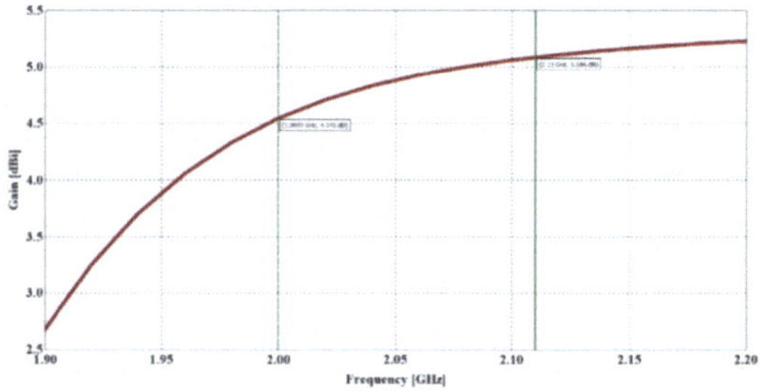

Fig. 21. Dependences of the gain of the wire collinear antenna on the frequency.

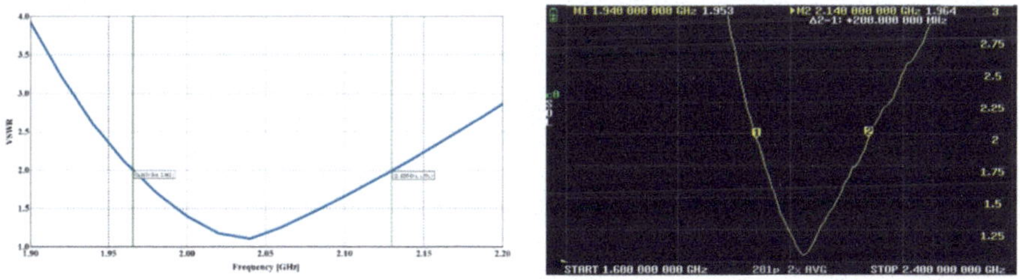

a *b*

Fig. 22. Dependences of the VSWR of the wire collinear antenna on the frequency: *a* – modeling; *b* – experiment.

Coaxial Collinear Antenna. The design of collinear antenna from sections of rigid coaxial cable (Fig. 23a) arose as a result of an attempt to create more technologically advanced designs. Each section of the coaxial cable is a dipole. The sections are connected crosswise, which ensures the currents are in phase on the outer surface of the tubular rods, provided that half wavelength is placed on each section. In-phase currents also flow on the central conductors of the coaxial cable. In this work, the coaxial collinear antenna (Fig. 23, *b*) using RG-402 coaxial cable was investigated for the frequency band of 5.5–5.8 GHz. The main results of the coaxial collinear antenna study are presented in Figs. 24, 25 and 26.

a *b*

Fig. 23. The coaxial collinear antenna.

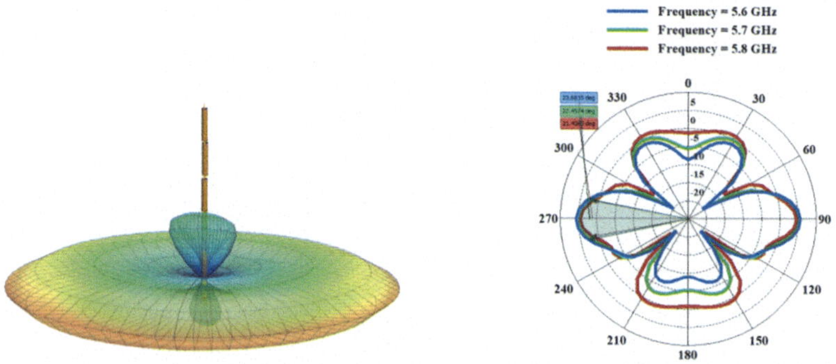

a *b*

Fig. 24. Radiation patterns of the coaxial collinear antenna: *a* – 3D; *b* – in the *E*-plane.

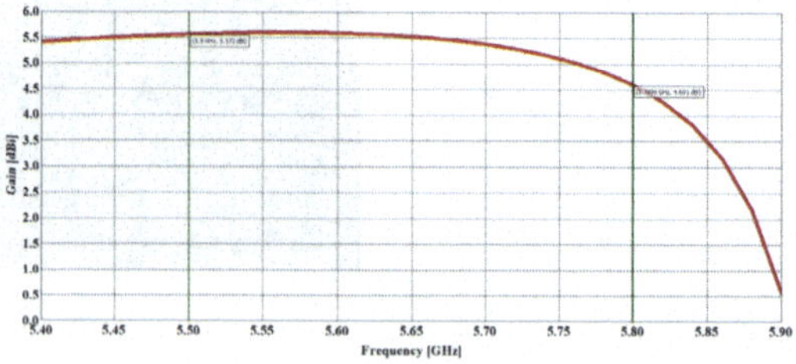

Fig. 25. Dependences of the gain of the coaxial collinear antenna on the frequency.

The experimental study showed that the antenna operating frequency band reaches more than 300 MHz at VSWR <2. The experimentally obtained frequency band is almost three times larger than that obtained in the simulation. This is due to the physical characteristics of the coaxial cable used for construction, since it is impossible to fully take them into account in the simulation.

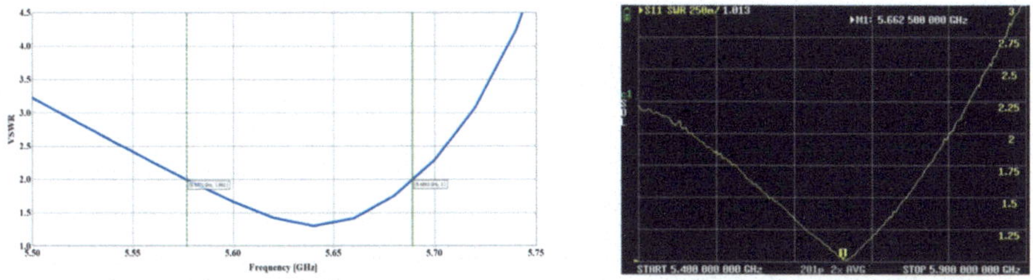

a *b*

Fig. 26. Dependences of the VSWR of the coaxial collinear antenna on the frequency: *a* – modeling; *b* – experiment.

2.4 Log-Periodic Antennas

UAV communication systems often require directional antennas. This allows not only to increase the range of the system, but also to provide spatial filtering. Antennas of this type can be used both in the ground and in the airborne segment of UAV communication. But they have found the widest application in the ground segment, since the airborne installation requires the use of a rotary device, which is impossible in many cases.

Log-periodic antennas (LPA) are non-equidistant unequal-amplitude AA with axial radiation. The process of constructing microstrip LPAs is described in detail in [4, 9, 18]. LPA belongs to the class of ultra-wideband antennas. Its range coverage coefficient reaches values greater than 4. Flat LPA has the form of a linear AA of symmetric dipoles, the length of which changes monotonically (Fig. 27).

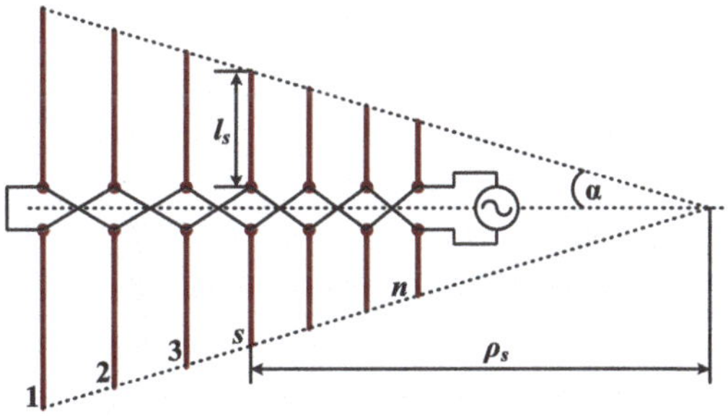

Fig. 27. LPA design.

The exact algorithm for calculating the LPA is quite complex. The main ratios of the antenna structure dimensions are:

- the tangent of the antenna opening angle: $l_1/\rho_1 = l_2/\rho_2 = \ldots = l_n/\rho_n = tg(\alpha/2)$;
- relative diameter of the *s*th dipole rod: $\sigma = a_s/\rho_s = const$;
- relative distance between adjacent dipoles: $\tau = \rho_{s+1}/\rho_s = const$.

In this paper, two samples of microstrip LPA are considered. The first sample of LPA (Fig. 28) was manufactured for a relatively narrow frequency range of 750–1150 MHz.

Dielectric substrate FR-4 was used for construction. The main results of the study of the LPA 750–1150 MHz are presented in Figs. 29, 30 and 31.

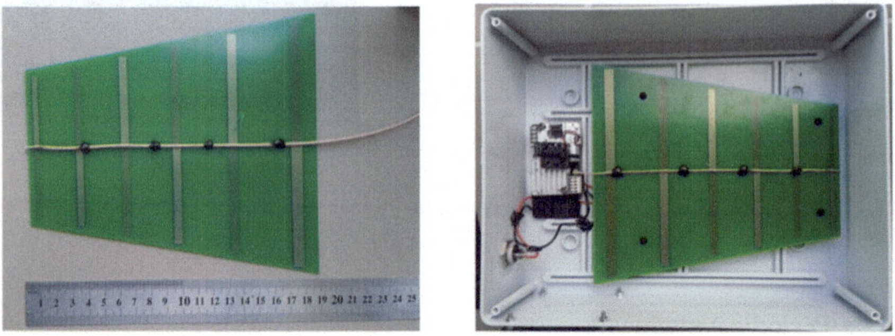

Fig. 28. The LPA 750–1150 MHz and the example of its placement in the ground antenna unit of the UAV communication system.

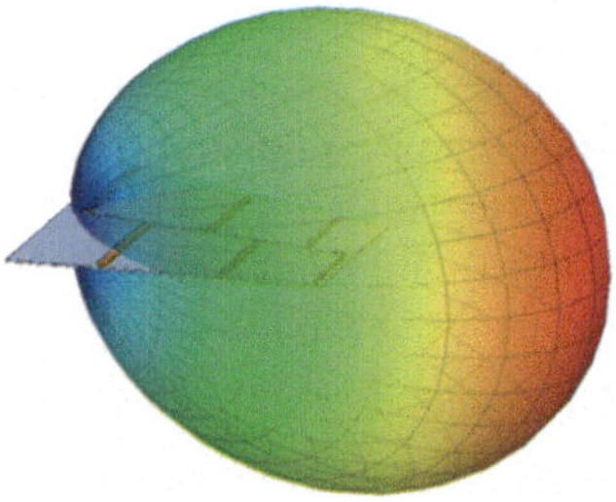

a

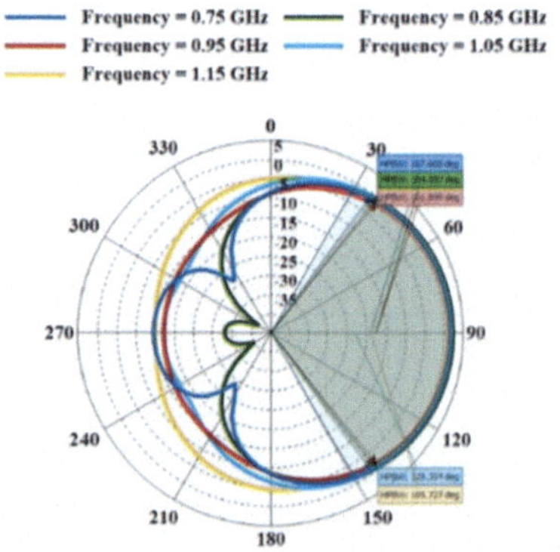

b

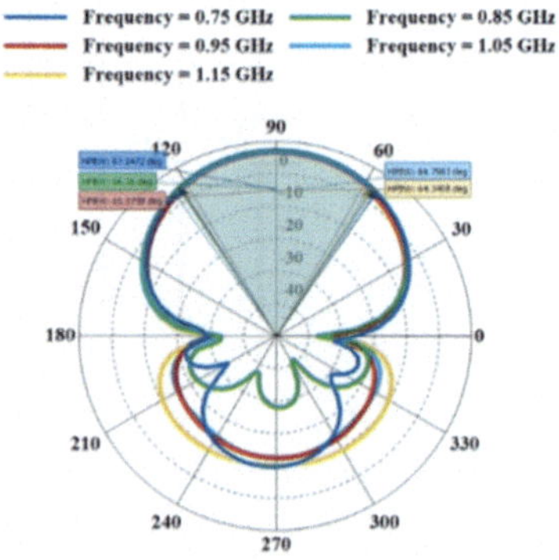

c

Fig. 29. Radiation patterns of the LPA 750–1150 MHz: *a* – 3D; *b* – in the *E*-plane; *c* – in the *H*-plane.

The second LPA sample was manufactured for a wider frequency range of 2–6 GHz. The Roger RO4003C dielectric substrate was used to construct the microstrip LPA

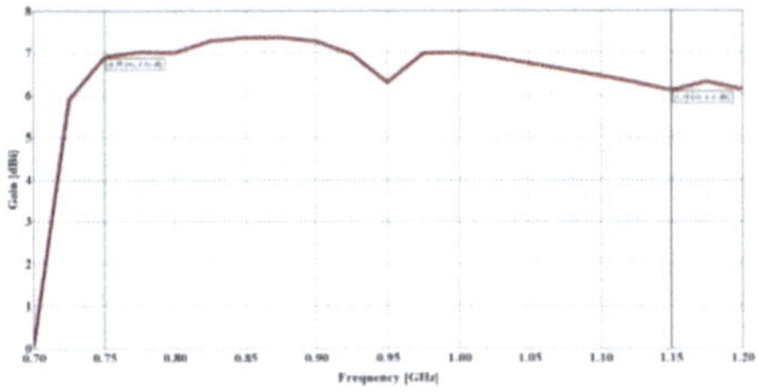

Fig. 30. Dependences of the gain of the LPA 750–1150 MHz on the frequency.

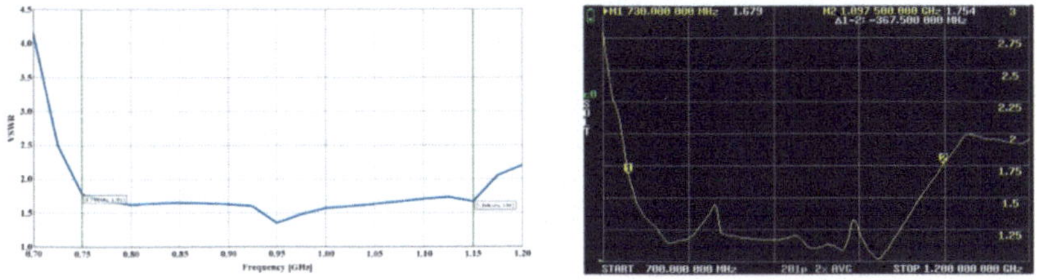

a *b*

Fig. 31. Dependences of the VSWR of the LPA 750–1150 MHz on the frequency: *a* – modeling; *b* – experiment.

(Fig. 32). The main results of the modeling and research are presented in Figs. 33, 34 and 35.

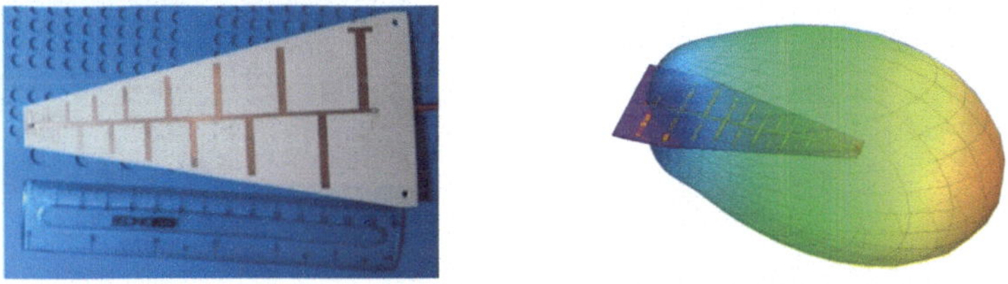

Fig. 32. The LPA 2–6 GHz and its 3D radiation pattern.

For the production of microstrip LPAs for frequencies above 3 GHz, it is recommended to use special ultra-high-frequency substrates (for example, from Roger), since standard substrates for printed circuit boards do not guarantee the stability of their parameters over wide frequency range.

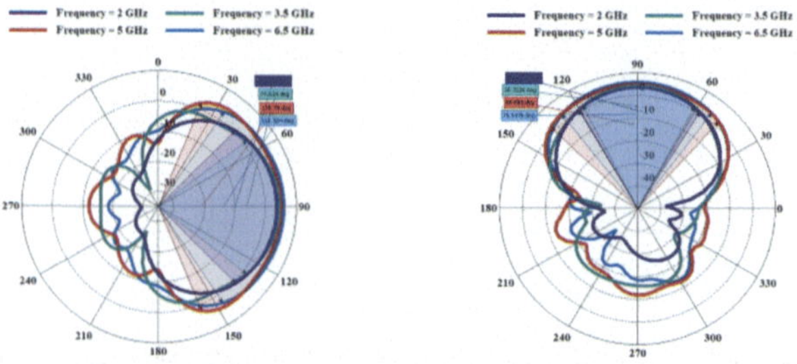

a *b*

Fig. 33. Radiation patterns of the LPA 2–6 GHz: *a* – in the *E*-plane; *b* – in the *H*-plane.

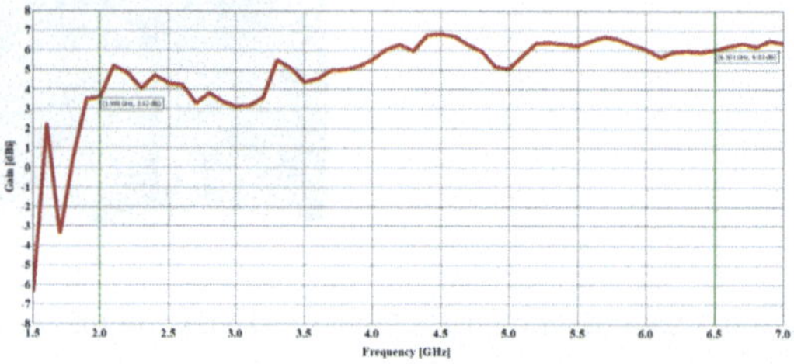

Fig. 34. Dependences of the gain of the LPA 2–6 GHz on the frequency.

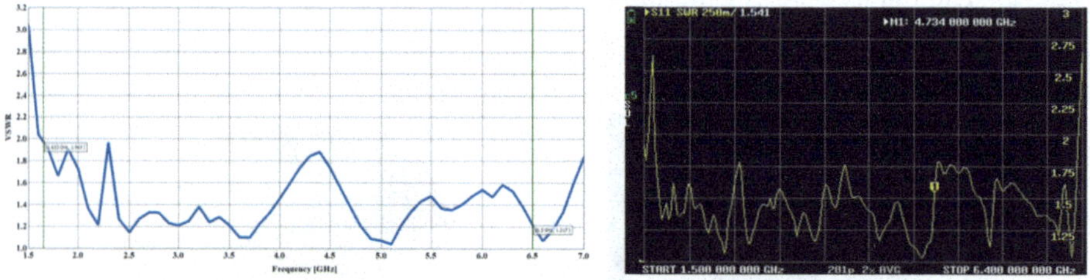

a *b*

Fig. 35. Dependences of the VSWR of the LPA 2–6 GHz on the frequency: *a* – modeling; *b* – experiment.

3 Conclusions

Antennas are among the most critical components in UAV communication systems, as their characteristics significantly impact the quality and reliability of communication between the aircraft and other systems. UAV antennas must have a simple, robust and compact design, and be easy to mount on both the onboard and ground segments of the UAV complex.

This article provides detailed discussion of the design principles of such simple antenna designs as dipole antennas and discone antennas for the onboard communication segment, as well as log-periodic antenna for the ground segment. The main quality characteristics of these antenna types, such as radiation pattern, gain, HPBW, and VSWR, were also modeled. The dependence of VSWR on the operating frequency was measured using antenna analyzers.

After theoretical substantiation, modeling and experimental study of each type of antenna, the authors of the article indicated several features and recommendations that can be useful for both beginners and professionals in the field of antenna design.

The future research directions are:

– experimental study of the communication system range with UAVs using the prototypes described in this work;
– study of increasing the range and accuracy of the communication system with UAVs using directional AA with electronic amplitude or phase control on the onboard segment;
– study of the combination of an automatic UAV tracking system on the onboard and ground segments of communication.

References

1. Ivashchuk, O., et al.: A configuration analysis of Ukrainian flight routes network. In: 16th International Conference on the Experience of Designing and Application of CAD Systems (CADSM), pp. 6–10. IEEE, Lviv (2021). https://doi.org/10.1109/CADSM52681.2021.9385263
2. Averyanova, Y., et al.: Polarimetric-radar drop size evaluation for wind speed estimate based on Weber criterion. In: Signal Processing Symposium (SPSympo), pp. 17–22. IEEE, Lodz (2021). https://doi.org/10.1109/SPSympo51155.2020.9593349
3. Mahfuz, M.M., Park, C.-W.: Review of patch antennas used in drone applications. IEEE Access. **11**, 58367–58388 (2023). https://doi.org/10.1109/ACCESS.2023.3284040
4. Nosrati, M., Jafargholi, A., Pazoki, R., Tavassolian, N.: Broadband slotted blade dipole antenna for airborne UAV applications. IEEE Trans. Antennas Propag. **66**(8), 3857–3864 (2018). https://doi.org/10.1109/tap.2018.2835524
5. Safaron, M.S., Majid, H.A., Esmail, B.A.F., Ab Ghafar, A.S., Saparudin, F.A., Ismail, M.F., Abdullah, M.A.: Directional cloverleaf antenna for unmanned aerial vehicle (UAV) application. Indones. J. Electr. Eng. Comput. Sci. **14**, 773–780 (2019). https://doi.org/10.11591/ijeecs.v14.i2
6. Shcherbyna, O., Zadorozhnyi, O., Stetsyshin, O.: Passive antenna arrays in UAV communication systems. Int. J. Comput. Netw. Inf. Secur. **16**(4), 31–51 (2024). https://doi.org/10.5815/ijcnis.2024.04.03
7. Cheong, P., Wu, K., Choi, W.-W., Tam, K.-W.: Yagi–Uda antenna for multiband radar applications. IEEE Antennas Wirel. Propag. Lett. **13**, 1065–1068 (2014). https://doi.org/10.1109/LAWP.2014.2328991
8. Yang, Y.-H., Guo, J.-L., Sun, B.-H., Huang, Y.-H.: Dual-band slot helix antenna for global positioning satellite applications. IEEE Trans. Antennas Propag. **64**(12), 5146–5152 (2016). https://doi.org/10.1109/TAP.2016.2623647
9. Shcherbyna, O., Tomai, O., Kozhokhina, O.: Quadrifilar helical antennas with different types of supply lines. In: International Conference on the Advances in Wireless and Optical Communications (RTUWO), pp. 167–170. IEEE, Riga (2018). https://doi.org/10.1109/RTUWO.2018.8587877
10. Shcherbyna, O., Kozhokhina, O.: Construction principles of quadrifilar helical antenna. Telecomm. Radio Eng. **79**(16), 1441–1453 (2020). https://doi.org/10.1615/TelecomRadEng.v79.i16.30

11. Shcherbyna, O., Zadorozhniy, R.: The log-periodic dipole array antenna for monitoring. In: 14th International Conference on Advanced Trends in Radioelectronics, Telecommunications and Computer Engineering (TCSET), pp. 583–586. IEEE, Lviv-Slavske (2018). https://doi.org/10.1109/TCSET.2018.8336270
12. Pojang, K., Rakluea, P.: The design of log periodic dipole array antenna for WLAN/LTE/UWB applications. In: 18th International Symposium on Communications and Information Technologies (ISCIT), pp. 66–69. IEEE, Bangkok (2018). https://doi.org/10.1109/ISCIT.2018.8587918
13. Lubis, M.A.K.S., Apriono, C., Zulkifli, F.Y., Rahardjo, E.T.: Design of narrow wall slotted waveguide antenna for X-band application. In: Progress in Electromagnetics Research Symposium–Fall (PIERS–FALL), pp. 2625–2628. IEEE, Singapore (2017). https://doi.org/10.1109/PIERS-FALL.2017.8293579
14. Effendi, M.R., Ernanto, R., Munir, A.: 4.2GHz compact collinear antenna for manpack satellite communication. In: International Workshop on Antenna Technology (iWAT), pp. 1–4. IEEE, Nanjing (2018). https://doi.org/10.1109/IWAT.2018.8379237
15. Sajjad, H., Sethi, W.T., Zeb, K., Mairaj, A.: Microstrip patch antenna array at 3.8 GHz for WiMax and UAV applications. In: International Workshop on Antenna Technology: Small Antennas, Novel EM Structures and Materials, and Applications (iWAT), pp. 107–110. IEEE, Sydney (2014). https://doi.org/10.1109/IWAT.2014.6958609
16. Chuang, T.-Y., Liao, W.-J., Ma, T.-G., Lee, Y., Ho, M.-C.: Compact directive array antenna design for UAV application. In: International Symposium on Antennas and Propagation (ISAP), pp. 1–2. IEEE, Phuket (2017). https://doi.org/10.1109/ISANP.2017.8228780
17. Ruiz, P.M., Begaud, X., Magne, F., Leder, E., Khy, A.: Microstrip antenna array design for unmanned aerial vehicles detection radar. Adv. Electromagn. **12**(3), 1–9 (2023). https://doi.org/10.7716/aem.v12i3.2066
18. Choi, Y.-S., Park, J.-S., Lee, W.-S.: Beam-reconfigurable multi-antenna system with beam-combining technology for UAV-to-everything communications. Electronics. **9**(6), 980 (2020). https://doi.org/10.3390/electronics9060980
19. Yang, Y.-L., Lin, D.-B.: A compact planar Wi-Fi antenna with optimized radiation patterns for small UAV applications. Appl. Sci. **13**(13), 7470 (2023). https://doi.org/10.3390/app13137470
20. Balanis, C.A.: Antenna Theory: Analysis and Design, 4th edn. John Wiley & Sons, New Jersey (2016)
21. Milligan, T.A.: Modern Antenna Design. John Wiley & Sons, New Jersey (2005)
22. Huang, Y.: Antennas: From Theory to Practice, 2nd edn. John Wiley & Sons, New Jersey (2021)
23. Shcherbyna, O., et al.: Diagnostic process efficiency analysis for block diagram of electric field parameters meter. In: IEEE XIIth International Conference on Electronics and Information Technologies (ELIT), pp. 5–9. IEEE, Lviv (2021). https://doi.org/10.1109/ELIT53502.2021.9501136

Frequency Synthesizer of X Band

Leonid Sibruk(✉) and Yaroslav Gritsev

National Aviation University, Liubomyra Huzara Ave., 1, Kyiv 03058, Ukraine
leonid.sibruk@npp.kai.edu.ua

Abstract. The paper introduces approaches to the development and implementation of an X-band frequency synthesizer, which is built on the basis of available components and is characterized by a small frequency change step and a satisfactory level of phase noise. The paper briefly deals with the features of various analog and digital phase-locked loop (PLL) devices, including those based on dielectric resonators. Main advantages and drawbacks of the digital signal synthesis (DSS) oscillators are considered. The functional diagram of the developed frequency synthesizer includes two PLLs and DDS circuit. The first PLL is DR series specifically for PLL use (PLDRO) with 8 GHz output oscillations and −132.85 dBc/Hz phase noise at 1 kHz offset, 1 kHz bandwidth and 10 MHz tuning band. The low noise output oscillations of Direct Digital Synthesis are in the lower band, but their bandwidth is 100 MHz and step is 400 Hz. In the second PLL the frequency convertor substitutes the high-N frequency divider. The synthesizer oscillation frequency varies in the range of 1...1.5 GHz with a step of 1 kHz. The phase noise is −98.32 dBc/Hz at 10 kHz offset. The paper also deals with frequency and standing wave ratio by voltage (VSWR) responses of used low pass and pass band filters.

Keywords: Frequency Synthesizer · Phase-locked Loop · Digital Signal Synthesis · Phase Noise · Dielectric Resonator oscillator · Phased Detector · Voltage Control Oscillator

1 Introduction

1.1 Introduction to the Problem

Frequency synthesizers are part of almost all transmitters and receivers for various applications. They generate signals in definite frequency band with crystal stability. Frequency synthesizers are built on the basis of phase-locked loop (PLL) or digital signal synthesis (DSS). This article deals with some aspects of frequency synthesizers design.

PLL synthesizers are widely used due versatility and relative simplicity. The generalized structure of synthesizer (Fig. 1) includes PLL and reference highly stable oscillator. The phases of waveforms from reference generator and oscillations, connected with synthesizer output via divider, are compared in phase detector (PFD). A linearized

I. Ostroumov et al. (Eds.): ACASD 2025, LNNS 1418, pp. 251–261, 2025.
https://doi.org/10.1007/978-3-031-91992-3_16

differential equation for the phase difference Φ_Δ of PFD input signals at condition of $\Phi_\Delta \lhd 90°$ [1, 2].

$$d\Phi_\Delta/dt = \omega_\Delta - K\Phi_\Delta, \tag{1}$$

where ω_Δ is the frequency difference of oscillations on the phase detector inputs; K is the loop factor, which value depends on PFD and voltage control oscillator (VCO) responses.

Solution of (1)

$$\Phi_\Delta(t) = e^{-Kt}(\Phi_{\Delta 0} - \omega_\Delta/K) + \omega_\Delta/K, \tag{2}$$

where $\Phi_{\Delta 0}$ is initial phase angle.

One of the main synthesizer parameters is a level of phase noise, which is understood as ratio of noise power to carrier power over one hertz. Noise level is measured at definite offsets from carrier frequency. In physical sense, phase noise is random small variations of the signal phase or its period in the time plane. This topic has been the subject of many studies, including the use of frequency conversion instead of its division, diminishing and avoiding frequency division in favor of multiplication, and other approaches that have improved the quality of operation [3–5]. In case of square waveform oscillations, the phase noise is diminished using charge-pump PLL (CP-PLL) [6].

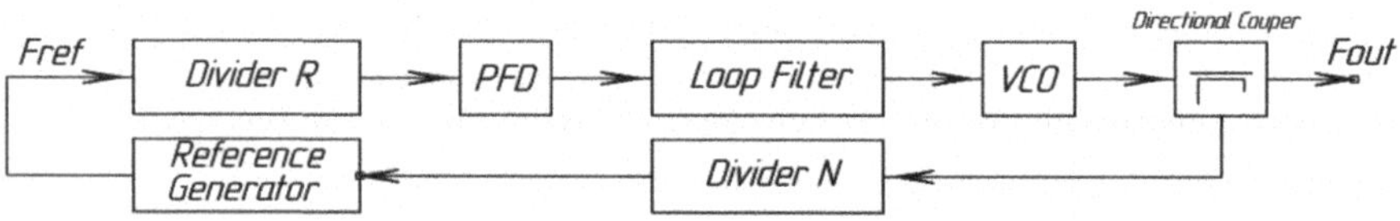

Fig. 1. The PLL synthesizer structure.

Phase noise causes the appearance of unwanted sidebands in the carrier spectrum, which reduces the ability to distinguish a useful signal from noise. This results in constellation rotating and higher BER levels in communication systems, masking of the reflected target signal and complicates the processing of Doppler data in radar technology, including Synthetic Aperture Radar [7–9].

Digital PLL devices have found wider application compared to analog ones. The frequency change in digital PLLs occurs discretely with a certain step, which is determined by the construction of the phase detector. When the tuning step decreases, the frequency division coefficients N grow and, accordingly, the complexity increase. The combined use of digital and analog circuits allows to reduce the coefficient N and obtain a high output frequency thanks to the analog PLL.

Sub-Sampling PLL (SS-PLL) oscillators [10] are a component of networks, which elements operate synchronously. The factor N is absent and the level of phase noise is low.

Bang-Bang PLL synthesizers [11] are digital PLLs, in which due to the application of quantization noise corrections it became possible to use a 1-bit phase detector known as Bang-Bang Phase Detector. Such synthesizers are characterized by low lock time.

The industry produces integer-N and fractional-N PLL synthesizers. In fractional ones, the reference frequency is higher, which allows to obtain high performance for analog and digital PLLs [12].

PLL with dielectric resonator will be discussed below.

PLLs belong to the indirect digital synthesis devices. Direct Digital Synthesis (DDS) is represented by oscillators [13, 14], which main parts include phase accumulator, phase-to-amplitude and digital-to-analog convertors (Fig. 2). The digital output of phase accumulator is proportional to the phase angle of the given oscillation. Phase-to-amplitude cascade transforms this number to the voltage sample, corresponding to the value of the phase angle, and digital-to-analog convertor (DAC) produces analog oscillation. Parameters and characteristics of DDS devices are rapidly evolving with the development of electronic technologies.

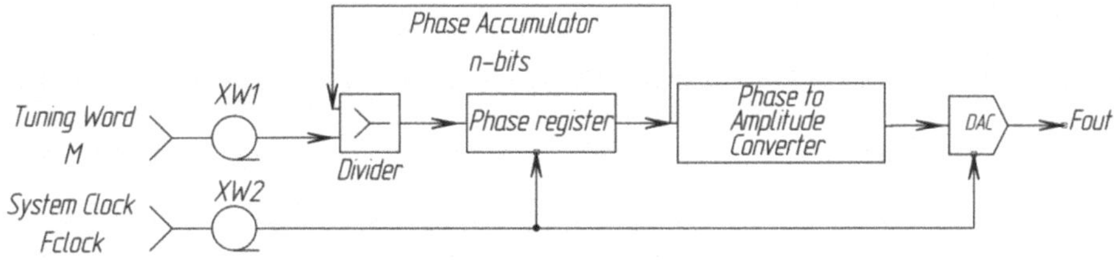

Fig. 2. The DDS synthesizer structure.

However, at the present time, the highest clock frequency of the DDS synthesizer produced by Analog Devices is equal to 12.3 GHz [15], which does not allow the use of DDS in the X band.

1.2 Motivation

In accordance with the material of Subsect. 1.1, frequency synthesizers can be analog or digital, built on the basis of phase-locked loop or digital signal synthesis. Besides exist many PLL variations. It is necessary to choose the optimal scheme of the synthesizer in accordance with the given frequency range, bandwidth, frequency step, phase noise level, cost and other parameters. The approach to the solution of the task was chosen to use, if possible, the advantages of existing technologies. The paper contains both results of mathematical modeling and measurements.

1.3 Contribution

The paper contains information about the state-of-the-art in the field of frequency synthesizers. The presented frequency synthesizer is an example of a specific development, which is characterized by accuracy characteristics confirmed by measurements.

1.4 The Organization of the Paper

The paper is organized as follows. The paper contains three sections. The first section describes introduction and state-of-the-art technology. The second section is devoted to design of frequency synthesizer, including dielectric resonators and oscillators on their basis, two PLLs and DSS oscillator, results of modeling and measurements. Finally, the third section is the conclusion.

2 Synthesizer Design

First, consider PLL circuits of the X range. For example, ADF5355 is fractional-N or integer-N PLL, operating with an external loop filter and a reference oscillator [16]. Due to data sheet the phase noise is equal to −107 dBc/Hz at 100 kHz offset from 10 GHz carrier. The simulating phase noise performance of the loop with ADF5355 in the ADIsimPLL software environment gave the results shown in Fig. 3. Also Fig. 4 presents a screenshot of the measurement device during the experimental study of the phase noise PLL with ADF5355.

The measuring results of phase noise differ from the mathematical modeling data in a worse way. So, at 100 kHz offset from carrier the noise value changes from −106 dBc/Hz to −84.75 dBc/Hz. At offsets of 1 kHz, 10 kHz, the level of phase noise varies from 80 dBc/Hz to 90 dBc/Hz that also unacceptable for high performance device.

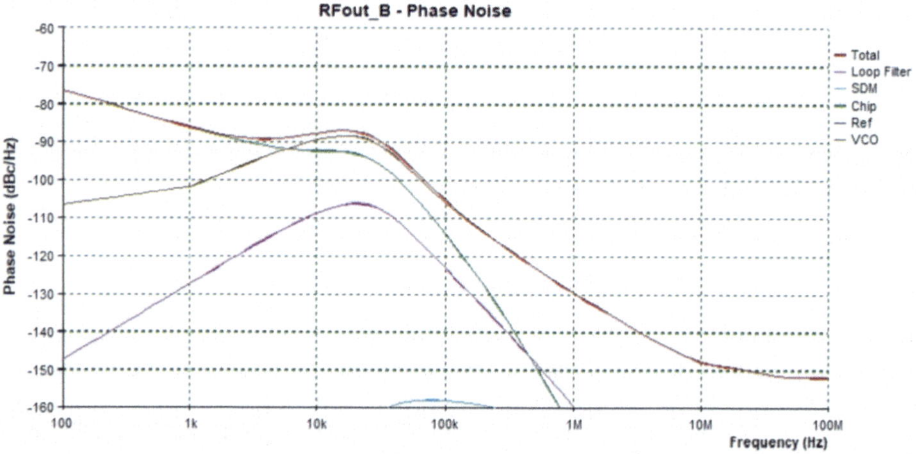

Fig. 3. Phase noise of the PLL on the basis of ADF5355 on frequency 10 GHz due to the modeling.

It is necessary to restrict the main sources of phase noise. Prediction of the phase noise can be made using the following formulas [4]. The component of synthesizer output phase noise N_{out_ref} due to the crystal oscillator noise N_{ref}

$$N_{out\_ref} = \frac{0.5N_{ref}}{f/f_{ref}} \frac{G_{loop}(f)}{R}, \qquad (3)$$

where $G_{loop}(f)$ is the closed loop gain (about 50 dB in X band); R is the division ratio for reference oscillator.

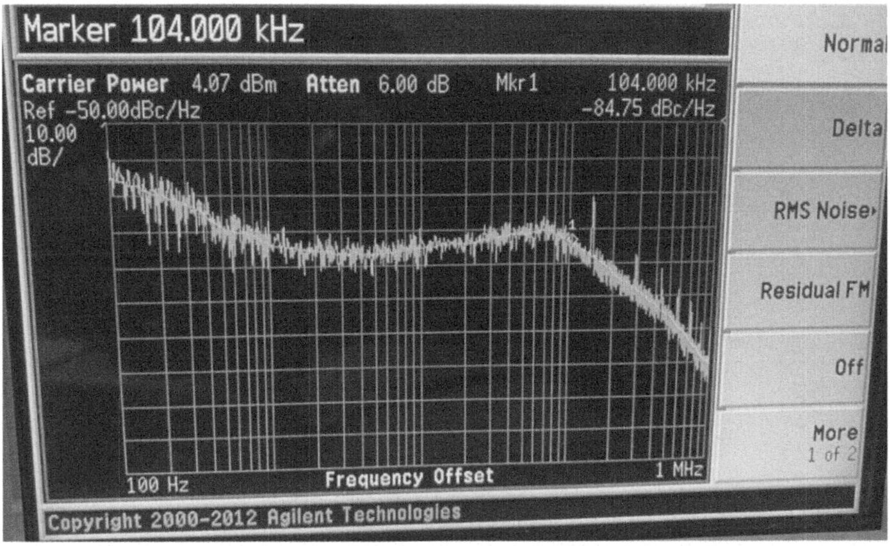

Fig. 4. The experimental study of the phase noise PLL with ADF5355.

The component of synthesizer output phase noise N_{out_pd} due to the phased detector

$$N_{out\_pd} = 10^{0.05(N_{pd\_ref} + 10\log f_{ref})} G_{loop}(f), \tag{4}$$

where N_{pd_ref} is the noise value, which can be taken from the data sheet of synthesizer circuit.

The third main source of the phase noise is VCO. Its value is evaluated using the level of noise N_{VCO_ref} at definite frequency f_{VCO_ref}

$$N_{VCO}(f) = \frac{0.5 N_{VCO\_ref}}{f/f_{VCO\_ref}} \cdot p \frac{1}{1 + G_{o\_loop}(f)}, \tag{5}$$

where $G_{o\_loop}(f)$ is open loop gain, which value is defined by loop parameters, except PFD.

In a well-known work [17], Leeson proved that the level of phase noise decreases in the oscillator circuit with high Q factor. Also known, that high Q factor resonators are built on the basis of DR and Yttrium Iron Garnet (YIR) devices. Sapphire dielectric resonant oscillators (DRO) operates with Q factor values 50k ... 70k and phase noise level up to −165 dBc/Hz at 1 kHz offset. But such values are reached using complex temperature stabilization appliance [18]. Manufacturers offer the DR series specifically for PLL use (PLDRO). In X band the PLDRO phase noise is typically equal to −85 dBc/Hz at 10 kHz offset. Studies of generators [19, 20] on the basis of DR confirm a low level of phase noise and jitter, but only in the region of the resonant frequency.

YIR oscillators offer even better parameters in comparison with DRO, but for many applications these advantages are compensated by greater complexity, mass, dimensions.

Mathematical modeling in ADIsimPLL confirms the low phase noise of PLDRO, for example PLDRO-10-7000-5P on 7 GHz (Fig. 5).

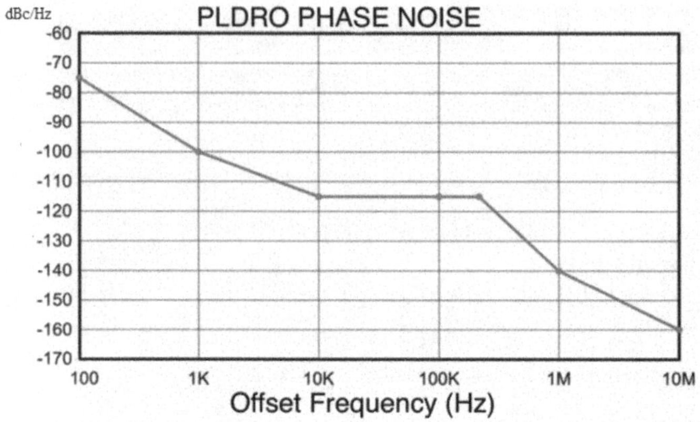

Fig. 5. Simulation results of PLDRO-10-7000-5P phase noise.

PLDRO was designed on the basis of self-made MgTiO3 DR. Magnesium titanium oxide is used for microwave applications. The DRO operates at 8 GHz with tuning in range 10 MHz by varicap. Figure 6 shows a DRO circuit.

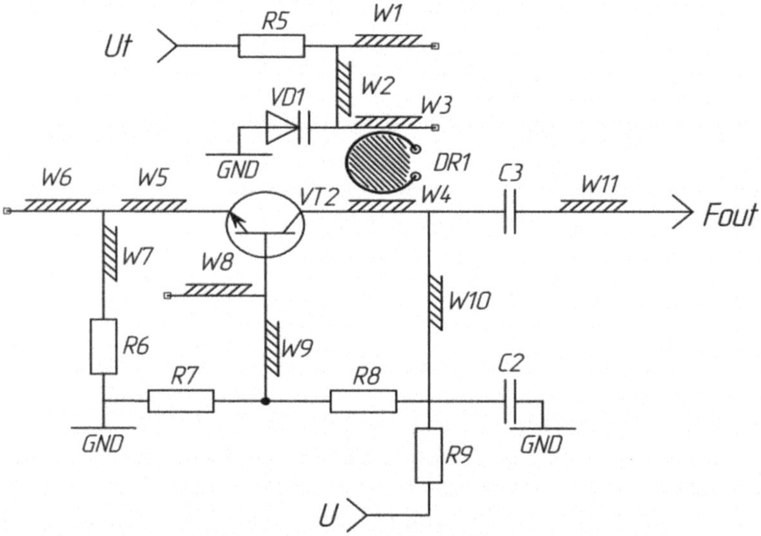

Fig. 6. The 8 GHz DR oscillator on the basis of MgTiO3 DR.

The DR operates on resonant frequency. Also, the DRO has narrow bandwidth of 1 kHz. This, in turn, creates conditions for minimizing the level of phase noise.

The DR design with microstrip excitation and control lines is shown in Fig. 7.

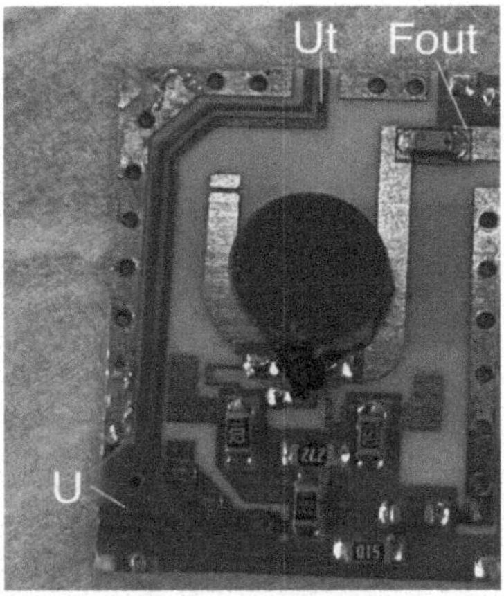

Fig. 7. Design of dielectric resonator.

Frequency synthesizer (Fig. 8) consists of two PLLs and one DSS. First loop is PLDRO, which includes the phase detector (HMC440), 1 kHz low pass filter on the low-noise operational amplifier ADR797AR, DR, directional coupler, the HMC434 divider by 8, 1000 MHz low pass filter and splitter. The operational amplifier ADR797AR sets the output voltage level of PLDRO. Besides phase detector the HMC440 also is used as frequency divider by a factor $N = 20$.

Phase noise response of PLDRO is shown in Fig. 9.

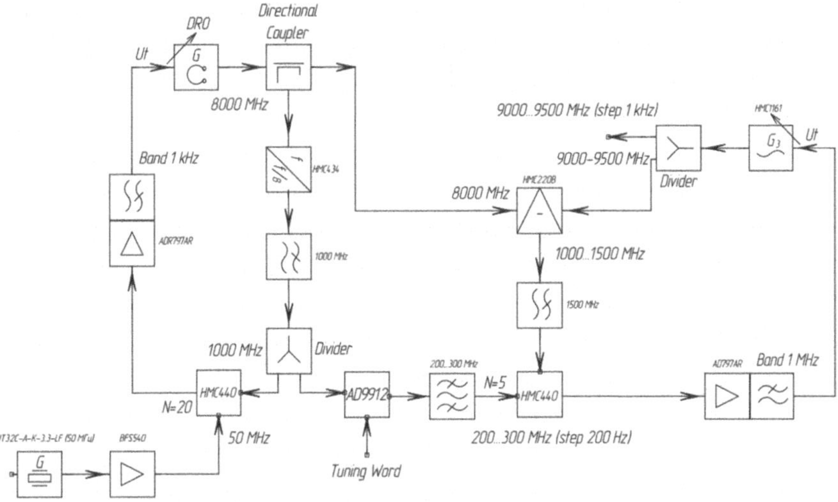

Fig. 8. The functional diagram of the X band frequency synthesizer.

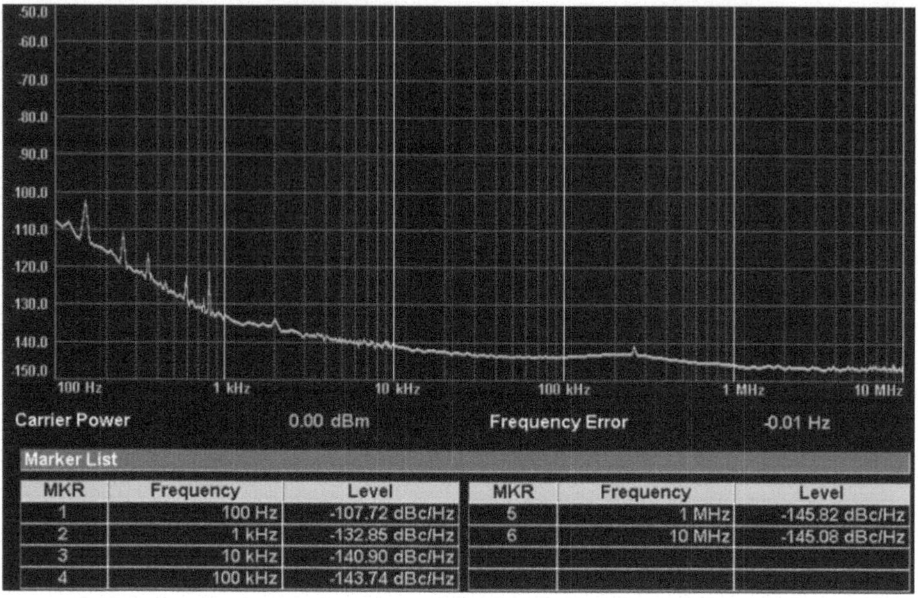

Fig. 9. Phase noise of PLDRO at resonant frequency 8 GHz.

The output oscillations of PLDRO are characterized by low phase noise and a frequency, which is close to the X range. That is, in the second PLL, you can avoid significant frequency division coefficients. But loop bandwidth of PLDRO is narrow.

The DDS synthesizer part, which includes DDS AD9912 with 14-bit DAC and pass band filter (PBF), provides low noise oscillations in the bandwidth from 200 MHz to 300 MHz with frequency step 200 Hz. The PDF is built on low pass filter and high pass filter (Fig. 10).

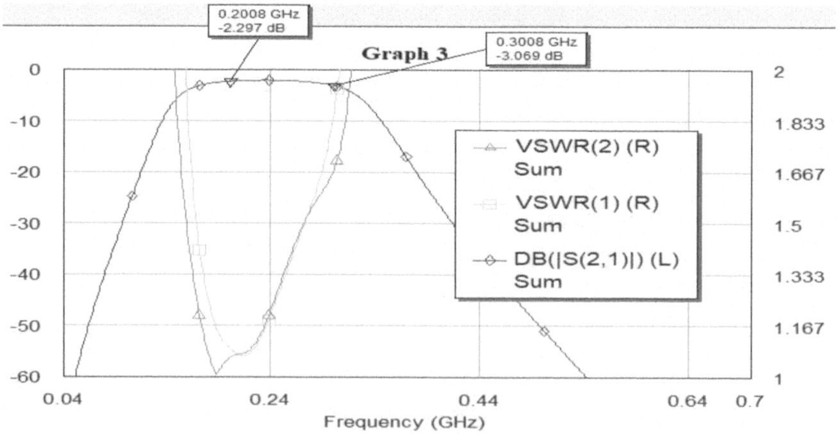

Fig. 10. Frequency responses of PBF. Left axis shows filter attenuation in dB and right axis – standing wave ratio by voltage (VSWR).

The double-balanced mixer HMC2208BMS in the second PLL down converted the 8 GHz oscillation from first PLL and output 9 GHz ... 9.5 GHz VCO oscillations on HMC1161. The spurs from the 1 GHz ... 1.5 GHz oscillations are deleted using low pass filter and is applied to the HMC440, which is used as divider $N = 5$ and phased detector. The second loop is closed by operational amplifier AD797AR and 1 MHz low pass filter (LPF). The frequency responses of low pass filter are shown in Fig. 11.

The phase noise response of frequency synthesizer is presented in Fig. 12.

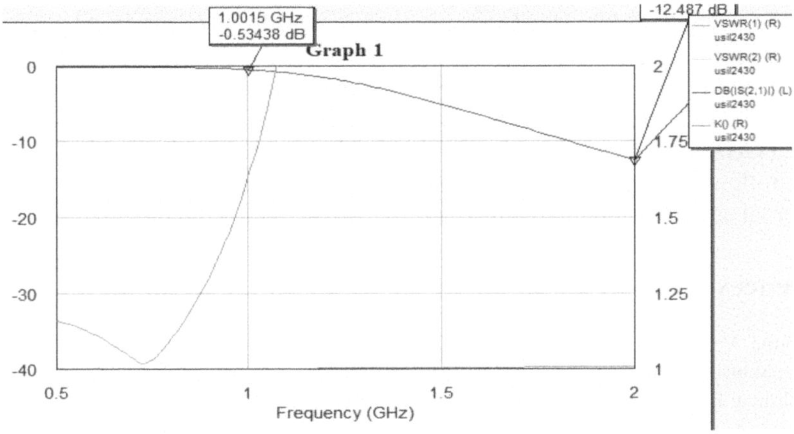

Fig. 11. Frequency responses of attenuation and VSWR LPF.

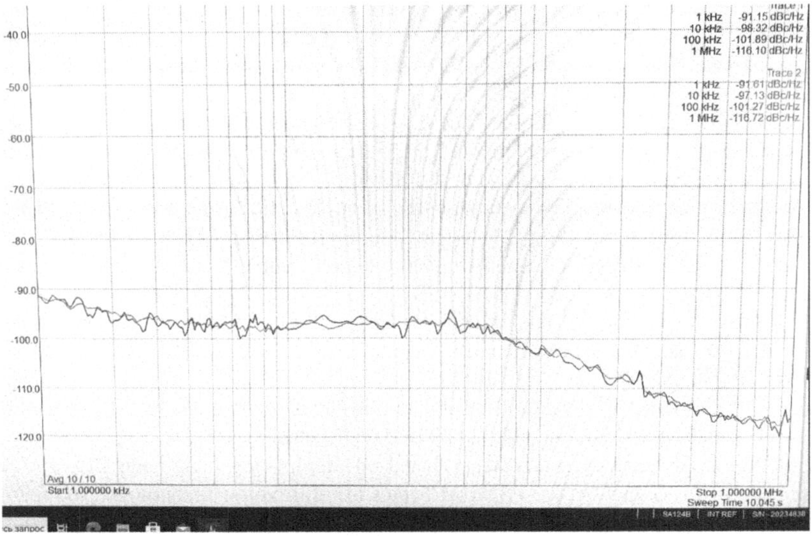

Fig. 12. Phase noise response of frequency synthesizer on 9 GHz.

3 Conclusions

According to the analysis, design of a frequency synthesizer using available components should rely on the capabilities of different technologies. In the presented synthesizer, PLDRO generates 8 GHz oscillations with low phase noise, 1 kHz bandwidth and 10 MHz tuning band. Stable, low-noise 8 GHz oscillations are required for the operation of second PLL.

The second necessary component of the synthesizer's output signal is obtained using a DDS. The output oscillations band of Direct Digital Synthesis circuit is from 200 MHz to 300 MHz with step 200 Hz. These oscillations are also characterized by low phase noise values.

In the second PLL, the high-N frequency divider is replaced by a frequency converter. As a result, at the output of the synthesizer, the oscillation frequency varies in the range of 1...1.5 GHz with a step of 1 kHz and at satisfactory level of phase noise.

The following studies will be devoted to conducting experiments on the influence of the external environment on the functioning of the frequency synthesizer.

References

1. Kroupa, V.: Phase Lock Loops and Frequency Synthesis. John Wiley, New York (2003)
2. Chenakin, A., Ojha, S., Shtin, N.: Phase noise suppression in PLL synthesizers. Anritsu Technical Bulletin, USA (2023)
3. Zhang, X., Du, Q., Liu, C., Ma, Y., Li, Y., Li, J.: A low phase noise frequency synthesizer with a fourth-order RLC. Electronics. **12**, 1–14 (2023). https://doi.org/10.3390/electronics12010224
4. Lascari, L.: Phase noise prediction in PLL synthesizers. Appl. Microw. Wirel. **12**, 30–38 (2000)
5. Yu-Chen, W., Mohammed, A., Dimitrios, P.: An L-band low phase noise evanescent-mode cavity-based frequency synthesizer. IEEE Trans. Circuits Syst. **65**(7), 2161–2168 (2018). https://doi.org/10.1109/TCSI.2017.2772872
6. Guo, M., Wang, L., Wang, S., Lu, J., Cui, M.: A low mismatch current charge pump applied to phased-locked loops. Micromachines. **15**, 913 (2024). https://doi.org/10.3390/mi15070913
7. Loria, E., Prager, S., Seker, I., Axmed, R., Hawkins, B., Lavalle, M.: Modelling the effects of oscillator phase noise and synchronization on multistatic SAR tomography. IEEE Trans. Geosci. Rem. Sens. **61**, 1–12 (2023). https://doi.org/10.1109/TGRS.2023.3249637
8. Doerry, A.: Radar Receiver Oscillator Phase Noise. Sandia National Laboratories, Albuquerque (2018)
9. Bai, L., Xu, W., Huang, P., Tan, W., Qi, Y., Chen, Y., Gao.: Z.: Phase noise compensation algorithm for space-borne azimuth multi-channel SAR. Sensors. **24**(14), 4494 (2024). https://doi.org/10.3390/s24144494
10. Gao, X., Klumperink, E., Nauta, B.: Sub-sampling PLL techniques. Inst. Eng. Technol. **21**, 583–603 (2020). https://doi.org/10.1049/PBCS064E_ch21
11. Bertulessi, L.: Frequency synthesizers based on fast-locking bang-bang PLL for cellular applications. In: Geraci, A. (ed.) Special Topics in Information Technology, Springer Briefs in Applied Sciences, pp. 27–36. Springer, Cham (2021). https://doi.org/10.1007/978-3-030-62476-7
12. Weltin-Wu, C., Zhao, G., Galton, I.: A 3.5 GHz digital fractional-N PLL frequency synthesizer based on ring oscillator frequency-to-digital conversion. IEEE J. Solid State Circuits. **50**(12), 2988–3002 (2015). https://doi.org/10.1109/JSSC.2015.2468712

13. Lei, T., Liang, L.: Research and design of digital unit for direct frequency synthesizer. J. Phys. Conf. Ser. **1907**, 012004 (2021). https://doi.org/10.1088/1742-6596/1907/1/012004
14. Leitner, S., Wang, H., Spyros, T.: Design techniques for direct digital synthesis circuits with improved accuracy over wide frequency ranges. J. Circuits Syst. Comput. **26**(2), 1750035 (2017). https://doi.org/10.1142/S0218126617500359
15. Direct Digital Synthesis. Analog Devices, Inc., https://www.analog.com/en/parametricsearch/11018#. Last accessed 16 Dec 2024
16. Microwave Wideband Synthesizer with Integrated VCO ADF5355. Analog Devices. https://www.analog.com/media/en/technical-documentation/data-sheets/ADF5355.pdf. Last accessed 16 Dec 2024
17. Leeson, D.: A simple model of feedback oscillator noise spectrum. Proc. IEEE. **54**(2), 329–330 (1966). https://doi.org/10.1109/PROC.1966.4682
18. Zhou, L., Yin, W., Wang, J., Wu, L.: Dielectric resonators with high Q-factor for tunable low phase noise oscillators. IEEE Trans. Compon. Packag. Manuf. Technol. **3**(6), 1008–1015 (2013). https://doi.org/10.1109/TCPMT.2013.2258465
19. Kaesbach, R., Delden, M., Musch, T.: A tunable dielectric resonator oscillator with phase-locked loop stabilization for THz time domain spectroscopy systems. IEICE Trans. Electron. **E106.C**(11), 718–721 (2023). https://doi.org/10.1587/transele.2023MMS0004
20. Van Delden, M., Pohl, N., Aufinger, K., Baer, C., Musch, T.: A low-noise transmission-type yttrium iron garnet tuned oscillator based on a SiGe MMIC and bond-coupling operating up to 48 GHz. IEEE Trans. Microw. Theory Tech. **67**(10), 3973–3982 (2019). https://doi.org/10.1109/TMTT.2019.2926293

Analysis of the Quality of Service Indicators for Aviation Communication Systems

Georgiy Konakhovych, Maksym Zaliskyi(✉), Serhii Tarasiuk, Anzhela Lelechenko, Veniamin Antonov, and Oleksandr Bondarev

National Aviation University, Liubomyra Huzara Ave., 1, Kyiv 03058, Ukraine
mzaliskyi@kai.edu.ua

Abstract. One of the key aspects of ensuring aviation safety is the communication system. This system allows transmitting useful and necessary information during aircraft flights and operations to control air routes, takeoff, and landing. The functioning of communication systems is aimed at meeting the needs of aviation enterprises and airlines. At the same time, communication equipment has various parameters that characterize its effectiveness in terms of functional purpose, technical capabilities, economic indicators, reliability, and operational properties. Some of these parameters relate to quality of service characteristics. This paper analyzes the characteristics of aviation communication equipment within the framework of the effectiveness of meeting the needs of consumers of air navigation services. The authors provide a complete list of quality of service characteristics and describe the features of their calculation and modeling. The main focus is on reliability characteristics, analysis of their models, and features of data processing for evaluating numerical values. The results of the study may be useful for aviation enterprise personnel when operating communication systems.

Keywords: Communication System · Operation · Efficiency Analysis · Quality of Service · Reliability · Aviation

1 Introduction

1.1 Introduction to the Problem

Today, civil aviation is undergoing rapid development. The main achievements of scientific thought are currently aimed at increasing the level of aviation safety [1, 2], reducing resource costs [3, 4], optimizing the organizational structure and production processes [5, 6], implementing achievements in the field of artificial intelligence [7, 8], and others.

Ensuring flight safety is the most important task in civil aviation [9]. The evolution of aviation safety has already gone through the technical era, the era of the human factor, and the organizational era. Currently, aviation safety is within the era of total systems. Indeed, civil aviation includes a wide variety of systems, each of which has its structure, content, resource provision, and options for interconnection with other systems [10, 11]. An integral part of the organizational structure of civil aviation is radio equipment

I. Ostroumov et al. (Eds.): ACASD 2025, LNNS 1418, pp. 262–275, 2025.
https://doi.org/10.1007/978-3-031-91992-3_17

[12]. This equipment is necessary for receiving radar and radio navigation information, as well as useful voice messages for control of flight activities [13]. According to the CNS/ATM concept, communication systems perform the function of transmitting useful and necessary information during aircraft flights and operations to control air routes, takeoff, and landing [14].

The useful functions of communication equipment are usually provided in the operation systems deployed at the aviation enterprise [15, 16]. The operation system is a set of organizational, regulatory, and management measures aimed at the effective utilization of equipment for its functional purpose [17]. The global goal of the operation system is to ensure compliance between the characteristics of the quality of equipment operation and the required level of air navigation service risks [18].

The main tasks of the communication equipment operation system are shown in Fig. 1.

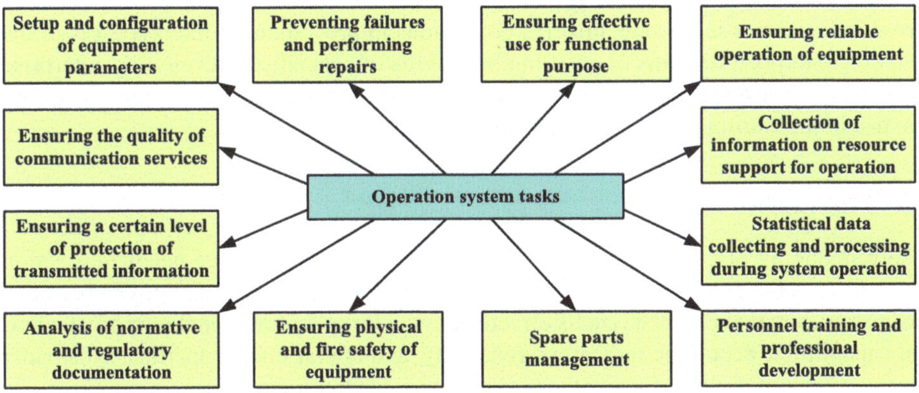

Fig. 1. The main tasks of the communication equipment operation system.

The communication system includes a large number of hardware and software modules [19, 20]. This is explained by the fact that the equipment is designed to meet a certain class of needs for service consumers [21]. From the point of view of economic optimization and safety, the use of communication equipment with limited functions is irrational. Therefore, the modern aviation communication system has different modes and functional features [22]. Therefore, during operation, it is necessary to perform settings and adjustments of the internal parameters of the communication system.

The task of failure prevention and repair is aimed at maintaining the serviceable condition of communication equipment [23, 24]. This task is implemented by performing the processes of monitoring the technical condition, diagnosing, and restoring serviceability. The need for this task is explained by objective factors of possible occurrences of failure, error, and malfunction [25].

Ensuring effective use for functional purposes involves solving the problem of optimizing the performance of the communication system. As productivity, we can usually understand the coefficient of equipment utilization, which is equal to the ratio of the real effect of equipment use to the maximum possible effect [26]. Productivity can be considered in the form of throughput or utilization factor.

Providing reliable operation of equipment is associated with optimizing the equipment structure with redundancy schemes for individual components, improving maintenance, repair processes, and others [27].

Obtaining required level of quality of services (QoS) is associated with achieving a compromise between the functionality of the communication system from the consumer's point of view and the costs of equipment and its operation [28, 29].

Ensuring a given level of protection of transmitted information includes a system of software and hardware information protection technologies in combination with a system of organizational and administrative actions to create the properties of secrecy, security, and noise immunity of the communication channel [30, 31].

Data collecting and processing contribute to the ability to make timely and accurate management decisions during equipment operation [32, 33].

The tasks of the operation system, related to documentation, personnel, resource provision, and fire safety, are aimed at continuous improvement of the processes of the aviation enterprise in terms of ensuring high skills of specialists, developing enterprise standards, ensuring proper working conditions, forming a stock of spare parts, justifying the use of measuring equipment, and others.

1.2 Motivation

The functioning of communication systems is aimed at meeting the needs of aviation enterprises and airlines [34, 35]. At the same time, communication equipment has various parameters that characterize its effectiveness in terms of functional purpose, technical capabilities, economic indicators, reliability, and operational properties. This paper analyzes the possibility of using these parameters as QoS indicators.

1.3 Contribution

This paper contributes to the theory of operation of communication systems in terms of substantiation of QoS parameters, features of their calculation, and modeling. The practical significance lies in the formation of recommendations for the selection of QoS parameters of aviation communication systems.

1.4 The Organization of the Paper

The paper consists of four sections. The first section considers basic information about using communication systems in civil aviation. The second section deals with the description of QoS indicators for aviation communication systems. The third section concentrates on the example of the calculation of QoS parameters. The fourth section discusses conclusions and future scope.

2 Materials and Methods

An aviation communication service can be defined by various properties. Considering it as an object of satisfying consumer requirements, it is possible to distinguish the properties of functionality, consumer need, quality, and costs.

Functionality determines the degree of utility for the consumer of the service, which characterizes the level of his satisfaction. The more functions the communication equipment can implement, the better the functionality.

Consumer needs for a communication service allow consumers to make decisions about the validity of using this service. According to international standards and recommendations in the field of communications, this property is characterized by effectiveness, security, ease of use, and safety. These properties can be described by such indicators as availability, continuity, and integrity.

An important characteristic of the communication system is its reliability. Usually, reliability can be characterized by serviceability, maintainability, and the possibility of conducting maintenance and repair.

Quality of service is a complex indicator that determines the beneficial effect resulting from the provision of the communication service. It characterizes the level of satisfaction of consumer needs. Since the consumer evaluates the beneficial properties of the service, the QoS can be a functional dependence on one or more indicators of the communication system. In operation practice, the use of sophisticated formulas is a difficult task, therefore, a multi-criteria approach to the analysis of the main indicators of the equipment can be used to assess the QoS. For example, for an aviation communication system, the following parameters can be introduced:

- the time frame for providing communication service parameters t_0;
- the service availability A as a ratio of the duration of successful provision of communication services to the total duration of operation;
- the delay t_d during information transmission;
- maximum value of the range t_{rv} of variation of the delay during information transmission;
- the probability p_t that guaranteed information transmission will be ensured during the observation interval;
- the level of noise immunity n of useful information:
- the proportion of lost packets in relation to the total number of transmitted packets;
- the proportion of packets with unreliable information distorted by natural or organized interferences;
- the level of load on the communication system, and others.

The cost of providing the service is an important aspect for the consumer when putting new equipment into operation. Usually, the price of communication equipment in civil aviation is quite high, but the cost of maintenance and repair during operation is much higher. Therefore, in practice, consumers of communication services take into account information on maintenance and repair tariffs, the cost of spare parts, the cost of logistics, and others.

Quality of service is usually considered from a statistical point of view. The process of information transmission is under the influence of random interferences. The process of establishing communication also occurs at random moments of time. In addition, during the operation of the equipment, failures, and malfunctions may occur in accordance with random factors. Therefore, in the general case, statistical characteristics of the QoS indicators should be used: probability density function, expected value, variance, and others.

In a geometric interpretation, the QoS indicator is a vector of a multidimensional space containing a certain set of indicators. In this case, it is necessary to establish the limit levels of the set of these indicators that meet the needs of consumers. The number of QoS levels in this case will be equal to the number of possible combinations of all values of the indicators of the multidimensional vector. In this case, each consumer can set his level of threshold values.

There are two types of QoS:

1. Guaranteed level of QoS. This level is determined at the design stage and declared in the operational documentation by the manufacturer of communication equipment. The guaranteed level is a functional dependence on the tactical, technical, and operational characteristics of the system.
2. Measured level of QoS. This level is determined at the operational stage by measuring the determining parameters, averaging, and processing to convert them into the value of the QoS indicator. Typically, the measurement is carried out over a certain observation interval, so the value of the QoS indicator is updated over time.

Figure 2 shows the structure of the system for QoS control. According to this structure, the equipment manufacturer ensures the level of QoS indicators at the design stage and develops maintenance recommendations to ensure the specified QoS levels during the operation stage.

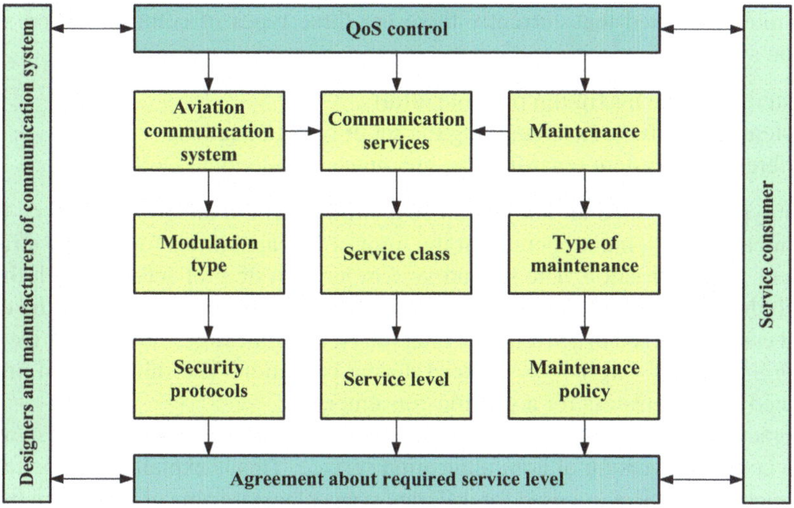

Fig. 2. The system for QoS control.

Practical experience in providing communication services shows that the number of equipment indicators of interest to consumers can reach 50. In addition, consumers require the ability to change these values in wide ranges. Therefore, in practice, classes of QoS parameters are established. The QoS class determines the levels of specific equipment indicators for each communication service.

Each class can be characterized by indicators, an example of which is given in Table 1. The following notations are made: MTDF is the mean time before failure; MTTR is the mean time to repair, MTRS is the mean time of recovering serviceability, MDV is the maximum of delay variation; MPTD is the maximum of packet transmission delay; MPLR is the maximum of packets lost rate; MIPER is the maximum of packet error rate. The minus sign in Table 1 indicates the absence of requirements for the QoS indicator.

Table 1. Example of QoS class characteristics.

QoS parameter	Class 0	Class 1	Class 2	Class 3	Class 4	Class 5
MTBF, hours	500	400	400	200	–	–
MTTR, hours	300	600	3600	–	–	–
MTRS, hours	600	1600	7200	–	–	–
MDV, ms	50	50	–	–	–	–
MPTD, ms	100	400	100	400	1000	–
MPLR	0.0001	0.0001	0.0001	0.0001	0.0001	–
MPER	0.000001	0.000001	0.000001	0.000001	0.000001	–

It should be noted that currently there are three types of communication service provision systems:

- a system with the maximum possible effort;
- a system with soft service regarding the set of QoS indicators;
- a differentiated system regarding the guarantee of individual QoS indicators.

The first type of system does not guarantee the fulfillment of all QoS indicators, but the communication equipment operates at the limit of its capabilities for a given structural scheme of its construction. The second system has a system of settings, with the help of which the values of certain QoS indicators that need to be guaranteed at the current moment can be changed in a given time interval. At the same time, the quantitative values of all QoS indicators are not guaranteed during operation. The third system provides guaranteed QoS indicators for a specific consumer.

In practical applications, attempts are made to combine these three systems and obtain a combined version of communication service. This is explained by the fact that communication services are used for different types of tasks, some of which can be given priority, and some of which can be assigned a secondary role.

Analysis of recent research also showed that currently there is no single methodology for substantiating, calculating, and determining the set of QoS indicators during the operation of communication equipment in civil aviation. This problem requires an immediate solution.

3 Results and Discussions

Let us consider the problem of calculating and modeling the decision-making process regarding the state of requirements satisfaction of the consumer of communication services. We assume that the consumer imposes requirements on three reliability indicators that characterize the quality of equipment operation. These indicators are:

- mean time between failures T_0;
- mean time between repairs T_R,
- steady-state availability A_{st}.

In this case, the given threshold values of these parameters are T_{0th}, T_{Rth}, and $A_{st\,th}$ respectively.

We assume that during the operation process, a scheme for evaluating the QoS indicators based on measurements is implemented. The necessary initial information for determining the estimates of the specified reliability indicators contains the operating time t_i between failures and the time t_{Ri} of repair procedures.

Since it is necessary to track the change in the QoS of indicators in real time, it is appropriate to perform the processing in the sliding window. The width of the sliding window is equal to n samples.

The estimates of reliability indicators in the sliding window are determined according to the equations

$$T_0 = \frac{1}{n}\sum\nolimits_{i=1}^{n} t_i. \tag{1}$$

$$T_R = \frac{1}{n}\sum\nolimits_{i=1}^{n} t_{Ri}. \tag{2}$$

$$A_{st} = \frac{T_0}{T_0 + T_R}. \tag{3}$$

The decision-making scheme for the quality of the communication system functioning is three-alternative and includes the following states:

- compliance with the requirements, when all three reliability indicators do not intersect the established threshold levels;
- partial compliance with the requirements, when one or two of the reliability indicators intersect the established threshold level;
- non-compliance with the requirements, when all three reliability indicators intersect the established threshold level.

Mathematically, we can write the decision in the form

$$D = \begin{cases} 1, \ \mathit{if} \ T_0 > T_{0th} \cap T_R < T_{Rth} \cap A_{st} > A_{st\ th}; \\ 0, \mathit{if} \ T_0 < T_{0th} \cap T_R > T_{Rth} \cap A_{st} < A_{st\ th}; \\ \quad 0.5, \ \mathit{otherwise}. \end{cases} \tag{4}$$

The process of evaluating QoS indicators will be carried out based on statistical simulation.

The modeling process includes the following stages:

- formation of random durations of failure occurrence according to an exponential distribution with a parameter 0.001;
- formation of random durations of repair procedure according to a normal distribution with a mean of 0.5 and a standard deviation of 0.1;
- modeling of random moments of occurrence of discrepancies that worsen the efficiency of operational processes;
- estimating reliability indicators in sliding window according to Eqs. (1)–(3);
- making a decision on the state of compliance of the QoS with the established requirements;
- modeling of the processes of performing control actions to restore compliance of the QoS with the established requirements.

The examples of realizations of the operating time between failures and the time of repair procedures are shown in Fig. 3.

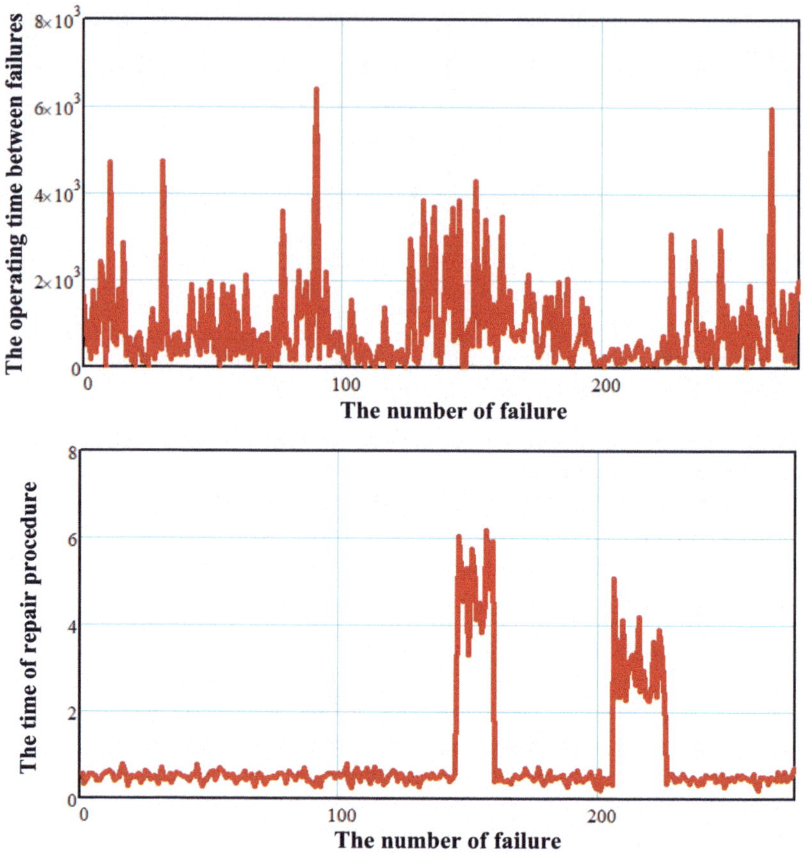

Fig. 3. The examples of initial parameters measurements.

Figure 4 shows the estimates of reliability indicators in the sliding window. In this simulation, we used the following threshold values: $T_{0th} = 500$, $T_{Rth} = 1.5$, $A_{st\,th} = 0.997$. We can see that each parameter contains an interval of non-compliance with the established requirements.

The decisions on compliance of the QoS with the established requirements are shown in Fig. 5. We see that at certain random moments in time when one of the QoS indicators deteriorates, the quality of communication services changes. In this case, the aviation enterprise performs settings, adjustments, and control measures to improve the efficiency of communication system operation processes. The state of non-compliance with consumer requirements is possible when the deterioration of two QoS indicators occurs at approximately the same time.

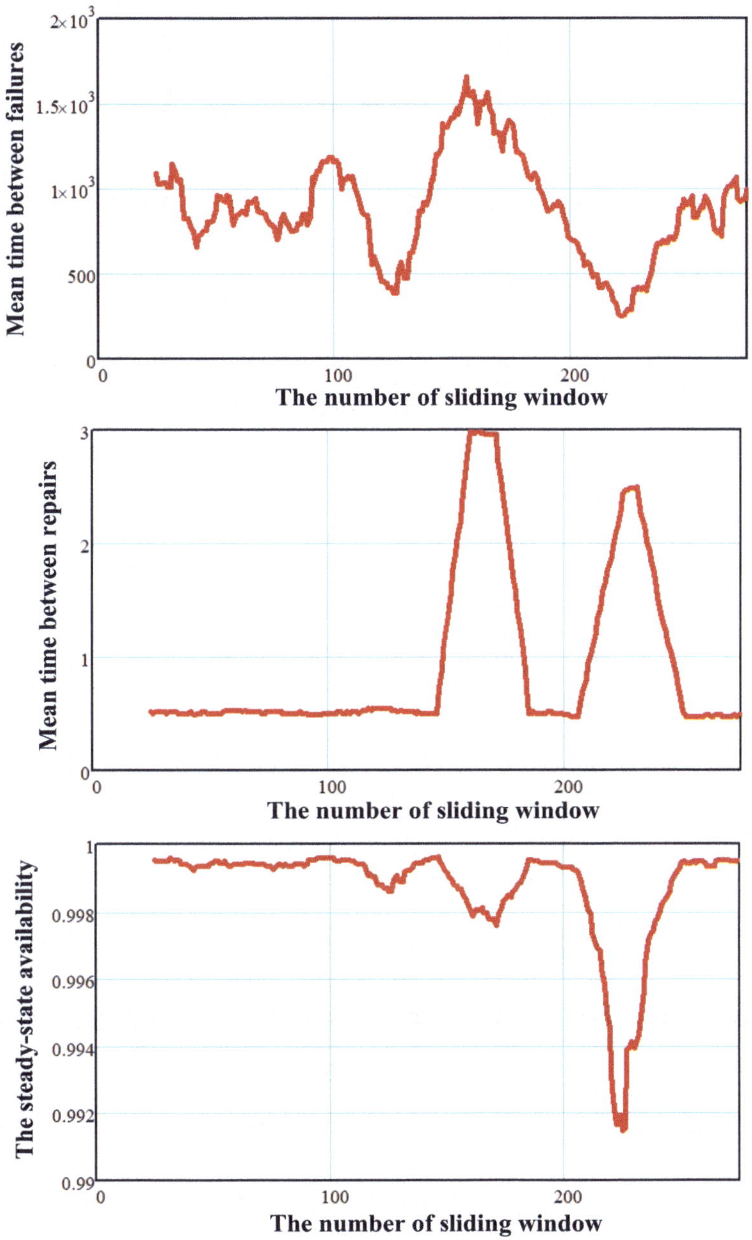

Fig. 4. The estimates of reliability indicators in sliding window.

It is quite natural to develop a new generalized QoS indicator that will take into account all the parameters of communication system that are of interest to the consumer.

An important issue is the determination of the functional dependence between the generalized parameter characterizing the QoS and the partial parameters. One type of

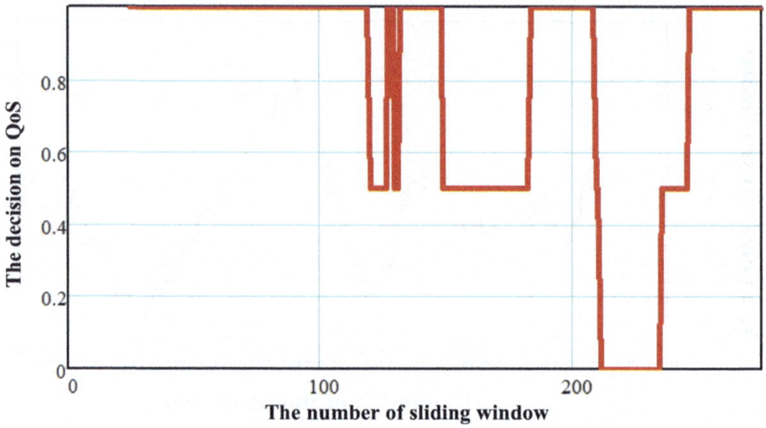

Fig. 5. The decisions on compliance of the QoS with the established requirements.

such dependence can be

$$G_{QoS} = \sum\nolimits_{i=1}^{m} w_i h_i^{(norm)}, \tag{5}$$

where w_i are the weighting coefficients calculated so that $\sum\nolimits_{i=1}^{m} w_i = 1$ and $0 < w_i < 1$, $h_i^{(norm)}$ are the normalized QoS indicators.

The weighting coefficients w_i can be determined based on consumer surveys and prioritization of QoS indicators. Normalized values $h_i^{(norm)}$ of QoS indicators can be converted to the range [0; 1] by one of the functional transformation methods, for example

$$h_i^{(\text{norm})} = \frac{h_i - h_i^{(\text{min})}}{h_i^{(\text{max})} - h_i^{(\text{min})}}, \tag{6}$$

where $h_i^{(min)}$ and $h_i^{(max)}$ are minimum and maximum value of the QoS indicator h_i.

Using Eqs. (5) and (6) will allow to obtain dimensionless QoS indicator in the range [0; 1], where values close to one will correspond to the best quality of communication services.

4 Conclusions

The paper discusses the characteristics of aviation communication equipment within the framework of the effectiveness of meeting the needs of consumers of air navigation services. For this purpose, the paper presents the complete list of QoS indicators and describes the features of their calculation and modeling. The QoS indicators determine the properties of equipment that characterize effectiveness, security, ease of use, and safety. The QoS indicators are usually random, so they can be described by statistical characteristics. The main focus of this research is on reliability characteristics, analysis of their models, and features of data processing for evaluating numerical values. The paper shows the example of QoS analysis in the case of establishing requirements for three reliability parameters. The quality of communication system operation is the result of three alternative classification with decisions: compliance, partial compliance, and non-compliance. The results of the study may be useful for aviation enterprise personnel when operating communication systems.

5 Future Scope

Further research will be aimed at: (a) development of models of spatial and temporal availability of the information transmission channel; (b) analysis of the efficiency of the communication channel in civil aviation within the framework of the possibility of packet loss; (c) development of recommendations for substantiating the list and limit values of QoS indicators for civil aviation enterprises.

Acknowledgments. This research is partially supported by the Ministry of Education and Science of Ukraine under the projects "Methods of building protected multilayer cellular networks 5G/6G based on the use of artificial intelligence algorithms for monitoring country's critical infrastructure objects" (№ 0124U000197).

References

1. Tachinina, O.M., Lysenko, O.I., Ponomarenko, S.O., Alekseeva, I.V.: Conceptual proposals for the creation of a fully reusable light-class aerospace system in Ukraine. In: IEEE 6th International Conference on Methods and Systems of Navigation and Motion Control, pp. 85–88. IEEE, Kyiv (2020). https://doi.org/10.1109/MSNMC50359.2020.9255504
2. Ostroumov, I.V., Kuzmenko, N.S.: Risk assessment of mid-air collision based on positioning performance by navigational aids. In: 6th International Conference on Methods and Systems of Navigation and Motion Control (MSNMC), pp. 34–37. IEEE, Kyiv (2020). https://doi.org/10.1109/MSNMC50359.2020.9255506
3. Galar, D., Sandborn, P., Kumar, U.: Maintenance Costs and Life Cycle Cost Analysis. CRC Press, Boca Raton (2017)
4. Poberezhna, Z.: Comprehensive assessment of the airlines' competitiveness. Econ. Ann. XXI. **167**(9–10), 32–36 (2017). https://doi.org/10.21003/ea.V167-07
5. Zaliskyi, M., et al.: Methodology for substantiating the infrastructure of aviation radio equipment repair centers. In: CEUR Workshop Proceedings, vol. 3732, pp. 136–148 (2024)
6. Solomentsev, O., Zaliskyi, M., Sushchenko, O., Bezkorovainyi, Y., Averyanova, Y., Ostroumov, I., et al.: Data processing through the lifecycle of aviation radio equipment. In: IEEE 17th International Conference on Computer Sciences and Information Technologies, pp. 146–151. IEEE, Lviv (2022). https://doi.org/10.1109/CSIT56902.2022.10000844
7. Andre, J.: Industry 4.0: Paradoxes and Conflicts. Wiley, New York (2019)
8. Averyanova, Y., Larin, V., Kuzmenko, N., Ostroumov, I., Zaliskyi, M., Solomentsev, O., et al.: Turbulence detection and classification algorithm using data from AWR. In: IEEE 2nd Ukrainian Microwave Week (UkrMW), pp. 518–522. IEEE, Lviv (2022). https://doi.org/10.1109/UkrMW58013.2022.10037172
9. Kharchenko, V.P., Kuzmenko, N.S., Ostroumov, I.V.: Identification of unmanned aerial vehicle flight situation. In: 5th International Conference Actual Problems of Unmanned Aerial Vehicles Development (APUAVD), pp. 116–120. IEEE, Kyiv (2017). https://doi.org/10.1109/APUAVD.2017.8308789

10. Sineglazov, V., Shildskyi, S.: Navigation systems based on GSM. In: International Conference on Methods and Systems of Navigation and Motion Control (MSNMC), pp. 95–98. IEEE, Kiev (2014). https://doi.org/10.1109/MSNMC.2014.6979740
11. Sushchenko, O., et al.: Algorithm of determining errors of gimballed inertial navigation system. In: Gervasi, O., et al. (eds.) Computational Science and Its Applications—ICCSA 2024 Workshops. ICCSA 2024 Lecture Notes in Computer Science, vol. 14816, pp. 206–218. Springer, Cham (2024). https://doi.org/10.1007/978-3-031-65223-3_14
12. Solomentsev, O., Zaliskyi, M., Zuiev, O.: Radioelectronic equipment availability factor models. In: Signal Processing Symposium 2013 (SPS 2013), pp. 1–4. IEEE, Serock (2013). https://doi.org/10.1109/SPS.2013.6623616
13. Zaliskyi, M., et al.: Heteroskedasticity analysis during operational data processing of radio electronic systems. In: Shukla, S., Unal, A., Varghese Kureethara, J., Mishra, D.K., Han, D.S. (eds.) Data Science and Security Lecture Notes in Networks and Systems, vol. 290, pp. 168–175. Springer, Singapore (2021). https://doi.org/10.1007/978-981-16-4486-3_18
14. Forouzan, B.A.: Data Communications and Networking. McGraw-Hill, New York (2007)
15. Dhillon, B.: Reliability, Quality, and Safety for Engineers. CRC Press, Boca Raton (2005)
16. Ostroumov, I., Kuzmenko, N., Bezkorovainyi, Y., Averyanova, Y., Larin, V., Sushchenko, O., et al.: Relative navigation for vehicle formation movement. In: 3rd KhPI Week on Advanced Technology, pp. 1–4. IEEE, Kharkiv (2022). https://doi.org/10.1109/KhPIWeek57572.2022.9916414
17. Anand, A., Ram, M.: System Reliability Management: Solutions and Techniques. CRC Press, Boca Raton (2021)
18. Modarres, M., Groth, K.: Reliability and Risk Analysis. CRC Press, Boca Raton (2023)
19. Kesidis, G.: An Introduction to Communication Network Analysis. Wiley, New York (2007)
20. Gnatyuk, S., Kinzeryavyy, V., Kyrychenko, K., Yubuzova, K., Aleksander, M., Odarchenko, R.: Secure hash function constructing for future communication systems and networks. In: Hu, Z., Petoukhov, S., He, M. (eds.) Advances in Artificial Systems for Medicine and Education II. AIMEE2018 2018, Advances in Intelligent Systems and Computing, vol. 902, pp. 561–569. Springer, Cham (2020). https://doi.org/10.1007/978-3-030-12082-5_51
21. Kasera, S., Narang, N., Narang, S.: Communication Networks: Principles and Practice. McGraw Hill Education, Delhi (2005)
22. Solomon, M.G., Kim, D.: Fundamentals of Communications and Networking. Jones & Bartlett Learning, Burlington (2021)
23. Jardine, A.K.S., Tsang, A.H.C.: Maintenance, Replacement, and Reliability: Theory and Applications, 2nd edn. CRC Press, Boca Raton (2017)
24. Zaliskyi, M., Solomentsev, O., Larin, V., Averyanova, Y., Kuzmenko, N., Ostroumov, I., et al.: Model building for diagnostic variables during aviation equipment maintenance. In: IEEE 17th International Conference on Computer Sciences and Information Technologies, pp. 160–164. IEEE, Lviv (2022). https://doi.org/10.1109/CSIT56902.2022.10000556
25. Larin, V., Solomentsev, O., Zaliskyi, M., Shcherban, A., Averyanova, Y., Ostroumov, I., et al.: Prediction of the final discharge of the UAV battery based on fuzzy logic estimation of information and influencing parameters. In: 3rd KhPI Week on Advanced Technology, pp. 1–6. IEEE, Kharkiv (2022). https://doi.org/10.1109/KhPIWeek57572.2022.9916490
26. Sushchenko, O., Bezkorovainyi, Y., Solomentsev, O., Kuzmenko, N., Averyanova, Y., Zaliskyi, M., et al.: Airborne sensor for measuring components of terrestrial magnetic field. In: 41st International Conference on Electronics and Nanotechnology (ELNANO), pp. 687–691. IEEE, Kyiv (2022). https://doi.org/10.1109/ELNANO54667.2022.9926760
27. Sushchenko, O., Bezkorovainyi, Y., Golitsyn, V., Kuzmenko, N., Averyanova, Y., Zaliskyi, M., et al.: Integration of MEMS inertial and magnetic field sensors for tracking power lines. In: XVIII International Conference on the Perspective Technologies and Methods in MEMS

Design, pp. 33–36. IEEE, Polyana (Zakarpattya) (2022). https://doi.org/10.1109/MEMSTECH55132.2022.10002907

28. Odarchenko, R., Iavich, M., Iashvili, G., Fedushko, S., Syerov, Y.: Assessment of security KPIs for 5G network slices for special groups of subscribers. Big Data Cogn. Comput. **7**(4), 169, 1–19 (2023). https://doi.org/10.3390/bdcc7040169
29. Solomentsev, O., Zaliskyi, M., Zuiev, O.: Estimation of quality parameters in the radio flight support operational system. Aviation. **20**(3), 123–128 (2016). https://doi.org/10.3846/16487788.2016.1227541
30. Odarchenko, R., Lavrynenko, O., Bakhtiiarov, D., Dorozhynskyi, S., Zharova, V.A.O.: Empirical wavelet transform in speech signal compression problems. In: IEEE 8th International Conference on Problems of Infocommunications, Science and Technology, pp. 599–602. IEEE, Kharkiv (2021). https://doi.org/10.1109/PICST54195.2021.9772156
31. Lavrynenko, O., Taranenko, A., Machalin, I., Gabrousenko, Y., Terentyeva, I., Bakhtiiarov, D.: Protected voice control system of UAV. In: IEEE 5th International Conference Actual Problems of Unmanned Aerial Vehicles Developments (APUAVD), pp. 295–298. IEEE, Kiev (2019). https://doi.org/10.1109/APUAVD47061.2019.8943926
32. Altundag, A., Wynn, M.: Advanced analytics and data management in the procurement function: an aviation industry case study. Electronics. **13**(8), 1554, 1–26 (2024). https://doi.org/10.3390/electronics13081554
33. Poberezhna, Z., Petrova, Y., Slimani, K.: Information technologies in logistics processes of enterprises in the aviation industry. In: CEUR Workshop Proceedings, vol. 3732, pp. 90–102 (2024). https://ceur-ws.org/Vol-3732/paper07.pdf
34. Ostroumov, I., Ivannikova, V., Kuzmenko, N., Zaliskyi, M.: Impact analysis of Russian-Ukrainian war on airspace. J. Air Transp. Manag. **124**, 102742 (2025). https://doi.org/10.1016/j.jairtraman.2025.102742
35. Smerichevska, S., Poberezhna, Z., Mykhalchenko, O., Shtyk, Y., Pokanevych, Y.: Modeling and evaluation of organizational and economic support for sustainable development of transport enterprises: innovative and ecological aspects. Financ. Credit Act. Probl. Theory Pract. **4**(51), 218–229 (2023). https://doi.org/10.55643/fcaptp.4.51.2023.4121

Antenna Feed by Several Sources of Waveforms

Ludvig Ilnitskiy, Ihor Zakutynskyi, Leonid Sibruk(✉), and Olga Shcherbyna

National Aviation University, Liubomyra Huzara Ave., 1, Kyiv 03058, Ukraine
leonid.sibruk@npp.kai.edu.ua

Abstract. The article is devoted to the study of the conditions under which the operation of two or more transceivers on one antenna is possible. At this radio lines acts on different frequencies and in-band full-duplex systems are not the subject of research. Conventional means of communication have been considered. After the introduction the theoretical foundations of the processes that occur in radio lines when working with a common antenna are presented. Also, calculation, design and modeling results of the band-pass filter are provided. During the open-area test, two modes of adjacent radio lines were investigated. The one of them involves reception and radiation modes in the adjacent radio lines, when working on a common antenna. Despite the relatively low selectivity of used Soft Designed Radio, the results of measuring the BER level indicate satisfactory operation of the radio lines at modulations BPSK, QPSK, 16QAM and 30 dB attenuation pass band filters. In the next test mode both radio lines radiate oscillations. At this mode the negative effect of interference is greater. Filters with 30 dB attenuation are not enough to obtain a satisfactory level of BER. It is necessary to use filters with attenuation at the level of 60 dB.

Keywords: Quadrupole · S-parameters · Intermodulation · Butterworth Filter · Selectivity · Modulation · Bit Error Rate · Balun · Receiver · Transmitter · Single Antenna

1 Introduction

1.1 Introduction to the Problem

In modern conditions, the reliable functioning of half-duplex or full-duplex (with frequency division) radio lines is ensured by the use of various methods technical and informational nature [1]. One of the simplest methods of combating intentional as well as unintentional interference is the transition from one wavelength to another, on which the electromagnetic interferences are absent. But at the same time, it should be taken into account that changing the wavelength (frequency) requires solving certain technical problems. The main ones are ensuring the effective emission of electromagnetic waves, maintaining the matching of the antenna (or several antennas) with the feeder, the independence of information streams in the radio lines, resolution of transceivers, time delays for changing frequency and etc.

I. Ostroumov et al. (Eds.): ACASD 2025, LNNS 1418, pp. 276–285, 2025.
https://doi.org/10.1007/978-3-031-91992-3_18

Some of these problems disappear when single antenna is fed with two or more sources of electrical oscillations. Such antenna simultaneously forms two (or more) fields of electromagnetic waves, independent of each other. This is especially positive when organizing radio communication lines with mobile objects, or communication between objects whose surface area has limitations for antenna location.

It should also be noted that the possibility of creating several radio lines with frequency division using one antenna simplifies the solution of the electromagnetic compatibility problem. Obviously, the use of a single radiator for each radio line can lead to a decreasing the number of antenna devices located in a limited space. Ensuring low levels of interference in each radio line requires optimization of the mutual location of radiators and minimization influences between devices that are structural elements of radio lines. The compatible operation of hulf-duplex radio lines using only one antenna as the final element does not widely studied during last years. But similar problem of in-band full-duplex, which solution gives impoving both spectral efficiency and data rate in communication and radar systems, is a subject of many works.

The first study of the problem of in-band full-duplex radio line considered the tranceiver with circular patch antenna, which provides 40 dB attenuation of the transmitting signal in receiving channel. Self-interference is a main problem of in-band full-duplex radio line [2, 3]. Self-interference compensation occurs in the analog or digital parts of the transceiver or in the antenna system [4–6]. Given that this work is devoted to radio lines operating at different frequencies, methods and means of in-band full-duplex duplex communication are not considered in detail. Obviously, that in-band full-duplex radio line are more complex compared to frequency-division ones.

1.2 Motivation

Using a separate radiator for each radio link can lead to significant increase in the number of antenna devices located in limited space. Ensuring low levels of interference in each radio link requires optimizing the mutual arrangement of antennas and minimizing the coupling effects between devices that are structural elements of radio links. Using the single radiator as the final element of several radio links, their joint operation can be ensured by selectivity of receivers and the low level of intermodulation interference in transmitters. The latter is achieved by electrical isolation of the transceivers outputs.

1.3 Contribution

The article contributes to the development of the theory and practice of antenna devices. The implementation of work results leads to the minimization of antenna number and improvement of electromagnetic compatibility.

1.4 The Organization of the Paper

The paper is organized as follows. The first section describes introduction, including a brief analysis of literature sources. The second section considers theoretical basic of feeder design for radio line, in which two transmitters operates on a single antenna. The third section is devoted to experimental studies. The last chapter presents conclusions and directions for future research.

2 Materials and Methods

When feeding the single radiating element with two transmitters (waveforms frequencies f_1 and f_2) it is necessary to create the feeder paths of both sources in such a way that in the output cascades of the transmitter generating oscillations with frequency f_1 will not penetrate the power of oscillations with frequency f_2 of the second transmitter, and vice versa. Besides, connecting the outputs of the transmitters to the antenna should not change the distribution or intensity of the currents in the antenna, i.e. not affect the set parameters and characteristics of the antenna itself. It is obvious that the biggest change in the design of the feeder, which can lead to the fulfillment of the stated requirements, is the sequential inclusion between the feeder output and the antenna input the quadrupole with frequency-dependent parameters. First of all, such a quadripole must be lossless and satisfy the conditions of reciprocity. The number of quadripoles corresponds to transmitters number.

The full decoupling of two transmitters is achieved when the power of transmitter 1 (frequency f_1) reaches the antenna without loss, and the current of generator 2 (frequency f_2) will be completely reflected from the quadripole 1, which is connected to transmitter 1. On the other hand, the quadripole 2 transfers the power of generator 2 without loss and prevents the flow of currents from transmitter 1 into the feeder of transmitter 2. Therefore, the elements of the scattering matrix of quadripole 1 should have the following values:

$$\left|\dot{S}_{11}(f_1)\right| \to 0; \quad \left|\dot{S}_{21}(f_1)\right| \to 1; \quad \left|\dot{S}_{22}(f_2)\right| \to 1; \quad \left|\dot{S}_{12}(f_2)\right| \to 0. \tag{1}$$

The elements of the scattering matrix of quadripole 2 should have similar values:

$$\left|\dot{S}_{11}(f_2)\right| \to 0; \quad \left|\dot{S}_{21}(f_2)\right| \to 1; \quad \left|\dot{S}_{22}(f_1)\right| \to 1; \quad \left|\dot{S}_{12}(f_1)\right| \to 0. \tag{2}$$

For passive quadripoles that obey the principle of reciprocity, the following equations are valid:

$$\dot{S}_{11} = \dot{S}_{22}; \quad \dot{S}_{21} = \dot{S}_{12}. \tag{3}$$

The given values of the scattering matrix elements are inherent to filters built on inductances and capacitors. The calculation and design of the band-pass filter (BPF) is considered below.

It is obvious that the radiator should be excited in two frequency bands with average frequencies f_1 and f_2. The frequency bands that ensure the necessary quality of radio channel functioning are limited by the lower frequency of the filter bandwidth f^{b1} and the upper frequency f^{b2}. The average frequencies of the filter's bandwidth are defined as

$$f_{01} = \sqrt{f_1^{b1} f_1^{b2}}; \qquad f_{02} = \sqrt{f_2^{b1} f_2^{b2}}. \tag{4}$$

At the boundaries of the bandwidth, the attenuation of electromagnetic power transmission takes the value A_{b1} for the first filter with the frequency f_1 and $A_{п2}$ for the second filter with the average frequency of the bandwidth f_2. All these four frequency parameters can be determined from the technical conditions on the radio line.

Filter calculation needs to define two more parameters. So, for a radio line the first filter A_{b1} is the attenuation in the bandwidth and the attenuation outside band A_{o1}, or in relative values L_{b1} and L_{o1}, where the attenuation is the ratio of the powers at the input and at the output of the filter $L = P_{in}/P_{out}$. The relationship between the attenuation and the values of the elements of the scattering matrix is quite simple:

$$\left.\begin{array}{l} L(f) = 1/\left|\dot{S}_{21}(f)\right|^2 \\ \left|\dot{S}_{11}\right|^2 = 1 - \left|\dot{S}_{21}\right|^2 \end{array}\right\}. \tag{5}$$

The element $\dot{S}_{11}$ of the scattering matrix is the reflection coefficient. As a rule, the reflection coefficient or voltage standing wave ratio (SWR) in the feeder is normalized, as well as the reflection coefficients from the inputs of certain devices included in the structure of the feeder tract. This makes it possible to calculate the permissible value of the SWR, or the reflection coefficient from the input of the filter:

$$\left|\dot{S}_{11}\right| = \left|\dot{\Gamma}\right| = \frac{\text{SWR} - 1}{\text{SWR} + 1}. \tag{6}$$

Therefore, the attenuation in the filter bandwidth is determined using the known structure of the feeder tract:

$$L_{b1} = 1/\left|1 - \left|\dot{\Gamma}\right|^2\right|, \tag{7}$$

where all values refer to the electrical processes in the feeder at frequency f_1 of transmitter 1, or to the path at frequency f_2 when calculating L_{b2}.

The attenuation outside bandwidth depends on two processes. One of them is the penetration of the wave power with a frequency f_1 into the feeder of second transmitter, and the second is the appearance of an electromagnetic energy flux from the output of the feeder of first transmitter to the output of the feeder of second transmitter along the radiator at separate points of the feeders connection to the antenna. The first process leads to the excitation of intermodulation oscillations in the output stages of the transmitters, and the second process - to the distortion of the antenna's directional pattern.

As is known, the power of side emissions of a radio transmitter, including the power of intermodulation emissions, is regulated. Therefore, based on the technical characteristics of the transmitter, it is possible to determine the maximum power of oscillations $P_{out}(f_2)$, which enter the feeder tracts without causing an intermodulation radiation.

It is rather difficult to calculate the power of currents with a frequency f_2, which penetrate via outside band of first filter at the output of first transmitter. This power will not exceed the threshold value of the output power of second transmitter ($P_{\text{th}\,f2}$). So, it can be assumed that

$$L'_{o1} = \text{P}_{\text{th}\,f2}/\text{P}'_{\text{out}}(f_2). \tag{8}$$

Another attenuation value L''_{o1} is calculated based on the radiation power of the wave currents, which flow from the output of the second transmitter to the point of connection of the feeder of the first transmitter to the antenna. This current can significantly change the distribution of total current and charges on the antenna and the change in parameters and characteristics of the antenna. It is possible to calculate the power, created by the current at fixed connection points and at known shape of the antenna. Thus, it is possible to obtain the second attenuation value in the stop bandwidth:

$$L''_{o1} = \mathrm{P}_{\mathrm{th}\,f2}/\mathrm{P}''_{\mathrm{out}}(f_2). \quad (9)$$

It is obvious that the attenuation L''_{o1} is determined only if the value $\left|\dot{S}_{11}(f_2)\right|$ of the scattering matrix is known for filter 1. Therefore, the attenuation L'_{o1} is used for the first stage of filter design, and at the second stage we specify its parameters. If it turns out that:

$$L''_{o1} < L'_{o1}, \quad (10)$$

then there is no need to redefine the filter parameters.

Using mathematical modeling and parameters described above, the frequency response of the antenna feeder with two frequency channels can be built. Such frequency response is shown in Fig. 1 in the attenuation plane (L, dB – frequency f) and has the form of two independent curves that overlap the specified frequency range. Each curve is the dependence of the band-pass filter attenuation on the frequency. The Butterworth filters separate the energy flows by frequency in order to feed the antenna device.

Butterworth filters are characterized by the flattest amplitude and linear phase responses, but the intervals between the bandwidth and outside band are quite large. Its approximation function [7]:

$$L(\Omega) = 1 + \gamma^2 \Omega^{2n}, \quad (11)$$

where n =1,2,3... is number of filter elements; $\Omega = \frac{f_0}{f^{b2} - f^{b1}}\left(\frac{f}{f_0} - \frac{f_0}{f}\right)$ is normalized frequency; γ is a factor, determined by the given value of attenuation L_1 at the boundary of the passband at $\Omega_{\mathrm{boun}} = 1$, namely $L = 10\,log\,(1 + \gamma^2)$.

The number of filter elements can be determined as follows:

$$n \geq \frac{\log\sqrt{(L_o - 1)/(L_b - 1)}}{\log(\Omega_2/\Omega_1)}, \quad (12)$$

$$\Omega_1 = \frac{f_0}{f^{b2} - f^{b1}}\left(\frac{f^{o1}}{f_0} - \frac{f_0}{f^{o1}}\right), \quad (13)$$

$$\Omega_2 = \frac{f_0}{f^{b2} - f^{b1}}\left(\frac{f^{o2}}{f_0} - \frac{f_0}{f^{o2}}\right). \quad (14)$$

The bandwidth of the first filter in Fig. 1 is 100 MHz and attenuation in the stop band is 30 dB. For the second filter these values are 50 MHz and 60 dB respectively.

3 Open-Area Test

In experimental study we use transceiver Analog Devices Pluto SDR [8], which is based on the AD9363. The AD9363 was designed for application in mobile communication [9], but also is recommended for experimental studies. The comparative test AD9363 proves [10], that this SDR transceiver do not belong to high performance devices. Nevertheless, its operation is better than Hack RF or RTL-SDR.

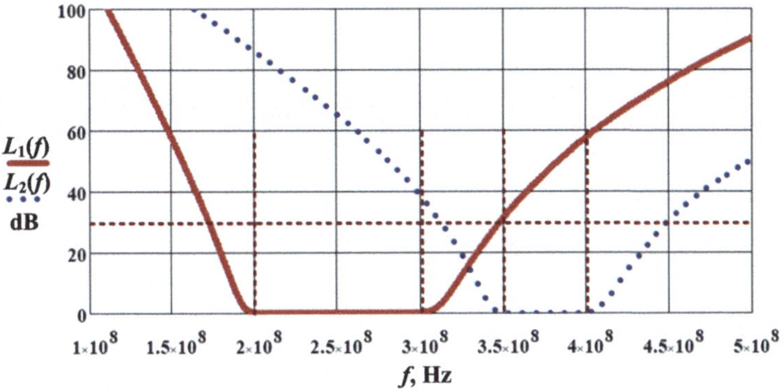

a

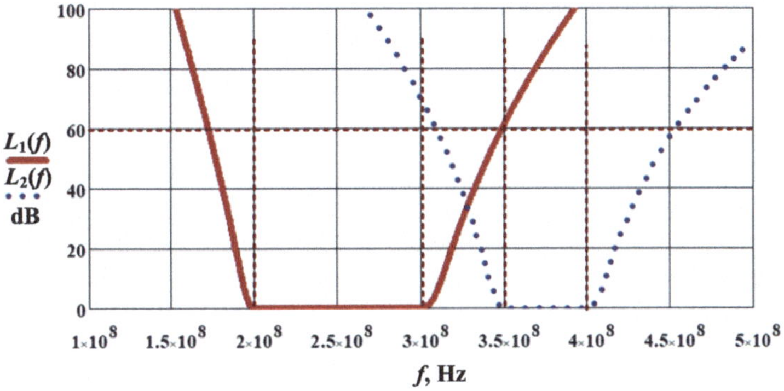

b

Fig. 1. Feeder frequency response: at SWR = 1.5, $L_o = 30$ dB ($n_1 = 9$, $n_2 = 5$) (*a*); at SWR = 1.5, $L_o = 60$ dB ($n_1 = 15$, $n_2 = 8$) (*b*).

In case of circuit with common antenna, the total selectivity consists of the receiver selectivity and the attenuation of the filter at the receiver input. The value of adjacent channel selectivity AD9363 is absent in data sheet. The mirror interference does not high: 18 dB. However, the AD9363 allows to work with three modulations, which include PSK and QAM. Also, the AD9363 calculates Bit Error Rate (BER).

The Fig. 2 shows a functional diagram of the radio line when interference directly affects the receiver. The case when interference affects the transmitter is in Fig. 3. The diagrams includes receivers and transmitters of three Pluto SDR, band-pass filters, which bandwidths (BW) correspond to the channel frequencies f1 and f2. The adder operates in both directions. The matching device also helps to expand the frequency range of the antenna.

The matching device is the self-made 1:1 Guanella balun. Obtained standing wave ratio by voltage (VSWR) in the 200 MHz BW with «fat» dipole is VSWR < 2.0. The number of winding turns was selected experimentally.

Three metal shielded filters of Mini-Circuits were used in study, namely RBP-253+ (BW from 253 MHz to 340 MHz), RBP-415+ (BW from 404 MHz to 426 MHz) and

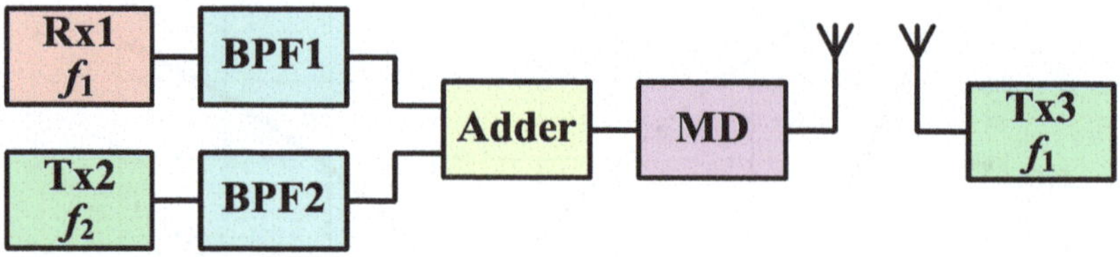

Fig. 2. The functional diagrams for test of a simultaneous transmitter and receiver operation on the single antenna.

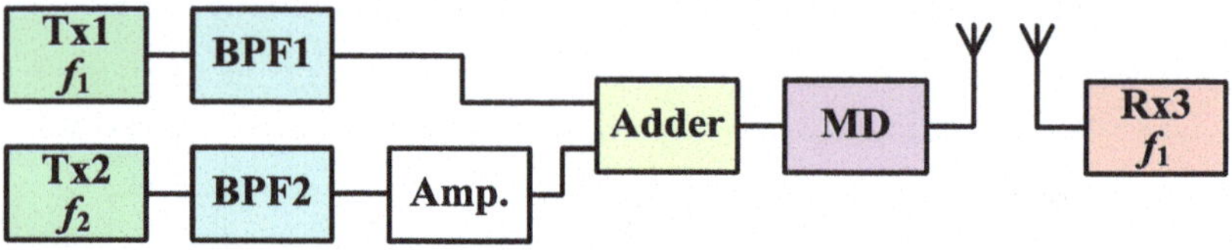

Fig. 3. The functional diagrams for test of two transmitters operation on the single antenna.

RBPF-435 (BW from 435 MHz to 535 MHz). Filters are characterized by relative small ripple in the bandwidth as can be seen from frequency responses (Figs. 4 and 5) and linear phase responses. Also filters operate at 1 W input power. The transition region between bandwidth and stopband at attenuation level about 30 dB due to the measured frequency responses is narrow.

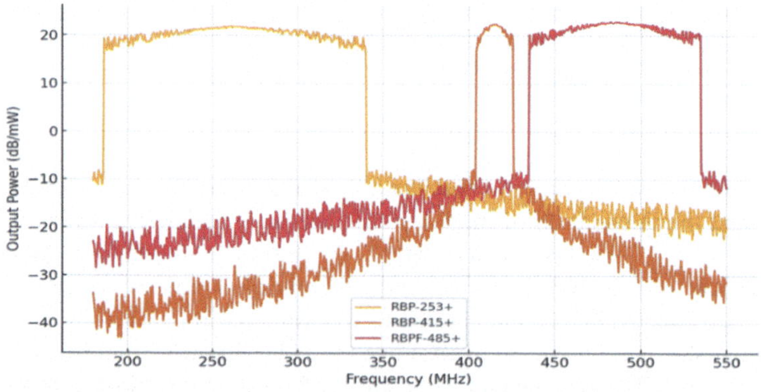

Fig. 4. Frequency responses of filters at input power 300 mW.

The dependence of BER in receiver Rx1 (Fig. 2) from ratios of signal power to interference power in case of pair transmitter-receiver simultaneously operating on single antenna is in Fig. 6. Here modulations are BPSK, QPSK and 16QAM. The filter RBP-415+ was connected to the receiver Rx1 and filter RBP-485+ to the transmitter Tx2.

The results of study the interference effect in case of pair neighboring transmitters, measured by Rx3, are presented in Fig. 7.

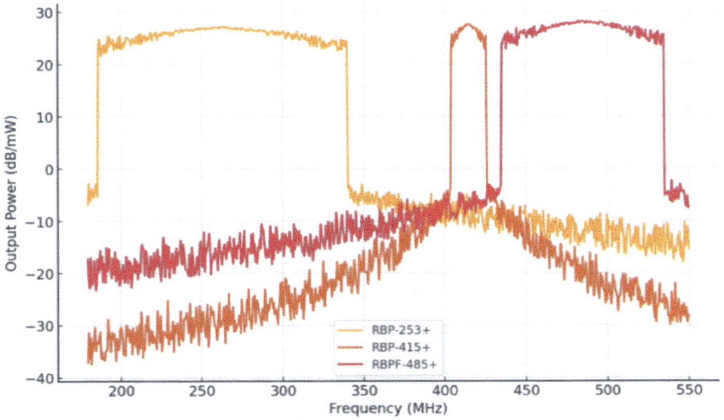

Fig. 5. Frequency responses of filters at input power 1000 mW.

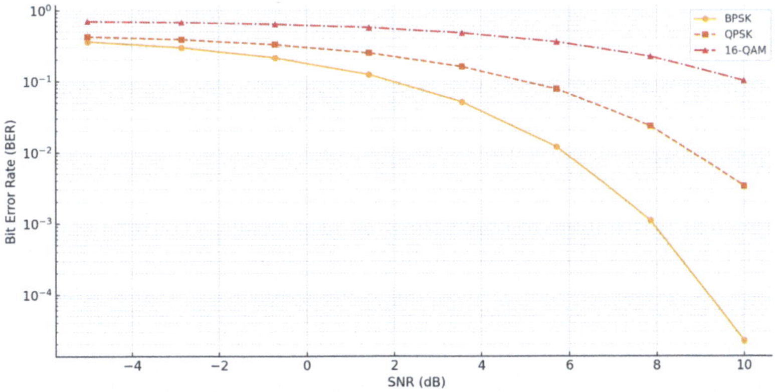

Fig. 6. Ratio of signal to interference power on receiver input (signal frequency is 415.0 MHz, interference frequency is 440 MHz).

The curve marked "single RBP-415+" in Fig. 6 belongs to the case of the application of filters RBP-415+ in both channels. The signal frequency is 410 MHz, and the interference is 424 MHz, that is near the boundary stop-transparency bands of the filter.

The curve marked "two RBP-415+" belongs to the case when in both channels' filters RBP-415+ operate in transparency band. At this two filters are connected in series in the signal channel. Successful operation is possible when the signal power exceeds the interference power by more than 10 dB.

The next case of operation two transceivers on a single antenna includes the filter RBP-415+ ($f_S = 410$ MHz) in the signal channel and the filter RBP-485+ ($f_i = 440$ MHz) in the interference channel. Both oscillations are inside filters bandwidth. As follows from Fig. 6 the quality of communication depends on the ratio of powers in the channels. For example, with equal powers at the outputs of the transceivers, 30 dB

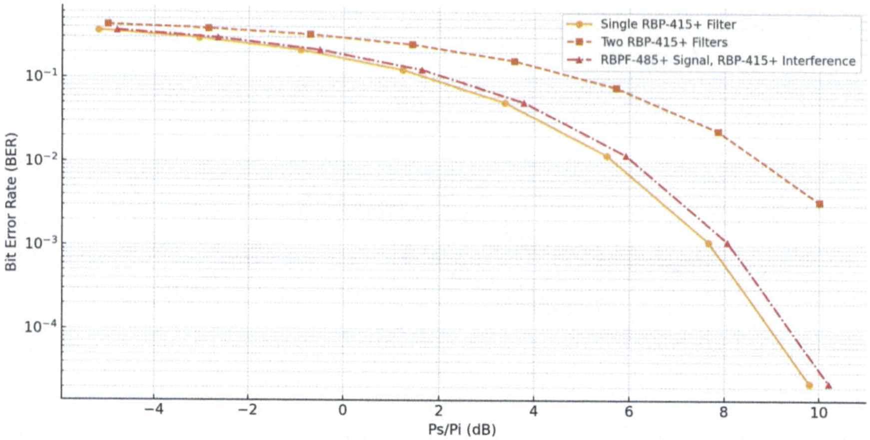

Fig. 7. Ratio of signal to interference power on receiver input (the modulation type is QPSK).

attenuation in the filters is not enough. It is necessary to provide approximately 60 dB attenuation in each channel due to filters.

4 Conclusions

The operation of two or more transceivers on single antenna is convenient, especially in mobile applications. One antenna is installed on the mast instead of several. There is no need to provide spatial diversity of antennas (usually at least 3 wavelengths).

The price for convenience is the inclusion of pass band filters between the outputs of the transceivers and the adder. As the study showed, the filters attenuation should be chosen taking into account the parameters of the equipment and application conditions. It is clear that operation with higher-level QAM modulation needs to be investigated. However, it should also be noted that the used Analog Devices Pluto SDR has low selectivity and the equipment with a typical value of selectivity at the level of 60...70 dB should bring better results.

The future research directions include experimental studies of communication systems using standard equipment to achieve specified BER probability values with PSK and QAM modulations.

References

1. Knox, M.E.: Single antenna full duplex communications using a common carrier. In: Wireless and Microwave Technology Conference (WAMICON), pp. 1–6. IEEE, Cocoa Beach, FL (2012). https://doi.org/10.1109/WAMICON.2012.6208455
2. Yalcinkaya, M.T., Sen, P., Fettweis, G.: Comparative analysis of antenna isolation characteristic with & without self-interference reduction techniques towards in-band full-duplex operation. IET Microw. Antennas Propag. **17**(5), 329–342 (2023). https://doi.org/10.1049/mia2.12355

3. Shyamala, C., Nithyashree, L.A., Sushma, S.J.: A full duplex SG wireless communication system using NI USRP2944. Int. J. Eng. **7**(9), 191–199 (2023). https://doi.org/10.33564/IJEAST.2023.v07i09.029
4. Tarek, N.A., Guerra, M.R., Nunez, A., Uddin, N., Elias, A.A.: Improving isolation in monostatic simultaneous transmit and receive systems using a quasi-symmetrical self-interference cancellation architecture. IEEE J. Microw. **3**(2), 665–664 (2023). https://doi.org/10.1109/JMW.2022.3227242
5. Zhang, L., et al.: A single-antenna full-duplex subsystem with high isolation and high gain. IEEE Open J. Antennas Propag. **5**(3), 620–625 (2024). https://doi.org/10.1109/OJAP.2024.3376568
6. Watkins, G.T., Thompson, W., Halls, D.: Single antenna full duplex cancellation network for ISM band. In: IEEE Radio and Wireless Symposium (RWS), pp. 21–24. IEEE, Anaheim, CA (2018). https://doi.org/10.1109/RWS.2018.8304935
7. Ilninskiy, L.Y., Savchenko, A.Y., Sibruk, L.V.: Antennas and Ultra High Frequency Devices. Ukrtelecom, Kyiv (2003) [in Ukrainian]
8. ADALM-PLUTO. Software-Defined Radio Active Learning Module. Analog Devices (2024). https://www.analog.com/en/resources/evaluation-hardware-and-software/evaluation-boards-kits/adalm-pluto.html#eb-documentation. Last accessed 15 Feb 2025
9. AD9363. RF Agile Transceiver. Analog Devices (2016). https://www.analog.com/en/products/ad9363.html#part-details. Last accessed 15 Feb 2025
10. Some Further Tests with ADALM-PLUTO SDR. RTL-SDR Com., v.1.0.0.1583, August (2017). https://www.rtl-sdr.com/some-further-tests-with-the-adalm-pluto-sdr. Last accessed 15 Feb 2025

Surveillance Data Processing in Civil Aviation

Object Coordinate Determination Based on Dempster-Shafer Evidence Theory

Igor Prokopenko[1], Sofiia Alpert[1(✉)], Maksym Alpert[2], and Yuliia Petrova[1]

[1] National Aviation University, Liubomyra Huzara Ave., 1, Kyiv 03058, Ukraine
sonyasonet87@gmail.com
[2] National Technical University of Ukraine "Igor Sikorsky Kyiv Polytechnic Institute", Poly-technichna Str., 37, Kyiv 03056, Ukraine

Abstract. Nowadays radars of air surveillance system are the main air traffic control system for the airspace around airports. They are applied for detection object's parameters, such as location, position, coordinates, angle, size, speed, and others. But a lot of airport surface applies multiple radar coverage, that's why fusion of information from different radars is one of the necessary procedures to provide a reliable object coordinate determination. The surveillance system employs data fusion algorithms to combine measurement data from radars for object coordinate determination and minimization of measurement error. Therefore, this paper focuses on researching the use of Dempster-Shafer evidence theory to combine surveillance data in order to process imprecise, inconsistent and partial data, received from various surveillance radars for object coordinate determination. The developed approach applies new algorithm of calculation of basic probabilities and mathematical expectation of coordinates of the object, integrating initial data of several surveillance radars and characteristics of their zones of ambiguity. The scientific novelty of this paper is the proposition of new function – basic probability density of basic probability (basic mass) for Dempster-Shafer evidence theory. Applying this new approach, we can determine most likely coordinates of the object. It also was considered three numerical examples of object coordinate determination applying evidence theory. This approach can be applied for solving various civil and military problems.

Keywords: Surveillance Radar · Dempster-Shafer Evidence Theory · Object Coordinate Determination · Basic Mass · Basic Probability Density · Monte Carlo Method

1 Introduction

1.1 Introduction to the Problem

Nowadays object detection is one of the most actual challenges of modern sensor systems. In civil applications modern sensor systems are used for air traffic control purposes, such as: the detection the presence and position of aircrafts, control and organization of the air traffic collision prevention, providing information about the aircrafts [1–3].

I. Ostroumov et al. (Eds.): ACASD 2025, LNNS 1418, pp. 289–304, 2025.
https://doi.org/10.1007/978-3-031-91992-3_19

The air surveillance system at large airports has integrated primary and secondary radar system. The primary surveillance radar consists of a large rotating parabolic antenna dish mounted on a tower. This radar antenna rotates and emits high power pulses of microwave radio waves in a narrow vertical fan-shaped beam around the airspace surrounding the airport. The primary radar pulses propagate at the speed of light. If the antenna of this radar rotates at a constant rate about a vertical axis, the radar beam scans the whole surrounding airspace about every 5 s. When the microwave beam strikes an aircraft, some of the energy of microwaves is reflected back to the antenna and is detected by the radar receiver. The radar receiving system determines the distance of the aircraft from the radar system by calculating the time delay between the transmitted and received reflected pulses.

The main function of primary radar is to determine the azimuth and range of the aircraft in respect of the antenna position. Then this received information is displayed on the workplace screen of the air traffic controller. Primary radar detects the position and range of aircraft by microwaves reflected back to the antenna from the aircraft's surface. A main advantage of the primary radar is that it provides good range information and doesn't demand any on-board equipment to locate aircraft.

A secondary radar system is known as secondary surveillance radar. It consists of a second rotating antenna mounted on the primary radar antenna. The secondary surveillance radar looks like a horizontal metal sheet. It requires an aircraft to be fitted with an onboard radar transmitter/receiver (transponder).

The transponder is an electronic device in the aircraft that produces a reply when it receives a radio-frequency interrogation. Transponder transmits a radio signal back containing the barometric altitude and the aircraft's identification.

The atmospheric phenomena cause echo signals that can mask the target echoes. Radar waves bend when they propagate through the atmosphere with increasing altitude and decreasing density of the atmosphere. Atmospheric absorption facilitates loss of radar energy [3, 4].

1.2 Motivation

Surveillance radars can differ in their credibility, precision, reliability, completeness, and zones of ambiguity (zone of confusion). Adaptations of different surveillance radars to the same environment are not the same. Some radars are better adapted to the environmental conditions than other radars. The zone of ambiguity is area where signal interference makes navigation uncertain. It is an area, where navigating is uncertain, because signals are precarious. The aircraft's navigation receiver cannot interpret the signals correctly in zones of confusion. Transmitted signals from the omnidirectional and directional antennas overlap in such a way that they cancel each other out. That's why, aerial surveillance information fusion is one of most efficient and important procedure for the applying air traffic control systems. Recorded data can be used together with the data fusion output information for object coordinate determination in modern radar systems. A lot of airport surface have multiple radar coverage, without the necessary synchronized sensors, so a data fusion scheme is essential to provide a stable and reliable output. The surveillance system employs fusion algorithms to process measurement data from radars (surveillance sources) for prediction of the future position of the aircraft and minimization of measurement error. Data fusion is also applied for target coordinate determination [4, 5].

Therefore, the data fusion techniques for target coordinate determination are still evolving, so this motivated us to synthesize a new data fusion approach based on modified Dempster-Shafer evidence theory in order to get a complex assessment faster and simplify the mathematical calculations.

1.3 Contribution

The scientific novelty of this paper is the improvement of the new data fusion approach based on the proposition of new function - basic probability density of basic mass for Dempster-Shafer evidence theory. The main contribution concerns the development of a new approach to solving the problem of object coordinate determination based on Dempster-Shafer evidence theory, information fusion of surveillance radars and characteristics of zones of ambiguity. This can be considered an extension of the Dempster-Shafer theory for solving various civil and military problems.

1.4 The Organization of the Paper

The paper consists of four sections. The first section presents the background information and actuality of the considered problem. The second section considers the basic concepts and main definitions of the Dempster-Shafer evidence theory. The third section contains a detailed algorithm of the proposed approach and numerical examples. The fourth section is conclusions.

2 Materials and Methods

2.1 Basic Definitions and Concepts of Dempster-Shafer Evidence Theory

Dempster-Shafer evidence theory is a mathematical approach for modeling uncertainty, it processes ambiguous, ignorance, contradictory and missing information. This theory also can be applied for data fusion, because it combines various sources of information to derive probabilistic conclusions. It processes different expert opinions and obtains an integral estimation. Evidence theory provides a multi-alternative solution of the problem and calculates degrees of belief.

The evidence theory is generalization of classical probability theory. But this theory does not require the strict assumptions of classic probability theory. Unlike known probability theory, Dempster-Shafer theory can process partial, ambiguous and ambivalent data. It can deal with probabilities of a many hypotheses, whereas a probability theory deals with only one single hypothesis. Let's consider main concepts and definitions of Dempster-Shafer evidence theory [6–9]. Ω is a frame of discernment, Ω is a set of all hypotheses; 2^{Ω} is number of all subsets of Ω. Set Ω and set $\emptyset$ are also included in this number. $m(A)$ is the basic mass or basic probability. It means the degree of belief allocated to the given hypothesis A. Basic mass satisfies the next conditions:

$$\sum\nolimits_{A\subseteq 2^{\Omega}} m(A) = 1. \quad (1)$$

$$m(\varnothing) = 0.$$

$$0 \leq m(A) \leq 1.$$

Subset A is called the focal subset, if basic mass $m(A) > 0$.

Basic differences between Dempster-Shafer evidence theory and probability theory. Necessary conditions are met for the classical probability theory

1. $P(\Omega) = 1$;
2. If $X \subset Y$, then the condition is obligatorily satisfied:$P(X) \leq P(Y)$;
3. $P(X) + P(\overline{X}) = 1$, where $\overline{X}$– the complement of the set : $X \cap \overline{X} = \varnothing$, $X \cup \overline{X} = \Omega$.

Necessary conditions are met for the Dempster-Shafer evidence theory:

1. $m(\Omega) \neq 1$;
2. If $X \subset Y$,then the condition is not necessarily satisfied: $m(X) \leq m(Y)$;
3. No relationship is required between $m(X)$ and $m(\overline{X})$, where $\overline{X}$ is the complement of the set X.

The classical probability theory demands complete knowledge of combined probabilities and the priory knowledge of probability distribution [10–12]. It also cannot model imprecision. Dempster-Shafer evidence theory can process incomplete and vague data, because it is more flexible approach than the probability theory. Dempster-Shafer theory can deal with probabilities of a collection of hypotheses, whereas a traditional probability theory deals with only one single hypothesis. Dempster-Shafer evidence theory is a generalization of probability theory [13, 14].

Belief function *Bel*(A) determines the minimum support for some hypothesis. *Bel*(A) is calculated by summing the basic probabilities over all nonempty subsets $B \leq A$. *Bel*(A) is defined as follows:

$$Bel(A) = \sum\nolimits_{B \subseteq A} m(B). \tag{2}$$

Plausibility function *Pl*(A) reflects the maximum or potential support for that hypothesis. The values of the *Pl*(A) are a set of basic probabilities of all nonempty subsets B intersecting with the subset A.

Pl(A) is defined as follows:

$$Pls(A) = \sum\nolimits_{B \cap A \neq \varnothing} m(B). \tag{3}$$

The interval [*Bel*(A), *Pl*(A)] is called the "belief interval". The length of "belief interval" shows the imprecision about the uncertainty value of A. Belief function *Bel*(A) and plausibility function *Pl*(A) show the level of hypothesis support. The Dempster's combination rule is applied for combining data from various experts or sources. Suppose that one expert assigned mass m_1 to the class A and another expert independently assigned mass m_2 to the same class. Then, the combined estimation of the mass of the class A is defined as follows:

$$m(A) = \frac{1}{1-K} \sum B_1 \cap B_2 = Am_1(B_1) \cdot m_2(B_2), \tag{4}$$

$$K = \sum B_1 \cap B_2 = \varnothing m_1(B_1) \cdot m_2(B_2), \tag{5}$$

where K is called conflict coefficient. This conflict coefficient K determines the degree of conflict among the sources or experts. It satisfies next condition: $0 \leq K \leq 1$.

The less contradictions we have, the closer is the K value to 0 [15, 16].

2.2 The Proposition of Dempster-Shafer Evidence Theory Enhancing by New Basic Probability Density of Basic Mass

The scientific novelty of this research is the improvement of the new data fusion approach based on the proposition of new function – basic probability density of basic mass (basic probability) for Dempster-Shafer evidence theory. Suppose, that the set A is the domain of a certain geometric shape (circle, rectangle, ellipse, etc.). The basic mass expresses the probability of an object falling into this domain.

The basic probability density is defined as follows:

$$p(A) = \frac{m_n(A)}{S_A}, \tag{6}$$

where $m_n(A)$ is normalized basic mass (basic probability), S_A is the area of a domain A.

Suppose, that we have 2 domains: A and B. These domains also overlap.

Then basic probability densities of basic masses for domain A, domain B, and intersection of these domains $A \cap B$ are defined as follows:

$$p(\{A\}) = \frac{m_n(\{A\})}{S_A}; \tag{7}$$

$$p(\{B\}) = \frac{m_n(\{B\})}{S_B}; \tag{8}$$

$$p(\{A \cap B\}) = p(\{A\}) + p(\{B\}), \tag{9}$$

where $S_A \cap S_B$is the area of the intersection of A and B, $m_n(A)$, $m_n(B)$, $m_n(A \cap B)$ are normalized basic probabilities.

Let's note, that the maximum density value will be in the intersection of domains. Thus, the most likely location of the object should be in the intersection $S_A \cap S_B$.

3 Applying the Modified Dempster-Shafer Evidence Theory to Object Coordinate Determination

We need to determine the most likely coordinates of an object by combining the information coming from two surveillance radars into a processing center. Let's note, that zones of ambiguity of different surveillance radars differ in size and shape [17].

3.1 First Numerical Example

Suppose we have 2 surveillance radars.

First surveillance radar has zone of ambiguity in the form of a rectangle with next vertices and sides: $A_1(1, 3)$, $B_1(1, 6)$, $C_1(8, 6)$, $D_1(8, 3)$, $a_1 = 3$, $b_1 = 7$.

Second surveillance radar has zone of ambiguity in the form of a rectangle with next vertices and sides: $A_2(5, 2)$, $B_2(5, 5)$, $C_2(16, 5)$, $D_2(16, 2)$, $a_2 = 11$, $b_2 = 3$.

These two zones of ambiguity in the form of a rectangle overlap (Fig. 1).

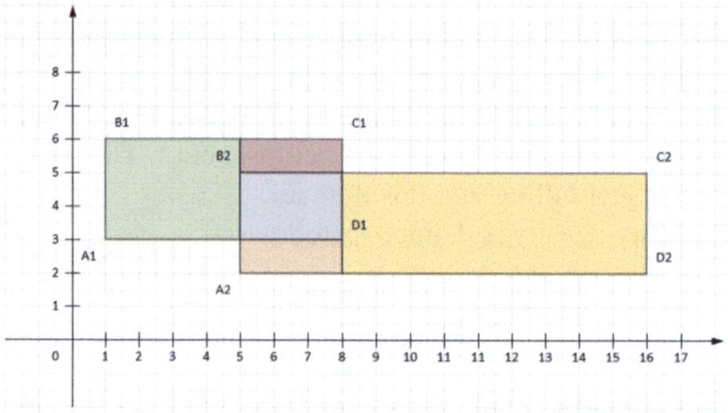

Fig. 1. The interconnection of two zones of ambiguity that have the shape of a rectangle.

The intersection of these zones of ambiguity is also a rectangle with next vertices and sides: $M(5, 3)$, $B_2(5, 5)$, $N(8, 5)$, $D_1(8, 3)$, $a_3 = 2$, $b_3 = 3$.

Let's define coordinates of the center of intersection of two zones of ambiguity:

$$x_0 = \frac{x_{B_2} + x_{D_1}}{2} = \frac{5+8}{2} = 6.5; \quad y_0 = \frac{y_{B_2} + y_{D_1}}{2} = \frac{5+3}{2} = 4.$$

The center of intersection of two zones of ambiguity is $O(6, 5, 4)$.

We need to determine the most likely coordinates of an object by applying Dempster-Shafer evidence theory, given characteristics of two zones of ambiguity and their intersection (vertices and sides of rectangles).

Then areas of two zones of ambiguity and their intersection are defined as:

- $S_1 = a_1 b_1 = 3 \bullet 7 = 21$-area of zone of ambiguity of first surveillance radar;
- $S_2 = a_2 b_2 = 11 \bullet 3 = 33$-area of zone of ambiguity of second surveillance radar;
- $S_3 = a_3 b_3 = 2 \bullet 3 = 6$-area of the intersection of zones of ambiguity.

$$S_1 - S_1 \cap S_2 = 21 - 6 = 15; \quad S_2 - S_1 \cap S_2 = 33 - 6 = 27.$$

Then basic masses (basic probabilities) for these focal subsets are defined as follows:

$$m(\{s_1\}) = \frac{S_1 - S_1 \cap S_2}{S_1} = \frac{15}{21} = 0.714; \quad m(\{s_2\}) = \frac{S_2 - S_1 \cap S_2}{S_2} = \frac{27}{33} = 0.818;$$

$$m(\{S_1 \cap S_2\}) = \frac{S_1 \cap S_2}{S_1} + \frac{S_1 \cap S_2}{S_2} = \frac{6}{21} + \frac{6}{33} = \frac{2}{7} + \frac{2}{11} = 0.468.$$

Then values of the basic probabilities are normalized by the next formula:

$$m_n(\{s_1\}) = \frac{m(\{s_1\})}{m(\{s_1\}) + m(\{ s_2\}) + m(\{S_1 \cap S_2\})};$$

$$m_n(\{s_1\}) = \frac{0{,}714}{0{,}714 + 0{,}818 + 0{,}468} = \frac{0{,}714}{2} = 0.357;$$

$$m_n(\{s_2\}) = \frac{m(\{s_2\})}{m(\{s_1\}) + m(\{ s_2\}) + m(\{S_1 \cap S_2\})} = \frac{0.818}{2} = 0.409;$$

$$m_n(\{S_1 \cap S_2\}) = \frac{0.468}{0.714 + 0.818 + 0.468} = \frac{0.468}{2} = 0.234.$$

We defined basic probabilities of an object falling into each of the zones of ambiguity respectively. Let's note, that the scientific novelty of this research is the improvement of the proposition of new function – basic probability density of basic probability.

Then basic probability densities for these zones of ambiguity are defined as follows:

$$p(\{s_1\}) = \frac{m_n(\{s_1\})}{S_1 - S_1 \cap S_2}; p(\{s_1\}) = \frac{0.357}{15} = 0.024;$$

$$p(\{s_2\}) = \frac{m_n(\{s_2\})}{S_2 - S_1 \cap S_2}; p(\{s_2\}) = \frac{0.409}{27} = 0.015;$$

$$p(\{S_1 \cap S_2\}) = \frac{m_n(\{S_1 \cap S_2\})}{S_1 \cap S_2}; p(\{S_1 \cap S_2\}) = \frac{0.234}{6} = 0.039.$$

The maximum density value will be in the intersection of zones of ambiguity. Thus, the most likely coordinates of the object should be in this intersection $S_1 \cap S_2$. Then we should calculate belief functions and plausibility functions for zones of ambiguity and their intersection:

1. Belief functions for S_1, S_2 and $S_1 \cap S_2$ are defined as follows:

$$Bel(\{S_1\}) = m_n(\{S_1\}) \approx 0.357; \;\; Bel(\{S_2\}) = m_n(\{S_2\}) \approx 0.409.$$

$$Bel(\{S_1 \cap S_2\}) = m_n(\{S_1 \cap S_2\}) \approx 0.234.$$

2. Plausibility functions for intervals S_1, S_2 and $S_1 \cap S_2$are defined as follows:

$$Pls(\{S_1\}) = m_n(\{S_1\}) = 0.357; \;\; Pls(\{S_2\}) = m_n(\{S_2\}) = 0.409;$$

$$Pls(\{S_1 \cap S_2\}) = m_n(\{S_1 \cap S_2\}) = 0.234.$$

The most probable coordinates of an object are the mathematical expectation of the two-dimensional continuous random variable falling within the area formed by the two rectangles. We would like to remind, that mathematical expectation of a two-dimensional random variable is defined by the next formulas:

$$E(X_1) = \int_{-\infty}^{+\infty}\int_{-\infty}^{+\infty} x_1 f(x_1, x_2) dx_1 dx_2; E(X_2) = \int_{-\infty}^{+\infty}\int_{-\infty}^{+\infty} x_2 f(x_1, x_2) dx_1 dx_2,$$

where $f(x_1, x_2)$ is probability density function. X_1 and X_2 are random variables with joint probability distribution function $f(x_1, x_2)$.

$$E(X_1) = \int_3^6 \int_1^5 x_1 \cdot 0{,}024 dx_1 dx_2 + \int_5^6 \int_5^8 x_1 \cdot 0{,}024 dx_1 dx_2 +$$

$$+ \int_3^5 \int_5^8 x_1 \cdot 0{,}039 dx_1 dx_2 + \int_2^3 \int_5^8 x_1 \cdot 0{,}015 dx_1 dx_2 + \int_2^5 \int_8^{16} x_1 \cdot 0{,}015 dx_1 dx_2;$$

1. $\int_3^6 \int_1^5 x_1 \cdot 0.024 dx_1 dx_2 = 0.024 \int_3^6 \int_1^5 x_1 dx_1 dx_2 = 0.864.$
2. $\int_5^6 \int_5^8 x_1 \cdot 0.024 dx_1 dx_2 = 0.024 \int_5^6 \int_5^8 x_1 dx_1 dx_2 = 0.468.$
3. $\int_3^5 \int_5^8 x_1 \cdot 0.039 dx_1 dx_2 = 0.039 \int_3^5 \int_5^8 x_1 dx_1 dx_2 = 1.521.$
4. $\int_2^3 \int_5^8 x_1 \cdot 0.015 dx_1 dx_2 = 0.015 \int_2^3 \int_5^8 x_1 dx_1 dx_2 = 0.2925.$
5. $\int_2^5 \int_8^{16} x_1 \cdot 0.015 dx_1 dx_2 = 0.015 \int_2^5 \int_8^{16} x_1 dx_1 dx_2 = 4.32.$

Additional calculations: $\int_8^{16} x_1 dx_1 = 96.$

$$E(X_1) = 0.864 + 0.468 + 1.521 + 0.2925 + 4.32 = 7.4655 \approx 7.5.$$

$$E(X_2) = \int_3^6 \int_1^5 x_2 \cdot 0{,}024 dx_1 dx_2 + \int_5^6 \int_5^8 x_2 \cdot 0{,}024 dx_1 dx_2 +$$

$$+ \int_3^5 \int_5^8 x_2 \cdot 0{,}039 dx_1 dx_2 + \int_2^3 \int_5^8 x_2 \cdot 0{,}015 dx_1 dx_2 + \int_2^5 \int_8^{16} x_2 \cdot 0{,}015 dx_1 dx_2;$$

1. $\int_3^6 \int_1^5 x_2 \cdot 0.024 dx_1 dx_2 = 0.024 \int_3^6 x_2 \int_1^5 dx_1 dx_2 = 1.296.$
2. $\int_5^6 \int_5^8 x_2 \cdot 0.024 dx_1 dx_2 = 0.024 \int_5^6 x_2 \int_5^8 dx_1 dx_2 = 0.396.$
3. $\int_3^5 \int_5^8 x_2 \cdot 0.039 dx_1 dx_2 = 0.039 \int_3^5 x_2 \int_5^8 dx_1 dx_2 = 0.936.$
4. $\int_2^3 \int_5^8 x_2 \cdot 0.015 dx_1 dx_2 = 0.015 \int_2^3 x_2 \int_5^8 dx_1 dx_2 = 0.1125.$
5. $\int_2^5 \int_8^{16} x_2 \cdot 0.015 dx_1 dx_2 = 0.015 \int_2^5 x_2 \int_8^{16} dx_1 dx_2 = 1.26.$

$$E(X_2) = 1.296 + 0.396 + 0.936 + 0.1125 + 1.26 = 4.0005 \approx 4.$$

The obtained mathematical expectation of the two-dimensional continuous random variable falling within the area formed by the two rectangles is the most probable coordinates of an object for this case: $O^*(E(X_1), E(X_2))$.

3.2 Second Numerical Example

Suppose we have 2 surveillance radars. First surveillance radar has zone of ambiguity in the form of a circle with next coordinate of center and radius: $O_1(3, 1)$, $R_1 = 1$. Second surveillance radar has zone of ambiguity in the form of a circle with next coordinate of center and radius:$O_2(5, 2)$, $R_2 = 2$. These two zones of ambiguity also overlap (Fig. 2).

1. We should find coordinates of the intersection points of these two circles and coordinates of the center of this intersection

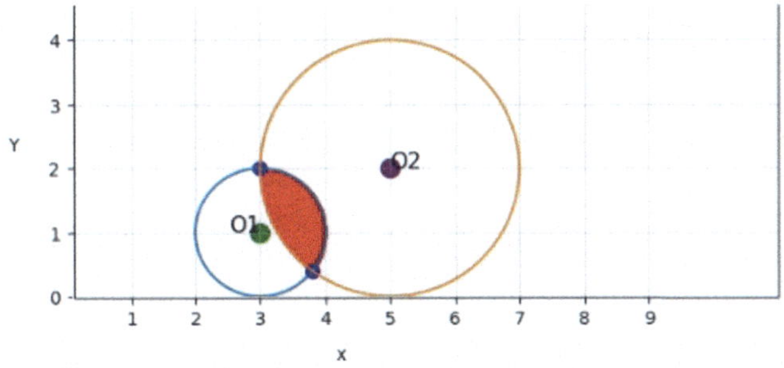

Fig. 2 The interconnection of two zones of ambiguity that have the shape of a circle.

The equation for the first circle is defined as: $(x-3)^2+(y-1)^2=1^2$.
The equation for the second circle is defined as: $(x-5)^2+(y-2)^2=2^2$.
Then the following system of equations is solved as:

$$\begin{cases} x^2-6x+9+y^2-2y=0; \\ x^2-10x+25+y^2-4y=0. \end{cases}$$

Therefore, we can get $-6x+9-2y=-10x+25-4y$ and $y=8-2x$. So

$$(x-3)^2+(8-2x-1)^2=1;$$

$$x_1=\frac{34-4}{2\cdot 5}=\frac{30}{10}=3;\ x_2=\frac{34+4}{10}=\frac{38}{10}=3.8.$$

$$y_1=8-2x_1=8-2\cdot 3=2;\ y_2=8-2x_2=8-2\cdot 3.8=0.4.$$

The coordinates of the intersection points are defined as follows: $K_1(3,2)$; $K_2(3.8,0.4)$.

The coordinates of the center of this intersection of these two circles are defined as:

$$O^*\left(\frac{3+3.8}{2},\frac{2+0.4}{2}\right)\rightarrow O^*(3.4,1.2).$$

Then areas of two circles (zones of ambiguity) and their intersection are defined. Area of the first circle: $S_1=\pi R_1^2=\pi\cdot 1^2=3.14$. Area of the second circle: $S_2=\pi R_2^2=\pi\cdot 2^2=12.56$.

Area of the intersection of two circles:

$$S_1\cap S_2=\int_3^{3.8}\int_{-\sqrt{4-(x-5)^2}+2}^{\sqrt{1-(x-3)^2}+1}dydx=\int_3^{3.8}(\sqrt{1-(x-3)^2}+1$$
$$+\sqrt{4-(x-5)^2}-2)dx=\int_3^{3.8}\left(\sqrt{1-(x-3)^2}+\sqrt{4-(x-5)^2}-1\right)dx.$$

Additional calculations:

$$\int_3^{3.8}\sqrt{1-(x-3)^2}dx=\left|\begin{array}{c}\text{we set } x-3=t,\\ \text{then } dx=dt\end{array}\right|=\int_0^{0.8}\sqrt{1-t^2}dt$$

$$= \frac{t}{2}\sqrt{1^2 - t^2} + \frac{1^2}{2}\arcsin\frac{t}{1}\Bigg|_0^{0.8} = \frac{t}{2}\sqrt{1 - t^2} + \frac{1}{2}arcsint\Bigg|_0^{0.8} \approx 0.70364$$

$$\int_3^{3.8}\sqrt{4 - (x-5)^2}dx = \int_3^{3.8}\sqrt{2^2 - (x-5)^2}dx = \left|\begin{array}{c}\text{we set } x - 5 = u,\\ \text{then } du = dt\end{array}\right|$$

$$= \int_{-2}^{-1.2}\sqrt{2^2 - u^2}du = \frac{u}{2}\sqrt{4 - u^2} + 2\arcsin\frac{u}{2}\Bigg|_{-2}^{-1.2} \approx 0.89459.$$

$$\int_3^{3.8} dx = 0.8.$$

In these calculations, the following formula and values of the arcsine were used:

$$\int \sqrt{a^2 - u^2}du = \frac{u}{2}\sqrt{a^2 - u^2} + \frac{a^2}{2}\arcsin\frac{u}{a} + C,$$

$$\arcsin\left(\frac{3}{5}\right) = 0.6435;\ \arcsin\left(\frac{4}{5}\right) = 0.9273;\ \arcsin(-1) = -1.571.$$

Then the area of the intersection of two circles:

$$S_1 \cap S_2 = \int_3^{3.8}\int_{-\sqrt{4-(x-5)^2}+2}^{\sqrt{1-(x-3)^2}+1} dydx = \int_3^{3.8}\left(\sqrt{1 - (x-3)^2} + 1\right.$$

$$\left.+\sqrt{4 - (x-5)^2} - 2\right)dx = \int_3^{3.8}\left(\sqrt{1 - (x-3)^2} + \sqrt{4 - (x-5)^2} - 1\right)dx \approx 0{,}798.$$

$$S_1 - S_1 \cap S_2 = 3.14 - 0.798 = 2.342;\ S_2 - S_1 \cap S_2 = 12.56 - 0.798 = 11.762.$$

2. Then basic masses (basic probabilities) for these focal subsets are defined as follows

$$m(\{s_1\}) = \frac{S_1 - S_1 \cap S_2}{S_1} = \frac{2.342}{3.14} = 0.746;\ m(\{\ s_2\}) = \frac{S_2 - S_1 \cap S_2}{S_2} = \frac{11.762}{12.56} = 0.936;$$

$$m(\{S_1 \cap S_2\}) = \frac{S_1 \cap S_2}{S_1} + \frac{S_1 \cap S_2}{S_2} = \frac{0.798}{3.14} + \frac{0.798}{12.56} = 0.318.$$

3. Then values of the basic probabilities are normalized by the next formula

$$m_n(\{s_1\}) = \frac{0.746}{0.746 + 0.936 + 0.318} = 0.373;\ m_n(\{s_2\}) = \frac{0.936}{0.746 + 0.936 + 0.318} = 0.468;$$

$$m_n(\{S_1 \cap S_2\}) = \frac{0.318}{0.746 + 0.936 + 0.318} = 0.159.$$

We defined basic probabilities of an object falling into each of the zones of ambiguity (circles) respectively.

4. These basic probability densities for these zones of ambiguity (two circles and their intersection) are defined as follows

$$p(\{s_1\}) = \frac{m_n(\{s_1\})}{S_1 - S_1 \cap S_2} = \frac{0.373}{2.342} = 0.159; p(\{s_2\}) = \frac{m_n(\{s_2\})}{S_2 - S_1 \cap S_2} = \frac{0.468}{11.762} = 0.04;$$

$$p(\{S_1 \cap S_2\}) = \frac{m_n(\{S_1 \cap S_2\})}{S_1 \cap S_2} = \frac{0.159}{0.798} = 0.199.$$

5. The maximum density value will be in the intersection of zones of ambiguity. Thus, the most likely coordinates of the object should be in this intersection $S_1 \cap S_2$.

Belief functions for S_1, S_2 and $S_1 \cap S_2$ are defined as follows:

$$Bel(\{S_1\}) = m_n(\{S_1\}) \approx 0.373; Bel(\{S_2\}) = m_n(\{S_2\}) \approx 0.468;$$

$$Bel(\{S_1 \cap S_2\}) = m_n(\{S_1 \cap S_2\}) \approx 0.159.$$

6. Plausibility functions for intervals S_1, S_2 and $S_1 \cap S_2$are defined as follows:

$$Pls(\{S_1\}) = m_n(\{S_1\}) = 0.373; Pls(\{S_2\}) = m_n(\{S_2\}) = 0.468;$$

$$Pls(\{S_1 \cap S_2\}) = m_n(\{S_1 \cap S_2\}) = 0.159.$$

The most probable coordinates of an object are the mathematical expectation of the two-dimensional continuous random variable falling within the area formed by the two circles, that intersect.

7. The values of the basic probability densities for zones of ambiguity (two circles and their intersection) are normalized by the next formula

$$p_n(\{s_1\}) = \frac{0.159}{0.159 + 0.04 + 0.199} = 0.399; p_n(\{s_2\}) = \frac{0.04}{0.159 + 0.04 + 0.199} = 0.101;$$

$$p_n(\{S_1 \cap S_2\}) = \frac{0.199}{0.159 + 0.04 + 0.199} = 0.5.$$

Let's note, that in this case, the mathematical expectation is the center of mass. The mean value is the location of the center of mass. We should determine the mathematical expectation of coordinate *X*:

$$E(X) = x_{O_1} \cdot p_n(\{s_1\}) + x_{O^*} \cdot p_n(\{S_1 \cap S_2\}) + x_{O_2} \cdot p_n(\{s_2\})$$

$$= 3 \cdot 0.399 + 3.4 \cdot 0,5 + 5 \cdot 0.101 = 1.197 + 1.7 + 0.505 = 3.402.$$

Then we determine the mathematical expectation of coordinate *Y:*

$$E(Y) = y_{O_1} \cdot p_n(\{s_1\}) + y_{O^*} \cdot p_n(\{S_1 \cap S_2\}) + y_{O_2} \cdot p_n(\{s_2\})$$

$$= 1 \cdot 0{,}399 + 1,2 \cdot 0,5 + 2 \cdot 0{,}101 = 0{,}399 + 0,6 + 0{,}202 = 1{,}201.$$

Problem answer: O^*(3,402, 1,201).

3.3 Third Numerical Example

Suppose we have 2 surveillance radars.

First surveillance radar has zone of ambiguity in the form of an ellipse with next coordinate of the center, semi-major and semi-minor axes: *a*, *b*.

$$O_1(0, 0);\ a_1 = 2;\ b_1 = 1.$$

Second surveillance radar has zone of ambiguity in the form of an ellipse with next coordinate of the center, semi-major and semi-minor axes:

$$O_2(4, 0);\ a_2 = 3;\ b_2 = 2.$$

These two zones of ambiguity also overlap (Fig. 3). We should find coordinates of the intersection points of these two ellipses and coordinates of the center of this intersection. The equation for the first ellipse is defined as: $\frac{x^2}{4} + \frac{y^2}{1} = 1$.The equation for the second ellipse is defined as: $\frac{(x-4)^2}{9} + \frac{y^2}{4} = 1$.

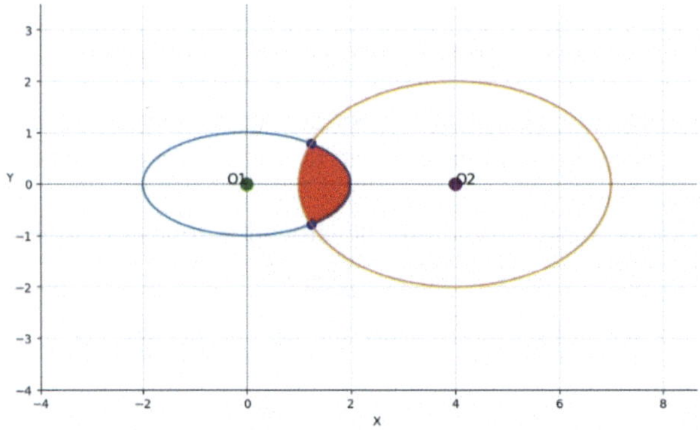

Fig. 3. The interconnection of two zones of ambiguity that have the shape of an ellipse.

Then the following system of equations is solved as:

$$\begin{cases} \frac{x^2}{4} + \frac{y^2}{1} = 1; \\ \frac{(x-4)^2}{9} + \frac{y^2}{4} = 1. \end{cases} \rightarrow \begin{cases} x^2 + 4y^2 = 4; \\ 4x^2 - 32x + 64 + 9y^2 = 36. \end{cases}$$

Therefore, we can get $y^2 = 1 - \frac{x^2}{4}$ and

$$y_1 = \sqrt{1 - \frac{x^2}{4}};\ y_2 = -\sqrt{1 - \frac{x^2}{4}}.$$

$$4x^2 - 32x + 64 + 9\left(1 - \frac{x^2}{4}\right) = 36 \text{ or } 1.75x^2 - 32x + 37 = 0;$$

$x_1 = x_2 = 1.24$ (Fig. 3).

$$y_1 = \sqrt{1 - \frac{(1.24)^2}{4}} = 0.784; y_2 = -\sqrt{1 - \frac{(1.24)^2}{4}} = -0.784.$$

The coordinates of the intersection points are defined as follows: $K_1(1, 24, 0{,}784)$, $K_2(1, 24, -0{,}784)$. The coordinates of center of this intersection of these two ellipses are defined as:

$$O^*\left(\frac{1.24 + 1.24}{2}, \frac{0.784 - 0.784}{2}\right) \to O^*(1.24, 0).$$

Then areas of two ellipses (zones of ambiguity) and their intersection are defined. Area of the first ellipse is defined as: $S_1 = \pi a_1 b_1 = \pi \bullet 2 \bullet 1 \approx 6.28$. Area of the second ellipse is defined as: $S_2 = \pi a_2 b_2 = \pi \bullet 3 \bullet 2 = 6\pi \approx 18.84$. Then we define the area of the intersection of two ellipses, applying Monte Carlo method and Python programming language.

The Monte Carlo algorithm for calculating the area of any figure is summarized as follows:

1. The figure is inscribed in a rectangle. Let's note, that the coordinates of the sides of the rectangle are known, so its area is known.
2. Then a lot of points are generated pseudo-randomly inside this rectangle. For each point, it is determined whether the point is inside the original figure or not.
3. Then the area of the original figure is calculated based on the next proportion: the ratio of the number of points that fell inside the figure to the total number of generated points is equal to the ratio of the figure area to the area of the rectangle bounding it.
4. From this proportion we find the area of the intersection of these two ellipses.

Let's note, that number of points that fell inside the figure, the total number of generated points and area of the rectangle are known values.

So, the area of the intersection of two ellipses is defined as:

$$S_1 \cap S_2 \approx 1.09.$$

$$S_1 - S_1 \cap S_2 = 6.28 - 1.09 = 5.19;$$

$$S_2 - S_1 \cap S_2 = 18.84 - 1.09 = 17.75.$$

1. Then basic masses (basic probabilities) for these focal subsets are defined as follows

$$m(\{s_1\}) = \frac{S_1 - S_1 \cap S_2}{S_1} = \frac{5.19}{6.28} = 0.826; m(\{ s_2\}) = \frac{S_2 - S_1 \cap S_2}{S_2} = \frac{17.75}{18.84} = 0.942;$$

$$m(\{S_1 \cap S_2\}) = \frac{S_1 \cap S_2}{S_1} + \frac{S_1 \cap S_2}{S_2} = \frac{1.09}{6.28} + \frac{1.09}{18.84} = 0.232.$$

2. Then values of the basic probabilities are normalized by the next formula

$$m_n(\{s_1\}) = \frac{0.826}{0.826 + 0.942 + 0.232} = 0.413; m_n(\{s_2\}) = \frac{0.942}{0.826 + 0.942 + 0.232} = 0.471;$$

$$m_n(\{S_1 \cap S_2\}) = \frac{0.232}{0.826 + 0.942 + 0.232} = 0.116.$$

We defined basic probabilities of an object falling into each of the zones of ambiguity (circles) respectively.

3. These basic probability densities for these zones of ambiguity (two circles and their intersection) are defined as follows

$$p(\{s_1\}) = \frac{m_n(\{s_1\})}{S_1 - S_1 \cap S_2} = \frac{0.413}{5.19} = 0.0796; p(\{s_2\}) = \frac{m_n(\{s_2\})}{S_2 - S_1 \cap S_2} = \frac{0.471}{17.75} = 0.0265;$$

$$p(\{S_1 \cap S_2\}) = \frac{m_n(\{S_1 \cap S_2\})}{S_1 \cap S_2} = \frac{0.116}{1,09} = 0.1064.$$

The maximum density value will be in the intersection of zones of ambiguity. Thus, the most likely coordinates of the object should be in this intersection $S_1 \cap S_2$.

4. Belief functions for S_1, S_2 and $S_1 \cap S_2$ are defined as follows:

$$Bel(\{S_1\}) = m_n(\{S_1\}) \approx 0.413; Bel(\{S_2\}) = m_n(\{S_2\}) \approx 0.471;$$

$$Bel(\{S_1 \cap S_2\}) = m_n(\{S_1 \cap S_2\}) \approx 0.116.$$

5. Plausibility functions for intervals S_1, S_2 and $S_1 \cap S_2$are defined as follows:

$$Pls(\{S_1\}) = m_n(\{S_1\}) = 0.413; Pls(\{S_2\}) = m_n(\{S_2\}) = 0.471;$$

$$Pls(\{S_1 \cap S_2\}) = m_n(\{S_1 \cap S_2\}) = 0.116.$$

The most probable coordinates of an object are the mathematical expectation of the two-dimensional continuous random variable falling within the area formed by the two circles, that intersect.

6. The values of the basic probability densities for zones of ambiguity (two ellipses and their intersection) are normalized by the next formula

$$p_n(\{s_1\}) = \frac{0.159}{0.159 + 0.04 + 0.199} = 0.399;$$
$$p_n(\{s_2\}) = \frac{0.04}{0.159 + 0.04 + 0.199} = 0.101;$$
$$p_n(\{S_1 \cap S_2\}) = \frac{0.199}{0.159 + 0.04 + 0.199} = 0.5.$$

7. Let's note, that in this case, the mathematical expectation is the center of mass. The mean value is the location of the center of mass.

8. We should determine the mathematical expectation of coordinate X:

$$E(X) = x_{O_1} \cdot p_n(\{s_1\}) + x_{O^*} \cdot p_n(\{S_1 \cap S_2\}) + x_{O_2} \cdot p_n(\{s_2\}) =$$

$$= 0 \cdot 0.3746 + 1.24 \cdot 0.5007 + 4 \cdot 0.1247 \approx 1.119668 \approx 1.12.$$

Then we determine the mathematical expectation of coordinate Y:

$$E(Y) = y_{O_1} \cdot p_n(\{s_1\}) + y_{O^*} \cdot p_n(\{S_1 \cap S_2\}) + y_{O_2} \cdot p_n(\{s_2\}) =$$

$$= 0 \cdot 0.3746 + 0 \cdot 0.5007 + 0 \cdot 0.1247 = 0.$$

Problem answer: $O^*(1.12, 0)$.

4 Conclusions

This paper presents the application of Dempster-Shafer evidence theory to an object coordinate determination. This theory deals with contradictory, ambiguous, uncertain and incomplete data. Evidence theory is generalization of classic probability theory, but it doesn't require the strict assumptions as a classic probability theory. Evidence theory can be applied for data fusion, because it can combine different sources of information and gives an integral estimation. The scientific novelty of this paper is the improvement of the new data fusion approach based on the proposition of new function – basic probability density of basic probability (basic mass) for Dempster-Shafer evidence theory. The proposed method is based on the new algorithm of calculation of basic probabilities, basic probability densities and mathematical expectation of coordinates of the object, integrating initial data of two surveillance radars and characteristics of their zones of ambiguity. It also was considered three numerical examples of object coordinate determination applying Dempster-Shafer theory and new proposed algorithm of calculation of basic probabilities, basic probability densities, combining information of two surveillance radars. This proposed approach can be applied for solving various civil and military problems.

The future research directions will be associated with object coordinate determination, integrating data of surveillance radars with zones of ambiguity of irregular geometric shape, Monte Carlo method and artificial intelligence.

References

1. Kim, E., Sivits, K.: Blended secondary surveillance radar solutions to improve air traffic surveillance. Aerosp. Sci. Technol. **45**, 203–208 (2015)
2. Crichton, J.: The NAV CANADA model. J. Air Traffic Control. **57**(2), 33–35 (2015)
3. Abdullah, A., Ismail, A., Badron, K., Rashid, N.: Improving radar detection by adaptation of automatic dependent surveillance-broadcast (ADS-B) technology. Adv. Sci. Lett. **22**(10), 2860–2864 (2016)

4. Zhang, C., Zhou, X., Zhuge, X., Xu, M.: Learnable optical flow network for radar echo extrapolation. IEEE J. Sel. Top. Appl. Earth Obs. Rem. Sens. **14**, 1260–1266 (2021). https://doi.org/10.1109/JSTARS.2020.3031244
5. He, W., Xiong, T., Wang, H., He, J., Ren, X., Yan, Y., Tan, L.: Radar echo spatiotemporal sequence prediction using an improved ConvGRU deep learning model. Atmosphere. **13**, 88 (2022). https://doi.org/10.3390/atmos13010088
6. Popov, M., Alpert, S., Podorvan, V., Topolnytskyi, M., Mieshkov, S.: Method of hyperspectral satellite image classification under contaminated training samples based on Dempster-Shafer's paradigm. Cent. Eur. Res. J. **1**, 86–97 (2015)
7. Denoex, T.: Decision-making with belief functions: a review. Int. J. Approx. Reason. **109**, 87–110 (2019). https://doi.org/10.1016/j.ijar.2019.03.009
8. Alpert, S.: The new approach to applying the Dezert–Smarandache theory in land-cover classification in UAV-based remote sensing. Manag. Dev. Complex Syst. **49**, 33–39 (2022). https://doi.org/10.32347/2412-9933.2022.49.33-39
9. Shi, F., Su, X., Qian, H., Yang, N., Han, W.: Research on the fusion of dependent evidence based on rank correlation coefficient. Sensors. **17**, 2362 (2017). https://doi.org/10.3390/s17102362
10. Deng, Y.: Generalized evidence theory. Appl. Intell. **43**(3), 530–543 (2015)
11. Zhang, W., Zhang, Z.: Belief function based decision fusion for decentralized target classification in wireless sensor networks. Sensors. **15**, 20524–20540 (2015)
12. Alpert, S.: Data combination method in remote sensing tasks in case of conflicting information sources. Ukr. J. Rem. Sens. **8**(3), 44–48 (2021). https://doi.org/10.36023/ujrs.2021.8.3.201
13. Zhang, Y., Liu, Y., Chao, H.C., Zhang, Z.: Classification of incomplete data based on evidence theory and an extreme learning machine in wireless sensor networks. Sensors. **18**, 1046 (2018). https://doi.org/10.3390/s18041046
14. Sun, G., Guan, X., Yi, X., Zhao, J.: Conflict evidence measurement based on the weighted separate union kernel correlation coefficient. Open Access J. **6**, 30458–30472 (2018). https://doi.org/10.1109/ACCESS.2018.2844201
15. Yuan, K., Xiao, F., Fei, L., Kang, B., Deng, Y.: Modeling sensor reliability in fault diagnosis, based on evidence theory. Sensors. **16**(1), 113 (2016). https://doi.org/10.3390/s16010113
16. Zhu, J., Wang, X., Song, Y.: Evaluating the reliability coefficient of a sensor based on the training data within the framework of evidence theory. IEEE Access. **6**, 30952–30601 (2018). https://doi.org/10.1109/ACCESS.2018.2816915
17. Prokopenko, I., Alpert, S., Petrova, Y.: Applications of Dempster–Shafer evidence theory to data processing in remote sensing. In: CEUR Workshop Proceedings, vol. 3895, pp. 65–76 (2025)

Estimation of Mid-Air Collision Risk Based on ADS-B Trajectory Data

Oleg Ivashchuk and Ivan Ostroumov(✉)

National Aviation University, Liubomyra Huzara Ave., 1, Kyiv 03058, Ukraine
ostroumov@ukr.net

Abstract. Civil aviation is grounded on ensuring the same safety level for each airspace user. The fast-growing market of air transportation requires to search for reliable methods of increasing capacity of airspace within the same volume. Increasing number of airspace users needs improving performance of navigation and accuracy of surveillance equipment. In this paper, we consider particular airspace safety analysis based on risk of mid-air collision estimated with the help of trajectory data transferred by Automatic Dependent Surveillance-Broadcast (ADS-B) technology. Coordinates of airspace users received under ADS-B could be used to identify each airplane's deviations from preplanned trajectory and processed with a statistical assessment tool to recover a model of airplane fluctuation along a flight path. Recovered model utilized errors in flight technique and navigation sensor used on board. Univariate generalizer error distribution is used to estimate probability of airplane deviations out of safety perils. Risk of a mid-air collision is estimated by each pair of airspace users with data visualization in a hierarchical hexagonal spatial indexing system. Airspace of Poland has been considered as an example of safety analysis.

Keywords: Air Navigation · Aviation · Airspace · Risk · Safety · CNS · Collision risk model · Air Traffic Management · Spatial Index

1 Introduction

Air transportation system is constantly moving forward, overcoming all the disagreements and problems which has arisen during a long period of civil aviation development. Today air transportation services have become an ordinary reality of human activity. In 2022 civil aviation includes 28 thousand aircraft in use globally with 37.7 million flights being made during 2023 [1]. In addition to commercial aviation, small sports aircraft and business jets are increasingly appearing. In 2019 number of lightweight aircraft exceeded 21 thousand. Future air transportation system will utilize services for a much wider number of airspace users which is caused by rapid integration of Unmanned Areal Vehicles (UAV) in controlled airspace [2, 3]. However, UAV integration faces a number of problems including limited airspace volume and low precision of navigation system [4, 5].

I. Ostroumov et al. (Eds.): ACASD 2025, LNNS 1418, pp. 305–318, 2025.
https://doi.org/10.1007/978-3-031-91992-3_20

Nowadays modern air transportation systems face a huge challenge of operation under wide airspace segregation, which is mostly caused by war in Ukraine. The Russian-Ukrainian war caused to closing of whole Ukrainian airspace and made airspaces of Russian Federation and Belorussia in limited in-use condition. This segregated airspace caused air traffic flow relocation to avoid entering restricted and riskier regions. Most flights between European airports and Asian have been reconfigured which increases the time of flight connection and length of flight path [6, 7].

In addition to segregated Northern side of Eurasia, military conflicts and local wars in countries of the Middle East region significantly limit traffic from Asia to the West side. Current situation made airspace of Turkish as the only safe path to reach Asian destinations. All air traffic flows were relocated to Turkish airspace which significantly increased workload of systems used for air traffic support [8].

Based on the global tendency of air traffic increase, airspace volume remains at the same size, which leads to an increased risk of mid-air collision in areas where the flow of aircraft has been redirected [9, 10]. In current air traffic situation, most neighboring airspaces to Ukraine utilize detour traffic flows caused by airspace segregation.

There are some approaches that are aimed to help in problem of limited size of airspace volume. Free route airspace (FRA) is one of the most promising approaches [11, 12], which allows pilots to use efficient routes within the selected airspace, thereby unloading existing air traffic routes and reducing the load on air traffic controllers. However, the disadvantage is the difficulty in determining possible conflict points, as well as the fact that according to this concept, the pilot must use a route in such a way as to avoid zones over which flight is prohibited. Conventual air traffic idea grounds on airspace users using predetermined routes and it helps to predict exact point of conflict. In FRA prediction and detection of overload regions by air traffic could be difficult and affect the safety of civil aviation.

Risk of mid-air collision could be estimated based on a particular risk model, which provides an estimation probability of airspace users collision. Also, there are a large number of scenarios should be initiated to model an overall risk value. For example in the case of flights in parallel routes, the risk of collision could be estimated in horizontal plane only based on the Reich model [13, 14]. However, in the case of intersecting routes, the Anderson model provides much more confident results [15].

In this paper, we analyzed models of mid-air collision estimation which could be useful in FRA for automating air traffic control system applications. Proposed model of mid-air collision grounds on Aldis model [16, 17], which is a variation of the Anderson collision risk model. Developed model is practically integrated automatic air traffic control software based on a global spatial indexing system. H3 library for Python is used to represent the airspace and implement visualization of mid-air risk in defined airspace volume.

2 Airspace Structure

All air transportation services are provided within the airspace. Airspace volume is characterized by vertical dimension which goes from relief height up to 100 km above sea level (Carmen line, beyond which space begins). Horizontal diminution usually coincides with administrative boundaries of a particular country over the land and is legislated by international agreements over ocean regions. Basic administrative boundaries may be too big to organize effective air traffic management, therefore airspace could be divided into zones, sectors, and areas to simplify air traffic control operations. State bodies are fully responsible for choosing configuration of available airspace division into zones and areas to ensure effective management and meet requirements of flight safety.

The core categories of airspace include the following ones [18]:

1. FIR (Flight Information Region) is a large area of airspace within which civil aviation authorities provide flight information and emergency notification services. Each country may have one or more FIRs, and these zones cover all of its airspace.
2. UIR (Upper Information Region) is a part of the FIR located at high altitudes. This area is specifically designated for air traffic control at altitudes where commercial airliners fly during cruise flights.
3. CTR (Control Zone) is airspace around an aerodrome where aircraft movements are controlled to ensure safe take-off and landing. The CTR usually extends from the relief surface to a certain height.
4. TMA (Terminal Control Area) is airspace surrounding one or more major airports. It is designed to control the movement of heavy aircraft in the areas adjacent to the airports and to ensure the safe approach and departure of aircraft.
5. ATZ (Aerodrome Traffic Zone) is airspace around an aerodrome designed to protect air traffic during take-off and landing. The ATZ is usually smaller and covers the space from the surface to a certain height.
6. CTA (Control Area) is a controlled area that is usually located above aerodromes and other land-based airspace and extends upwards. It is intended for the control and management of air traffic at altitudes higher than the CTR (Controlled Area Around an Aerodrome) zones.

Airspace volume is also classified into different classes, which are highlighted by letters from A to G, depending on the level of control and equipment and communication requirements. For example, Class A is controlled airspace for commercial and military flights, while Class G is uncontrolled airspace where pilots have sole responsibility for avoiding collisions. Dividing airspace into these zones and areas helps to ensure effective air traffic management, prevent collisions, and improve flight safety.

Effective airspace structure is an important element of flight safety. But, using airspace is also required from airplanes to ensure the exact following a cleared flight path. If each airspace user follows initially predefined paths risk of mid-air collision could be minimized to zero level. Unfortunately, different factors introduce bias in the system. Appeared uncertainty in airplane departure time or weather fluctuation may seriously disturb nominal air transportation system operation.

Ensuring air traffic operation at required level of flight safety is also grounds on level of navigation services provided by air navigation service provider. Navigation services are guaranteed based on particular level of precision to ensure safe airplane arrive from departure to destination airport.

Air traffic management system grounds commands provided by air traffic controller (ATC) to pilots to ensure safe airplane separations. ATC uses advanced technologies of communication, surveillance, and radar systems to coordinate the actions of crews in the airspace segment to maintain a safe, effective, and economical level of air traffic flow. The quality of ATC service depends on elements of system work properly and professional skills of person involved in ATC.

Initially, air traffic was organized based on a conventional navigation method, under which aircraft had to move from one radio beacon to another, to ensure its presence at a particular flight route. Then, with the development of technology, a Performance-Based Navigation (PBN) was implemented. PBN made it possible to use any possible flight path with no required to move through radio beacons. PBN caused to change in the idea of airspace use to wide navigation by coordinates (latitude and longitude) [19, 20]. ATC uses a specific surveillance system to identify aircraft position. Ground radar is an efficient system however which could not provide altitude measurements. Also, common radars have significant errors of azimuth measuring that reduce overall system performance. Automatic Dependent Surveillance (ADS) technology is an airplane-to-ground data link to hare measured on-board parameters with ground ATC. Precision of on-board positioning system is much higher than radar system. Thus measured by global navigation satellite system and shared by ADS airplane position is more accurate than measured by ground radar network [21].

Modern concept of vision of future air traffic management system grounds on ADS technology as a primary source of data. ADS opens new horizons to further airspace development based on improving accuracy of onboard positioning sensor. Integration of positioning in multi networks (GNSS, communication wireless networks, and navigation aids networks) will help to increase positioning performance in 0.1 m precision and guarantee required level of reliability. All these improvements will help to increase number of airspace users at the same airspace volume while ensuring the same level of safety [22, 23]. Also, precise navigation will simplify further development of FRA to ensure a high capacity level.

PBN considers maintaining airplane position within 95% confidence level [20]. Improving performance of navigation system may be a reason to made confidence level significantly wide. It is also important that not only one primary positioning technology should be used to maintain air traffic. Based on reliability theory at least two positioning approaches should be utilized at each point of airspace to ensure a safe level of air traffic in PBN [24, 25].

Air traffic management authority have to provide required level of safety and security within a particular sector, which allows users to independently create its flight route within sector boundaries, flying around areas where flights are prohibited, as well as using fixed entry and exit points from a particular zone. FRA is actively developing within Eurocontrol airspace as well as around the globe in a variety of flight regions.

3 Model of Mid-Air Collision Risk Estimation

In this study, we use the risk of mid-air collision as a parameter distributed over airspace area to classify airspace. This classification helps to identify areas with increased risk which should be taken into account during airspace structure improvement. As a main source of surveillance data, we use ADS technology. There are plenty of web-based services that provide easy and flexible access to airplane trajectory data obtained by

ADS over API. We work with actual airplane positions obtained by ADS and shared by dynamic API with less than 7 s delay from real traffic. Input data includes actual coordinates of airspace users within defined airspace obtained in JSON data format.

Position of each airspace user is used to estimate probability of horizontal deviation based on error model and airplane dynamic model. The Aldis model of mid-air collision risk estimation is a variation of the Anderson model, in which the time limits of the integrals become infinite. This model ignores uncertainty of velocities for both involved airplanes and distance between them. Simplification in three integrals significantly increases computation performance of model. Mid-air collision risk for a pair of two involved airplanes going on intersecting paths could be estimated as follows [16]:

$$\mathrm{R} = n\int_{-\infty}^{\infty} P_h P_v \left(\frac{2V}{\pi\lambda_{xy}} + \frac{V_H}{2\lambda_z}\right)dt, \tag{1}$$

where n is number of airplane pairs; P_h is probability of horizontal collision; P_v is probability of airplane getting vertical deviation; V is relative speed between two airplanes in pair; λ_{xy} is average horizontal airplane size; λ_z is vertical safety size of airplane; V_H is a mean relative vertical speed.

Probability of airplane exceeding the horizontal limits of flight route could be used to utilize uncertainty introduced by on onboard positioning system. In general case each airplane trajectory should be recorded during some period of time and position fluctuation around flight route center line could be used to get an error model of a particular airspace user. It requires accumulating statistics of deviation in some array which is continuously changing. That statistical analysis could be applied for accumulated data parameters of mean value and dispersion of deviation could identified and used in the risk model. Also, Kernel probability density function could be used to get a precise model of airplane deviation.

Assuming a particular probability density function (PDF) is required due to the limited amount of learning sample size. Analysis of discrete values in the form of frequency of airplane deviation analysis has less confidence than probability estimated based on known PDF. Also, different PDF models could be used based on the required precision. An univariate generalized error distribution (UGED) could be used to accurately tune the shape of function from double exponential to normal [26, 27]:

$$\rho(x) = \frac{1}{2^{k+1}\sigma\Gamma(k+1)} exp\left(-\frac{1}{2}\left|\frac{x-\mu}{\sigma}\right|^{\frac{1}{k}}\right), \tag{2}$$

where σ is mean standard deviation; k is parameter of control the skewness; μ is a mean value; $\Gamma(x)$ is an Eelier-Gamma function.

UGED is changing its shape similar to normal PDF if $k = 0.5$, and to double exponential if $k = 1$.

The risk (probability) of airplane going out of horizontal perils of the route is estimated based on model (2) as follows:

$$P_h = \int_{-\infty}^{-l} \rho(x)dx + \int_{\infty}^{l} \rho(x)dx, \tag{3}$$

or

$$P_h = 1 - \int_{-l}^{l} \rho(x)dx, \tag{4}$$

where l is a separation minimum for latitudinal deviation.

Probability of airplane vertical deviation could be estimated as follows [11]:

$$P_z = \int_{-\lambda_z}^{\lambda_z} \int_{-\infty}^{+\infty} L(z_1)L(h_z + z_1 - z)dz_1 dz, \tag{5}$$

where $L(x)$ is Laplace PDF.

The scale parameter of this distribution is 38 ft for altitudes below 290 FL (410 FL if the airplane is within airspace where reduced vertical separation minimums are applied) and 76 ft above.

The International Civil Aviation Organization (ICAO) specifies target safety level values (TSL) for controlled airspace at a level of 5×10^{-9}. Thus, the risk of mid-air collision estimated by (1) for each pair of airplanes should be compared with TSL.

In a practical application of a risk assessment model for the whole airspace, it makes sense to use Global Spatial Indexing System (GSIS) which provides a comprehensive solution for flexible processing, visualization, and storage of data related to geographic coordinates. GSIS treats the ellipsoidal model of the Earth as a series of cells of varying sizes, defined by resolution levels. Hierarchical indexing assigns unique addresses to each cell in the system. There are three common types of GSIS based on the geometry of the cell: triangular, square (or rectangular), and hexagonal. Each of the GSIS types offers certain advantages for different applications. Rectangular cells are particularly well suited for mapping and addressing locations, while hexagonal cells have numerous advantages for data processing and navigation.

A spatial index with hexagonal cells facilitates navigational calculations by providing equal distances to the nearest hexagons. The centers of adjacent cells are located at the vertices of a larger hexagon. This regular hexagonal shape allows the coordinates of each vertex to be calculated economically from a given reference point. Accordingly, calculating a parental hexagon requires locations of all children's hexagons. The hierarchical hexagonal spatial index (H3 software library [28]) provides a global spatial index system that is useful not only for statistical processing of geographic data but also for vehicle navigation and traffic control (including air traffic) [29, 30]. The main advantage of H3 is its ability to speed up calculations due flexible use cell indexes in calculation instead of coordinates [31, 32].

We use GSIS for safety analysis of particular airspace. One of the main advantages of a hierarchical indexing system is speed data fusion between levels of precision. Due to collecting all the children's data based on particular prefix in cell address. A hexagonal hierarchical spatial index is used to visualize estimated risks over analyzed airspace area.

4 Numerical Demonstration

In the numerical demonstration, we use geometry of EPWW flight information region which is an airspace of Poland. Airspace is analyzed with ADS data obtained by API from surveillance database [33]. Analyzed data sets include all airspace users from

03.12.2024 to 10.12.2024. In the initial data request a rectangular area specified as 48°33′19.1″ 14°14′48.6″E and 54°48′40.7″N 24°41′56.3″E is used. During this time, data from more than 4 thousand flights were recorded. Data processing has been done in specially developed software in Python. Example of data collected in the air traffic are given in Table 1.

Table 1. Example of air traffic data collected.

ICAO code, hex	Callsign	Departure airport	Arrival airport	ETD	ETA	Airplane type
471f6a	WZZ12CF	EBCI	EPWA	12/10/24 10:22	12/10/24 12:11	A321
48adb9	LOT5RW	LKPR	EPWA	12/10/24 10:47	12/10/24 12:13	E75L
471f38	WZZ3	LHBP	EKCH	12/10/24 11:52	12/10/24 13:21	A21N
48ae82	LOT4CA	EPWA	LEMD	12/10/24 12:14	12/10/24 15:07	B38M
48ae86	LOT3FG	EPWA	LKPR	12/06/24 12:15	12/06/24 13:18	B738
48ada4	LOT646	LROP	EPWA	12/08/24 18:40	12/08/24 20:06	E190
48ad0c	LOT646	LROP	EPWA	12/09/24 18:49	12/09/24 20:18	E75L
48ad82	LOT5KW	EPWR	EPWA	12/09/24 19:31	12/09/24 20:16	E190
48ad08	LOT776	EYVI	EPWA	12/08/24 19:17	12/08/24 20:05	E190
48ad86	LOT3923	EPWA	EPKK	12/04/24 17:39	12/04/24 18:21	E190
48c226	RYR551E	EPMO	LIRA	12/04/24 17:40	12/04/24 19:50	B738

API is used for connection with ADS database of actual flights. API response includes JSON format of airplane location data within specified rectangular region. Data in JSON is structured by each airspace user and includes a unique airplane code (ICAO registration code), coordinates in altitude, latitude, longitude, and some specific parameters. Each API response includes an air traffic state at a particular time and in order to process data by each airplane we collect all responses in a software database. We will need to use the parameters of the aircraft model to calculate the risk. We will use the aircraft model E190 (height 10.55 m, length 36.24 m) and B737-800 (height 12.55 m, length 39.48 m) [34, 35].

Unique airplane identifier is used to get all flight data regarding the aircraft's location during the movement through considered airspace [36, 37]. Coordinates of airplanes are compared with a flight-routes network to identify airplane deviations. Results of deviations of airplanes from a flight-route network in the airspace of Poland are estimated for each unique ICAO code based on all available data related to this airplane during the considered time. The 511 unique flights have been processed. As an example, results of statistical analysis of deviations of SPTVA, WZZ3, and RYR551E are given in Figs. 1, 2 and 3. Also, figures include results of UGED (2) fitting by the maximum likelihood method.

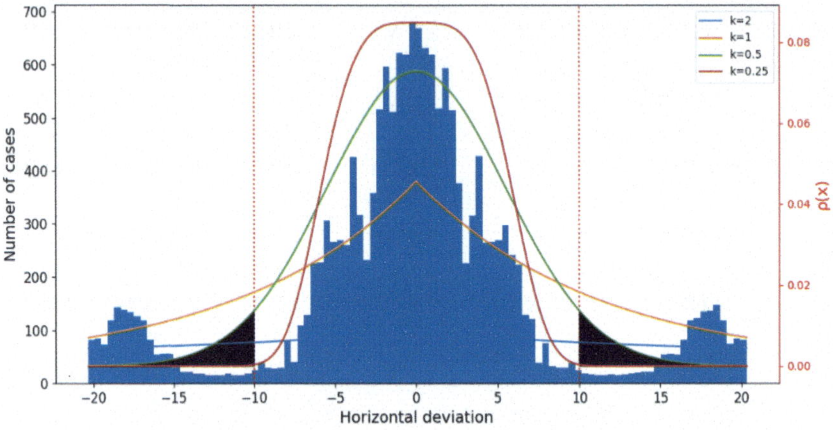

Fig. 1. PDF of SPTVA deviations from flight routes network.

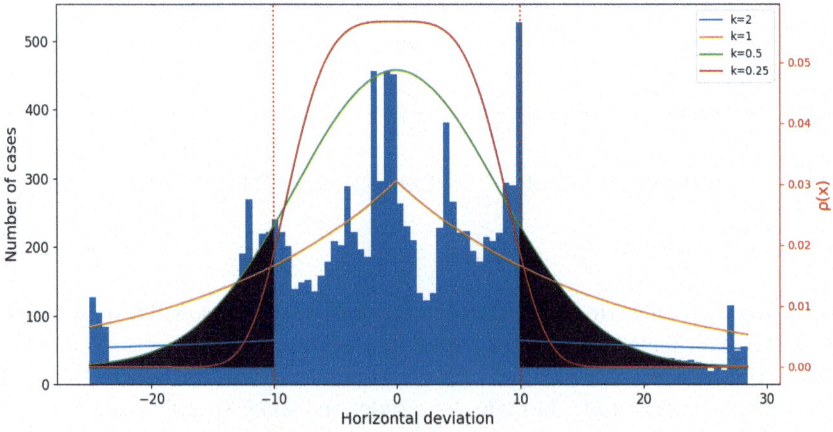

Fig. 2. PDF of WZZ3 deviations from flight routes network.

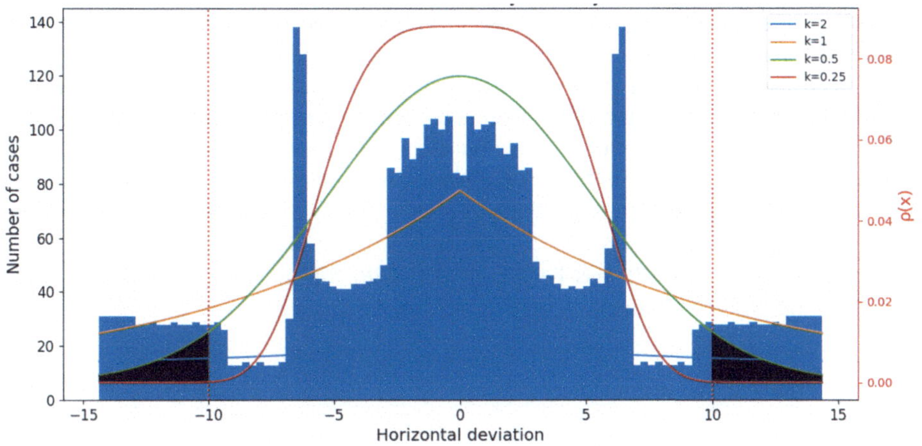

Fig. 3. PDF of RYR551E deviations from flight routes network.

Obtained results show that aircraft mostly maintain a center of flight-route, but there are several cases when excessive deviations are present. Results of PDF fitting are used for estimation probability of airplane deviation out of safety boundaries by (4). Results of P_h estimation are given in Table 2.

Table 2. Results of estimation probability of airplane deviation out of safety boundaries.

Callsign	Probability of horizontal perils lost
SPTVA	6.5×10^{-3}
SPSPO	0.77
WZZ12CF	0.98
SPSMP	0.61
LOT5RW	0.79
WZZ3	7.1×10^{-2}
LOT4CA	0.63
LOT3FG	0.71
LOT646	5.55×10^{-16}
LOT776	0.48
LOT5KW	3.3×10^{-3}
LOT3923	0.5
RYR551E	0.12
THY3LZ	7.1×10^{-2}

Coordinates of each airspace user have been associated with a particular hexagonal cell of GSIS based on a specified level of precision. Based on time criteria all visible airplanes are considered by pairs to estimate risk of a mid-air collision between them by (1). Based on risk model a probability of mid-air collision is estimated for each hexagon in considered space for each pair. Results of mid-air collision estimation between selected pairs of airplanes are given in Table 3 for hexagonal cells with the highest risk value. Then picture of risk is collected by all the airplane pairs in considered airspace to get an overall mid-air collision analysis.

Data processing in GSIS has been provided with folium library which visualizes results on OpenStreetMap maps.

Table 3. Results of mid-air collision risk estimation for selected pairs.

Date time	Call sign of first aircraft	Call sign of second aircraft	Index of conflict cell	Risk of mid-air collision	Probability of airplane vertical deviation
12/03/24 13:41:34	SPTVA	SPSPO	871f523a8ffffff	8.3×10^{-4}	0.19
12/10/24 11:39:04	WZZ12CF	SPSMP	871e240e0ffffff	1.2×10^{-111}	2.51×10^{-111}
12/10/24 12:05:01	WZZ12CF	LOT5RW	871f5230bffffff	1.1×10^{-7}	1.74×10^{-7}
12/10/24 12:35:03	WZZ3	LOT4CA	871e20a93ffffff	7.5×10^{-3}	0.19
12/06/24 12:26:12	SPSPO	LOT3FG	871e21b70ffffff	2.6×10^{-80}	5.77×10^{-80}
12/08/24 20:03:14	LOT646	LOT776	871f52308ffffff	1.3×10^{-19}	5.9×10^{-4}
12/09/24 20:12:28	LOT646	LOT5KW	871f52301ffffff	3.0×10^{-19}	0.19
12/04/24 17:53:36	LOT3923	RYR551E	871e2cc54ffffff	6.2×10^{-43}	1.17×10^{-41}
12/09/24 20:25:21	EXS8KF	THY3LZ	871e05a45ffffff	1.1×10^{-84}	1.88×10^{-84}
12/09/24 20:31:39	EXS8KF	LOT3508	871e2e642ffffff	1.3×10^{-7}	1.7×10^{-7}
12/04/24 11:14:34	LOT5RW	SPJOL	871e20411ffffff	1.4×10^{-111}	2.51×10^{-111}
12/09/24 20:13:45	WZZ19WS	LOT5KW	871f52353ffffff	8.26×10^{-128}	2.94×10^{-125}

Due to all airplanes in pairs moving at different flight levels with significant separation, it reduces the overall risk of mid-air collision to extremely low values. It makes obtained results significantly lower than the TSL value.

Results of analysis show that there are some hexagons in which risk is higher than TLS. It happened due to different safety boundaries that should be applied for airplanes moving in standard schemes in airport vicinity. As an example, several high-risk regions were identified in the Warsaw airport vicinity. ATC could direct the airplanes and they move with a very small separation interval. Also, increased risk is obtained due to airplanes placed in the holding scheme near destination airport.

Results of mid-air risk estimation in EPWW airspace for December 12, 2024 are given in Fig. 4. Concentration of trajectory data for a particular time period is shown with smaller cell size. Hexagons with a risk higher than TLS are highlighted by red color. Figure 5 shows how often planes flew through hexagons during the studied period. The highest number of flights was observed near Warsaw and Krakow airports.

Results of safety analysis demonstrate a high level of safety based on model (1). Risk of airplane simultaneous presence in one hexagonal region is much more than TLS. Results of the safety study during a considered period show absent of conflict situations based on comparison risk value with TLS for an en-route phase of flight. Indicated conflicts are resulted of approved lover separation minimums which is used in the study to show the conflict identification logic.

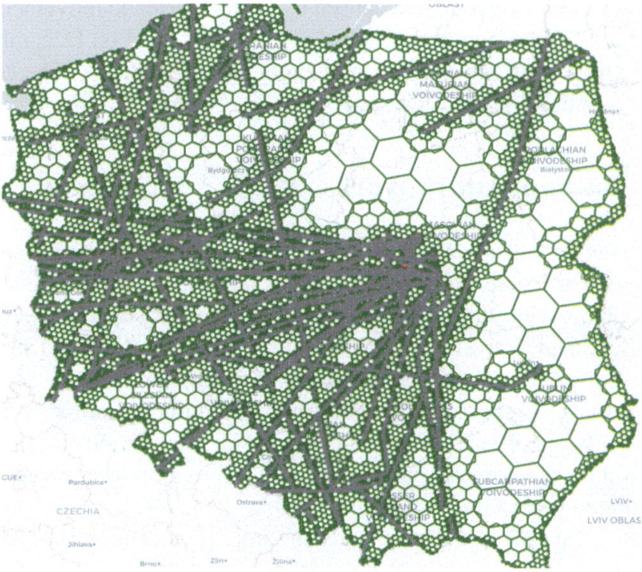

Fig. 4. Results of risk analysis on December 12, 2024.

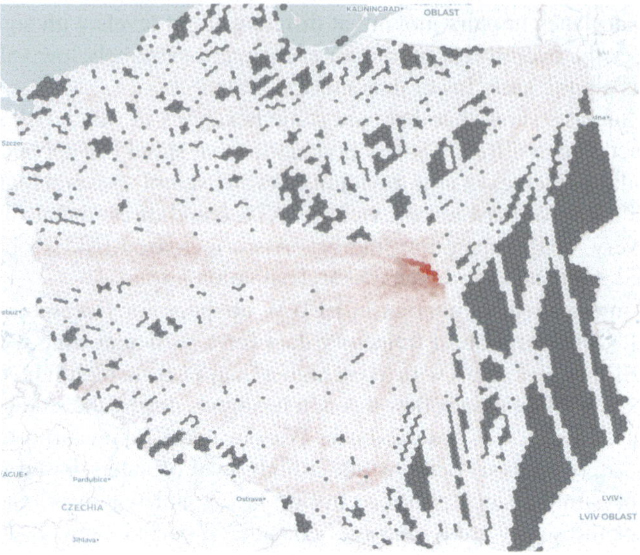

Fig. 5. Density of aircraft use of airspace in the hexagon.

Identified hexagons with a high level of mid-air collision have resulted of small vertical separations in the airport area caused by correct airplane movement on standard instrument schemes. Maximum risk reached value 8.3×10^{-4} on December 3, 2024 at 13:41:34 between SPTVA and SPSPO in hexagon 871f523a8ffffff. The case was fixed near Fryderyk Chopin Airport in Warsaw.

5 Conclusions and Future Works

Risk analysis of airspace is an important component of air transportation safety ensuring system. Modern surveillance technology provides reach data for the safety study of each flight region on planet with a high level of precision. Also, vary of ADS databases provide easy access to these data via API interface. Results of safety analysis of a particular region based on the hexagonal airspace model give reliable results for studying total safety levels and could be useful for flight routes network modernization or implementation of FRA.

Results of safety analysis show a low level of mid-air collision risk in most airplane pairs which could be neglectable. As an example for a pair of airplanes WZZ19WS and LOT5KW estimated risk of mid-air collision was 8.26×10^{-128}.

In future works, it makes sense to study the impact of different models of PDF on result of safety estimation in order to identify models with low computation time and reliable results.

References

1. Annual Review 2023. IATA (2023)

2. Tuncal, A., Erol, U.: Integrating unmanned aerial vehicles in airspace: a systematic review. J.Aviat. Res. **6**(1), 89–115 (2024). https://doi.org/10.51785/jar.1393271
3. Kouretas, K., Kepaptsoglou, K.: Planning integrated unmanned aerial vehicle and conventional vehicle delivery operations under restricted airspace: a mixed nested genetic algorithm and geographic information system-assisted optimization approach. Vehicles. **5**(3), 1060–1086 (2023). https://doi.org/10.3390/vehicles5030058
4. Main Report for the 2005/2012 Integrated Risk Picture for Air Traffic Management in Europe, EUROCONTROL (2006)
5. State of Global Aviation Safety. ICAO safety report. ICAO (2019)
6. Ostroumov, I., Ivashchuk, O., Kuzmenko, N.: Preliminary estimation of war impact in Ukraine on the global air transportation. In: 12th International Conference on Advanced Computer Information Technologies (ACIT), pp. 281–284. IEEE, Ruzomberok (2022). https://doi.org/10.1109/ACIT54803.2022.9913092
7. Ostroumov, I., Ivannikova, V., Zaliskyi, M., Kuzmenko, N.: Impact analysis of Russian-Ukrainian war on airspace. J. Air Transp. Manag. **124**, 102742 (2025). https://doi.org/10.1016/j.jairtraman.2025.102742
8. Ennen, D., Wozny, F.: Airspace closures following the war of aggression in Ukraine: the impact on Europe-Asia airfares. Transp. Res. Procedia. **78**, 103–110 (2024). https://doi.org/10.1016/j.trpro.2024.02.014
9. Su, Y., Xu, Y.: A risk assessment method for mid-air collisions in urban air mobility operations. IEEE Trans. Intell. Veh., 1–15 (2024). https://doi.org/10.1109/TIV.2024.3426915
10. Dy, L.R., Mott, J.H.: Airspace saturation and midair collision risk: a case study at a class D airport. Int. J. Aviat. Aeronaut. Aerosp. **11**(1), 4 (2024). https://doi.org/10.58940/2374-6793.1882
11. Simorgh, A., Soler, M., Dietmüller, S., Matthes, S., Yamashita, H., Castino, F., Yin, F.: Robust 4D climate-optimal aircraft trajectory planning under weather-induced uncertainties: free-routing airspace. Transp. Res. Part D: Transp. Environ. **131**, 104196 (2024). https://doi.org/10.1016/j.trd.2024.104196
12. Teutsch, J., Petersen, C.: Dynamic airspace re-configuration for manned and unmanned operations in shared airspace. In: 2024 Integrated Communications, Navigation and Surveillance Conference (ICNS), pp. 1–14. IEEE, Herndon, VA (2024). https://doi.org/10.1109/ICNS60906.2024.10550738
13. ERNIP Part 1. European Airspace Design Methodology Guidelines—General Principles and Technical Specifications for Airspace Design, EUROCONTROL (2022)
14. The Longitudinal Reich Collision Risk Model. SASP-WG/WHL/20-IP/10. ICAO (2012)
15. Anderson, D., Lin, X.: A collision risk model for a crossing track separation methodology. J. Navig. **49**, 337–349 (1996). https://doi.org/10.1017/S0373463300013576
16. Aldis, G., Barry, S.: Fast evaluation of the Anderson-Hsu collision risk for a crossing track scenario. In: Proceedings of the ICAO SASP-WGH-27 Meeting, Oklahoma City, OK, USA (2015)
17. Kallinen, V.: Collision risk quantification for pairs of recorded aircraft trajectories. J. Navig. **76**(2–3), 181–199 (2023). https://doi.org/10.1017/s0373463323000085
18. Air traffic management. Procedures for Air Navigation Services, Doc. 4444. ICAO (2016)
19. Ostroumov, I.V., Kuzmenko, N.S.: An area navigation (RNAV) system performance monitoring and alerting. In: First International Conference on System Analysis & Intelligent Computing (SAIC), pp. 1–4. IEEE, Kyiv (2018). https://doi.org/10.1109/SAIC.2018.8516750
20. Performance-Based Navigation (PBN) Manual. Doc 9613. ICAO (2008)
21. Ostroumov, I., Kuzmenko, N.: Statistical analysis and flight route extraction from automatic dependent surveillance-broadcast data. In: Integrated Communication, Navigation and Surveillance Conference (ICNS), pp. i–ix. ION, Dulles, VA (2022). https://doi.org/10.1109/ICNS54818.2022.9771515

22. Ostroumov, I.V., Kuzmenko, N.S.: Accuracy estimation of alternative positioning in navigation. In: 4th International Conference Methods and Systems of Navigation and Motion Control: MSNMC 2016, pp. 291–294. IEEE, Kyiv (2016). https://doi.org/10.1109/MSNMC.2016.7783164
23. Kuzmenko, N.S., Ostroumov, I.V.: Performance analysis of positioning system by navigational aids in three dimensional space. In: First International Conference on System Analysis & Intelligent Computing (SAIC), pp. 1–4. IEEE, Kyiv (2018). https://doi.org/10.1109/SAIC.2018.8516790
24. Okoro, O.C., Zaliskyi, M., Dmytriiev, S., Solomentsev, O., Sribna, O.: Optimization of maintenance task interval of aircraft systems. Int. J. Comput. Netw. Inf. Secur. **14**(2), 77–89 (2022). https://doi.org/10.5815/ijcnis.2022.02.07
25. Zaliskyi, M., Odarchenko, R., Gnatyuk, S., Petrova, Y., Chaplits, A.: Method of traffic monitoring for DDoS attacks detection in E-health systems and networks. In: CEUR Workshop Proceedings, vol. 2255, pp. 193–204. https://ceur-ws.org/Vol-2255/paper18.pdf (2018)
26. Giller, G.: A Generalized Error Distribution (2005)
27. Ostroumov, I.V., Marais, K., Kuzmenko, N.S., Fala, N.: Triple probability density distribution model in the task of aviation risk assessment. Aviation. **24**(2), 57–65 (2020). https://doi.org/10.3846/aviation.2020.12544
28. Hierarchical hexagonal spatial index. H3. https://h3geo.org/docs/comparisons/s2. Last accessed 15 Jan 2025
29. Ma, Y., Li, G., Yao, X., Cao, Q., Zhao, L., Wang, S., Zhang, L.: A precision evaluation index system for remote sensing data sampling based on hexagonal discrete grids. ISPRS Int. J. Geo Inf. **10**(3), 194 (2021). https://doi.org/10.3390/ijgi10030194
30. Li, M., McGrath, H., Stefanakis, E.: Geovisualization of hydrological flow in hexagonal grid systems. Aust. Geogr. **2**(2), 227–244 (2022). https://doi.org/10.3390/geographies2020016
31. Sahadevan, N., Ali, H.: Enhancing trajectory-based operations for UAVs through hexagonal grid indexing: a step towards 4D integration of UTM and ATM. Int. J. Aviat. Aeronaut. Aerosp. **10**(2), 5 (2023). https://doi.org/10.58940/2374-6793.1815
32. Aini, A.N., Dewantari, O.A., Mandala, D.P., Bisri, M.B.: An enhanced earthquake risk analysis using H3 spatial indexing. IOP Conf. Ser. Earth Environ. Sci. **1245**(1), 012014 (2023)
33. The OpenSky Network. https://opensky-network.org. Last accessed 15 Jan 2025
34. FlyEmbraer. APM 190. https://www.flyembraer.com/irj/go/km/docs/download_center/Anonymous/Ergonomia/Home%20Page/Documents/APM_190.pdf. Last accessed 15 Jan 2025
35. 737 Airplane Characteristics for Airport Planning (Rev. A). Boeing. https://www.boeing.com/content/dam/boeing/boeingdotcom/commercial/airports/acaps/737_RevA.pdf. Last accessed 15 Jan 2025
36. Ivashchuk, O., et al.: A configuration analysis of Ukrainian flight routes network. In: 16th International Conference on the Experience of Designing and Application of CAD Systems, pp. 6–10. IEEE, Lviv (2021). https://doi.org/10.1109/CADSM52681.2021.9385263
37. Averyanova, Y., Larin, V., Kuzmenko, N., Ostroumov, I., Zaliskyi, M., Solomentsev, O., et al.: Turbulence detection and classification algorithm using data from AWR. In: IEEE 2nd Ukrainian Microwave Week (UkrMW), pp. 518–522. IEEE, Lviv (2022). https://doi.org/10.1109/UkrMW58013.2022.10037172

Classification of Flight Situation Based on Risk of Flight Level Loss for Improved Airspace Safety

Ivan Ostroumov[1](✉), Yurii Bezkorovainyi[1], Oleksii Holubnychyi[1], Olha Sushchenko[1], Oleksandr Solomentsev[1], Maksym Zaliskyi[1], Yuliya Averyanova[1], Kostiantyn Cherednichenko[1], Olena Sokolova[1], Viktoriia Ivannikova[2], Borys Kuznetsov[3], Ihor Bovdui[3], Tatyana Nikitina[4], and Roman Voliansky[5]

[1] National Aviation University, Liubomyra Huzara Ave., 1, Kyiv 03058, Ukraine
ostroumov@ukr.net

[2] Dublin City University, Collins Ave. Ext., Dublin, Ireland

[3] Anatolii Pidhornyi Institute of Mechanical Engineering Problems of the National Academy of Sciences of Ukraine, Pozhars'koho Str., 2/10, Kharkiv 61046, Ukraine

[4] Educational Scientific Professional Pedagogical Institute Ukrainian Engineering Pedagogical Academy, University Str. 16, Kharkiv 61003, Ukraine

[5] National Technical University of Ukraine "Igor Sikorsky Kyiv Polytechnic Institute", Ber-esteiskyi Ave., 37, Kyiv 03056, Ukraine

Abstract. The safety of air transportation depends on the level of airplane deviations from preplanned trajectory and cleared flight level. Minimums of vertical separation are used as perils for safe airplane deviation from required altitude of flight level. Vertical separation minimums are designed to take into account errors in pressure altitude measurements. In the paper, we consider airplane flight situation classification based on airplane deviation from cleared flight level. We analyze a full group of classes which include normal flight, high deviation, complicated, emergency, and catastrophic situations. A maximum posterior probability method is used for classification. A Gaussian function is used as a conditional probability density function of each class. A prior probability of each class is estimated by statistical data analysis of historical flight realizations. Proposed classifier could be used in the Flight Management System for safety monitoring and alerting or could be integrated into algorithms of surveillance data processing in automatic air traffic control systems.

Keywords: Flight Level · Risk · Posterior Probability · Air Navigation · Air Traffic · CNS

1 Introduction

Global air transportation system is designed based on criteria of ensuring the minimum required level of flight safety. Usage of airspace is planned in advance by each user. During the flight each airspace user has to maintain a preplanned trajectory and cleared

I. Ostroumov et al. (Eds.): ACASD 2025, LNNS 1418, pp. 319–334, 2025.
https://doi.org/10.1007/978-3-031-91992-3_21

flight level to guarantee the safety of air transportation [1, 2]. Deviation of airplane from predefined trajectory could indicate reduced safety and increase the risks of mid-air collision with other airspace users [3]. Performance-based navigation (PBN) considers requirements for maintaining horizontal deviations from the centerline of flight route [4]. Due to noise action in airplane position, PBN takes into account navigation system error and flight technical error [5, 6]. Navigation system error is an error of on-board positioning system, which depends on a particular system that is used for measuring latitude and longitude of airplane position. Flight technical error utilize errors of maintaining predefined trajectory in autopilot or manual modes.

Each flight route is divided by altitude into flight levels (FL). FL is based on minimum of vertical separations between airspace users [7] which could be different with altitude to utilize error of pressure altitude measurements. Direction of flight levels could be mixed (for example odd FL could be directed to one side and even FL could be directed to opposite) to guarantee an equal opportunity for different airflows within one flight route. Airplane deviation from the center altitude of FL could indicate about increased risk of mid-air collision with air traffic ahead.

Air traffic management authority uses specific surveillance equipment to identify airspace users and current weather [8]. Air traffic controllers (ATC) obtain information about airspace user's position from the network of ground radars, multilateration systems, and Automatic Dependent Surveillance-Broadcast (ADS-B) [9, 10]. Unfortunately, in civil aviation, no sensor could measure altitude of airspace users from the ground. Primary surveillance radars and weather radars couldn't provide airplane altitude data. However, secondary surveillance radars and ADS-B support pressure altitude measurement transmission from the airplane side to the ground. Airplane pressure altitude received by the ground equipment of ATC is used to organize and control air traffic.

Pressure altitude is used in vary of studies to classify flightphase in both real-time system and post-processing modes. Pressure altitude is calculated from an isobaric surface of constant pressure (which is a "zero" altitude level). There are several studies of pressure altitude conversion to altitude from WGS84 ellipsoidal model which makes it possible to hard airplane fixed with a relief model and environment. However, such transformation requires actual weather data along a flight path [11, 12]. Also, pressure altitude measurements include errors of different types [13]. Pressure altitude is calculated from static pressure measurements [14]. Static pressure sensing on board of speedy moving object introduces aerodynamic error in addition to common error components [15, 16]. Pressure altitude measurement is a complicated process including catching a part of air masses outside the airplane body in multiple places and moving air static pressure through a specific tube system to the pressure sensor. Measured static pressure is sent to the air data system for data processing and calculation of pressure altitude. Multiple components of pressure altitude measurement introduce a bias in the obtained results.

Nowadays algorithms of on-board Traffic Collision and Avoidance System (TCAS) use a static metric of critical relative altitude between airspace users which does not consider error action in pressure altitude data [17, 18]. Therefore, results of a vertical test in TCAS may be affected by noise of pressure altitude measurements that reduce total level of flight safety.

In the paper, we propose a probability-based method for classification of flight situations by deviation of actual pressure altitude from the altitude of cleared flight level. A risk-based classifier provides a posterior probability of each class and gives results with minimum error. Also, we propose a metric of classes based on the level of dangerous deviation from cleared FL.

2 Dangerous Levels of Airplane Deviations

Airspace structure includes a set of FL within each flight route. A title of FL includes three numbers of cleared flight level in hundreds of feet. A vertical separation minimum below 41,000 ft (410FL) is 1000 ft (300 m) and 2000 ft (600 m) above [19]. For instrumental flight, FL begins from 110FL and finishes at 650FL. Above 650FL a free-route airspace is introduced in many regions around the globe.

We consider degradation of flight situation with airplane deviations from cleared FL to a dangerous stage. The normal flight could be specified by perils of ±200 ft (±60 m) from an altitude of cleared FL below 410FL and ±300 ft (±90 m) above. Conditions of normal flight could be as follows:

$$H - h_{NF} \leq h \leq H + h_{NF}, \tag{1}$$

where H is pressure altitude of cleared FL; h_{NF} is a peril value; h is actual pressure altitude.

When airplane pressure altitude within perils (1) flight is nominal and deviation does not affect the safety of air transportation. Any vertical deviations out of these perils affect the safety of airflow in the particular flight route.

It makes sense to consider four stages of dangerous deviation which sequentially increase the risk of mid-air collision based on level of deviation.

Dangerous stages:

- High deviation (HD).
- Complicated situation (CS).
- Emergency situation (ES).
- Catastrophic situation (CaS).

Metrics of dangerous stages are presented in Fig. 1.

Dangerous stage HD could be initiated when an airplane descends or climbs out of the safety perils of Normal flight (NF) to the middle point between neighbor FLs. Conditions of HD for both cases of climbing and descending modes are the following:

$$H \mp h_{NF} < h \leq H \mp 0.5h_{FL}, \tag{2}$$

where h_{FL} is the minimum of vertical separation for a particular FL.

In the case, of increasing deviation from cleared FL, an airplane is crossing middle point between FLs which indicates about beginning of CS. Maximum limits for CS are NF perils of neighbor FLs. The condition of CS could be as follows:

$$H \mp 0.5h_{FL} < h \leq H \mp h_{FL} \mp 0.5h_{NF}. \tag{3}$$

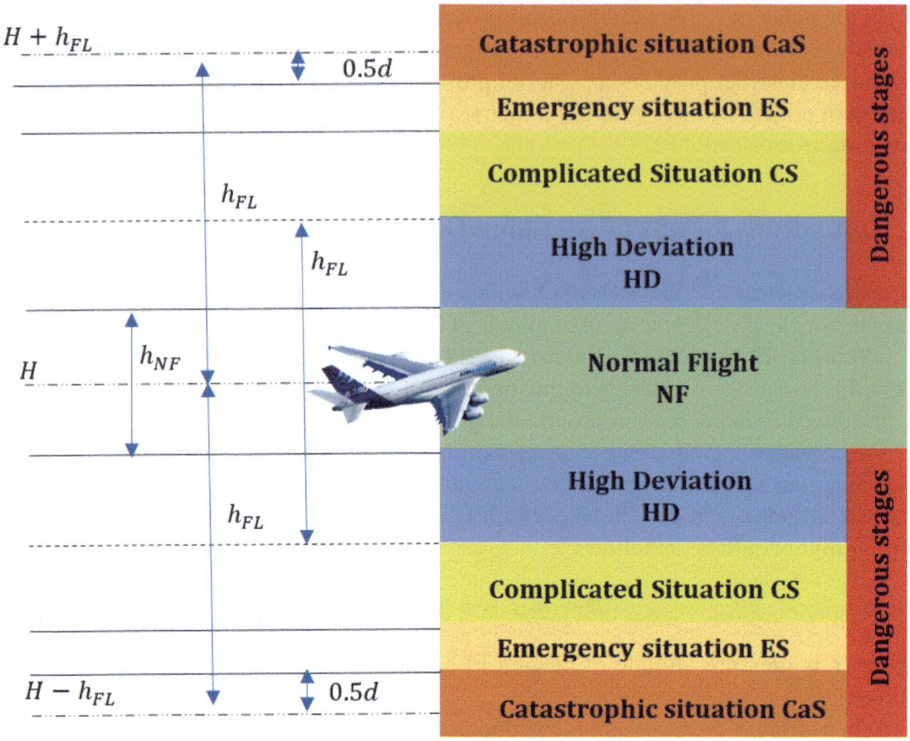

Fig. 1. Metric of dangerous stages.

ES begins when airplane is inside the perils of neighbor FL:

$$H \mp h_{FL} \mp 0.5h_{NF} < h \leq H \mp h_{FL} \mp 0.5d. \tag{4}$$

where d is the geometrical height of airplane.

Finally, CaS indicates about presence of airplane at the center of neighboring FL:

$$H \mp h_{FL} \mp 0.5d < h \leq H \mp h_{FL} \mp 0.5d. \tag{5}$$

Degradation of flight situations is a sequential process of switching from one stage to the next one based on level of deviation. Stages could not be missed in the process of deviation. Appearance of a particular stage requires all previous stages to pass through. The process of stages development could be represented as a directed graph, which is presented in Fig. 2.

From each stage, there are only two options for switching to the side of increasing or decreasing the risk of mid-air collision. A catastrophic situation is only a case of intruding in FL of another airplane with no information of presents airplane there. Therefore, in the case of neighbor FL busy with another airspace user, CaS stage leads to catastrophe, if no it is a violation of vertical separation minimums only.

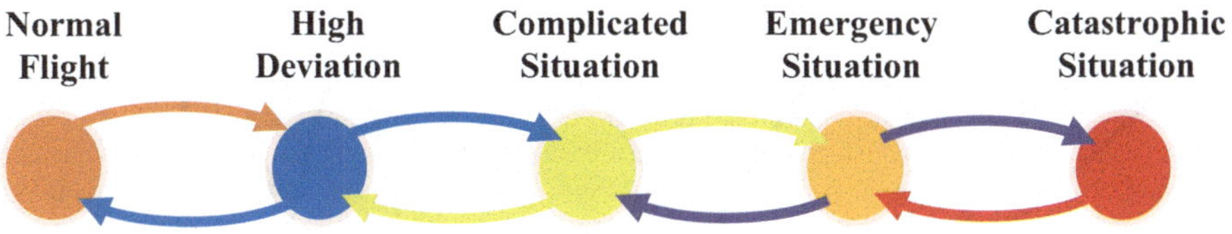

Fig. 2. Flight stages development process.

3 Classification

Errors actions in results of pressure altitude measurements require to use probability-based classifier which takes into account error model and provides better performance in comparison with generic logic [20]. We propose to use a classification method based on the maxim of posterior probability. In this case, a posterior probability should be estimated for each stage. Each stage is accompanied by a particular hypothesis of this stage's appearance:

- H_{NF} is hypothesis of Normal Flight;
- H_{HD} is hypothesis of Height Deviation;
- H_{CS} is hypothesis of Complicated Situation;
- H_{ES} is hypothesis of Emergency Situation;
- H_{CaS} is hypothesis of Catastrophic situation.

A prior probability of each hypothesis is known: P_{NF}, P_{HD}, P_{CS}, P_{ES}, P_{CaS}. These probabilities compose a full group of evidence:

$$P_{NF} + P_{HD} + P_{CS} + P_{ES} + P_{CaS} = 1. \tag{6}$$

Each stage is described by a conditional probability density function. Different functions could be used to specify conditional probability. We use the Gaussian probability density function due to normal error distribution in pressure altitude measurements:

$$\rho_i(x) = \frac{1}{\sqrt{2\pi}\sigma_i}\exp\left(-\frac{(x-\mu_i)^2}{2\sigma_i^2}\right), \tag{7}$$

where σ_i is a standard deviation; μ_i is a mean value; i is the identification of hypothesis $i \in [NF, HD, CS, ES, CaS]$.

Conditional probability density functions (7) are fixed at the middle of each stage (Fig. 3). A standard deviation of conditional probability density function can be estimated based on criteria of a particular level of confidence. A confidence level of 95% corresponds to $\pm 2\sigma$ band in stage perils. Also, the level of 99.7% of confidence could be used which meets stage perils in $\pm 3\sigma$ band.

Formulas for estimation mean value (μ) and standard deviation (σ) for each conditional probability density function in the case of upward deviation are presented for each stage in Table 1.

Each dangerous stage has two possible cases: downwards and upwards based on the side of airplane deviation. Thus, there are two basic cases of setting conditional probability density functions:

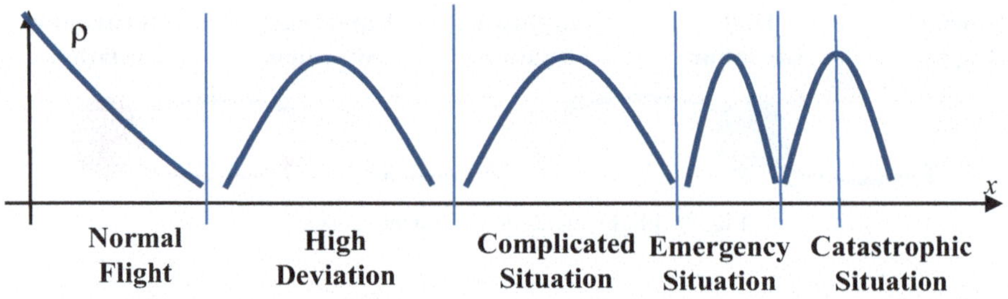

Fig. 3. Conditional probability density functions.

Table 1. Parameters of conditional probability density functions.

Flight situation	Mean value	Standard deviation case 95%	Standard deviation case 99.7%
Normal flight	H	$\frac{h_{NF}}{4}$	$\frac{h_{NF}}{6}$
High deviation	$H + 0.25(h_{FL} + h_{NF})$	$\frac{h_{FL}-h_{NF}}{4}$	$\frac{h_{FL}-h_{NF}}{6}$
Complicated situation	$\frac{2H+1.5h_{FL}-0.5h_{NF}}{2}$	$\frac{h_{FL}-h_{NF}}{4}$	$\frac{h_{FL}-h_{NF}}{6}$
Emergency situation	$\frac{2H+2h_{FL}-0.5h_{NF}-0.5d}{2}$	$\frac{-h_{NF}-d}{4}$	$\frac{-h_{NF}-d}{6}$
Catastrophic situation	$H + h_{FL}$	$\frac{d}{2}$	$\frac{d}{3}$

- Consider a case of nine classes which include: Normal flight, 4 dangerous stages of downward, and 4 dangerous stages of upward maneuvers.
- Consider a case of five classes with specific conditional PDFs.

We use a case of five classes with specific conditional PDFs. A normal flight is specified with PDF (7). However, other dangerous stages are classified with a sum of two Normal PDFs which utilize downward and upward manners.

Conditional PDF for dangerous stages based on data in Table 1 are as follows:

- High Deviation

$$\rho_{HD}(x) = \frac{4}{\sqrt{2\pi}(h_{FL} - h_{NF})}\exp\left(-\frac{8(x + 0.25(h_{FL} + h_{NF}))^2}{(h_{FL} - h_{NF})^2}\right) + \\ + \frac{4}{\sqrt{2\pi}(h_{FL} - h_{NF})}\exp\left(-\frac{8(x - 0.25(h_{FL} + h_{NF}))^2}{(h_{FL} - h_{NF})^2}\right); \quad (8)$$

- Complicated Situation

$$\rho_{CS}(x) = \frac{4}{\sqrt{2\pi}(h_{FL} - h_{NF})}\exp\left(-\frac{8(x + 0.25(3h_{FL} - h_{NF}))^2}{(h_{FL} - h_{NF})^2}\right) +$$

$$+\frac{4}{\sqrt{2\pi}(h_{FL}-h_{NF})}\exp\left(-\frac{8(x-0.25(3h_{FL}-h_{NF}))^2}{(h_{FL}-h_{NF})^2}\right); \tag{9}$$

- Emergency Situation

$$\rho_{ES}(x) = \frac{4}{\sqrt{2\pi}(h_{NF}+d)}\exp\left(-\frac{8(x+0.5(2h_{FL}-0.5h_{NF}-0.5d))^2}{(h_{NF}+d)^2}\right)+$$
$$+\frac{4}{\sqrt{2\pi}(h_{NF}+d)}\exp\left(-\frac{8(x-0.5(2h_{FL}-0.5h_{NF}-0.5d))^2}{(h_{NF}+d)^2}\right); \tag{10}$$

- Catastrophic Situation

$$\rho_{CS}(x) = \frac{2}{d\sqrt{2\pi}}\exp\left(-\frac{2(x+h_{FL})^2}{2(h_{FL}+h_{NF})^2}\right)+\frac{2}{d\sqrt{2\pi}}\exp\left(-\frac{2(x-h_{FL})^2}{(h_{FL}+h_{NF})^2}\right). \tag{11}$$

A configuration of conditional PDFs for altitudes above 410FL are presented in Fig. 4. Also, we use 30 m of geometrical height of airplane (d = 30 m).

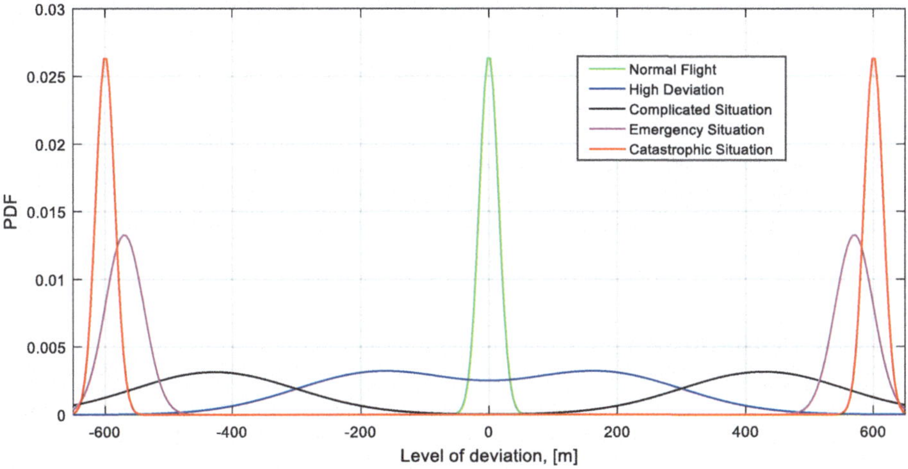

Fig. 4. Conditional PDF.

A posterior probability of each stage could be estimated based on Baes' formula:

$$q_i = \frac{p_i\rho_i}{\sum_{k=1}^{N} p_k\rho_k}, \tag{12}$$

where N = 5 is the number of stages used.

Results of posterior probability estimation for each stage along the possible deviation levels are presented in Fig. 5.

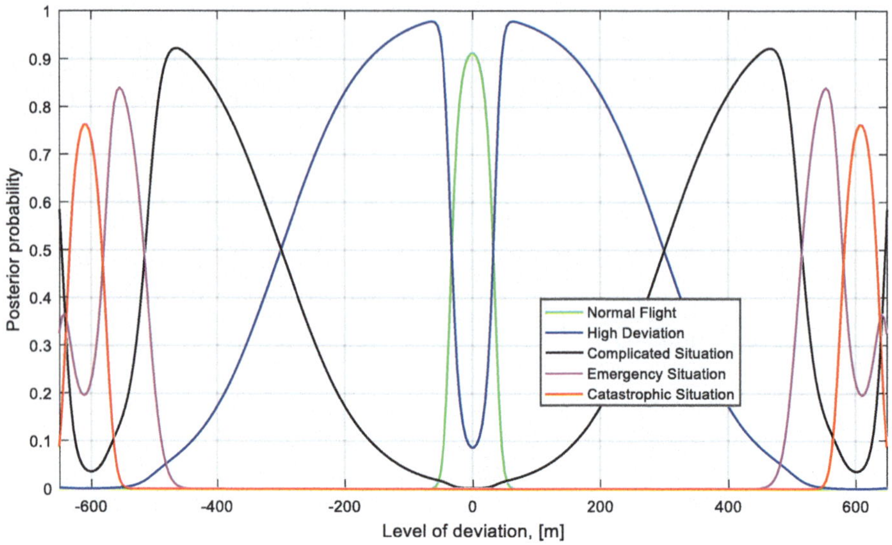

Fig. 5. Posterior probabilities for different situations.

Detection of flight situation or appearance of airplane deviation in a particular stage could be based on a maximum of posterior probability q:

$$n = \max_{i=1,N} q_i, \tag{13}$$

where n is an identification of flight situation.

Maximum posterior probability classifier supports obtaining a decision in the condition of a noisy environment with simultaneously providing a probability of correct detection which is equal to posterior probability.

4 Prior Probabilities

Detection of flight situations based on a maximum of posterior probability requires input of prior probability of each class. Prior probabilities could be calculated based on available statistical data. Prior probability could not be equal, because most time airplane is at the cleared altitude. Valuable deviations from cleared FL to the neighbor FL are classified as rare events. Due to limited statistical data from previous flights, a rare event may not happen, therefore prior probability of each stage could not be calculated as a frequency of this event [21, 22].

In this case, a previous flight realization could be useful with a particular model that takes into account rare events. For each dataset of historical flight a specific algorithm of en-route flight phase identification could be initiated to cut only horizontal flight at cleared flight level [23, 24]. Then, deviations from cleared FL could be calculated. Calculated deviations are used in statistical data analysis to estimate a probability density function (PDF).

Different functions could be useful to analyze PDF of airplane deviation from cleared FL. A normal PDF is mostly used in statistical analysis of results of measurements with errors:

$$\rho(h) = \frac{1}{\sigma\sqrt{2\pi}}\exp\left(-\frac{(H-h-\mu)^2}{2\sigma^2}\right), \tag{14}$$

where H is altitude of cleared FL; h is the actual pressure altitude.

A Double Exponential PDF is mostly used in analysis of airplane deviations from cleared FL [25, 26]:

$$\rho(h) = \frac{(1-\alpha)}{2b_1a_1\Gamma(b_1)}exp\left(-\left|\frac{H-h-\mu}{a_1}\right|^{1/b_1}\right)+ \\ \frac{\alpha}{2b_2a_2\Gamma(b_2)}exp\left(-\left|\frac{H-h-\mu}{a_2}\right|^{1/b_2}\right), \tag{15}$$

$$\Gamma(y) = \int_0^\infty \exp(-x)x^{y-1}dx, \tag{16}$$

where α is a coefficient of mixture of two Exponential PDF; a are scale factors of both PDFs; b are shape coefficients of both PDFs; $\Gamma(b)$ is a Gamma-Euler function.

The main benefit of using Double Exponential PDF is the possibility of controlling "tails". One component is the core of PDF and the second one is used to tune PDF shape based on risk requirements in rare events. As an example, Double Exponential PDF has been used to validate possibility of implementation of reduced vertical separation minimum airspace below 410FL.

However, a Triple Univariate Generalized Error Distribution (TUGED) PDF is used for safety study in the horizontal plane [27]. Three components are used to utilize navigation system error, flight technical system error, and rare events:

$$\rho(h) = \alpha f(b_1, a_1, h) + \beta f(b_2, a_2, h) + (1-\alpha-\beta)f(b_3, a_3, h), \tag{17}$$

$$f(b, a, h) = \frac{1}{2ab\Gamma(b)}\exp\left(-\left|\frac{H-h-b}{a}\right|^{b^{-1}}\right), \tag{18}$$

where β is weight coefficient; $f(b, a, h)$ is a generalized error PDF.

Also, in the case of long statistical dataset, a kernel PDF could be used. It is not recommended to use kernel PDF with a small length of learning sample due to bad modeling of tails. That leads to wrong risk values estimation for emergency and catastrophic stages.

A maximum likelihood method (MLM) is used to estimate parameters of PDF based on input statistical data set of airplane deviations. MLM gets first derivative from PDF by each unknown parameter and makes result equal to zero. Obtained system of equations could be solved by one of the available methods based on level of complexity. Estimated parameters of PDF are result of tuning PDF to the input dataset.

Prior probability of each stage is estimated by the area below PDF within geometrical perils of a particular stage (Fig. 6).

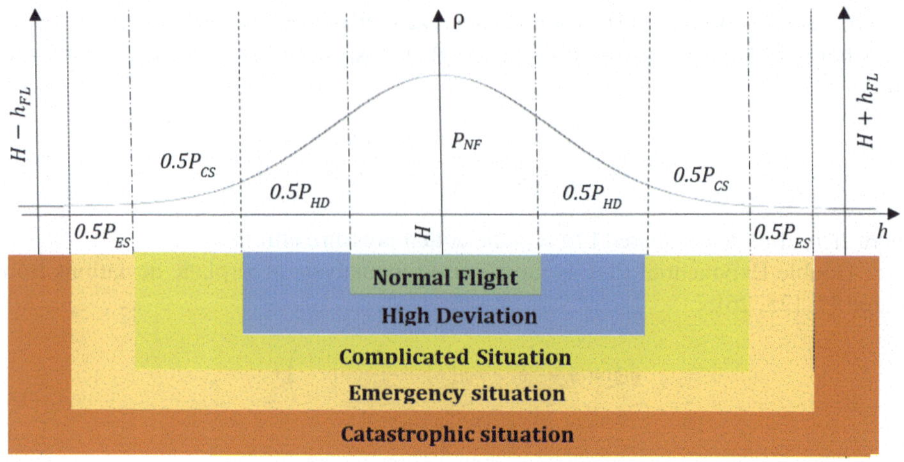

Fig. 6. Prior probabilities of stages.

Based on the model PDF which is used prior probabilities are estimated as follows:

$$P_i = P_i\{\xi \in [c_i, d_i]\} = \int_{c_i}^{d_i} \rho(h)dh, \tag{19}$$

where c_i, d_i are geometrical perils of a particular stage calculated by (1)–(5).

5 Numerical Study

In a numerical study, we use surveillance data obtained under ADS-B to estimate prior probabilities of each class and validate classifier based on maximum posterior probability. A trajectory data of flight AAL 169 for flight connection between Los Angeles International Airport (KLAX) and Tokyo International Airport (RJTT) has been used. Trajectory variation of AAL 169 during June 2023 is presented in Fig. 7. Trajectory variation of AAL 169 direct and AAL 170 opposite flights is caused by weather variation. Airlines use effective trajectory planning based on weather distribution in the region. Thus, trajectory variation is a result of an effective trajectory selection process within a network of flight routes and wind parameters forecast for the time of flight.

Trajectory data have been obtained from the database of surveillance data fixed by the global network of software-defined radios under ADS-B mode 1090ES [28, 29]. Unfortunately, ADS-B data are not synchronized in time and depend on the ground network configuration. Therefore, many data are received in different timestamps and may include long-length gaps. A vertical profile of AAL-169 based on pressure altitude is presented in Fig. 8.

Also, we do not provide data recovery for gaps, to avoid adding errors of interpolation. Therefore, statistical data processing is performed with raw pressure altitudes available in ADS-B reports. Different size of available raw data for each flight gives variation in number of iterations in Fig. 8. Figure 8 shows that cleared FL is changed during the

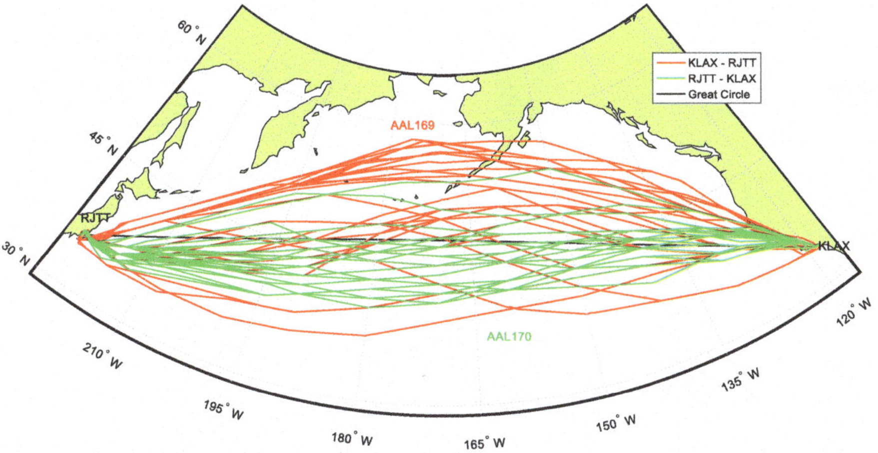

Fig. 7. Trajectory of flight AAL 169 for June 2023.

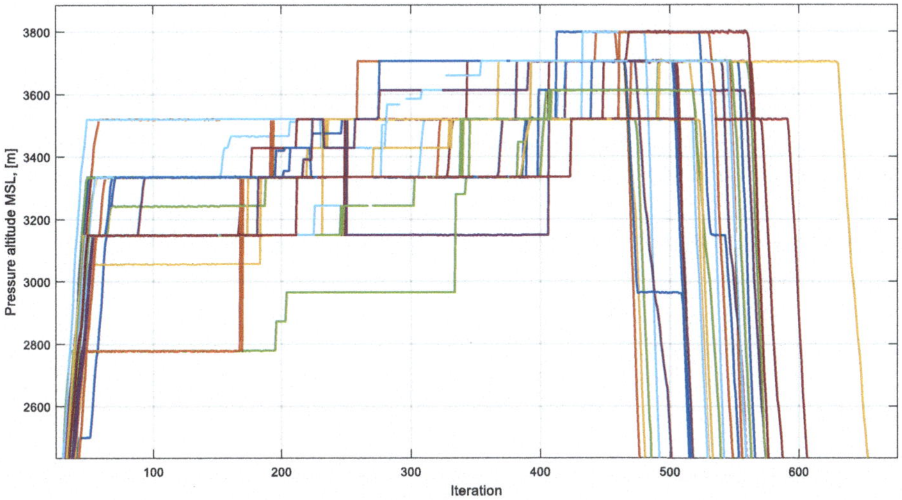

Fig. 8. Pressure altitude of flight AAL 169 during June 2023.

flight several times and FL could be different daily. Active changing FL during en-route phase of the flight is a result of multiple airspace changes due to the long length of flight path (9201 km mean) that takes about 11 h (mean).

Cleared FL for each flight data sequence is detected by comparison of pressure altitude with FL scheme for particular airspace during en-route phase. Phases of flight for each flight are detected by a probability-based classifier by pressure altitude, ground, and vertical speeds. Difference between altitude of cleared FL and actual pressure altitude gives deviations. A Histogram of AAL 169 deviation from the centerline of cleared FL based on 29 flight datasets performed during June 2023 is presented in Fig. 9.

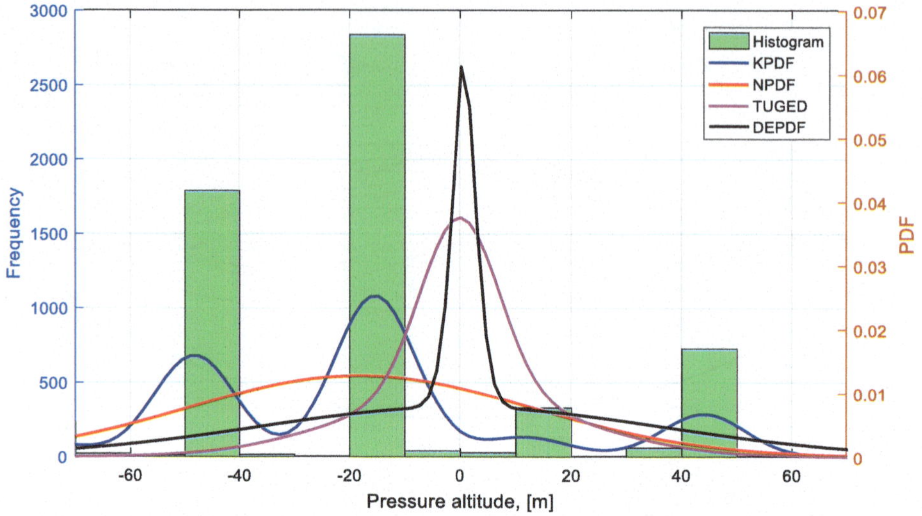

Fig. 9 Statistical analysis of airplane deviations from cleared FL for AAL 169 during June 2023.

Results of statistical analysis of datasets show airplane fluctuations around cleared FL altitude in the range from −80 m up to +60 m which mostly corresponds to Normal flight. A frequency of airplane deviations has been used for tuning PDF. Results of tuning Normal PDF, Double Exponential PDF, TUGED, and Kernel PDF are presented in Fig. 9.

The following parameters of PDFs have been estimated based on MLM:

- Normal PDF
  $\mu = -18.04$ m; $\sigma = 31.16$ m
- Double Exponential PDF
  $\alpha = 0.3$; $a_1 = 2.98$ m; $a_2 = 51.17$ m; $b_1 = 0.61$ m; $b_2 = 0.5$ m; $\mu_1 = 0.65$ m; $\mu_2 = 0$ m;
- TUGED
  $\alpha = 0.5$; $\beta = 0.4$; $a_1 = 30$ m; $a_2 = 10$ m; $a_3 = 2$ m; $b_1 = 0.5$ m; $b_2 = 0.5$ m; $b_2 = 1$ m; $\mu_1 = 0$ m; $\mu_2 = 0$ m; $\mu_3 = 0$ m.

Results of prior probabilities of classes, estimated by (19) for different PDF models are presented in Table 2.

Estimated prior probabilities of classes by (19) are used to calculate posterior probability by (12) for each class. Prior probabilities and conditional probabilities of classes (8)–(11) are used in the classifier by maximum posterior probabilities. Classifier could be used in real-time for immediate signalization of flight class to pilot or could be used in post-processing mode during analysis of flight data. As an example, result of flight stage classification for AAL 169 on June 02, 2023 for en-route phase of flight is presented in Fig. 10.

Results of classification show that most time airplane was in Normal Flight however a Hight deviation situation has also happened. Probability of correct detection was above

Table 2. Prior probabilities of classes.

Probability density function	Normal PDF	Double exponential PDF	TUGED
Normal flight	0.996	0.99	0.99
High deviation	3.8×10^{-3}	4.5×10^{-3}	1.1×10^{-5}
Complicated situation	3.06×10^{-12}	4.47×10^{-17}	1.0×10^{-22}
Emergency situation	2.9×10^{-20}	2.6×10^{-21}	3.5×10^{-27}
Catastrophic situation	1.02×10^{-47}	4.5×10^{-3}	2.9×10^{-15}
Dangerous stage	3.8×10^{-3}	9.0×10^{-3}	1.1×10^{-5}

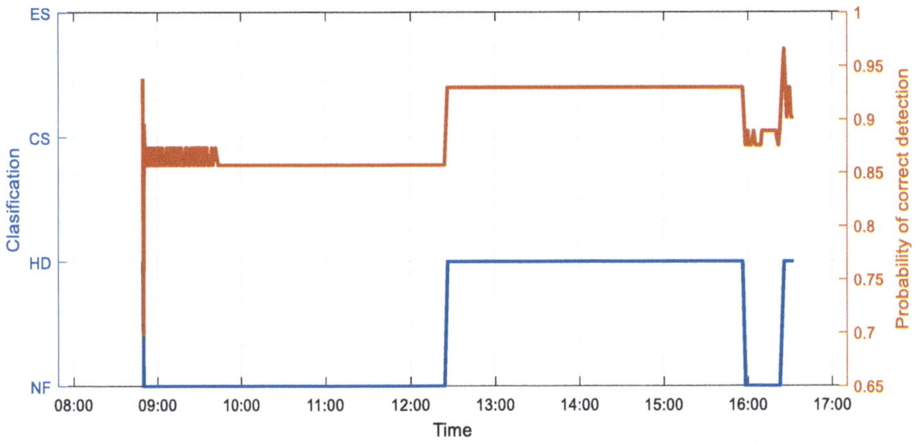

Fig. 10. Results of flight stage classification for AAL 169 on June 02, 2023.

85% most of the time. Also, proposed classifier supports results with a minimum risk of misdetection error.

6 Conclusions

Unplanned deviations from pressure altitude of cleared flight level are a critical parameter for the safety of air transportation. Appearance and development of deviation is a result of degradation factors action in the aviation system. These deviations could be result of flight technical errors in maintaining predefined values or they could be a result of any critical system fault that is connected with airplane stabilization or flight support. Thus, detection of deviation appearance could indicate about beginning of some critical stages that can develop in catastrophe. Proposed in the paper group of five classes, helps to identify particular flight stage based on safety criteria. Method of maximum posterior probability supports classification of current state with a minimum error of misdetection. Also, a probability-based classifier is well-tuned for pressure data with noise action. Classifier in parallel to the decision provides a probability of correct detection which is an important parameter for safety systems.

Detected dangerous flight situations at the beginning helps to understand the level of flight safety and find solutions to avoid further dangerous situations development. Immediate alarming of dangerous stages presence to pilots helps to identify degradation factors that reduce flight safety and orient pilots to minimize its action. Also, a sequence of actions could be provided to pilot via an electronic flight instrument system to back airplane to normal flight.

Classifier requires prior probabilities which are estimated based on previous flight realization at the same flight connection. Prior probabilities could be estimated differently: for a particular airplane, for a particular pilot team, by an airline, or by a particular flight route. Proposed classifier could be integrated into internal algorithms of Flight Management System on board airplanes for probability-based alarming of dangerous stages in real-time. Also, a classifier could be used in software of air traffic control support for alarming violations of cleared flight level and total safety of air transportation estimation. Proposed group of classes and classifier could be used to operate with another metric of class to support algorithms of Traffic Collision Avoidance System and Terrain Awareness and Warning System.

Acknowledgments. Research has been supported by grants from the Ministry of Education and Science of Ukraine under no. 425-DB22 "Methods of increasing the operational efficiency of telecommunication and radio-electronic systems" and no. 243-DB19 "Design of integrated drone guidance, navigation, and flight control system".

References

1. Ale, B.J., Bellamy, L.J., Cooke, R.M., Goossens, L.H., Hale, A.R., Roelen, A.L., Smith, E.: Towards a causal model for air transport safety—an ongoing research project. Saf. Sci. **44**(8), 657–673 (2006). https://doi.org/10.1016/j.ssci.2006.02.002
2. Lin, X., Fulton, N., Westcott, M.: Target level of safety measures in air transportation—review, validation and recommendations. In: Proceedings of the International Conference IASTED, vol. 662, p. 222 (2009)
3. A Unified Framework for Collision Risk Modelling in Support of the Manual on Airspace Planning Methodology for the Determination of Separation Minima, Doc. 9689. ICAO (2009)
4. Performance-Based Navigation (PBN) Manual, Doc 9613. ICAO (2008)
5. Ostroumov, I.V., Kuzmenko, N.S.: An area navigation RNAV system performance monitoring and alerting. In: 1st International Conference System Analysis & Intelligent Computing: SAIC 2018, pp. 211–214. IEEE, Kyiv (2018). https://doi.org/10.1109/SAIC.2018.8516750
6. Tserne, E., Pavlikov, V., Zhyla, S., Dergachov, K., Havrylenko, O., Popov, A., Volosyuk, V., Ruzhentsev, N., Shmatko, O.: Method of optimal threshold calculation in case of radio equipment maintenance. In: Data Science and Security Lecture Notes in Networks and Systems, vol. 462, pp. 69–79. Springer, Singapore (2022)
7. Air traffic management. Procedures for Air Navigation Services, Doc. 4444. ICAO (2016)
8. Kuzmenko, N.S., Ostroumov, I.V.: Performance analysis of positioning system by navigational aids in three dimensional space. In: First International Conference on System Analysis & Intelligent Computing (SAIC), pp. 1–4. IEEE, Kyiv (2018). https://doi.org/10.1109/SAIC.2018.8516790
9. Hun, K.C.: Development of an algorithm for correlation of aircraft positioning data from radar and ADS-B sensors. Doctoral dissertation, University of Malaya (Malaysia) (2019)

10. Zhyla, S., Volosyuk, V., Pavlikov, V., Ruzhentsev, N., Tserne, E., Popov, A.: Practical imaging algorithms in ultra-wideband radar systems using active aperture synthesis and stochastic probing signals. Radioelectron. Comput. Syst. **105**(1), 55–73 (2023)
11. Simonetti, M., Crespillo, O.G.: Robust modeling of geodetic altitude from barometric altimeter and weather data. In: 34th International Technical Meeting of the Satellite Division of The Institute of Navigation (ION GNSS+ 2021), pp. 1176–1189 (2021)
12. Bolanakis, D.E., Kotsis, K.T., Laopoulos, T.: Temperature influence on differential barometric altitude measurements. In: 8th International Conference on Intelligent Data Acquisition and Advanced Computing Systems: Technology and Applications (IDAACS), pp. 120–124 (2015). https://doi.org/10.1109/IDAACS.2015.7340711
13. Callegari, S., Zagnoni, M., Golfarelli, A., Tartagni, M., Talamelli, A., Proli, P., Rossetti, A.: Experiments on aircraft flight parameter detection by on-skin sensors. Sens. Actuators A Phys. **130**, 155–165 (2006). https://doi.org/10.1016/j.sna.2005.12.026
14. Grigorie, T.L., Dinca, L., Corcau, J.I., Grigorie, O.: Aircrafts' altitude measurement using pressure information: barometric altitude and density altitude. WSEAS Trans. Circuits Syst. **9**(7), 503–512 (2010)
15. Rodi, A.R., Leon, D.C.: Correction of static pressure on a research aircraft in accelerated flight using differential pressure measurements. Atmos. Meas. Tech. **5**(11), 2569–2579 (2012). https://doi.org/10.5194/amt-5-2569-2012
16. Reasor, D.A., Bhamidipati, K.K., Woolf, R.K.: Numerical predictions of static-pressure-error corrections for a modified T-38C aircraft. J. Aircr. **52**(4), 1326–1335 (2015). https://doi.org/10.2514/1.C032925
17. Aeronautical Telecommunication. Annex 10 to the convention on International Civil aviation, vol. IV. Surveillance and Collision Avoidance System (2007)
18. Lin, C.E., Hung, T.W., Chen, H.Y.: TCAS algorithm for general aviation based on ADS-B. Proc. Inst. Mech. Eng. Part G J. Aerosp. Eng. **230**(9), 1569–1591 (2016). https://doi.org/10.1177/0954410016631974
19. Manual on a 300 m (1 000 ft) Vertical Separation Minimum Between FL 290 and FL 410 Inclusive, Doc 9574, AN/934. ICAO (2012)
20. Bendarkar, M.V., Behere, A., Briceno, S.I., Mavris, D.N.: A Bayesian safety assessment methodology for novel aircraft architectures and technologies using continuous FHA. In: AIAA Aviation 2019 Forum, p. 3123 (2019). https://doi.org/10.2514/6.2019-3123
21. Ostroumov, I.V., Kuzmenko, N.S.: Risk analysis of positioning by navigational aids. In: International Conference of IEEE Signal Processing Symposium: SPSympo-2019, pp. 92–95 (2019). https://doi.org/10.1109/SPS.2019.8882003
22. Kuzmenko, N., Ostroumov, I., Bezkorovainyi, Y., Averyanova, Y., Larin, V., Sushchenko, O.: Airplane flight phase identification using maximum posterior probability method. In: 3rd International Conference on System Analysis & Intelligent Computing, pp. 1–5. SAIC, Kyiv (2022). https://doi.org/10.1109/SAIC57818.2022.9922913
23. Fala, N., Georgalis, G., Arzamani, N.: Study on machine learning methods for general aviation flight phase identification. J. Aerosp. Inf. Syst. **20**(10), 636–647 (2023). https://doi.org/10.2514/1.I011246
24. Ostroumov, I., Ivannikova, V., Zaliskyi, M., Kuzmenko, N.: Impact analysis of Russian-Ukrainian war on airspace. J. Air Transp. Manag. **124**, 102742 (2025). https://doi.org/10.1016/j.jairtraman.2025.102742
25. Ostroumov, I., Ivashchuk, O., Kuzmenko, N.: Preliminary estimation of war impact in Ukraine on the global air transportation. In: 12th International Conference on Advanced Computer Information Technologies (ACIT), pp. 281–284. IEEE, Ruzomberok (2022). https://doi.org/10.1109/ACIT54803.2022.9913092
26. Review of the General Concept of Separation Panel (RGCSP), Doc. 9536, Montreal, Canada. ICAO (1988)

27. Mori, R.: Identifying the ratio of aircraft applying SLOP by statistical modeling of lateral deviation. Trans. Jpn Soc. Aeronaut. Space Sci. **54**(183), 30–36 (2011). https://doi.org/10.2322/tjsass.54.30
28. Ostroumov, I.V., Marais, K., Kuzmenko, N.S., Fala, N.: Triple probability density distribution model in the task of aviation risk assessment. Aviation. **24**(2), 57–65 (2020). https://doi.org/10.3846/aviation.2020.12544
29. Flightaware. International database of ADS-B. https://www.flightaware.com

Methods of Operational Efficiency Improvement for Civil Aviation Systems

Algorithms for Selecting the Optimal Filter in the Architecture of a Multi-channel Passive Interference Compensation System

Igor Prokopenko and Anastasiia Dmytruk(✉)

National Aviation University, Liubomyra Huzara Ave., 1, Kyiv 03058, Ukraine
gyhyre@gmail.com

Abstract. The practical implementation of adaptive algorithms in radar systems for detecting moving targets is a relevant and promising direction of modern signal processing technologies. One of the possible approaches to hardware implementation of adaptive algorithms is developing a multi-channel passive interference compensation system, which may be appropriate for improving the performance of radar systems in complex and dynamic environments. This paper considers the problem of selecting the optimal filter in a multi-channel architecture to maximize the signal-interference ratio. To solve the problem, it is proposed to use the optimal filter selection criterion stage in the system architecture. Three criteria are proposed and considered, namely the criterion of the ratio of ordinal statistics, which is based on the ratio between the maximum value of the modulus of the discretized signal samples and the median of the modulus of this same vector, the criterion based on the minimum of the variance, since the variance characterizes the variability of the signal, and the criterion based on the Fourier transform, which is based on the analysis of the frequency components of the signal. To evaluate their effectiveness, computer modeling was conducted, leading to conclusions about the feasibility of their use. This emphasizes the connection between theoretical justification and practical implementation possibilities.

Keywords: Moving Target Detection · Adaptive Algorithm · Autoregressive Interference · Maximum Likelihood Ratio · Multi-Channel System

1 Introduction

The problem of radar detection of moving targets against the background of interference arose in the 1940s and became an impetus for the development of moving target selection systems, the fundamental idea of which is the use of the Doppler effect [1, 2]. However, further scientific and practical development and improvement of existing systems and methods continues to this day, since in the conditions of constant technological development of mankind there is a need to increase the requirements for the characteristics of radar detection systems, which are forced to function under conditions of interference of various nature, which mask the useful signal and can significantly complicate the receipt of useful information [3, 4].

I. Ostroumov et al. (Eds.): ACASD 2025, LNNS 1418, pp. 337–347, 2025.
https://doi.org/10.1007/978-3-031-91992-3_22

Since interference has a priori uncertainties of statistical characteristics and dynamically changes in space and time, classical approaches are often ineffective, therefore there is a need to create specialized methods for processing received signals, among which adaptive detection methods are of great interest, which can change their parameters directly taking into account the interference parameters, allowing to improve the efficiency and accuracy of modern radar systems for selecting moving targets [5, 6]. A significant advantage of adaptive methods is their ability to ensure optimal system performance under conditions of unpredictable changes in the environment, which is important for radar systems. The paper considers the problem of detecting moving targets against the background of broadband-correlated interference that occurs when a signal is reflected from various environmental objects (water surface, vegetation, local objects, etc.) and the possibility of practical implementation of the synthesized solution using the parallel architecture of programmable logic integrated circuits [7, 8]. These problems are characterized by a priori uncertainty in the statistical characteristics of the interference. Under these conditions, the likelihood ratio maximization method and the empirical Bayesian approach are used to synthesize algorithms [9, 10].

2 Formulation of the Problem

Given that the signal-interference mixture at the input of the receiving device is by its nature the implementation of a random process, statistical methods are the main means for solving the problem of moving targets against the background of passive problems. Regarding the implementation of the input sample x_i, the appropriate probability density functions for two hypotheses are considered: H_0 assumes the presence of only interference in the sample, while H_1 is an alternative hypothesis that the sample contains both interference and signal and is described through the corresponding probability distribution densities$f_0\left(\overline{x, \vartheta}|H_0\right), f_1\left(\overline{x}, \overline{\vartheta}|H_1\right)$ [11, 12]. Taking into account the condition of insufficient a priori data, it is advisable to use the Neyman-Pearson criterion, according to which, based on the described probability distribution densities $f_0\left(\overline{x, \vartheta}|H_0\right), f_1\left(\overline{x}, \overline{\vartheta}|H_1\right)$of hypotheses H_0 and H_1, the likelihood ratio L is calculated (1).

$$L(\overline{x}) = \frac{f_1\left(\overline{x}, \overline{\vartheta}, H_1\right)}{f_0\left(\overline{x, \vartheta}, H_0\right)}, \tag{1}$$

where $f_0\left(\overline{x}, \overline{\vartheta}, H_0\right)$, $f_1\left(\overline{x}, \overline{\vartheta}, H_1\right)$ – probability density functions correspond to the statistical hypotheses, $\overline{\vartheta}$ – the parameter vector for the mixture of signal and interference.

When forming a detection algorithm taking into account the selected criterion, the maximum likelihood method (MLM) is used, which consists of choosing a hypothesis that maximizes the likelihood ratio $L(\overline{x})$ (2).

$$L(\overline{x}) = \frac{\max\limits_{\overline{\vartheta}} f_1\left(\overline{x}, \overline{\vartheta}, H_1\right)}{\max\limits_{\overline{\vartheta}} f_0\left(\overline{x, \vartheta}, H_0\right)}. \tag{2}$$

Due to the lack of a priori data, a concomitant problem arises in estimating the vector of unknown interference parameters. In this regard, for the further synthesis of adaptive detection algorithms [13, 14], the work uses an empirical Bayesian approach, which allows to refinement of parameter estimates based on posterior information, using the maximum posterior probability density method. The above-described procedure for synthesizing an adaptive detection algorithm can be generalized in the form of an adaptive detector structure (Fig. 1).

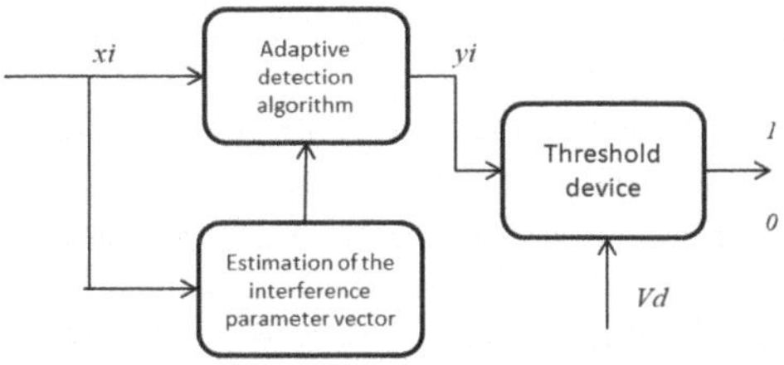

Fig. 1. Block diagram of a robust adaptive detector of a harmonic signal with an unknown phase against a mixture of correlated autoregressive and impulse interference.

When considering the detection problem, an important step is to formalize the signal and interference model. According to this a mixture of harmonic signal and autoregressive interference is proposed.

$$x_i = bS_i + y_i, i = \overline{1, n}, \tag{3}$$

where S_i – samples of a harmonic signal, b - signal parameter that determines the amplitude of the received signal, $y_i = \sum_{j=1}^{p} a_j y_{i-j} + \eta_i$, – an autoregressive model of interference with η_i generating process, p – model order.

In works [15, 16], an adaptive algorithm for detecting a deterministic signal against the background of autoregressive interference was synthesized. Considering the above, the conditional density distribution of the mixture in case H_0 and H_1 will be expressed as (4).

$$\begin{gathered} f\left(x_{p+1}, \ldots, x_n \middle| x_1, \ldots, x_p, H_0\right) = \frac{1}{\sqrt{2\pi\sigma_3^2}} exp^{-\frac{x_1{}^2}{2\sigma_3^2}} \times \\ \times \frac{1}{\sqrt{2\pi\sigma_3^2(a_1,\ldots a_{\min(i,p)})}} \prod_{i=2}^{n} \left[exp\left\{ -\frac{\left(x_i - \sum_{j=1}^{\min(i,p)} a_j x_{i-j}\right)^2}{2\sigma_3^2(a_1,\ldots a_{\min(i,p)})} \right\} \right], \\ f\left(x_{p+1}, \ldots, x_n \middle| x_1, \ldots, x_p, H_1\right) = \frac{1}{\sqrt{2\pi\sigma_3^2}} exp^{-\frac{(x_1 - S_1)^2}{2\sigma_3^2}} \times \\ \times \frac{1}{\sqrt{2\pi\sigma_3^2(a_1,\ldots a_{\min(i,p)})}} \prod_{i=2}^{n} \left[exp\left\{ -\frac{\left((x_i - S_i) - \sum_{j=1}^{\min(i,p)} a_j (x_{i-j} - S_{i-j})\right)^2}{2\sigma_3^2(a_1,\ldots a_{\min(i,p)})} \right\} \right]. \end{gathered} \tag{4}$$

According to local-optimal conception [15] the derivative of logarithm of the likelihood ratio (2) on the signal parameter b in the point $b = 0$ will have the following form (5).

$$\frac{\partial}{\partial b}\ln(L(x_1, \ldots, x_n, b))_{|b=0} =$$

$$= \sum\nolimits_{i=2}^{n}\left(x_i - \sum\nolimits_{j=1}^{\min(i,p)} a_j{}^* x_{i-j}\right)\left(S_i - \sum\nolimits_{j=1}^{\min(i,p)} a_j{}^* S_{i-j}\right) > V\left(\overline{S}, \sigma_i, \overline{a^*}, \alpha\right), \quad (5)$$

where $V\left(\overline{S}, \sigma_i, \overline{a^*}, \alpha\right)$ – decision threshold, $\overline{a^*}$ are the estimated autoregressive coefficients, σ_i – the variance of a random process, α – false alarm probability.

According to (2), the structure of the algorithm (4) is formed, as illustrated in Fig. 2. The detector consists of two identical channels – two rejection filters: the channel for processing the sequence of values of the mixture of signal and interferencex_i, $i = \overline{1, n}$, and the channel of the sequence of signal valuesS_i, $i = \overline{1, n}$.

Each of the notch filters contains delay elements for the sampling interval τ, amplifiers with transmission coefficients $(-a_1^*, \ldots, -a_p^*)$, the values of which are calculated based on the interference sample in the parameter estimation block and the adder. As a result of the operation of the upper adaptive filter, discrete samples of the filtered signal are formed. The outputs of both channels are fed to a multiplication circuit, and then to an adder, the output of which is fed to a comparison circuit with a decision threshold, the value of which is determined according to the Neyman-Pearson criterion. An important element of the structure is an adaptive filter, the adaptability of which is ensured by estimating the interference parameters according to the synthesized estimation algorithm, which, due to the insufficiency of a priori data, is based on an empirical Bayesian approach, which allows obtaining parameter estimates based on a posteriori information, using the maximum likelihood method.

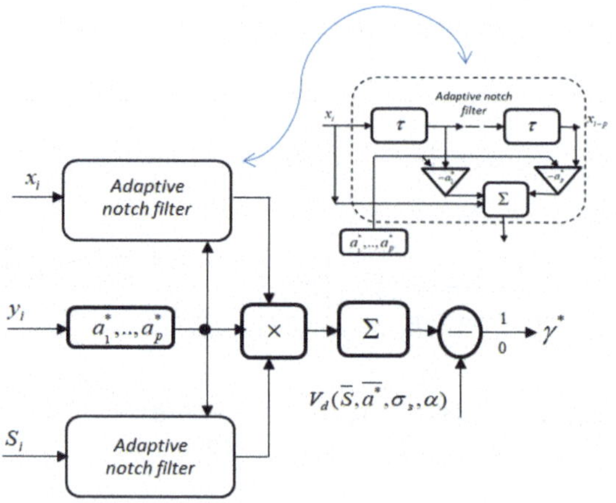

Fig. 2. The structural diagram of a synthesized adaptive algorithm (5).

According to the obtained results, the corresponding problem of estimating the coefficients of autoregressive interference is formulated. The solution can be implemented using a multichannel system consisting of a set of notch filters tuned to the given coefficients of autoregression. The input mixture is processed by these filters, and the filter output with the maximum signal-to-interference ratio is selected.

3 Criteria for Choosing the Optimal Filter

The implementation of synthesized structures, which, taking into account the specifics of the task and algorithm, can be effectively executed on FPGA since the use of FPGA provides the execution of computationally complex algorithms in real time with high performance and flexibility [17, 18].

One of the possible implementation options for the detection algorithm described above involves building a multi-channel system, where each channel uses predefined coefficients to process input signals (Fig. 3). The coefficients of adaptive filters are stored in the FPGA memory, which provides quick access to them during calculations. The parallel architecture of the FPGA allows us to implement a system that includes the required number of parallel filters, each of which operates based on its unique coefficients, which allows the simultaneous processing of several signals.

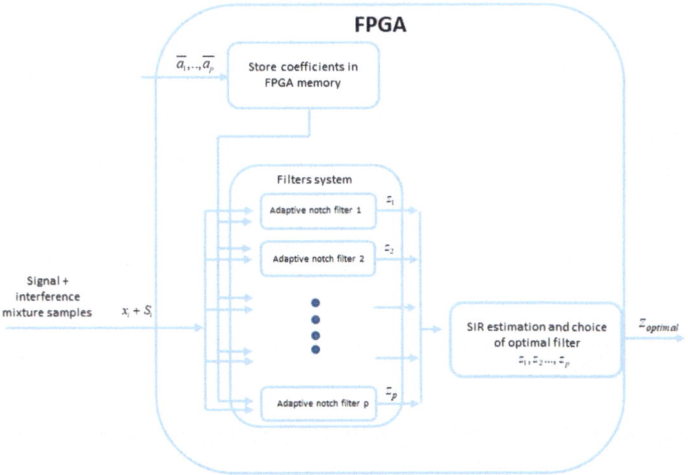

Fig. 3. Architecture of a multi-channel passive interference compensation system.

The task is to develop an algorithm for selecting the optimal filter at the output of which the maximum signal-to-interference ratio is formed. This structure implements the calculation of the maximum likelihood ratio (2).

To solve the problem, it is proposed to consider three algorithms (criteria) for estimating the signal-to-interference ratio: (1) the ordinal statistics ratio algorithm; (2) the minimum variance algorithm; (3) the algorithm based on the Fourier transform.

The criterion based on the ratio of ordinal statistics is defined as the ratio between the maximum value of the modulus of the sampled signal samples at the filter output and the median of the modulus of this same vector (6):

$$k1 = \frac{\max(|\overline{y}|)}{median(|\overline{y}|)}. \tag{6}$$

As a result of the work, the optimal filter is considered to be the one for which the value of the criterion $k1$ is minimal. This is explained by the fact that minimizing the criterion reduces the influence of interference, since in the case of weak interference compensation, the value of the criterion $k1$ will increase.

An increase in this value indicates the presence of an uneven distribution of signal energy, which is associated with the predominance of interference in the filter output signal.

The criterion of minimum dispersion is that the optimal filter is chosen to provide the minimum value of the dispersion of the output signal (7):

$$k2 = \min(\text{var}(\overline{y})). \tag{7}$$

Dispersion characterizes the variability of the output signal. A large dispersion indicates a strong presence of interference. Thus, a filter at the output of which a minimum dispersion is formed rejects interference better.

The signal-to-interference ratio maximization criterion based on the Fourier transform is an effective tool for extracting a useful signal in the presence of interference by analyzing the frequency components of the signal (8):

$$k3 = \frac{\max(\text{abs}(\text{FFT}(\overline{y})))}{\text{mean}(\text{abs}(\text{FFT}(\overline{y})))}. \tag{8}$$

4 Computer Simulation

Computer simulation is an important step to evaluate the performance of the proposed algorithms. It allows you to verify the effectiveness of the proposed criteria in practice, as well as to study the behavior of the algorithms in various experimental conditions. This allows us not only to confirm the theoretical validity of the criteria but also to determine the optimal conditions for their application. The results of the computer simulation of the algorithms for selecting the optimal filter in the multichannel system (Fig. 3) are shown in Figs. 4, 5, and 6.

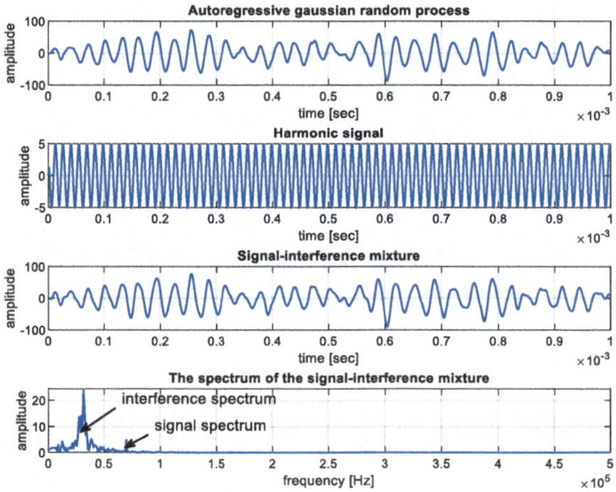

Fig. 4. Spectral characteristics of the filtered signal sequence at the output of each of the filters of the multichannel structure.

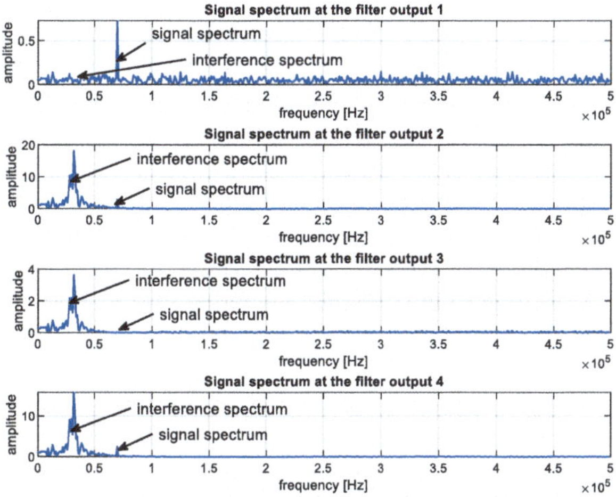

Fig. 5. Spectral characteristics of the filtered signal sequence at the output of each of the filters of the multichannel structure.

Figure 4 shows the time plots and spectral representation of the principal components used in the simulation. The first plot shows the time implementation of the simulated autoregressive interference model with a Gaussian generating process. The second plot shows the harmonic signal used as the useful signal model. The third plot illustrates the time implementation of the signal-interference mixture obtained by adding the harmonic signal and the interference model. The fourth plot shows the spectral characteristic of the signal-interference mixture, which allows us to demonstrate the frequency distribution of the components.

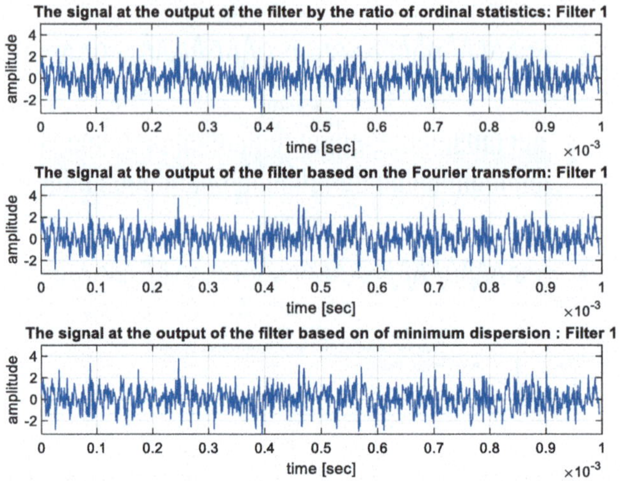

Fig. 6. Results of simulation of the operation of the used criteria for selecting the optimal filter.

Figure 5 presents the results of the multichannel filter system structure, which demonstrates the filtered signal sequence's spectral characteristics at each filter's output. According to the obtained results, it is noticeable that at the output of the first filter, the interference is completely rejected, thus this filter under the specified modeling conditions will be considered optimal.

In turn, Fig. 6 presents the results of the criteria for selecting the optimal filter. The presented results allow us to compare the effectiveness of different criteria, assess their sensitivity to changing conditions, and determine the most appropriate criterion for specific tasks. According to the obtained modeling results, it is clear that under the specified conditions, all criteria demonstrate the required level of effectiveness, that is, the optimal filter is selected.

To determine the most appropriate selection criterion in the context of a specific problem statement, it is necessary to analyze the effectiveness of their work. To do this, the probability of choosing the correct filter is determined based on the results of statistical tests - that is, the one that, according to the simulation results, provides the best signal-to-interference ratio of each of the proposed criteria.

The obtained analysis results are given in Table 1. They allow us to compare the effectiveness of the criteria and determine which of which is most appropriate for use in the problem being solved, taking into account the changes in the specifics of the signal-interference environment.

During the analysis of efficiency, a four-channel system of notch filters was highlighted. To assess the efficiency of the criteria, a series of statistical tests were conducted with the number of repetitions $N = 1000$ and changing the modeling parameters, namely the parameters of the interference model a, amplitude A, and frequency f of the useful signal. As a result, the probability of choosing the correct filter for each of the criteria was determined.

The results of the study showed that the best performance indicators were demonstrated by the minimum variance criterion and the criterion based on the ratio of ordinal

Table 1. The probability of choosing the correct filter depending on the criteria used.

Simulation parameters	Probability of choosing the correct filter (criterion 1)	Probability of choosing the correct filter (criterion 2)	Probability of choosing the correct filter (criterion 3)
$A = 5; a = [1.95; -0.99];$ $f = 70\ kHz.$	$P1 = 1$	$P2 = 1$	$P3 = 0.959$
$A = 5; a = [1.5; -0.75];$ $f = 70\ kHz.$	$P1 = 0.997$	$P2 = 1$	$P3 = 0.9964$
$A = 1; a = [1.9; -0.99];$ $f = 51\ kHz.$	$P1 = 1$	$P2 = 1$	$P3 = 0.87$
$A = 1; a = [1.8; -0.95];$ $f = 51\ kHz.$	$P1 = 1$	$P2 = 1$	$P3 = 0.78$
$A = 1; a = [1.4; -0.9];$ $f = 40\ kHz.$	$P1 = 0.813$	$P2 = 1$	$P3 = 0.0$
$A = 1; a = [1.7; -0.95];$ $f = 40\ kHz.$	$P1 = 0.938$	$P2 = 1$	$P3 = 0.0$

statistics. The simultaneous criterion based on the Fourier transform turned out to be more sensitive to the spectral features of the signal and interference, which can be an advantage in conditions of narrow-band interference or signals with a complex spectral structure.

Thus, the choice between these criteria depends on the specifics of the signal-interference environment and the requirements for adaptive filtering in a particular task.

5 Conclusions

Since interference has a priori uncertainties of statistical characteristics and dynamically changes in space and time, classical approaches are often ineffective, therefore there is a need to create specialized methods for processing received signals, among which adaptive detection methods are of great interest, which can change their parameters directly taking into account the interference parameters, allowing to improve the efficiency and accuracy of modern radar systems for selecting moving targets.

Given the specifics of the problem under consideration, a generalized scheme of an adaptive detection algorithm was developed, which serves as the basis for further synthesis of such structures. The algorithms are based on the use of the Neumann-Pearson criterion, according to which the optimality index of the synthesized algorithms is formed, which consists of ensuring the maximum probability of detection at a given probability of false alarm.

The use of autoregressive interference models and adaptive parameter estimation methods allows forming of locally optimal algorithms, which allows to a significant increase in the accuracy of signal detection in complex signal-interference situations. The algorithms are synthesized using a statistical approach, which includes a statistical interference model and takes into account a priori uncertainty in the parameters of both the interference model and the signal. An important feature of such algorithms is the structure of interference parameter estimation, which characterizes their ability to dynamically adapt to changes in the statistical properties of the interference.

The paper also considered the effectiveness of several criteria for selecting the optimal filter based on a multi-channel system of notch filters. The best results were shown by the minimum dispersion criterion and the criterion based on the ratio of ordinal statistics. The minimum dispersion criterion provides stable efficiency in a wide range of conditions, which makes it a universal choice for tasks with different characteristics of the signal-noise environment. The criterion based on the Fourier transform is more sensitive to the spectral characteristics of the signal and noise, which makes it effective in cases where the spectral features are clearly expressed. At the same time, the use of a multi-channel filter structure allows for adaptive signal processing in a complex signal-noise environment, improving the quality of filtering due to the possibility of simultaneous analysis and noise suppression in several channels.

Future research may focus on implementing these algorithms on high-performance hardware platforms such as FPGAs to achieve real-time processing capabilities for practical applications, making them suitable for modern radio-electronic systems operating in challenging environments, for example, to protect radio-electronic systems from countermeasures.

References

1. Richards, M.: Fundamentals of Radar Signal Processing. McGraw-Hill, New York (2005)
2. Emslie, A., McConnell, R.: Moving target indication. In: Radar Systems Engineering. McGraw-Hill, New York (1947)
3. Prokopenko, I., Omelchuk, I., Maloyed, M.: Synthesis of signal detection algorithms under conditions of aprioristic uncertainty. In: IEEE Ukrainian Microwave Week, pp. 418–423. IEEE, Kharkiv (2020). https://doi.org/10.1109/UkrMW49653.2020.9252687
4. Prokopenko, I., Omelchuk, I., Chyrka, Y.: RADAR signal parameters estimation in the MTD tasks. Int. J. Electron. Telecommun. **58**(2), 159–164 (2012)
5. Kay, S.: Fundamentals of Statistical Signal Processing: Detection Theory. Prentice-Hall, Upper Saddle River (1998)
6. Neyman, J., Pearson, K.: On the Problem of the Most Efficient Test of Statistical Hypotheses. Springer, New York (1992). https://doi.org/10.1007/978-1-4612-0919-5_6
7. Meyer-Baese, U.: Digital Signal Processing with Field Programmable Gate Arrays. Springer, Berlin (2007)
8. Hauck, S., DeHon, A.: Reconfigurable Computing, the Theory and Practice of FPGA-Based Computation. Morgan Kaufmann, San Francisco (2008)
9. Forbes, C., Evans, M., Hastings, N., Peacock, B.: Statistical Distributions. Hoboken, New Jersey (2010)

10. Solomentsev, O., Zaliskyi, M., Sushchenko, O., Bezkorovainyi, Y., Averyanova, Y., Ostroumov, I., Larin, V., Kuzmenko, N.: Data processing through the lifecycle of aviation radio equipment. In: IEEE 17th International Conference on Computer Sciences and Information Technologies, pp. 146–151. IEEE, Lviv (2022). https://doi.org/10.1109/CSIT56902.2022.10000844
11. Ross, S.: Stochastic Processes. Wiley, Montreal (1996)
12. Shirman, Y.: Radio Electronic Systems: Fundamentals of Construction and Theory. (1998)
13. Haykin, S.: Adaptive Filter Theory. Pearson Education, London (2007)
14. Kelly, E.: An adaptive detection algorithm. IEEE Trans. Aerosp. Electron. Syst. **22**(2), 115–127 (1986). https://doi.org/10.1109/TAES.1986.310745
15. Prokopenko, I., Dmytruk, A.: Implementation of adaptive algorithms in the task of MTDI filtration. In: IEEE International Conference on Information and Telecommunication Technologies and Radio Electronics, pp. 226–231. IEEE, Odesa (2021). https://doi.org/10.1109/UkrMiCo52950.2021.9716711
16. Prokopenko, I., Prokopenko, K., Dmytruk, A.: Comparison of neural network and statistical approaches to the problem of signal detection. In: Ostroumov, I., Zaliskyi, M. (eds.) Proceedings of the 2nd International Workshop on Advances in Civil Aviation Systems Development. ACASD 2024, Lecture Notes in Networks and Systems, vol. 992, pp. 224–241. Springer, Cham (2024). https://doi.org/10.1007/978-3-031-60196-5_18
17. Li, S., Wang, Y.: The design of radar signal processor based on FPGA. In: Zhang, Y. (ed.) Future Wireless Networks and Information Systems Lecture Notes in Electrical Engineering, vol. 143, pp. 283–288. Springer, Berlin (2012). https://doi.org/10.1007/978-3-642-27323-0_36
18. Shannon, L., Cojocaru, V., Dao, C., Leong, P.: Technology scaling in FPGAs: trends in applications and architectures. In: IEEE 23rd Annual International Symposium on Field-Programmable Custom Computing Machines, pp. 1–8. IEEE, Vancouver, BC (2015). https://doi.org/10.1109/FCCM.2015.11

Predictive Analytics of Decision-Making for Aviation Operation Enterprise

Oleksandr Solomentsev, Maksym Zaliskyi(✉), Zarina Poberezhna, and Oleksiy Zuiev

National Aviation University, Liubomyra Huzara Ave., 1, Kyiv 03058, Ukraine
mzaliskyi@kai.edu.ua

Abstract. At the current stage of development of science and technology, as well as in conditions of limited financial and natural resources, there are problems of effective use of equipment for its intended purpose. Efficiency is usually understood as ensuring a given level of reliability with minimal resource consumption. It is known that equipment is characterized by the presence of four stages of the life cycle. The longest and most important stage from the point of view of using the equipment functions for its intended purpose is operation. The high cost of the operation stage is due to the presence of maintenance and repair processes. These processes perform the functions of providing high-quality equipment usage and restoration. Maintenance and repair are based on the procedures of monitoring, technical condition control, data processing, diagnostics, decision-making, and implementation of preventive and control actions. This paper considers the problem of analyzing the efficiency of maintenance and repair processes, which are implemented using statistical data processing algorithms that are aimed at reducing average operating costs. The paper presents structural diagrams of the interconnection of processing operations and conducts a comparative analysis based on statistical simulation.

Keywords: Operation · Data Processing · Prediction · Decision-making · Radio Equipment · Maintenance · Repair

1 Introduction

1.1 Introduction to the Problem

The task of saving resources has historically always been relevant [1, 2]. At the current stage of development of science and technology, a large number of new industries and spheres of human activity have appeared [3]. In this regard, for specific applications, there are limitations on material, human, intellectual, and natural resources [4, 5]. In the aviation industry, the most important issues are the safety and regularity of flights [6, 7]. To ensure them, organizational, technical, and technological measures are used [8, 9]. This applies to improving infrastructure, harmonizing regulatory support, personnel training, introducing new technologies and equipment, and others [10, 11].

I. Ostroumov et al. (Eds.): ACASD 2025, LNNS 1418, pp. 348–363, 2025.
https://doi.org/10.1007/978-3-031-91992-3_23

Aircraft, avionics, and ground radio equipment interact to effectively perform flight tasks and maximize the satisfaction of users of aviation services [12].

An objective process is the occurrence of failures and malfunctions during the use of aviation radio equipment [13, 14]. One of the ways to eliminate failures, malfunctions, and their consequences is to implement a set of maintenance, repair, flight checks, and resource extension processes [15, 16]. Aviation enterprises implement these processes in operation systems (OS) [17].

The main components of the OS are:

- organizational structure;
- personnel and workplaces;
- documentation;
- expendable and information resources;
- equipment in operation;
- technological processes;
- measuring equipment, and others [18, 19].

A promising direction for improving the OS is the use of new methods, models, and data processing technologies [20]. They are aimed at reducing the uncertainty of the OS elements, assessing compliance with the requirements for these elements, adapting to operating conditions and new circumstances, and widely automating actions, measurements, and decision-making [21].

The use of new information technologies guarantees optimal solutions, efficient processes of using equipment for its intended purpose, and reduced operating costs.

Life cycle periods have different levels of costs. The analysis shows that during the operation period, the costs of electricity, maintenance and repair, ground and flight inspections, creation of design documentation for commissioning, personnel training, spare parts, and others many times exceed the costs of design and manufacturing [22]. Taking into account these features, maintenance and repair issues have a strong impact and significance in terms of optimizing operational resources [23].

It should be noted that within the aviation industry and individual aviation enterprises, data hubs have not yet been created for the components of resource costs throughout the entire life cycle of radio equipment.

Experience shows that in many countries, the aviation industry has a well-established triad consisting of a state regulatory body, operating enterprises, and research institutions [24, 25]. These organizations, within the framework of systematically defined powers and functions, timely identify problematic issues, put forward proposals, carry out developments, and implement them [26].

Maintenance and repair processes include technological procedures. This applies to both manual and automated implementation. Today, a number of procedures in the field of repair cannot be performed without a person [27].

In the OS, there are monitoring procedures that represent observations of diagnostic variable trends over time using built-in or external measuring equipment [28]. This procedure can provide information about the achievement of threshold levels of parameters. For reliability indicators, monitoring is performed after the actual calculation of the numerical value of the reliability indicator [29].

Another operation process is technical condition control. In a certain number of different states, it is possible to classify belonging to a particular state: (1) serviceable and faulty for a two-alternative classification; (2) serviceable, partially faulty, and completely faulty for a three-alternative classification. Classification issues should take into account the capabilities of personnel, the degree of influence of radio equipment on the effectiveness of its use, material costs, and others. The difference between control and monitoring operations is shown in Fig. 1.

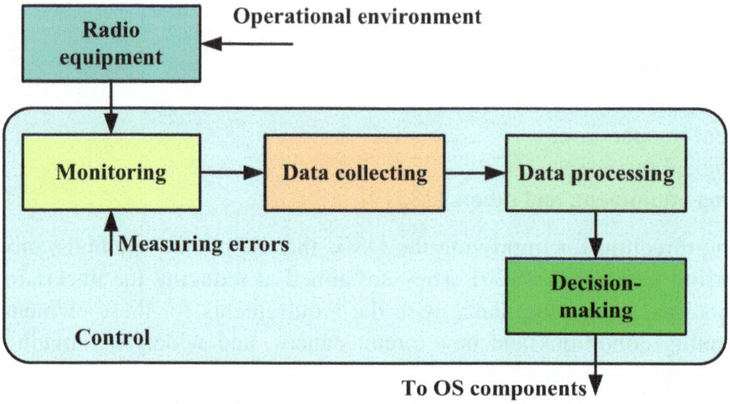

Fig. 1. The monitoring and control.

Figure 1 shows that monitoring is an integral part of control. Control can be implemented as current and predictive. Data processing algorithms can be associated with calculating mathematical expectations, identifying changepoints, forecasting trends for implementing preventive maintenance, analyzing heteroscedasticity, determining the type and parameters of the distribution law, and others [30, 31].

1.2 Motivation

Today, the industry is aimed at total resource saving in all areas of human activity. In the field of aviation equipment operation, there are four stages of the life cycle [32]. The operation stage is the longest and most expensive. Therefore, at this stage, it is necessary to apply new methods, approaches, models, and data processing technologies, including the use of artificial intelligence [33, 34]. In this case, it is advisable to pay attention to the processes of technical maintenance and repair as the most expensive processes.

1.3 Contribution

The main contribution of this paper is the development of new data processing methods using prediction procedures, which allow for preventive actions to be taken to avoid the negative consequences of possible events in the equipment operation system. In addition, the developed methods take into account possible changepoints in the trends of diagnostic parameters that can objectively arise in the process of deterioration of the technical condition of the equipment. The practical significance lies in the fact that there is an opportunity to improve the systems for forming and implementing control and preventive actions in the aviation equipment OS.

1.4 The Organization of the Paper

The paper consists of five sections. The first section considers background information related to the maintenance of aviation radio equipment. The second section deals with the development of data processing structures for the maintenance process. The third section deals with the statistical simulation of data processing algorithms. The last two sections are conclusions and future research directions.

2 Materials and Methods

The paper provides a comparative analysis of the efficiency of the following options for organizing maintenance and repair procedures:

1. Procedures are performed by personnel in accordance with a predetermined frequency and a list of control and measuring procedures. In this case, the personnel use built-in or external control and measuring equipment and make the following decisions, taking into account the known tolerances for diagnostic variables: (a) if the parameter at the time of measurement is within the operating tolerances, then a decision is made to continue operation; (b) if the parameter goes beyond the operating tolerances, then a decision is made to perform adjustment or current repair procedures.
2. The procedures are performed by personnel in accordance with the condition-based maintenance strategy with control of diagnostic variables or with control of the reliability level. It should be noted that the diagnostic variables control option is simpler. For its implementation, two preventive tolerances are used inside the operating tolerances. If the diagnostic variable is within the preventive tolerances, then a decision is made to continue the radio equipment operation. In the case of diagnostic variable value outside the preventive tolerances, but within the operating tolerances, a decision is made to plan and perform preventive maintenance. In the third case, repair procedure is performed. The maintenance strategy with reliability level control is more complex in terms of organization and execution. Based on the estimates of the values of the radio equipment reliability indicators, a decision is made as to which parameters of the entire maintenance system should be changed or adjusted (volume of work, frequency of work, threshold levels of operating tolerances, and others).
3. Maintenance using additional data processing on diagnostic variables and reliability indicators by forming predictive estimates of these parameters. In general, the condition-based maintenance strategy is a simplified version of the application of forecasting procedures for the technical condition of radio equipment. The numerical values of the predictive estimates in a more complete form can be obtained using special data processing algorithms that implement certain forecasting methods (maximum likelihood, least squares method, exponential smoothing method, method of moments, and others).

For the predictive analytics option, two processing variants are also possible: with two and four decision thresholds. Options with a larger number of preventive thresholds are also possible, which allow for more detailed tracking of changes in trends of diagnostic variables.

In general, the considered list of options for organizing and implementing maintenance can be considered in two aspects. The first aspect is associated with the use of manual control operations without automation tools. The second aspect involves the use of automated processing based on special computing tools (built into the radio equipment or external devices). The advantage of automation is obvious. It allows for increased efficiency, and the possibility of implementing complex additional processing, for example, forecasting technical conditions.

When considering the issues of initial design or modernization of the OS, it is necessary to determine a number of relevant features:

1. The content of synthesis and analysis tasks. We will assume that these tasks are implemented as follows. First, it is necessary to synthesize a number of alternative options for maintenance strategies. Next, for each alternative option, an efficiency analysis should be performed. After that, based on a criterion or a list of efficiency criteria, we select the best option. We believe that synthesis and analysis are processes, therefore they have a certain sequence of actions.
2. Selection of efficiency indicators and criteria. We believe that efficiency indicators allow to assess the degree of achievement of the goal of the maintenance system. The criterion is a rule for comparing alternative options and choosing the best one. There are single-criteria and multi-criteria analysis tasks. In this paper, we will consider the single-criteria variant of efficiency analysis. The efficiency indicator is the average costs of implementing maintenance procedures.
3. Selection of the model used to describe trends of changes in reliability indicators and diagnostic variables. In this paper, for simplicity of calculations, we assume that the condition of radio equipment is characterized by one diagnostic variable. The trend model is stochastic. This assumption is adequate for the physical processes of power supply instability, random occurrence of failures and malfunctions, unstable electromagnetic compatibility, and incorrect actions of operating personnel.

Within the framework of the trend model, we will distinguish two options: option 1 without changepoint and option 2 with changepoint. We understand the changepoint as the process of the presence of several intervals of quasi-stationarity of the parameters of the probability density function (PDF).

We assume that within one interval of quasi-stationarity, the PDF does not change and its parameters remain stable. For option 1, we will use a normal PDF over the entire observation interval. For option 2, a trend with PDF parameters different from the previous one begins at a random point in time.

For simple variants of diagnostic variables model, the problems of efficiency analysis can be solved analytically. However, for more complex variants of models and decision-making schemes, it is proposed to use statistical simulation.

An important issue during the implementation of the prediction procedure is the choice of the training sample. In this study, it is formed in a sliding window. The accuracy of prediction is determined by the coincidence of models on the training interval and the forecasting interval.

It should be noted that the probabilities of decision-making will be dependent since the statistics at two neighboring points in the decision process will be updated gradually. Analytically, this effect can be taken into account by using single, double, and multi-connected Markov chains. This analysis is quite complex, so it is appropriate to use statistical simulation.

During the synthesis of the data processing algorithm with prediction, it is necessary to detect the moment of the changepoint in time. We believe that the prediction will be effective if the models of the primary process and the models in the forecasting interval are the same. If the changepoint objectively occurred, then all previous sample values before its occurrence must be discarded from the sliding window and a new window must be formed. Until a homogeneous sample is formed, the forecasting results may contain an error regarding future trend values.

When synthesizing data processing algorithms, three options are possible:

- strict analytical synthesis of a complex data processing structure;
- heuristic synthesis;
- mixed synthesis.

The heuristic approach is based on intuition, experience in solving similar problems, the use of artificial intelligence, and others. In this paper, a third approach was used, which combines strict synthesis and heuristic procedures.

We believe that the synthesis problem is solved if a structural diagram of the interconnection of individual data collecting and processing operators is formed. In the processing procedure, erroneous decisions regarding the technical condition of radio equipment are possible. Errors can occur because the initial data are stochastic, and the algorithms are not optimal.

From a methodological point of view, it is possible to form a step-by-step procedure for the synthesis of data processing algorithms in the OS, which includes the following sequence of actions:

- definition of the maintenance object (a set of radio equipment, one sample, or part);
- definition of the purpose of the functioning of the data processing system in the OS;
- justification of efficiency indicators and the required levels of their numerical values;
- selection of efficiency criteria;
- selection of parameter models that characterize the functioning of the maintenance object;
- definition of requirements for parameters and characteristics of data processing structures;
- development of a scheme for forming alternative variants of data processing structures;
- definition of a list and groups of data processing and decision-making operators;
- justification of the list and characteristics of control influences on radio equipment to restore or maintain its serviceability;
- development of a scheme for inspecting radio equipment (presence of redundant elements, built-in and external measuring equipment, backup power supply, personnel

skills, regulatory documents on operation, conditional probabilities of error α and β, and others);
- selection of maintenance strategy (scheduled, condition-based, predictive).

The step-by-step procedure for the analysis of data processing algorithm contains:

- determination of efficiency indicators and criteria within the following hierarchical levels: data processing structure, maintenance system, equipment reliability;
- selection of numerical values of parameters that will characterize the implementation of data processing operators;
- establishment of threshold levels of parameters;
- selection of analysis methods (analytical, simulation-based, mixed);
- implementation of analysis methods;
- formation of conclusions and recommendations for operation enterprises, design organizations, educational institutions, and others;
- creation of guidelines, standards, recommendations, and methods;
- organization of feedback for developers of data processing algorithms from operation enterprises, design organizations, educational institutions, and others.

Scientific justification and implementation of the synthesis and analysis stages will allow optimizing the costs of their usage and will contribute to increasing the efficiency of the operation of aviation radio equipment.

Let us consider the model of sample formation. We will assume that the control is discrete. In the case of scheduled maintenance, the control frequency is Δ. The measurement is manual.

The automatic control system also has the same time discreteness, but the data accumulation is performed with a greater discreteness τ. Therefore, during the period Δ, the accumulation and processing of samples are performed, and the sample size is

$$N = \frac{\Delta}{\tau}.$$

During automatic control, we obtain a sample of size N, so the mathematical expectation and standard deviation are immediately estimated.

The observation period L during which estimates of average costs will be performed contains K manual measurements, then $L = KN$.

Diagnostic variable model

$$y(t) = x(t) + \theta(t) + \gamma(t), \tag{1}$$

where $x(t)$ is the signal component of the trend, for example $x(t) = a_0 + a_1 t + a_2 t^2$ with parameters a_0, a_1, and a_2 that are constant for one observation period L and stochastic for different periods; $\theta(t)$ is the stochastic signal component that takes into account inaccuracy of trend description; $\gamma(t)$ is the noise component caused by measuring equipment errors.

We assume for this research that during observation period signal component $x(t) = a_0$ and it can be described by normal PDF.

We also believe that the repair operations were performed effectively. After the repair procedures was performed, the diagnostic variables are monitored. The probabilities of type I and type II errors were assumed to be zero. The model of the signal component $x(t)$ of diagnostic variable after repair will be equal to a constant but random value a_0. Then there may be several jumps in the value a_0 during the observation period.

Possible processing options during maintenance and repair are shown in Fig. 2.

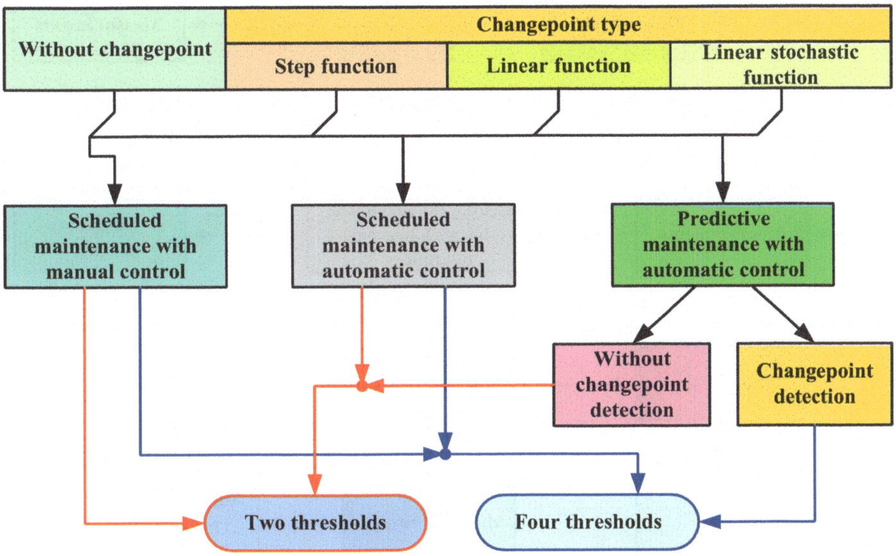

Fig. 2. Data processing for maintenance strategies.

Let us consider data processing options for different types of maintenance, presented in Fig. 2.

In the case of scheduled maintenance with two thresholds and manual control, the average operating costs are

$$C_{\Sigma}(L) = KC_{meas} + n_r(C_r + C_{meas}), \tag{2}$$

where C_{meas} is the cost of measurement, C_r is the cost of repair, n_r is the quantity of repairs during observation period.

The data processing flowchart for the scheduled maintenance strategy in case of two thresholds is shown in Fig. 3. The different actions in Fig. 3 are marked with different colors: blue is procedures, yellow is data processing, and green is decision-making.

In the case of scheduled maintenance with four thresholds and manual control, the average operating costs are

$$C_{\Sigma}(L) = KC_{meas} + n_r(C_r + C_{meas}) + n_{pm}\left(C_{pm} + C_{meas}\right), \tag{3}$$

where C_{pm} is the cost of preventive maintenance, n_{pm} is the quantity of preventive maintenances during observation period.

The data processing flowchart for the scheduled maintenance strategy in case of four thresholds is shown in Fig. 4. The type of elements in this diagram is marked with colors similar to Fig. 3.

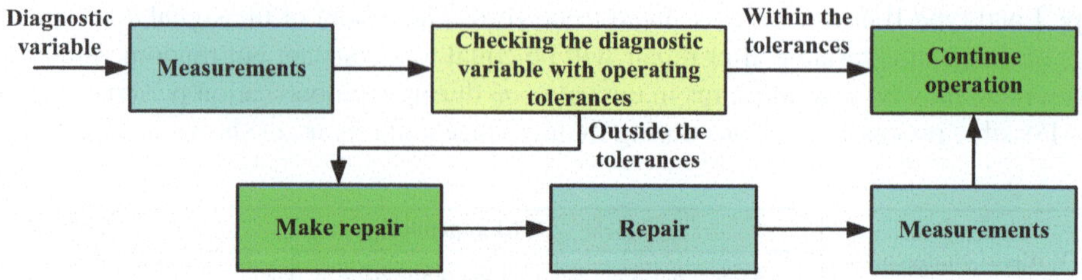

Fig. 3. The data processing flowchart for the scheduled maintenance in case of two thresholds.

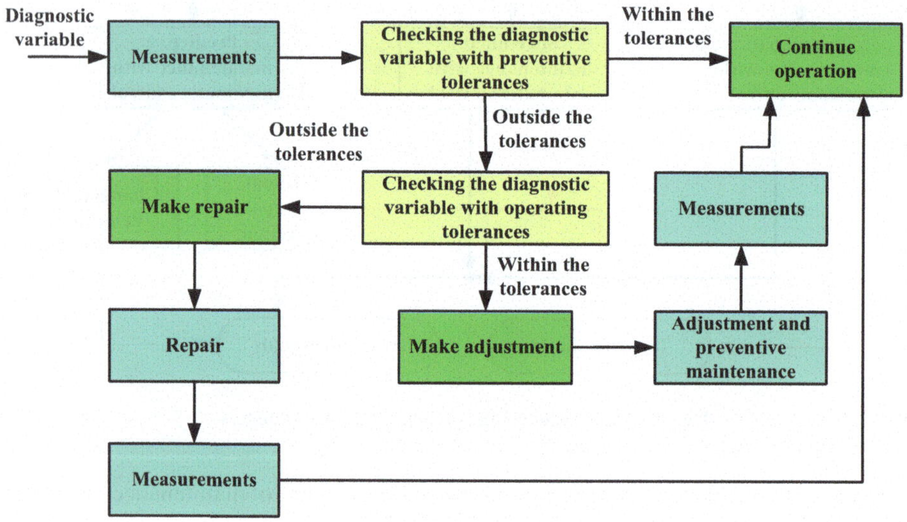

Fig. 4. The data processing flowchart for the scheduled maintenance in case of four thresholds.

Let us consider the features of the formation of the trend of the diagnostic variable for this maintenance strategy. If there is no changepoint, then the nature of the trends changes after adjustments and repairs taking into account the Gaussian law of the PDF for the parameter a_0. If the changepoint occurs, then the nature of the trends changes according to the law of deterioration of the technical condition of the equipment. In this case, the expected value of the parameter a_0 will not be equal to zero but will change according to the random duration of the quasi-stationarity interval and the random value of the parameter a_0.

The data processing flowcharts for the automatic control maintenance strategy options will be the same as shown in Figs. 3 and 4. In this case, the costs of measurement will be much lower since there will be no operational personnel. The average operating costs in this case will be determined similarly to Eqs. (2) and (3), however, the volume of measurements in them will be equal to L, and the cost of measurement will be equal to the cost of automatic control.

If data processing is performed without the sliding window, then it will be impossible to organize optimal detection of the changepoint. At the same time, operating costs will still be reduced by performing automatic measurements and calculating trend parameters.

After repairs and adjustments, we can calculate the average trend value. Note that this procedure can also be performed by operating personnel in cases where there is no automatic control scheme. In this case, the quality of processing is the same but the labor intensity of the procedures is different.

The sliding window is necessary to implement the prediction algorithm. In this case, prediction can be performed regardless of whether the changepoint is detected or not. However, detection results impact on the accuracy of prediction.

The data processing flowchart for the predictive maintenance strategy in case of two thresholds is shown in Fig. 5. The type of elements in this diagram is marked with colors similar to Fig. 3.

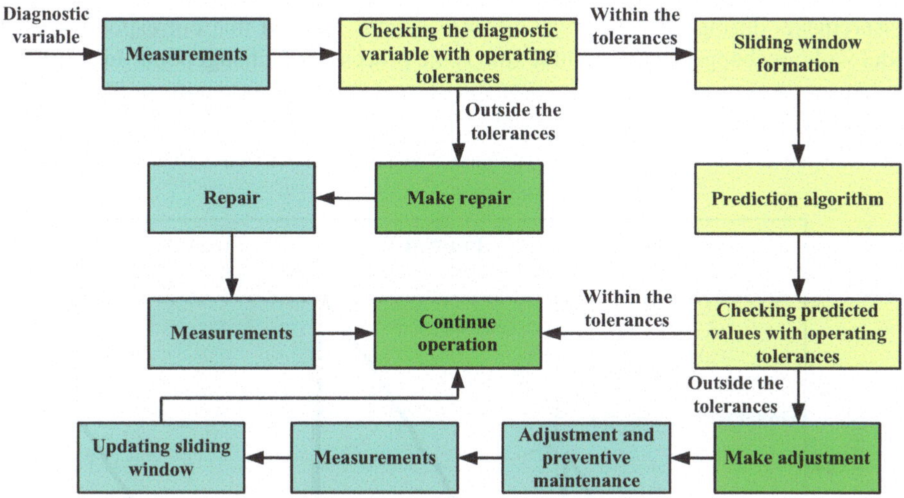

Fig. 5. The data processing flowchart for the predictive maintenance in case of two thresholds.

In the case of predictive maintenance strategy, the average operating costs are

$$C_{\Sigma}(L) = LC_{ac} + n_r(C_r + C_{ac}) + n_{ppm}\left(C_{pm} + C_{ac}\right), \tag{4}$$

where C_{ac} is the cost of automatic control, n_{ppm} is the quantity of predicted preventive maintenances during observation period.

The quantity of predictive maintenance procedures will depend on the quality of the prediction algorithm and the probability of correct detection of the changepoint. This value can be either comparable to the quantity of preventive maintenances in the four-threshold scheduled maintenance strategy or much smaller it.

3 Results and Discussions

Let us consider the basic issues of statistical simulation. For this purpose, the imitational model was developed. The imitational model gives the possibility to calculate the average operating costs for different variants of initial data. In this research, we used the following values: $C_{meas} = 100$ USD, $C_{ac} = 0.01$ USD, $C_r = 5000$ USD, $C_{pm} = 300$ USD, $K = 30$, $N = 1440$, expected value and standard deviation of parameter a_0 are 220 and 2, operating thresholds are 260 and 180, and level of noise is 0.5. The imitational model simulates three types of maintenance: scheduled with manual measurements, scheduled with automatic control, and predictive. All types of maintenance correspond to two threshold case.

At the first step of imitation, we obtained the realization of diagnostic variable. After every interval of observation, the parameter a_0 of the diagnostic variable will be equal to the random value. The same happens after the predictive maintenance and repair. Figure 6 presents examples of diagnostic variable trends for scheduled maintenance with manual measurements during three intervals of observation. The red line corresponds to the model without changepoint. The blue line is the process with changepoint. In this case, the changepoint occurs at a random moment of time (according to uniform PDF) with a constant inclination angle of linear trend.

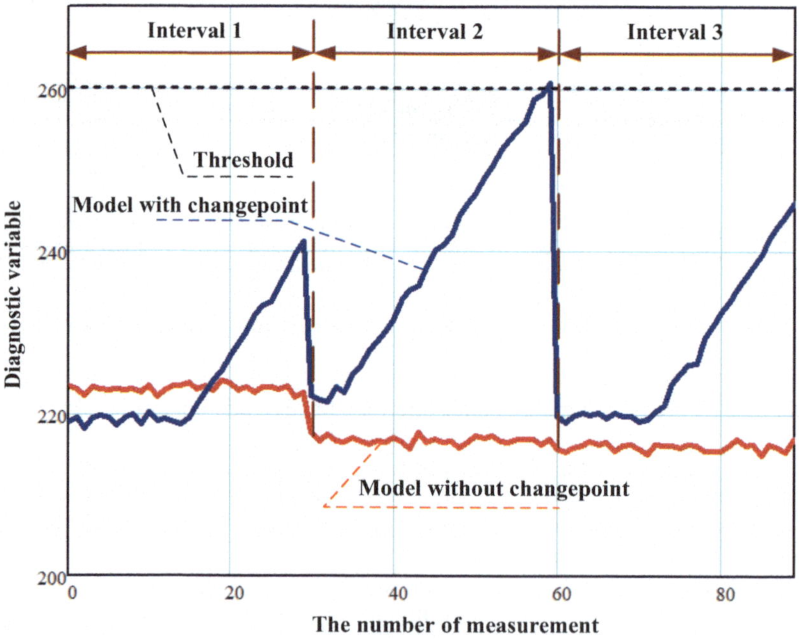

Fig. 6. The examples of diagnostic variable trends for scheduled maintenance with manual measurements during three intervals of observation.

The prediction algorithm was implemented in the sliding window. For simplicity, no preprocessing was performed to detect the changepoint. The width of the sliding window was chosen to be 50. The forecasting algorithm was reduced to a linear approximation of the trend of the diagnostic variable using the least squares method and determination of the moment of crossing the operational threshold. Figures 7 and 8 show the dependence of average operating costs on the cost of measurement and cost of repair. The dependencies were obtained for 10,000 iteration of simulation process.

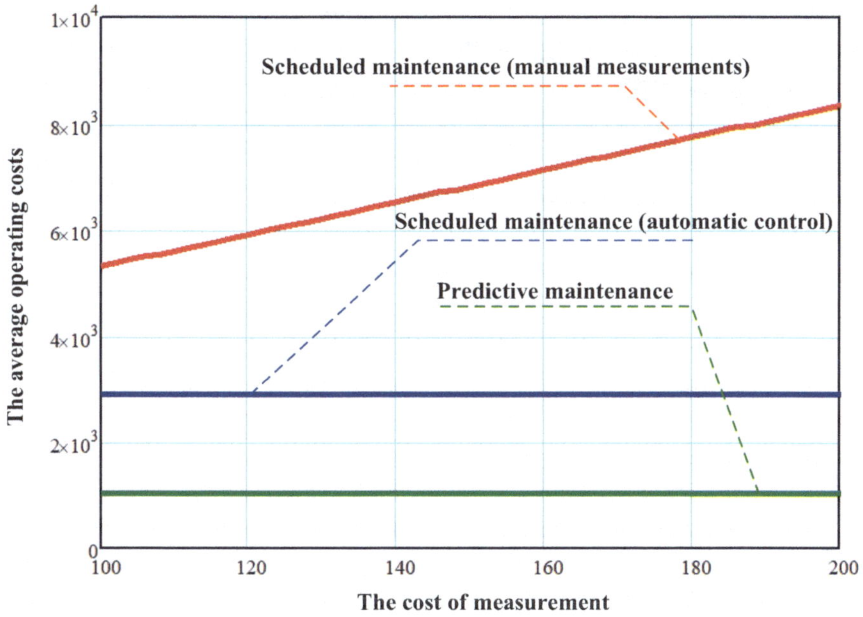

Fig. 7. The average operating costs (dependence on the cost of measurement).

Analysis of the graphs in Fig. 7 showed that for different values of the measurement cost, the average operating costs do not change. The best option is predictive maintenance. The gain of predictive maintenance compared to maintenance with automatic control is 64.1%. At the same time, the use of automatic control reduces the cost of scheduled maintenance by at least 45.2%. Moreover, this gain increases as the measurement cost increases. It should be noted that it is possible to calculate the marginal cost of measurement that provides the same average costs as the maintenance with automatic control. In this case, this value is 14.2 USD.

Analysis of the graphs in Fig. 8 showed that predictive maintenance has a gain in average operating costs compared to maintenance with automatic control of at least 60.9%. This gain increases with increasing repair costs. Automatic control improves the efficiency of scheduled maintenance by at least 34.7%. This gain increases with decreasing repair costs.

The conducted simulation proved the efficiency of predictive maintenance in terms of operating cost minimization.

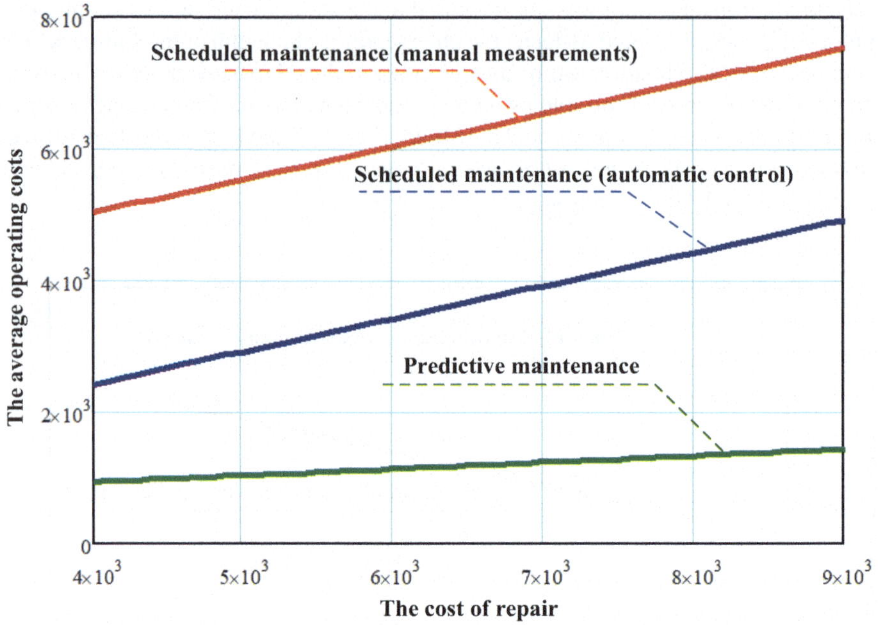

Fig. 8. The average operating costs (dependence on the cost of repair).

4 Conclusions

The paper discusses the comparative analysis of maintenance and repair strategies for aviation radio equipment. The main attention is paid to data processing algorithms during the application of predictive maintenance.

The simulation results showed that automatic control increases the efficiency of the OS in terms of minimizing operating costs. Data pre-processing to detect the changepoint at the same time improves the accuracy of predicting future equipment states. The comparative analysis of different maintenance strategies was performed for the linear model of deterioration of equipment condition. Sliding window processing for such a model of the diagnostic variable is effective in terms of timeliness and accuracy of detecting the changepoint.

The presented results can be used when improving the OS of aviation radio equipment in terms of the application of information technologies for algorithmic support of management decision-making in aviation enterprises.

5 Future Scope

Future research will be aimed at:

- analysis of the efficiency of predictive maintenance for different models of changepoint;
- development of sequential data processing procedures;

- development of a methodology for the simultaneous application of analytical, computational, and modeling approach to OS research;
- improvement of the structure of data hubs for collecting, processing, and using statistical data;
- research of the impact of measurement equipment errors on the efficiency of maintenance and repair processes.

Acknowledgments. This research is partially supported by the Ministry of Education and Science of Ukraine under the projects "Methods of building protected multilayer cellular networks 5G/6G based on the use of artificial intelligence algorithms for monitoring country's critical infrastructure objects" (# 0124U000197) and is partially supported by EURIZON project # 871072 (Project EU #3035 EURIZON "Research and development of Ukrainian ground network of navigational aids for increasing the safety of civil aviation").

References

1. Galar, D., Sandborn, P., Kumar, U.: Maintenance Costs and Life Cycle Cost Analysis. CRC Press, Boca Raton (2017)
2. Acemoglu, D., Laibson, D., List, J.A.: Economics. Pearson, London (2018)
3. Eriksen, E.L., Isaksen, A.: The emergence of new industries at the regional level: alignment of organizational and regional industrial culture. Reg. Stud. Reg. Sci. **8**(1), 387–401 (2021). https://doi.org/10.1080/21681376.2021.1987976
4. Andre, J.: Industry 4.0: Paradoxes and Conflicts. Wiley, New York (2019)
5. Melo, L.d.S., et al.: Industry 4.0 and sustainability: empirical validation of constructs of industry technology and sustainable development. Platforms. **2**, 150–164 (2024). https://doi.org/10.3390/platforms2040010
6. Goncharenko, A.V.: Several models of artificial intelligence elements for aircraft control. In: 4th International Conference on Methods and Systems of Navigation and Motion Control, pp. 224–227. IEEE, Kiev (2016). https://doi.org/10.1109/MSNMC.2016.7783148
7. Ostroumov, I.V., Kuzmenko, N.S.: Compatibility analysis of multi signal processing in APNT with current navigation infrastructure. Telecommun. Radio Eng. **77**(3), 211–223 (2018). https://doi.org/10.1615/TelecomRadEng.v77.i3.30
8. Stroeve, S., Smeltink, J., Kirwan, B.: Assessing and advancing safety management in aviation. Safety. **8**(20), 1–26 (2022). https://doi.org/10.3390/safety8020020
9. Holubnychyi, O., et al.: Self-organization technique with a norm transformation based filtering for sustainable infocommunications within CNS/ATM systems. In: Ostroumov, I., Zaliskyi, M. (eds.) Proceedings of the 2nd International Workshop on Advances in Civil Aviation Systems Development. ACASD 2024, Lecture Notes in Networks and Systems, vol. 992, pp. 262–278. Springer, Cham (2024). https://doi.org/10.1007/978-3-031-60196-5_20
10. Bartulovic, D., Steiner, S.: Conceptual model of predictive safety management methodology in aviation. Aerospace. **10**(268), 1–46 (2023). https://doi.org/10.3390/aerospace10030268
11. Ivashchuk, O., et al.: A configuration analysis of Ukrainian flight routes network. In: 16th International Conference on the Experience of Designing and Application of CAD Systems, pp. 6–10. IEEE, Lviv (2021). https://doi.org/10.1109/CADSM52681.2021.9385263
12. Zyluk, A., Cur, K., Tomaszewska, J., Czerwinski, T.: Analysis of aircraft operation system regarding readiness—case study. Aerospace. **9**(14), 1–10 (2022). https://doi.org/10.3390/aerospace9010014

13. Habib, K.A., Turkoglu, C.: Analysis of aircraft maintenance related accidents and serious incidents in Nigeria. Aerospace. **7**(178), 1–28 (2020). https://doi.org/10.3390/aerospace7120178
14. Zaliskyi, M., et al.: Methodology for substantiating the infrastructure of aviation radio equipment repair centers. In: CEUR Workshop Proceedings, vol. 3732, pp. 136–148 (2024). https://ceur-ws.org/Vol-3732/paper11.pdf
15. Solomentsev, O., et al.: Efficiency analysis of current repair procedures for aviation radio equipment. In: Ostroumov, I., Zaliskyi, M. (eds.) Proceedings of the 2nd International Workshop on Advances in Civil Aviation Systems Development. ACASD 2024, Lecture Notes in Networks and Systems, vol. 992, pp. 281–295. Springer, Cham (2024). https://doi.org/10.1007/978-3-031-60196-5_21
16. Goncharenko, A.: Multi-optional hybridization for UAV maintenance purposes. In: 5th International Conference Actual Problems of Unmanned Aerial Vehicles Developments, pp. 48–51. IEEE, Kiev (2019). https://doi.org/10.1109/APUAVD47061.2019.8943902
17. Solomentsev, O., Zaliskyi, M., Sushchenko, O., Bezkorovainyi, Y., Averyanova, Y., Ostroumov, I., et al.: Data processing through the lifecycle of aviation radio equipment. In: IEEE 17th International Conference on Computer Sciences and Information Technologies, pp. 146–151. IEEE, Lviv (2022). https://doi.org/10.1109/CSIT56902.2022.10000844
18. Solomentsev, O., Zaliskyi, M., Zuiev, O.: Estimation of quality parameters in the radio flight support operational system. Aviation. **20**(3), 123–128 (2016). https://doi.org/10.3846/16487788.2016.1227541
19. Jardine, A.K.S., Tsang, A.H.C.: Maintenance, replacement, and reliability: theory and applications, 2nd edn. CRC Press, Boca Raton (2017)
20. Tachinina, O., Lysenko, O., Alekseeva, I., Guida, O., Novikov, V., Sushyn, I.: Methodology of energy-efficient algorithmic modernization of dynamic characteristics of digital electric drives of mechatronic devices for small civil aviation. In: Ostroumov, I., Zaliskyi, M. (eds.) Proceedings of the International Workshop on Advances in Civil Aviation Systems Development. ACASD 2023, Lecture Notes in Networks and Systems, vol. 736, pp. 152–165. Springer, Cham (2023). https://doi.org/10.1007/978-3-031-38082-2_12
21. Sushchenko, O., et al.: Algorithm of determining errors of gimballed inertial navigation system. In: Gervasi, O., et al. (eds.) Computational Science and Its Applications—ICCSA 2024 Workshops. ICCSA 2024 Lecture Notes in Computer Science, vol. 14816, pp. 206–218. Springer, Cham (2024). https://doi.org/10.1007/978-3-031-65223-3_14
22. Ostroumov, I.V., Kuzmenko, N.S.: Accuracy improvement of VOR/VOR navigation with angle extrapolation by linear regression. Telecommun. Radio Eng. **78**(15), 1399–1412 (2019). https://doi.org/10.1615/TelecomRadEng.v78.i15.90
23. Tachinina, O.M., Lysenko, O.I., Alekseeva, I.V.: Algorithm for operational optimization of two-stage hypersonic unmanned aerial vehicle branching path. In: IEEE 5th International Conference on Methods and Systems of Navigation and Motion Control (MSNMC), pp. 11–15. IEEE, Kiev (2018). https://doi.org/10.1109/MSNMC.2018.8576319
24. Pop, G.I., Titu, A.M., Pop, A.B.: Enhancing aerospace industry efficiency and sustainability: process integration and quality management in the context of Industry 4.0. Sustainability. **15**, 16206, 1–20 (2023). https://doi.org/10.3390/su152316206
25. Bakhtiiarov, D., Konakhovych, G., Lavrynenko, O.: Protected system of radio control of unmanned aerial vehicle. In: 4th International Conference on Methods and Systems of Navigation and Motion Control (MSNMC), pp. 196–199. IEEE, Kiev (2016). https://doi.org/10.1109/MSNMC.2016.7783141
26. Meeus, J., Dewulf, W., Macario, R.: Management systems in aviation: challenges and opportunities to upgrade to an integrated management system. Sustainability. **15**(10424), 1–23 (2023). https://doi.org/10.3390/su151310424

27. Dhillon, B.: Reliability, Quality, and Safety for Engineers. CRC Press, Boca Raton (2005)
28. Sushchenko, O., Bezkorovainyi, Y., Golitsyn, V., Kuzmenko, N., Averyanova, Y., Zaliskyi, M., et al.: Integration of MEMS inertial and magnetic field sensors for tracking power lines. In: XVIII International Conference on the Perspective Technologies and Methods in MEMS Design, pp. 33–36. IEEE, Polyana (Zakarpattya) (2022). https://doi.org/10.1109/MEMSTECH55132.2022.10002907
29. Solomentsev, O., Zaliskyi, M., Zuiev, O.: Radioelectronic equipment availability factor models. In: Signal Processing Symposium 2013 (SPS 2013), pp. 1–4. IEEE, Serock (2013). https://doi.org/10.1109/SPS.2013.6623616
30. Prokopenko, I.: Nonparametric change point detection algorithms in the monitoring data. In: Hu, Z., Petoukhov, S., Dychka, I., He, M. (eds.) Advances in Computer Science for Engineering and Education IV. ICCSEEA 2021, Lecture Notes on Data Engineering and Communications Technologies, vol. 83, pp. 347–360. Springer, Cham (2021). https://doi.org/10.1007/978-3-030-80472-5_29
31. Zaliskyi, M., et al.: Heteroskedasticity analysis during operational data processing of radio electronic systems. In: Shukla, S., Unal, A., Varghese Kureethara, J., Mishra, D.K., Han, D.S. (eds.) Data Science and Security Lecture Notes in Networks and Systems, vol. 290, pp. 168–175. Springer, Singapore (2021). https://doi.org/10.1007/978-981-16-4486-3_18
32. Ostroumov, I., Ivannikova, V., Kuzmenko, N., Zaliskyi, M.: Impact analysis of Russian-Ukrainian war on airspace. J. Air Transp. Manag. **124**, 102742 (2025). https://doi.org/10.1016/j.jairtraman.2025.102742
33. Larin, V., Solomentsev, O., Zaliskyi, M., Shcherban, A., Averyanova, Y., Ostroumov, I., et al.: Prediction of the final discharge of the UAV battery based on fuzzy logic estimation of information and influencing parameters. In: 3rd KhPI Week on Advanced Technology, pp. 1–6. IEEE, Kharkiv (2022). https://doi.org/10.1109/KhPIWeek57572.2022.9916490
34. Averyanova, Y., Larin, V., Kuzmenko, N., Ostroumov, I., Zaliskyi, M., Solomentsev, O., et al.: Turbulence detection and classification algorithm using data from AWR. In: IEEE 2nd Ukrainian Microwave Week (UkrMW), pp. 518–522. IEEE, Kharkiv (2022). https://doi.org/10.1109/UkrMW58013.2022.10037172

The Determination of the Probability of Failure-Free Operation for System Operated in Two Modes

Petro Yablonskyi[1], Oleksandr Avramenko[1], Valentyn Dyptan[1], Pavlo Openko[1], Serhii Pustovyi[2], and Volodymyr Klymchuk[3](✉)

[1] The National Defence University of Ukraine Named After Ivan Cherniakhovskyi, Povitryanikh Sil Av., 28, Kyiv 03049, Ukraine
[2] LLC "Adron-DAM", Vyshniakivska Str., 9, Kyiv 02140, Ukraine
[3] National Aviation University, Liubomyra Huzara Ave., 1, Kyiv 03058, Ukraine
klimchuk1591@ukr.net

Abstract. There are many technical systems that operate with varying usage modes. The variable usage modes of technical systems in aircraft involve flexible adjustment and adaptation of systems to different flight conditions and types of tasks. This, in particular, allows for optimizing fuel consumption, improving flight efficiency, and increasing flight safety. Special attention is paid to the reliability of such systems. It is known that reliability refers to the ability of a technical system to maintain within specified limits all parameters that characterize its ability to perform required functions under given modes and conditions of use, maintenance, storage, and transportation. This paper discusses the technical systems of aircraft that operate in a two-shift mode, assuming that the time spent in each mode is non-random. It is also assumed that the exponential distribution law applies to each of the two modes. Taking into account these constraints and assumptions, and using a semi-Markov random process, a mathematical model has been developed for calculating the reliability of technical systems in aircraft that operate in variable operating mode.

Keywords: Technical System · Usage Mode · Reliability · Distribution Function · Probability of Failure-Free Operation · Semi-Markov Process · Mathematical Model

1 Introduction

1.1 Introduction to the Problem

For any technical system, one of the most important indicators of its quality is reliability [1, 2]. It is known that reliability refers to the ability of a technical system to maintain within specified limits all parameters that characterize its ability to perform required functions under given modes and conditions of use, maintenance, storage, and transportation [3, 4]. Reliability is a complex property, which, depending on the purpose of

I. Ostroumov et al. (Eds.): ACASD 2025, LNNS 1418, pp. 364–375, 2025.
https://doi.org/10.1007/978-3-031-91992-3_24

the technical system and its operating conditions, may include failure-free operation, durability, maintainability, and preservation, or certain combinations of these properties [5, 6].

At various times, a significant amount of scientific research has been dedicated to the reliability of technical systems, including those for military purposes [7–9]. The development of reliability models for military technical systems, which take into account the typical usage cycle, is discussed in [10–12]. In this case, reliability characteristics are considered with respect to a fixed calendar duration of operation, typically during the stage of normal operation, which is characterized by a constant failure rate [13, 14]. However, during prolonged operation, when military equipment exhausts its technical resource, issues of repair or replacement with new systems arise, which requires taking into account the dependence of reliability indicators on the calendar duration of operation and other factors [15, 16]. In article [17], the impact of the reliability of mobility assets in military technical systems on the assessment of the efficiency preservation coefficient of military technical systems is considered. In article [18], a model of the effectiveness of military technical systems is presented, which incorporates the parameters of the typical usage cycle. It is assumed that the reliability model, which describes failure and recovery processes of military technical systems during the stage of combat readiness, is well-known. The impact of periodic maintenance and checks on the functioning of military technical systems on reliability is not considered.

The study of the reliability of aircraft, including military ones, is a fundamental direction in modern science. The reliability of aircraft is not just a desirable attribute, but a vital requirement. Clearly, the failure of any of the aircraft's systems can lead to catastrophic consequences.

A significant number of technical systems in aircraft operate in variable usage modes. Variable usage modes involve flexible adjustment and adaptation of systems to different flight conditions and types of tasks. This, in particular, allows for optimizing fuel consumption, improving flight efficiency, and enhancing flight safety. Typical examples of technical systems in aircraft with variable usage modes include: engine control systems (takeoff mode, cruising mode, climb mode, descent mode, landing approach mode), power supply systems (normal mode, emergency power mode, reduced consumption mode), air conditioning systems (cooling mode, heating mode, ventilation mode), navigation systems (autopilot mode, manual mode, approach mode to the airport), and others. Each of the usage modes of technical systems in aircraft is characterized by specific time intervals of operation and reliability indicators.

Recently, semi-Markov random processes have been widely used to study the reliability of technical systems, including aircraft [11, 19]. A semi-Markov random process extends the concept of a Markov process and allows for more flexible modeling of situations where transitions between states in the model can be complex and variable.

The results of studying the existing scientific and methodological framework in the field of aircraft reliability allow for the conclusion that the development of a mathematical model for calculating the reliability of technical systems in aircraft operating in variable modes, using a semi-Markov random process, is a relevant scientific task.

1.2 Motivation

Failure-free operation of the radio-electronic equipment system is critically important for the safety and efficiency of aircraft flights. It not only determines the successful completion of the flight mission but also the lives of people on board and on the ground.

Scheduled maintenance is an integral part of the safe operation of any aircraft, regardless of its purpose or complexity. Timely performed scheduled maintenance helps prevent potential failures of the aircraft. According to the aircraft's technical maintenance schedule, scheduled maintenance is carried out at non-random time intervals. Preventive and repair work is performed according to technological charts and, in most cases, also lasts for a non-random time interval.

In the paper [20], the problem of determining the probability of failure-free operation of technical systems operating in two variable modes is addressed using the exponential distribution law, where the system stays in each mode for a random amount of time. In the paper [21], the issues of assessing the reliability of technical systems operating in variable modes were partially considered. It was assumed that in each of the operating modes, the technical system works for a non-random time interval, and the distribution function of the failure-free operating time follows an exponential distribution.

This paper is dedicated to the development of a mathematical reliability model using a semi-Markov random process, which will allow for calculating the probability of failure-free operation for the radio-electronic equipment systems of aircraft operating in variable modes, with two operating modes and specific values of time and failure intensities in each mode.

1.3 Contribution

This paper contributes to the development of the theory of using semi-Markov random processes for reliability calculations of aircraft technical systems. The practical value of the research lies in the development of an algorithm for determining the probability of failure-free operation of radio-electronic equipment systems in aircraft operating in two-shift modes, with an exponential distribution law and non-random time spent in each mode. This will enable the operators of aircraft radio-electronic equipment systems to assess their reliability effectively.

1.4 The Organization of the Paper

The first section of this paper presents the main provisions and references to studies in the field of aircraft reliability, provides examples of aircraft systems that operate in variable modes, and justifies the appropriateness of using semi-Markov random processes in mathematical modeling. The second section contains information about the main formulaic dependencies and characteristics of the developed mathematical model. The third section focuses on the errors made in the preceding studies and presents the results of the mathematical calculations carried out. The fourth section provides conclusions and summarizes the research results.

2 Materials and Methods

We will assume that the distribution function of the failure-free operating time for the radio-electronic equipment systems of aircraft, according to the State Standard of Ukraine [22], follows an exponential form.

Let the failure intensity of the radio-electronic equipment system of the aircraft operating in two variable modes be denoted as λ_0 for the first mode and λ_1 for the second mode. The discrepancy in the values of λ_0 and λ_1 is explained by the differing number of elements in each operating mode. We assume that the system, when operating in the first mode, can function without failure for a random time η_0, and in the second mode, for a random time η_1.

In the paper, it is assumed that the random variables η_0 and η_1 are distributed according to an exponential distribution.

$$P(\eta_i \leq t) = \begin{cases} 1 - e^{-\lambda_i t}, \text{if } t > 0, \\ 0 \text{ otherwise.} \end{cases} \quad (1)$$

We will assume that the system is in the first mode for a time ξ_0, and then switches to the second operating mode for a time ξ_1. After that, the system switches back to the first mode, and so on.

Let $P(\xi_0 < t) = F_0(t)$ and $P(\xi_1 < t) = F_1(t)$ be the distribution functions of the random variables ξ_0 and ξ_1, respectively. The Laplace transforms for the functions $F_0(t)$ and $F_1(t)$ are given by:

$$\varphi_0(s) = \int_0^\infty e^{-stdF_0(t)}.$$

$$\varphi_1(s) = \int_0^\infty e^{-stdF_1(t)}.$$

To determine the probability of failure-free operation of such a system, we apply a semi-Markov random process.

According to [23], a semi-Markov random process with a finite number of states $E = (e_1, e_2, \ldots e_i \ldots e_n)$ is defined by the duration of stay in states e_i, $i = \overline{1, n}$ with distribution functions $H_i(t) = P(\xi_i \leq t)$ and conditional transition probabilities $q_{ij}(t)$ from state e_i to state e_j, given that the semi-Markov process was in state e_i for time t.

Let E_1 be a fixed subset of the states of the semi-Markov random process $E_i < E$, $E = E_1$.

In [23], it is shown that the expected value $\overline{\tau_i} = M\tau_i$ satisfies the system of algebraic equations:

$$\overline{\tau} = M\xi_i + \sum\nolimits_{j \in E_i} P_{ij}\tau_j, \quad (2)$$

where

$$P_{ij} = \int_0^\infty q_{ij}(t) dH_i, \quad (3)$$

$$\Psi_i(s) = P_i(s) + \sum\nolimits_{j \in E_i} P_{ij}(s)[\Psi_i(s) - 1], \tag{4}$$

$$P_{ij}(s) = \int_0^\infty e^{-st} dP_{ij}(t) = \int_0^\infty e^{-st} q_{ij}(t) dH_i(t), \tag{5}$$

$$P_i(s) = \int_0^\infty e^{-st} dH_i(t) = \sum_{j=1}^{n} P_{ij}(t), \tag{6}$$

$$P_{ij}(t) = \int_0^t q_{ij}(u) dH_i(u). \tag{7}$$

In formula (2), τ_j represents the average time that the semi-Markov process spends in the subset of operational states. Let us introduce the following state definitions for the semi-Markov process:

- e_0 – the system is in the first operating mode, and all components are operational. The time spent in state e_0 is $\xi_0 = \min(\xi_0, \eta_0)$, meaning the system remains in this mode and functional until it either switches to the second operating mode or a failure occurs. One of these events leads to the transition of the system from state e_0;
- e_1 – the system is in the second operating mode, and all components are operational. The system stays in this state until it either switches back to state e_0, or a failure occurs. The time spent by the semi-Markov process in state e_1 is $\xi_1 = \min(\xi_1, \eta_1)$;
- e_2 – the system is in a failure state. The time spent by the semi-Markov process in state e_2 is ξ_2.

The probability of failure-free operation can be written as:

$$P(\tau_1 > T) = R_i(T), i = 0, 1. \tag{8}$$

We will assume that at the initial time, the system is in the first operating mode. Let's introduce some notations:

$$P(\tau_i < T) = Q_0(T), P(\tau_1 < t) = Q_1(T), \tag{9}$$

$$R_0(T) = 1 - Q_0(T); R_1(T) = 1 - Q_1, \tag{10}$$

$$\Psi_0(s) = \int_0^\infty e^{-st} dQ_0(t); \Psi_1(s) = \int_0^\infty e^{-st} dQ_1(t). \tag{11}$$

The conditional transition probabilities $q_{ij}(t)$ are given by:

$$q_{01}(t) = e^{-\lambda_0 t}; q_{10}(t) = e^{-\lambda_1 t}; q_{02}(t) = 1 - e^{-\lambda_0 t}; q_{12}(t) = 1 - e^{-\lambda_1 t}.$$

We define the transition probabilities $P_{ij}(t)$ from (7) as:

$$P_{01}(t) = \int_0^t e^{-\lambda_0 t} dF_0(u); P_{10}(t) = \int_0^t e^{-\lambda_1 t} dF_1(u).$$

According to (6):

$$P_{01}(s) = \int_0^\infty e^{-st} dP_{01}(t) = \int_0^\infty e^{-(s+\lambda_0)t} dF_0(t) = F_0(s+\lambda_0).$$

$$P_{10}(s) = \int_0^\infty e^{-st} dP_{10}(t) = \int_0^\infty e^{-(s+\lambda_1)t} dF_1(t) = F_1(s+\lambda_1).$$

From Eq. (6), we define$P_i(s)$ as:

$$P_0(s) = \int_0^\infty e^{-st} dP[\min(\xi_{01}, \eta_0)].$$

$$P_0(s) = \int_0^\infty e^{-st\lambda_0} e^{-\lambda_0 t}(1 - F_0(t))dt.$$

$$P_0(s) = \lambda_0 \int_0^\infty e^{-(s+\lambda_0)t}[1 - F_0(t)]dt + F_0.$$

In the last expression, we integrate by parts and obtain:

$$P_0(s) = \frac{sF_0(s+\lambda_0)}{s+\lambda_0}.$$

Similarly, it can be obtained:

$$P_1(s) = \frac{sF_1(s+\lambda_1) + \lambda_1}{s+\lambda_1}.$$

Therefore:

$$P_i(s) = \frac{sF_i(s+\lambda_i) + \lambda_i}{s+\lambda_i}$$

for $i = 0, 1$.

After substituting the values for $P_{ij}(s)$, $P_i(s)$, and $\Psi_i(s)$ into the system of Eq. (4), we obtain:

$$\Psi_0(s) = \frac{sF_0(s+\lambda_0) + \lambda_0}{s+\lambda_0} + F_0(s+\lambda_0)[\Psi_1(s) - 1].$$

$$\Psi_1(s) = \frac{sF_1(s+\lambda_1) + \lambda_1}{s+\lambda_1} + F_1(s+\lambda_1)[\Psi_0(s) - 1].$$

From these equations we find that:

$$\Psi_0(s) = \frac{\lambda_0[1 - F_0(\lambda_0 + s)]}{(\lambda_0 + s)[1 - F_0(\lambda_0 + s)F_1(\lambda_1 + s)]} + \frac{\lambda_1[1 - F_1(\lambda_1 + s)]F_0(\lambda_0 + s)}{(\lambda_1 + s)[1 - F_0(\lambda_0 + s)F_1(\lambda_1 + s)]}. \tag{12}$$

Formula (12) holds under the condition that the time the system spends in the first and second modes is arbitrary [24].

Next, we will assume that the system stays in the first mode for a fixed time T_0 and in the second mode for a fixed time T_1, meaning we will assume that

$$F_i(t) = \begin{cases} 0 \text{ for } t < T_i, \\ 1 \text{ for } t \geq T_i. \end{cases} \text{ for } i = 0, 1. \tag{13}$$

After substituting (13) into (12), we will obtain:

$$\Psi_0(s) = \frac{\lambda_0\left(1 - e^{-(s+\lambda_0)T_0}\right)}{(s+\lambda_0)\left[1 - e^{-(s+\lambda_0)T_0}e^{-(s+\lambda_1)T_1}\right]} + \frac{\lambda_1\left[1 - e^{-(s+\lambda_1)T_1}\right]e^{-(s+\lambda_0)T_0}}{(s+\lambda_1)\left[1 - e^{-(s+\lambda_0)T_0}e^{-(s+\lambda_1)T_1}\right]}. \tag{14}$$

With the presence of the Stieltjes-Laplace transformation, we can find the Laplace transform, namely:

$$LP(t) = \int_0^\infty e^{-st}P(t)dt,$$

which is expressed through $\Psi_0(s)$ as:

$$LP(t) = \frac{1 - \Psi_0(s)}{s}. \tag{15}$$

After substituting (14) into (15), we will obtain:

$$LP(t) = \frac{1}{(s+\lambda_0)\left[1 - e^{-[(s+\lambda_0)T_0 + (s+\lambda_1)T_1]}\right]} + \frac{(\lambda_0+\lambda_1)e^{-(s+\lambda_0)T_0}}{(s+\lambda_0)(s+\lambda_1)\left[1 - e^{-[(s+\lambda_0)T_0 + (s+\lambda_1)T_1]}\right]} - \frac{e^{-[(s+\lambda_0)T_0 + (s+\lambda_1)T_1]}}{(s+\lambda_1)\left[1 - e^{-[(s+\lambda_0)T_0 + (s+\lambda_1)T_1]}\right]}. \tag{16}$$

Using the properties of linearity, the originals of $LP(t)_1$ can be found as the sum of the originals of the corresponding images.

$$LP(t)_1 = \frac{1}{(s+\lambda_0)\left[1 - e^{-[(s+\lambda_0)T_0 + (s+\lambda_1)T_1]}\right]}. \tag{17}$$

Let us denote $1 - e^{-[s(T_0+T_1)+\lambda_0T_0+\lambda_1T_1]}$ as $1 + \beta e^{-s}$, where $\beta = e^{-(\lambda_0T_0+\lambda_1T_1)}$ and $\alpha = T_0 + T_1$; $\frac{1}{s+\lambda_0} = q(s)$ – an arbitrary function, where $q(s) = \int_0^\infty e^{-st}F(t)dt$.

In formula (17), $[1 + \beta e^{-\alpha s}]^{-1}$ represents a geometric progression, which can be written as:

$$\frac{1}{1+\beta e^{-\alpha s}} = \beta e^{-\alpha s} + \beta^2 e^{-2\alpha s} - \beta^3 e^{-3\alpha s} + \beta^4 e^{-4\alpha s} + \ldots (-1)^n \beta^n e^{-n\alpha s}.$$

Thus, formula (17) can be written as:

$$LP(t)_1 \to q(s)\left[1 - \beta e^{-\alpha s} + \beta^2 e^{-2\alpha s} - \beta^3 e^{-3\alpha s} + \ldots (-1)^n \beta^n e^{-n\alpha s}\right].$$

Using the properties of superposition, we obtain the original form of (17) as:

$$LP(t)_1 \to \sum\nolimits_{0 \le n < \frac{t}{T_0+T_1}} (-1)^n \left(-e^{-(\lambda_0 T_0 + \lambda_1 T_1)}\right)^n f(K_i) e^{-\lambda_0\left(t-\left(T_0+T_1\right)n\right)}, \tag{18}$$

where $f(K_i) = \begin{cases} 0 \text{ for } t < \left(T_0 + T_1\right)n, \\ 1 \text{ for } t \ge \left(T_0 + T_1\right)n. \end{cases}$

The second component of the expression (16) can be written as:

$$LP(t)_2 = \frac{(\lambda_0 - \lambda_1)e^{-(s+\lambda_0)T_0}}{(s+\lambda_0)(s+\lambda_1)\left[1 - e^{-[(s+\lambda_0)T_0 + (s+\lambda_1)T_1]}\right]}. \tag{19}$$

Let in the last expression

$$q(s) = \frac{(\lambda_0 - \lambda_1)e^{-(s+\lambda_0)T_0}}{(s+\lambda_0)(s+\lambda_1)}. \tag{20}$$

To determine the original form of (20), we use [25] and obtain:

$$q(s) \to e^{-\lambda_0 T_0}\left(e^{-\lambda_1(t-T_0)} - e^{-\lambda_0(t-T_0)}\right).$$

The values obtained from (19) can be represented as:

$$LP(t)_2 \to \sum\nolimits_{0 \le n < \frac{t}{T_0+T_1}} (-1)^n \left(-e^{-(\lambda_0 T_0 + \lambda_1 T_1)}\right)^n f(K_j) \times$$
$$\times \left(e^{-\lambda_1\left(t-T_0-\left(T_0+T_1\right)n\right)} - e^{-\lambda_1\left(t-T_0-\left(T_0+T_1\right)n\right)}\right). \tag{21}$$

$$f(K_j) = \begin{cases} 0 \text{ for } t < T_0 + \left(T_0 + T_1\right)n, \\ 1 \text{ for } t \ge T_0 + \left(T_0 + T_1\right)n. \end{cases} \tag{22}$$

The third component of the expression (16) can be written as:

$$LP(t)_3 \to -\frac{e^{-[(s+\lambda_0)T_0 + (s+\lambda_1)T_1]}}{(s+\lambda_1)\left[1 - e^{-[(s+\lambda_0)T_0 + (s+\lambda_1)T_1]}\right]}. \tag{23}$$

Let's introduce the notation:

$$q(s) \to \frac{e^{-(\lambda_0 T_0 + \lambda_1 T_1)} - s(T_0 + T_1)}{s + \lambda_1}. \tag{24}$$

Next, we will use the delay theorem and the table of integral transforms as discussed in [26, 27], after which we will obtain:

$$q(s) \to e^{-(\lambda_0 T_0 + \lambda_1 T_1)} e^{-\lambda_1\left(t-\left(T_0+T_1\right)\right)}. \tag{25}$$

From here, we will obtain:

$$LP(t)_3 \to -\sum\nolimits_{0<n<\frac{t}{T_0+T_1}} (-1)^n\left(-e^{-(\lambda_0T_0+\lambda_1T_1)n}e^{-(\lambda_0T_0+\lambda_1T_1)}\right)$$
$$\times f(K_z)e^{-\lambda_1\left(t-\left(T_0+T_1\right)(n+1)\right)}. \quad (26)$$

$$f(K_z) = \begin{cases} 0 \text{ for } t < \left(T_0+T_1\right)(n+1), \\ 1 \text{ for } t \geq \left(T_0+T_1\right)(n+1). \end{cases} \quad (27)$$

Thus, the probability of failure-free operation of a technical system operating in two variable modes (in the first mode for a non-random time T_0 and in the second mode for a non-random time T_1) can be expressed as:

$$R(t) = \sum_{0\leq n<\frac{t}{T_0+T_1}} (-1)^n\left(-e^{-(\lambda_0T_0+\lambda_1T_1)n}f(K_i)e^{-\lambda_0\left(t-\left(T_0+T_1\right)n\right)}+\right.$$
$$\left.+e^{-\lambda_0T_0}f\left(K_j\right)\left(e^{-\lambda_1\left(t-T_0-\left(T_0+T_1\right)n\right)}--e^{-\lambda_0\left(t-T_0-\left(T_0+T_1\right)n\right)}\right)\right)-$$
$$-e^{-\lambda_0T_0}f(K_z)e^{-\lambda_1T_1}e^{-\lambda_1\left(t-\left(T_0+T_1\right)(n+1)\right)}. \quad (28)$$

$$f(K_i) = \begin{cases} 0 \text{ for } t < \left(T_0+T_1\right)n, \\ 1 \text{ for } t \geq \left(T_0+T_1\right)n. \end{cases} \quad (29)$$

$$f\left(K_j\right) = \begin{cases} 0 \text{ for } t < T_0+\left(T_0+T_1\right)n, \\ 1 \text{ for } t \geq T_0+\left(T_0+T_1\right)n. \end{cases} \quad (30)$$

$$f(K_z) = \begin{cases} 0 \text{ for } t < \left(T_0+T_1\right)(n+1), \\ 1 \text{ for } t \geq \left(T_0+T_1\right)(n+1). \end{cases} \quad (31)$$

3 Results and Discussions

It should be noted that formula (28) differs from formula published in the work [24].

The difference between formulas lies in the fact that in formula published in the work [24], in the terms $f(\kappa_\iota)e^{-\lambda_0\left(t-(T_0+T_1)^n\right)}$, $f\left(\kappa_j\right)\left(e^{-\lambda_1\left(t-T_0-(T_0+T_1)^n\right)}\right)$, and $f(\kappa_z)e^{-\lambda_1T_1}e^{-\lambda_1(t-(T_0+T_1)(n+1))}$, the elements under the exponent are raised to the powers n or $n+1$, whereas they should be multiplied by the time intervals T_0+T_1, which is a mistake made in work [24], and needs to be corrected.

If in formula (28) $\lambda_1 = 0$ and $T_1 = 0$ (i.e., the system is assumed to operate in only one mode), then we get:

$$R(t) = \sum\nolimits_{0 \le n < \frac{t}{T_0}} (-1)^n \left(-e^{-\lambda_0 T_0}\right)^n \left(\left[f(K_i)e^{-\lambda_0 (t-nT_0)} - e^{-\lambda_0 T_0} f\left(K_j\right) - \right.\right.$$

$$\left.\left. -e^{-\lambda_0\left(t - T_0 - \left(1+n\right)\right)}\right)\right). \tag{32}$$

$$f(K_i) = \begin{cases} 0 \text{ for } t \le nT_0, \\ 1 \text{ for } t \ge nT_0. \end{cases} \tag{33}$$

$$f\left(K_j\right) = \begin{cases} 0 \text{ for } t < (1+n)T_0, \\ 1 \text{ for } t \ge (1+n)T_0. \end{cases} \tag{34}$$

In Fig. 1, the probability of the failure-free operation of the avionics system of the aircraft, which operates in a variable mode and is calculated on a computer according to formula (28), is shown, as well as the probability of the failure-free operation of the system operating in a single mode, calculated using formula (32).

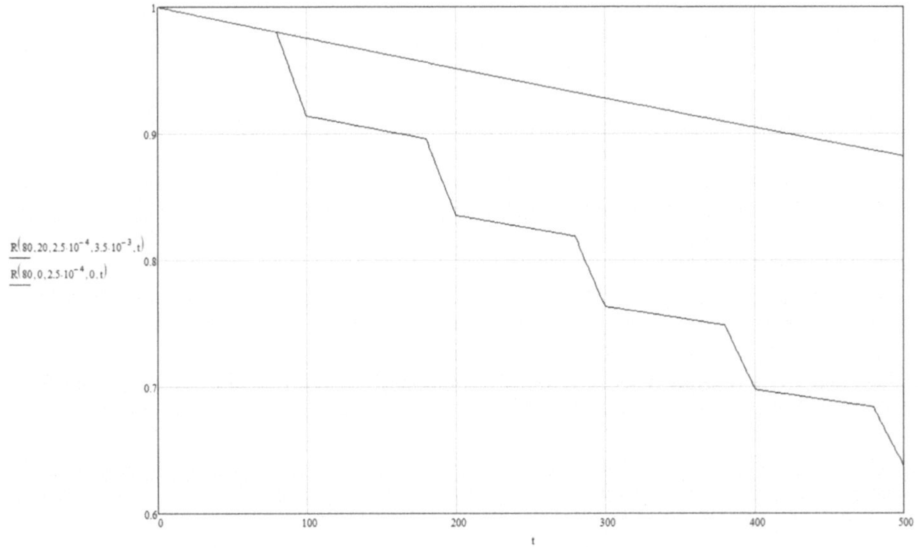

Fig. 1. Probability of failure-free operation of the aircraft avionics system operating in a variable mode, and the probability of failure-free operation of the system operating in a single mode

The graphs shown are constructed with constant values of $T_0 = 80$ h, $T_1 = 20$ h, $\lambda_0 = 2.5 \times 10^{-4}$ h^{-1}, $\lambda_1 = 3.5 \times 10^{-3}$ h^{-1}.

From the graphs, it is evident that as time increases, the probability of failure-free operation decreases, and the greater the failure intensity, the faster the decline. Specifically, when $\lambda_1 = 3.5 \times 10^{-3}$ h^{-1}, the probability of failure-free operation decreases much faster than when $\lambda_0 = 2,5 \times 10^{-4}\ \text{h}^{-1}$. It should be emphasized that the time the system spends in each mode is non-random, which is impossible with a Markov process. For a system operating in a single mode, the probability of failure-free operation is described by an exponential law.

Thus, formula (32) is a partial form of formula (28) with $T_1 = 0$ and $\lambda_1 = 0$.

4 Conclusions

The developed mathematical model and the obtained results enable the calculation of the probability of failure-free operation of the radio-electronic equipment system of an aircraft operating in a variable mode, with specific values of time and failure intensity in each mode. The proposed mathematical model allows for the operation of two systems, alternating in equal, non-random time intervals, following the pattern: the first set operates for one day, and then the second set is activated, also operating for one day. This situation repeats multiple times. The failure intensity for each set can be different. An important assumption is the exponential distribution of failure-free operation time in each mode.

References

1. DSTU 2860-94: 1994: Reliability of Equipment. Terms and Definitions. Official Edition. Institute of Reliability of Machines and Structures, Kyiv (1994)
2. ADMP-01: 2014. Guidance for Developing Dependability Requirements. Edition A, Version 1.-14 (2014)
3. ADMP-02 :2014. Guidance for Dependability In-Service. Edition A, Version 1.-15 (2014)
4. Tobias, P.A., Trindade, D.C.: Applied Reliability, 3rd edn. CRC Press, Boca Raton (2012)
5. Drejper, N., Smit, G.: Applied Regression Analysis, 3rd edn. Wiley, Hoboken (1998)
6. Beichelt, F., Franken, P.: Reliability and Maintenance-Mathematical Methods. Verlag Technik, Berlin (1983)
7. Avramenko, O.V., Openko, P.V., Pyavchuk, O.O., Tselishchev, Y.P., Yablonsky, P.M.: Determining the maintenance frequency of military equipment operated by technical condition using the diffusion-monotone distribution law of the failure. Sci. Herit. **3**(66), 41–47 (2021)
8. Dyptan, V.P., Pyavchuk, O.O., Openko, P.V., Yablonsky, P.M.: Mathematical model of storing aviation destruction means using a diffusion-nonmonotonic failure distribution. Honor Law. **3**(82), 99–107 (2022)
9. Yablonskyi, P., Maistrov, O., Koskov, Y., Sarapin, Y.: Determination of the probability of states of the doppler plane velocity meter and the drift angle of an aircraft using a Markov random process. Air Forces Ukr. **1**(6), 87–92 (2024)
10. Al-Ammouri, A., Tsukanov, I.M., Al-Ammori, H.A.: Information technologies for research semi-Markov reliability model of complex systems. Inf. Process. Technol. Syst. Transp. **4**, 3–16 (2016)
11. Avramenko, O., Sarapin, Y., Yablonskyi, P., Fedorov, O.: Determination of the optimum periodicity of maintenance of the fire alarm system and notification at military storage facilities of ammunition (aviation attack means). Air Forces Ukr. **2**(7), 96–105 (2024). https://doi.org/10.33099/2786-7714-2024-2-7-96-105
12. Ostroumov, I., Ivannikova, V., Kuzmenko, N., Zaliskyi, M.: Impact analysis of Russian-Ukrainian war on airspace. J. Air Transp. Manag. **124**, 102742 (2025). https://doi.org/10.1016/j.jairtraman.2025.102742

13. Cui, L., Li, Y., Shena, J., Cong, L.: Reliability evaluation based on a dependent two-stage failure process with competing failures. Appl. Math. Model. **64**, 699–712 (2018)
14. Solomentsev, O., Zaliskyi, M., Zuiev, O.: Estimation of quality parameters in the radio flight support operational system. Aviation. **20**(3), 123–128 (2016). https://doi.org/10.3846/16487788.2016.1227541
15. Lukyanchuk, V.V., Lanetskii, B.M., Koval, I.V., Tyurin, V.V., Open'ko, P.V., Khoma, V.V.: Model of the surface-to-air missile system reliability during long-term operation. Adv.Mil. Technol. **16**(2), 277–288 (2021). https://doi.org/10.3849/aimt.01506
16. Open'ko, P.V., Drannyk, P.A., Kobzev, V.V., Brovko, M.B., Zalevsky, G.S.: Substantiation of reliability requirements for mobility means of surface-to-air missile systems. Adv. Mil. Technol. **12**(1), 91–99 (2017). https://doi.org/10.3849/aimt.01122
17. Open'ko, P.V., Tyurin, V.V., Mironyuk, M.Y., Drannyk, P.A., Kobzev, V.V.: Reliability and maintainability requirements for mobility means of surface-to-air missile systems. Adv. Mil. Technol. **15**(2), 425–434 (2020). https://doi.org/10.3849/aimt.01402
18. Lanetskii, B., Lukyanchuk, V., Khudov, H., Fisun, M., Zvieriev, O., Terebuha, I.: Developing the model of reliability of a complex technical system of repeated use with a complex operating mode. East.-Eur. J. Enterp. Technol. **4**(107), 55–65 (2020). https://doi.org/10.15587/1729-4061.2020.214995
19. Pohoruy, A.O., Chemeris, O.A.: Introduction to the Theory of Random Processes: Textbook. Zhytomyr State University named after Ivan Franko, Zhytomyr (2020)
20. Brody, S.M., Vlasenko, O.N.: Assessment of the reliability of systems with variable usage modes. Tech. Cybernet. **1**, 35–38 (1967)
21. Yablonskyi, P.M., Avramenko, O.V., Ivanov, V.I., Pyavchuk, O.O., Tselishchev, Y.P.: Assessment of the reliability of a system operating in two-shift usage modes. Air Forces Ukr. **1**(1), 77–79 (2021). https://doi.org/10.33099/2786-7714-2021-1-1-77-79
22. DSTU 3433-96: 1996. Reliability of Equipment. Failure Models. Main Provisions. Official Edition. Institute of Reliability Problems of Machines and Structures, Kyiv (1996)
23. Korolyuk, V.S.: The Sojourn time of a semi-Markov process in a fixed set of states. Ukr. Math. J. **3**(7), 123–128 (1965)
24. Yablonskyi, P.M.: Assessment of the reliability of systems with variable usage modes. Autom. Comput. Technol. **2**, 26–28 (1977)
25. Bateman, H., Erdelyi, A.: Tables of Integral Transforms. Volume 1. Fourier, Laplace. Mellin Transforms, Nauka (1969)
26. Yuzhakova, H.O., Kireyenko, O.V.: Fourier Transforms. Problem Collection. Educational Guide. NTU "KPI named after I. Sikorsky", Kyiv (2022)
27. Hegeza, V.P., Oleshchenko, L.M.: Operational Calculus. Practical Course. Educational Guide. NTU "KPI named after I. Sikorsky", Kyiv (2018)

Sensors in Civil Aviation

Features of Stability Analysis of Complex Nonlinear Dynamic Systems in Phase Space

Petro Snitsarenko[1], Rimvidas Khrashchevskyi[2], Olga Ivanets[2](✉), Maryna Arkhyrei[2], and Nataliia Bilak[2]

[1] National Defence University of Ukraine Named After Ivan Cherniakhovskyi, Povitryanikh Sil Ave., 28, Kyiv 03049, Ukraine

[2] National Aviation University, Liubomyra Huzara Ave., 1, Kyiv 03058, Ukraine

olchik2104@ukr.net

Abstract. The development of the theory of stability of dynamic systems is actively taking place in the direction of studying chaotic dynamic systems. The class of such systems, along with others, includes biological systems (objects), among which the highest importance is given to a person as a complex biological object that requires appropriate study. Special attention for research is required by a person-operator of complex production (technological) processes, who is under the influence of destabilizing factors, in particular, this is a significant part of the personnel of the aviation industry. Ensuring a stable state of such specialists, functional stability of their body is an urgent task. Professions related to aviation security need to develop technologies that allow assessing the state of the operator who makes decisions under the influence of destabilizing factors. Therefore, the functional health of aviation operators is an important trigger in a proactive approach to the entire risk management system for the aviation industry. The paper presents studies that make it possible to obtain additional information from existing biomedical data to assess the functional state of the operator. This approach makes it possible to predict the possibility of a violation of the functional state and, as a consequence, become a reason for failure to fulfill professional duties. The approach is based on studying the stability of a dynamic system to disturbances and analyzing the signals produced by this system.

Keywords: Attractor · Phase Space · Non-Linear Process · Diagnostics · Stochastic Influence · Dynamic System · Analysis

1 Introduction

1.1 Introduction to the Problem

The modern theory of stability of dynamic systems is actively developing in the direction of studying chaotic dynamic systems, in particular biological ones, within the framework of the tasks of their analysis by data processing using nonlinear dynamics methods. In the phase space of a biological system (object), specific sets of states arise that characterize the complex process of homeostasis, which constantly returns the system to a stable state

I. Ostroumov et al. (Eds.): ACASD 2025, LNNS 1418, pp. 379–388, 2025.
https://doi.org/10.1007/978-3-031-91992-3_25

after compensating for the action of destabilizing factors. This is especially characteristic of biological objects, which are people-operators of complex processes, for example, in the aviation industry, who, due to their professional activities, are constantly faced with the action of destabilizing factors. Ensuring a stable state of such specialists, the functional stability of their organism as a complex biological object, is an urgent task. To solve it, an objective assessment of this state is necessary based on the analysis of a set of relevant survey data based on modern methods and technologies.

1.2 Motivation

The stimulus for the development of the theory of chaotic dynamical systems and their applications was the work [1]. Many works are devoted to the analysis of chaotic phenomena that arise mainly in dissipative systems [2], and in recent decades these approaches have been actively used to analyze data from biological systems [3, 4]. Today, the influence of stochastic factors on biomedical data from complex biological objects is actively studied [5, 6]. Although interest in the study of chaotic phenomena arose quite a long time ago [1], with the discovery of exponential instability it turned out that almost any typical nonlinear system with more than one degree of freedom can exhibit chaotic properties [7]. In addition, one of the main issues related to this pattern is the question of the measure of chaotic trajectories. At the same time, it is known that in nonlinear systems, chaotic trajectories always border on a set of coexisting regular (non-chaotic) orbits, and the measure of such regular trajectories can be zero. As the analysis results show, the answer to the question of the role of such regular trajectories in biological systems may be far from simple [8, 9]. The behavior of a biological system is much more complex than the behavior of technical and artificial systems, so not all patterns can be analyzed and taken into account by the methods and technologies of analyzing these systems. Because of this, a biological system requires a separate approach to the analysis and study of the transformations it undergoes during its functioning. Here, studies are needed that analyze the phase trajectories of a biomedical signal in order to make a decision about the stability of the behavior of a certain biological system, in particular, a human operator. Such research allows obtaining additional prognostic information that can be used to assess the predictor (impairment of the functional state of operators), in particular in the methodological provision of risk reduction in the flight safety management system [8].

1.3 Contribution

This study develops the theory of stability of dynamic systems within the framework of the problems of their analysis by signal processing using nonlinear dynamics methods. The practical value is associated with the possibility of reducing the level of risks when assessing the functional state of a human operator of complex processes on the basis of obtaining additional information based on the results of data analysis.

1.4 The Organization of the Paper

The paper is organized as follows. The first section describes the features of biomedical data processing (for analyzing the stability of biological systems) in phase space and presents an analysis of scientific works in this field. The second section justifies the choice of cardiovascular system data for analyzing the stability of the entire dynamic system. The third section deals with the processing of experimental heart rate variability data using nonlinear dynamics methods in phase space and the calculation of key criteria for making decisions regarding the stability of the system. The fourth section is devoted to the conclusions and stages of the study.

2 Materials and Methods

Chaotic processes present in the states of a biological system as a complex dynamic object, in particular a human operator, contain an additional informative component about the functioning of this system and the regularities of internal regulation processes – their study is the task of this work. This requires analysis using nonlinear methods and approaches that take into account the chaotic component of these processes [10].

To analyze the state of a biological system, it is necessary to consider the phase trajectories of its state for a certain period of time and study its phase trajectories, which will allow determining the stability of its functioning and taking into account the chaotic component to obtain additional information about the dynamic behavior of this system. In this case, the analysis of variability can give an answer to how the biological system reacts to the influence of constant destabilizing factors and reveal the regularity of long-term and short-term variability that occurs during daily, seasonal and other biological rhythms.

The human cardiovascular system can be considered as a complex dynamic system with a self-organized hierarchy of autonomous subsystems, which in turn is an indicator of the functional stability of the entire organism, i.e. the stability of a complex biological system (object). In general, the heart rate variability signal is non-stationary, has both periodic and chaotic components [10]. During the analysis of literature sources [6, 8] it was established that when studying dynamic biological systems, the issue of the emergence of chaotic boundary sets in such systems is of great interest. Moreover, the unpredictability that arises in such systems sometimes depends exclusively on the internal properties of the dynamic system, and not on certain external chaotic excitations of this system. However, in this unpredictability (chaoticity), upon closer examination, a number of regularities in the behavior of the system can be identified, which distinguishes this phenomenon from classical random processes. In view of the above, the analysis of the stability of the functioning of a biological system (human operator) was carried out by studying the state of the cardiovascular system.

The paper considers a solution that allows to increase the informativeness of already collected data by using information technology for data processing regarding such a biological system. For this purpose, the phase plane method is used to describe a complex nonlinear dynamic system (cardiovascular system), when in its phase space the transverse sections of the phase trajectory are considered, and the time interval of these sections is taken as the period of external influence on the system.

Using the idea of phase space to study a biological system allows you to use the sections of its phase trajectories with a plane as a two-dimensional picture containing a sequence of points of intersection of phase trajectories with different planes [9]. This area shows the coordinates of the initial point after one, two or more periods of external influence on the system, which allows you to determine short-term and long-term variability. Usually, the time interval of these system intersections is considered as the period of external influence [11]. In this case, the process of obtaining additional useful information from already measured biomedical signals to reduce uncertainty in assessing the functional state of the biological system of the organism [5] is considered using the following approach: analysis of time series of biomedical signals; construction of an attractor and a colored three-dimensional histogram, a colored Poincaré plot, their visual analysis and a preliminary conclusion about the stability of the functioning of the biological system [12, 13]; calculation of quantitative values of the Hurst index, Kolmogorov entropy, approximate and sample entropy. This approach allows you to visualize points in the phase space of a biological system and determine the degree of its stability [14, 15]. The calculated values of the Hurst index, Kolmogorov entropy, as well as approximate and sample entropy were used to confirm the conclusions obtained by visual analysis.

3 Results and Discussions

In the work, to assess the functional stability of the study participants in the phase space of states, the RR intervals of the electrocardiogram states were selected, represented by time series obtained from the open database physionet.org [16]. Visual graphs were constructed to study the states of the phase space of the time series of RR intervals. As an example, Figs. 1 and 2 present such graphs for operators No. 10 and No. 11.

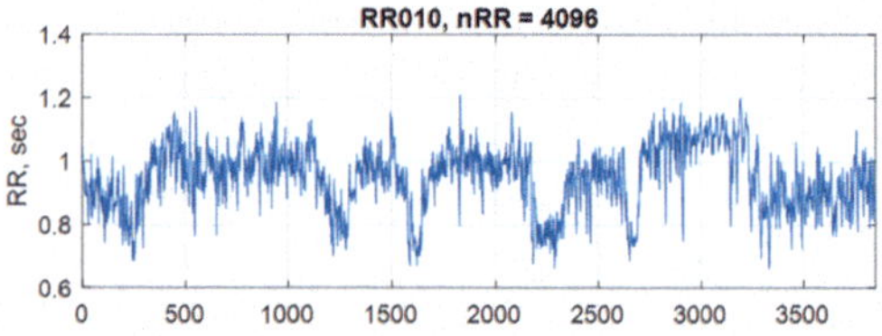

Fig. 1. The data of heart rate variability subject number 10.

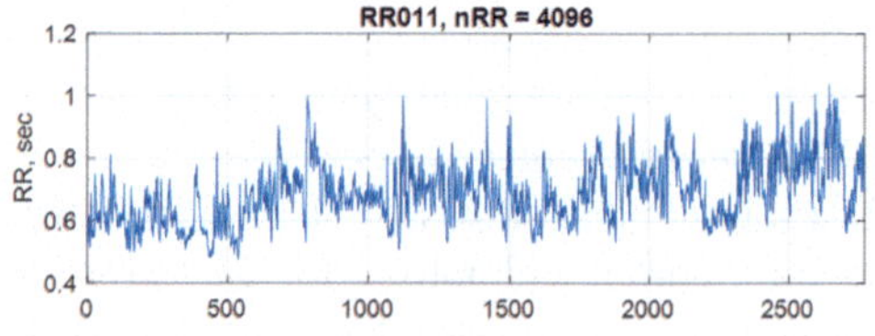

Fig. 2. The data of heart rate variability subject number 11.

Then, computer processing of these data was carried out. The method under consideration is able to reflect the nonlinear features of the time series. The first step was to construct Poincaré plots for the same data. Poincaré plots, which are also called scatterograms, can be used to make a diagnostic conclusion about the stability of the system according to the shape of the resulting ellipse [15, 17]. Such diagrams (plots) based on the results of the time series (Figs. 1 and 2) are presented in Figs. 3 and 4.

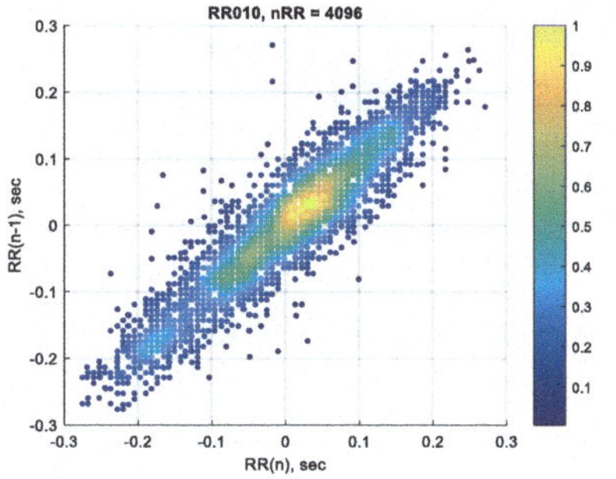

Fig. 3. The Poincare plot of heart rate variability of the research participant number 10.

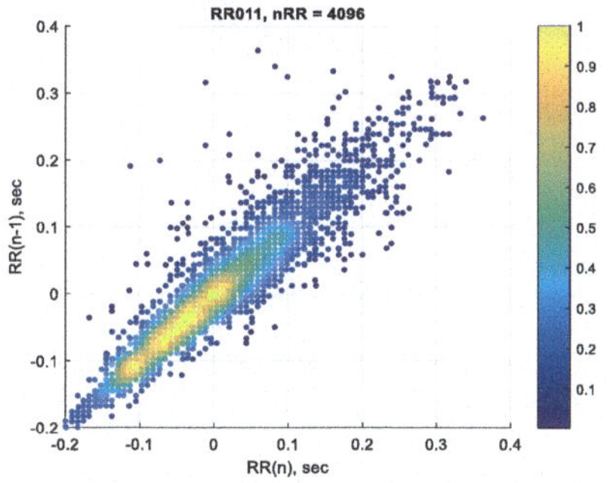

Fig. 4. The Poincare plot of heart rate variability of the research participant number 11.

As can be seen from Fig. 3, the time series of RR intervals of subject No. 10 is characterized by the "typical" form of the Poincare plot, i.e. it has the form of a "comet" of a cigar-shaped shape and a cloud of points oriented along the identity line [12, 17]. At the same time, in subject No. 11, the center of the scatterogram cloud is shifted to the lower left corner and has the form of a narrowed ellipse, but already resembles a "blurred" cigar shape (Fig. 4), which indicates a violation of the state of the cardiovascular system of subject No. 11.

The second step (to confirm the hypothesis of a violation of the state of the phase space of the RR intervals of subject No. 11) was to construct attractors of the time series data (Figs. 5 and 6). According to Takens' theorem [3], the main properties of the attractor will be the same as those of the object under study, and its characteristics can be determined by similarity.

Figure 6 shows that individual points of the phase space that "fall out" from the general attractor cloud form a larger number of them and have a greater distance from the general cloud (the upper part of the cloud) than in Fig. 5. As a third step, for a more visual display of such dynamics in a three-dimensional coordinate space to visualize the

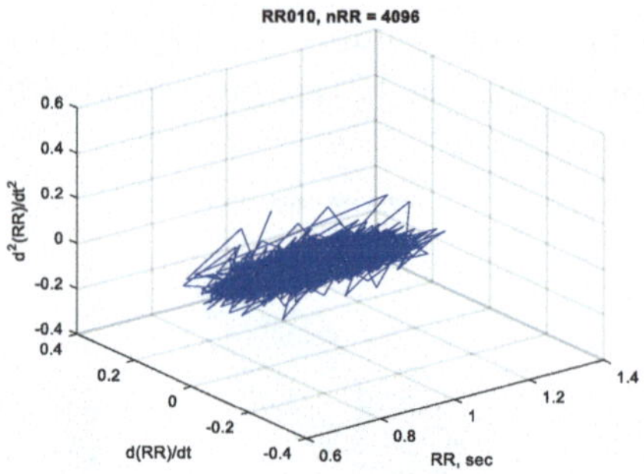

Fig. 5. The attractor of the signal of the research participant number 10.

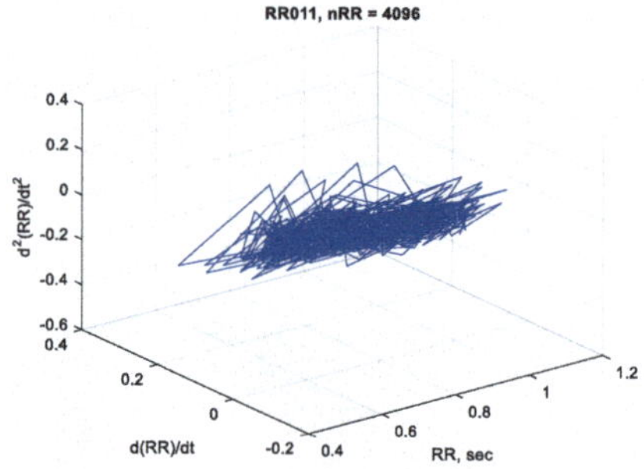

Fig. 6. The attractor of the signal of the research participant number 11.

location of pairs of consecutive RR intervals, three-dimensional histograms (models) were constructed for the corresponding RR intervals (Figs. 7 and 8).

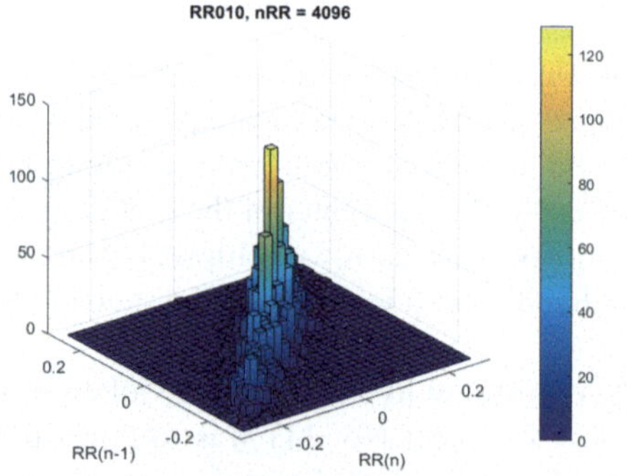

Fig. 7. The attractor of the signal of the research participant number 10.

To test this hypothesis, quantitative values obtained by nonlinear dynamics methods were calculated in the work, namely: Hurst's index, Kolmagorov's entropy, approximate and sample entropy [16, 17]. This list of characteristics is not complete, but it was selected in view of the study [8].

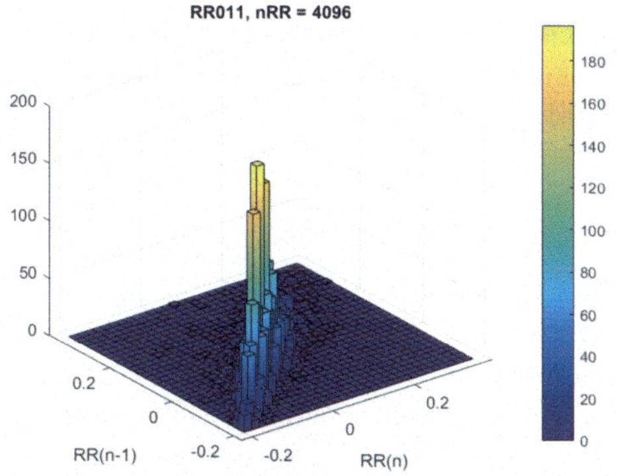

Fig. 8. The attractor of the signal of the research participant number 11.

In Fig. 8 we can see that the histogram is shifted to the lower corner and is not in the center of the line, i.e. the deviation of the points from the center of the bisector makes it possible to determine by how much the ith interval between consecutive R waves is longer or shorter than the $(i + 1)$th interval, which may be a sign of a violation of the functional stability of the cardiovascular system.

And the last step to test this hypothesis was to calculate descriptors for quantitative assessment of changes in the time series of RR intervals. The calculated quantitative values were obtained by methods of nonlinear dynamics, namely: Hurst index, Kolmogorov entropy, approximate and sample entropy. This list of characteristics is not complete, but it was selected taking into account the research [6].

Kolmogorov entropy allows us to estimate the rate of loss of information about the processes occurring in a biomedical system over time [6]. The Kolmogorov entropy as the sum of all positive Lyapunov exponents:

$$\langle \xi\ (t) \rangle_{\tau} = \frac{1}{\tau} \sum\nolimits_{t=1}^{\tau} \xi(\tau). \tag{1}$$

The use of the classical method of the normalized range (R/S-method) for the analysis of time series using the Hurst method consists in calculating the average value $\langle \xi(t) \rangle$ on the time interval τ, which has the same dimension as the time t for the existing time series $\xi(t)$ [18]. To calculate the dependence of the accumulated deviation $X(t, \tau)$ on the time interval τ, there is a dependence:

$$X(t, \tau) = \sum\nolimits_{u=1}^{\tau} \left\{ \xi(u) - \langle \xi\ (t) \rangle_{\tau} \right\}. \tag{2}$$

To calculate the absolute range function R corresponding to the accumulated deviation:

$$R(\tau) = \max_{1 \le t \le \tau} X(t, \tau) - \min_{1 \le t \le \tau} X(t, \tau). \tag{3}$$

The range should remain until the end of the interval and may increase due to its increase. The extent of the dimensionless function R/S after an hour interval τ is indicated by the section under R on the standard deviation S row $\xi(t)$:

$$s(\tau) = \sqrt{\frac{1}{\tau} \sum\nolimits_{u=1}^{\tau} \left\{ \xi(u) - \langle \xi\ (t) \rangle_{\tau} \right\}^2}. \tag{4}$$

According to the results of the study of many natural processes, an empirical relationship between the normalized range of R/S and the length of the interval τ has been established through the indicator H [8]:

$$\frac{R}{S} = \left(\frac{\tau}{2}\right)^H, \tag{5}$$

where H can accept values from 0 to 1.

The last step is to estimate the approximate and sample entropy. Sampling entropy (SampEn) is a modification of approximate entropy (ApEn), which is used to assess the complexity of physiological time series of signals, to diagnose a functional state disorder. SampEn has two advantages over ApEn: independence from data length and ease of implementation [2]. The results of calculations of the Hurst index, Kolmagorov entropy, approximate and sample entropy for the time series considered above are presented in Table 1.

Table 1. The results of the calculation.

Signal file	Kolmogorov entropy	Hurst index	ApEn	SaEn
RR10	4.7587	0.0764	1.1477	1.0701
RR11	4.4775	0.2469	0.8569	0.6562

Thus, the study of the time series of RR intervals of participants No. 10, 11, conducted using the described information technology, which is based on the application of nonlinear dynamics methods, provides additional information, in particular numerical information, which allows us to make the following assessments.

1. In participant No. 10, the shape and appearance of the Poincaré and attractor plot, as well as the values of the sample and approximate entropy of 1.076 and 1.148, respectively, and the value of the Hurst index of 0.08 may indicate normotonia and vague-sympathy of the balance of the cardiovascular system, that is, the expected stability (stability) of the functioning of this operator.
2. The appearance of the Poincaré plot, the attractors of the participant of the study No. 11 and the quantitative values of the sample and approximate entropy, which are 0.656 and 0.857 respectively, and the value of the Hurst index of 0.25 indicate a possible violation of the central, peripheral and humoral mechanisms of regulation of this biological system, that is, a possible significant violation of the stability (stability) of the operator's functioning, hence the risk of performing his professional duties.

4 Conclusions

Traditional time series graphs cannot fully reflect the characteristics of a biological system, in particular a human operator, that change over time, that is, a complex dynamic system, and with manifestations of a chaotic state. Therefore, alternative methods of visualizing the dynamics of complex time series, which primarily characterize such a system, are appropriate for assessing the level of stability of such a biological system. Therefore, additional information is obtained about the real state of the system, which increases the level of reliability of its assessment.

This research substantiates a methodological approach to the implementation and application of information technology for analyzing the stability of a dynamic system based on medical biodata processing using nonlinear methods of time series analysis. This confirms the feasibility of studying the level of stability of chaotic dynamic systems in the manner shown, in particular, in [18], which should be considered the main scientific result of this study. The obtained result in practical application allows to implement a proactive mechanism for reducing the frequency of crisis situations associated with the work of a human operator, due to the possibility of a more objective assessment of his functional state [19].

Further research should be focused on solving technical issues of implementing the developed information technology in the practice of a risk management system in complex socio-technical systems, in particular in the aviation industry.

References

1. Kovalenko, S.O., Kudiy, L.I.: Heart rate variability. Methodical Aspects. Cherkasy National University Named After B. Khmelnytskyi, Cherkasy (2016)
2. Ivanets, O.B., Burichenko, M.Y., Arkhyrei, M.V., Bratko, V.Y.: Peculiarities of using nonlinear dynamics methods for biomedical data processing. Sci. Technol. **4**(56), 317–334 (2022). https://doi.org/10.18372/2310-5461.56.17131
3. Takens, F.: Detecting strange attractors in turbulence. In: Rand, D., Young, L.S. (eds.) Dynamical Systems and Turbulence Lecture Notes in Mathematics, vol. 898, pp. 366–381. Springer, Berlin (1981). https://doi.org/10.1007/BFb0091924
4. Deka, B., Deka, D.: Nonlinear analysis of heart rate variability signals in meditative state: a review and perspective. Biomed. Eng. Online. **22**, 35 (2023). https://doi.org/10.1186/s12938-023-01100-3
5. Lisun, Y.B., Uglev, E.I.: Heart rate variability, use and methods of analysis. Pain Anaesth. Intensive Care. **4**(93), 83–89 (2020). https://doi.org/10.25284/2519-2078.4(93).2020.220693
6. Nayak, S.K., Bit, A., Dey, A., Mohapatra, B., Pal, K.: A review on the nonlinear dynamical system analysis of electrocardiogram signal. J. Healthc. Eng. **6920420**, 1–19 (2018). https://doi.org/10.1155/2018/6920420
7. Ivanets, O., Morozova, I.: Features of evaluation of complex objects with stochastic parameters. In: 11th International Conference on Advanced Computer Information Technologies (ACIT 2021), pp. 159–162. IEEE, Deggendorf (2021). https://doi.org/10.1109/acit52158.2021.9548579
8. Henriques, T., Ribeiro, M., Teixeira, A., Castro, L., Antunes, L., Costa-Santos, C.: Nonlinear methods most applied to heart-rate time series: a review. Entropy. **22**(3), 309 (2020). https://doi.org/10.3390/e22030309
9. Faes, L., Gómez-Extremera, M., Pernice, R., Carpena, P., Nollo, G., Porta, A., Bernaola-Galván, P.: Comparison of methods for the assessment of nonlinearity in short-term heart rate variability under different physiopathological states. Chaos. **29**(12), 123114 (2019). https://doi.org/10.1063/1.5115506
10. Shchapov, P.F., Ivanets, O.B., Sevryukova, O.S.: Dynamic properties of the time series of results of biomedical measurements. Sci. Technol. **2**(46), 236–244 (2020). https://doi.org/10.18372/2310-5461.46.14811

11. Ivanets, O., Burichenko, M., Schapov, P., Morozova, I., Arkhyrei, M., Kulakov, P.: Features of evaluation of risks for complex objects. In: 13th International Conference on Advanced Computer Information Technologies (ACIT 2023), pp. 63–69. IEEE, Wrocław (2023). https://doi.org/10.1109/ACIT58437.2023.10275575
12. Billeci, L., Marin, D., Insana, L., Vatti, G., Varanini, M.: Patient-specific seizure prediction based on heart rate variability and recurrence quantification analysis. PLoS One. **13**, e0204339 (2018). https://doi.org/10.1371/journal.pone.0204339
13. Karmakar, C.K., Khandoker, A.H., Voss, A., Palaniswami, M.: Sensitivity of temporal heart rate variability in Poincaré plot to changes in parasympathetic nervous system activity. Biomed. Eng. Online. **10**, 17 (2011). https://doi.org/10.1186/1475-925X-10-17
14. Golińska, A.K.: Poincar'e plots in analysis of selected biomedical signals. Stud. Log. Grammar Rhetor. **35**(48), 117–127 (2013). https://doi.org/10.2478/slgr-2013-0031
15. Satti, R., Abid, N.U., Bottaro, M., De Rui, M., Garrido, M., Raoufy, M.R., Montagnese, S., Mani, A.R.: The application of the extended Poincaré plot in the analysis of physiological variabilities. Front. Physiol. **10**, 116 (2019). https://doi.org/10.3389/fphys.2019.00116
16. The MIT-BIH normal sinus rhythm database, https://archive.physionet.org/physiobank/database/nsrdb/. Last accessed 27 Dec 2024
17. Rahman, S., Habel, M., Contrada, R.J.: Poincaré plot indices as measures of sympathetic cardiac regulation: responses to psychological stress and associations with pre-ejection period. Int. J. Psychophysiol. **3**, 79–90 (2018). https://doi.org/10.1016/j.ijpsycho.2018.08.005
18. Kamath, C.: Analysis of heart rate variability signal during meditation using deterministic-chaotic quantifiers. J. Med. Eng. Technol. **37**(7), 436–448 (2013). https://doi.org/10.3109/03091902.2013.828106
19. Burichenko, M., Ivanets, O., Arkhyrei, M., Khrashchevskyi, R., Melnykov, O.: Using time series for biomedical signal processing under uncertainties. In: Ostroumov, I., Zaliskyi, M. (eds.) Proceedings of the 2nd International Workshop on Advances in Civil Aviation Systems Development. ACASD 2024. Lecture Notes in Networks and Systems, vol. 992, pp. 385–394. Springer, Cham (2024). https://doi.org/10.1007/978-3-031-60196-5_28

Design of Robust Tracking System for Combined Control of Information and Measuring Devices of Aircraft

Olha Sushchenko(✉)

National Aviation University, Liubomyra Huzara Ave., 1, Kyiv 03058, Ukraine
sushoa@ukr.net

Abstract. The paper describes the principles of designing a robust tracking system to control aviation information and measuring devices operated on moving vehicles in difficult real operation conditions accompanied by parametrical and coordinate disturbances. Expressions describing the connections of input and output signal vectors are represented. The simulation results of the synthesized system are shown. A relevant task of creating robust tracking systems providing combined control of aviation information-measuring devices, considering external coordinate disturbances and measurement interference, is set and solved. A mathematical description is obtained for the integrated function estimating the quality of the developed system. It considers the system's accuracy and robustness, and control expenses under conditions of internal parametric and external coordinate disturbances. An algorithm for designing robust tracking systems of the type under study is presented based on structural synthesis and the results of modeling a robust tracking system for measuring and observing units functioning on aircraft. The represented outcomes could be applied to instrumentation assigned for operation on moving vehicles of the wide class.

Keywords: Robust Tracking Systems · Aviation Information-Measuring Devices · Combined Control · Robust Structural Synthesis

1 Introduction

Currently, the complexity of control processes that accompany the operation of moving objects in general and aircraft in particular is increasing. It should be noted that high precision characteristics characterize modern information and measuring devices. When operating on moving objects, these advantages cannot be fully realized without using new approaches to designing systems that ensure their stabilization and control of the orientation of the measuring axes [1]. The design of such systems is carried out under conditions of uncertainty due to many factors, including inaccuracies in the mathematical description, significant changes in some parameters during operation, the action of intense external coordinate disturbances, and the presence of measurement interference [2–4]. One of the modern approaches to solving this problem is the design of robust tracking systems capable of ensuring stability and providing system quality indicators within

I. Ostroumov et al. (Eds.): ACASD 2025, LNNS 1418, pp. 389–400, 2025.
https://doi.org/10.1007/978-3-031-91992-3_26

acceptable limits under the influence of both parametric and coordinate disturbances [5–7].

An error signal, which characterizes the difference between the command and output signals, can control information-measuring devices installed on a movable base [8–10]. Nevertheless, in this case, it is impossible to ensure high accuracy of the control processes for the position of the measuring axes of information devices. The solution to this problem can be implemented based on combined control, which uses control on both error and reference commands. In this case, the designed system will include two controllers implementing control in direct and feedback [11–13].

The choice of the design method for tracking systems for information-measuring devices used on mobile objects is determined by the following circumstances: First, the parameters of these devices can change significantly during operation on mobile objects [14–16]. Second, the functioning of the systems of the type under study is carried out under conditions of intense external influences.

Considering these circumstances, the problem of designing tracking systems for information-measuring devices used on aircraft should be solved as a problem of robust structural synthesis of a system with combined control [17–19].

In such systems, the preliminary (corrective) filter controls according to the reference action, and the controller in the feedback ensures robust stability and suppression of disturbances [20–22].

Such an approach ensures the high accuracy of stabilization and tracking of information and measuring devices operated on moving vehicles in difficult conditions of parametric and coordinate disturbances [23].

Robust control systems demonstrate the ability to provide stable operation even under significant external disturbances, such as wind gusts, turbulence, or sudden load changes. They effectively compensate for internal factors, including sensor errors, dynamic changes in system parameters, and other uncertainties [24–26].

This paper aims to solve the problem of designing robust tracking systems with combined control based on synthesis, taking into account the impact of internal parametric and external coordinate disturbances and the interference of the measuring system.

The paper's structure can be described like this: in Sect. 2, the design of a robust tracking system with combined control is represented. Section 3 deals with implementing the robust system assigned for tracking and stabilizing observation and information units consigned for mobile objects.

2 Design of Robust Tracking System with Combined Control

In modern foreign problems of control systems, systems with combined control are called two-degree-freedom ones [5, 27, 28]. The structural synthesis of robust two-degree-freedoms systems can be implemented in different ways. The method proposed in [13] and developed in [5, 29, 30] is based on robust stabilization and assigning parametric disturbances using normalized left mutually prime factorization. This method is grounded on the synthesis algorithm, which is based on achieving the desirable frequency responses of a robust closed-loop control system using the expansion of the transfer function of an open system [5, 31, 32]. An expanded stabilization object is formed by introducing weight transfer functions (the so-called pre- and post-compensators).

Nevertheless, for many applications, primarily for tracking systems that control information-measuring devices used on aircraft, the influence of external coordinate disturbances of a certain nature that act on the stabilization object is of great importance [3, 6, 9].

The novelty of the algorithm for designing a robust tracking system presented in this article lies in taking into account the impact of external disturbances and interference with the measuring system. For this purpose, the sensitivity function of the closed system concerning external disturbances and measurement interference is introduced into the optimization function of the problem under consideration. The structural diagram of the designed system is shown in Fig. 1 [5].

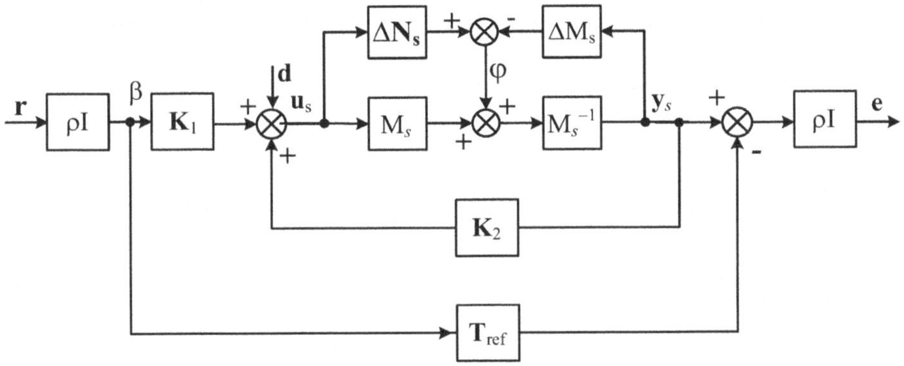

Fig. 1. Structural diagram of a robust tracking system with combined control: ρ is the design parameter.

To implement the H_∞−synthesis procedure, it is advisable to specify external disturbances acting on the system of the type under study in the form of moments that arrive at the output of the stabilization object.

While designing a robust tracking system with combined control, the structural synthesis of the feedback controller K_2 is carried out to ensure the robust stability and suppression of parametric structured disturbances that act on the system [5, 26].

The preliminary filter K_1 ensures the correspondence of the command signal and specified reference model T_{ref} [5, 26].

Thus, the problem of structural synthesis of a robust system with two degrees of freedom with additional control by disturbance consists of finding a stabilizing controller

$$K = [K_1 \ K_2] \tag{1}$$

for the stabilization object, expanded by weight transfer functions to the form $G_s = W_2GW_1$ and presented as a result of normalized left mutually prime factorization $G_s = M_s^{-1}N_s$ [5, 13]. In this case, the controller K_1 is in a direct connection, and the controller K_2 is in a reverse connection.

Following the diagram represented in Fig. 1, the connection between the vectors of the input and output signals of the system taking into account the expression (1) can be represented as follows:

$$\begin{bmatrix} u_s \\ y_s \\ e \end{bmatrix} =$$

$$\begin{bmatrix} \rho(I+K_2G_s)^{-1}K_1 & K_2(I+G_sK_2)^{-1}G_d & K_2(I+G_sK_2)^{-1} & K_2(I+G_sK_2)^{-1}M_s^{-1} \\ \rho(I+G_sK_2)^{-1}G_sK_1 & (I+G_sK_2)^{-1}G_d & (I+G_sK_2)^{-1} & (I+G_sK_2)^{-1}M_s^{-1} \\ \rho^2\left[(I+G_sK_2)^{-1}G_sK_1 - T_{ref}\right] & \rho(I+G_sK_2)^{-1}G_d & \rho(I+G_sK_2)^{-1} & \rho(I+G_sK_2)^{-1}M_s^{-1} \end{bmatrix} \begin{bmatrix} r \\ d \\ n \\ \phi \end{bmatrix}. \quad (2)$$

or $z = \Phi w$, where $z^T = \left[u_s \; y_s \; e \right]$ is the output vector that allows you to estimate the quality of the system's operation and includes control, output, and error signals, respectively; is the input vector and includes the command signal, as well as external coordinate disturbances, measurement noise, and internal parametric disturbances, respectively; Φ is the matrix transfer function that corresponds to the closed-loop system. H_∞ is the norm of the matrix transfer function Φ included in expression (2) and is a generalized functional of the quality of the designed system, since the elements of this matrix, by [5], determine such characteristics of the system as accuracy, robustness, and control costs. In this case, the components $(I + G_sK_2)^{-1}$, $(I + K_2G_s)^{-1}$ are the input and output of the system's sensitivity functions.

The presented problem of designing a robust tracking system should be solved from the point of view of ensuring γ-optimal control [5]. With this approach, a search is made for a suboptimal controller that ensures the fulfillment of the condition $\|\Phi\|_\infty < \gamma$, where γ is the minimum achievable number, called the tolerance level.

The procedure for synthesizing a robust tracking system with combined control is completed by replacing the preliminary filter K_1 with a weighted filter K_1W_i, which ensures the selection of the controller parameters in a steady state [5].

To be able to use modern automated means of implementing structural H_∞ synthesis, it is necessary to define the matrix P of the generalized system, which ensures the connection between its input and output signals in the form

$$\begin{bmatrix} z \\ y \end{bmatrix} = \begin{bmatrix} P_{11} & P_{12} \\ P_{21} & P_{22} \end{bmatrix} \begin{bmatrix} w \\ u \end{bmatrix}. \quad (3)$$

For the system under consideration with combined control, expression (3) takes the form

$$\begin{bmatrix} u_s \\ y_s \\ e \\ \beta \\ y_s \end{bmatrix} = \begin{bmatrix} P_{11} & P_{12} \\ P_{21} & P_{22} \end{bmatrix} \begin{bmatrix} r \\ d \\ \varphi \\ u_s \end{bmatrix}, \quad (4)$$

here $y^T = \left[\beta \; y_s \right]$, $u = u_s$.

To determine the matrix of a generalized system, one can use the relationships between the input and output signals of an open-loop system, which for the case under consideration are as follows:

$$\begin{gathered} u_s = Iu_s; \\ y_s = G_d d + M_s^{-1}\varphi + G_s u_s + n; \\ e = -\rho^2 T_{ref} r + \rho G_d d + \rho M_s^{-1}\varphi + \rho G_s u_s; \\ \beta = \rho I r. \end{gathered} \tag{5}$$

Based on relations (4), (5), the matrix of the generalized system takes the form:

$$P = \begin{bmatrix} 0 & 0 & 0 & 0 & I \\ 0 & G_d & M_s^{-1} & I & G_s \\ -\rho^2 T_{ref} & \rho G_d & \rho M_s^{-1} & 0 & \rho G_s \\ \rho I & 0 & 0 & 0 & 0 \\ 0 & G_d & M_s^{-1} & I & G_s \end{bmatrix} \tag{6}$$

In the state space, matrix (6) can be written as:

$$P = \begin{bmatrix} A_s & 0 & 0 & B_d & 0 & \left(B_s D_s^T + Z_s C_s^T\right) D_{M_s} & B_s \\ 0 & A_r & B_r & 0 & I & 0 & 0 \\ 0 & 0 & 0 & 0 & 0 & 0 & I \\ C_s & 0 & 0 & D_d & 0 & D_{M_s^{-1}} & D_s \\ \rho C_s & -\rho^2 C_r & -\rho^2 D_r & \rho D_d & 0 & \rho D_{M_s^{-1}} & \rho D_s \\ 0 & 0 & \rho I & 0 & 0 & 0 & 0 \\ C_s & 0 & 0 & D_{d1} & I & D_{M_s^{-1}} & D_s \end{bmatrix} \tag{7}$$

where A_s, B_s, C_s, D_s are the state, control, observation, and disturbance matrices that define the description of the stabilization object in the state space; A_r, B_r are the state and control matrices that define the description of the reference model T_{ref} in the state space.

Model (7) can be used as a foundation for the implementation of robust structural synthesis. Such an approach allows us to design control systems resistant to internal and external disturbances.

At the same time, the synthesized system keeps accuracy characteristics in the range of given requirements. In other words, a compromise between accuracy and robustness is achieved.

By the approaches to the representation in the state space of mathematical models obtained as a result of normalized left mutually prime factorization described in [5, 13], the following relationships hold:

$$D_{M_s} = R^{-1/2}, D_{M_s^{-1}} = R^{1/2},$$

here $R = I + D_s D_s^T$.

The component included in the matrix (7) is a positive definite solution of the algebraic Riccati equation [5, 13].

$$\left(A_s - B_s S^{-1} D_s^T C_s\right) Z + Z\left(A_s - B_s S^{-1} D_s^T C_s\right)^T - Z C_s^T R^{-1} C_s Z + B_s S^{-1} B_s^T = 0,$$

where $S = I + D_s^T D_s$.

3 Applied Implementation of Robust Tracking System for Information-Measuring Devices Operated on Moving Vehicles

The application of the proposed robust structural synthesis procedure can be considered using the example of a tracking system with combined control used to control the orientation of information-measuring devices operated on moving objects [9]. The system under study includes measuring equipment, which in this case represents the control object, DC motors, and gyroscopic angular velocity meters. In this case, the control signal from the controller is fed to the motor through a pulse-width modulator, which forms a sequence of impulses. The width of impulses is defined by the control signal coming from the controller.

The most critical parameters of the system of the type under study include the moment of inertia of the control object and the rigidity of the elastic connection between the actuator and the stabilization object. During operation, these parameters can change within 50%.

By the recommendations set out in [9], it is advisable to implement control of the information-measuring devices of the type under consideration based on the integral of the difference between the command control signal and the feedback signal coming from the gyroscopic sensor measuring the absolute angular velocity of the controlled object.

To improve the quality of control in systems of the type under consideration, additional feedback is usually used, for example, based on the current signal in the motor armature circuit, which is proportional to the external disturbance moment [3].

As for external disturbances, among them, we can distinguish the moment of system imbalance and moments caused by the angular motion of the object due to the parametrical and coordinate disturbances.

The first step in creating a procedure for the structural synthesis of a robust tracking system for information-measuring devices operated on moving objects is the creation of a mathematical description.

The main features and assumptions of the mathematical description of the control system of the type under consideration include the need to take into account the elastic connection between the control object and the actuator, neglect of the dynamics of the gyroscopic angular velocity sensor, and the linearization of the main nonlinearities of the model, including the pulse-width modulator. The linearized model of one channel of the tracking system for information-measuring devices used on mobile objects, based on [6], takes the form:

$$\dot{x}_1 = r - k_{ars} x_5; \tag{8}$$

$$\dot{x}_2 = x_5; \tag{9}$$

$$\dot{x}_3 = x_6; \tag{10}$$

$$\dot{x}_4 = -\frac{1}{T_a} x_4 - \frac{c_e}{T_a} x_6 + k_{PWM} \frac{U_{PWM}}{T_a}; \tag{11}$$

$$\dot{x}_5 = -\frac{c_r}{J_{co}}x_2 + \frac{c_r}{n_r J_{co}}x_3 - \frac{f_{co}}{J_{co}}x_5 - \frac{M_{dist}}{J_{co}}; \tag{12}$$

$$\dot{x}_6 = \frac{c_r}{n_r J_m}x_2 - \frac{c_r}{n_r^2 J_m}x_3 + \frac{c_m}{R_w J_m}x_4 - \frac{f_m}{J_m}x_6. \tag{13}$$

where k_{ars} is the angular velocity device gain; T_a is the constant time of the motor anchor circuit; c_e is the proportionality coefficient of the motor angular velocity and the emf; k_{PWM} is the gain of the linearized PWM; U_{PWM} is the PWM input; J_{co} is the inertia moment of the controlled object; c_r is the gearbox rigidity; n_r is the gearbox ratio; f_{co} is the factor of the linearized friction torque of the controlled object; J_m is the motor moment of inertia; f_{co} is the factor of the linearized friction torque of the motor; c_m is the factor of the loading torque on the motor shaft; R_w is the resistance in the motor anchor windings.

The state variables included in the system of Eqs. (8)–(13) are the integral of the difference between the command signal and the voltage corresponding to the absolute angle speed of the controlling object; the angle of turn of the control object; the angle of rotation of the engine; the voltage in the anchor circuit of the engine; the angular velocity of the controlled object and the angular velocity of the engine. For the set of differential Eqs. (8)–(13), matrices of state, control, observation, and perturbation, allowing the mathematical representation of the controlled object to be represented in the state space, become:

$$A = \begin{bmatrix} 0 & 0 & 0 & 0 & k_{ars} & 0 \\ 0 & 0 & 0 & 0 & 1 & 0 \\ 0 & 0 & 0 & 0 & 0 & 1 \\ 0 & 0 & 0 & -\frac{1}{T_a} & 0 & -\frac{c_e}{T_a} \\ 0 & -\frac{c_r}{J_{co}} & \frac{c_r}{n_r J_{co}} & 0 & -\frac{f_{co}}{J_{co}} & 0 \\ 0 & 0 & -\frac{c_r}{n_r^2 J_m} & \frac{c_m}{R_w J_m} & 0 & -\frac{f_m}{J_m} \end{bmatrix}; \tag{14}$$

$$B^T = \begin{bmatrix} 0\,0\,0 & U_{PWM} & 0 & 0 \\ 0\,0\,0 & 0 & M_{dist} & 0 \end{bmatrix}; \tag{15}$$

$$C = \begin{bmatrix} 1\,0\,0 & 0 & 0 & 0 \\ 0\,0\,0 & 1/R_w & 0 & 0 \\ 0\,0\,0 & 0 & k_{ars} & 0 \end{bmatrix}. \tag{16}$$

The next important phase in the structural synthesis of a robust tracking system is the choice of a reference model [5].

It must be said that the selection of a reference signal depends on the requirements for the transient process of processing the command signal. However, this process is sufficiently difficult and unambiguous. Therefore, it is connected with heuristic techniques.

In the case under consideration, it is most appropriate to use a model of the type:

$$T_{ref} = \frac{k}{T^2 p + 2\xi T p + 1}, \tag{17}$$

where $k = 1$, $T = 0.2$, $\xi = 0.7$.

The next step is to select the design parameter ρ. Its recommended value by [5] is in the range of 1...3. For the system of the type under study, it is advisable to use the maximum value from this range.

The definition of transfer functions for pre and post-compensators is the most difficult step in the procedure of structural H_∞−synthesis of a robust tracking system since it is mainly based on the trial and error method. By recommendations of [5], it is advisable to take the value of the post-compensator W_2 equal to 1. As for the pre-compensator, as a result of the studies, the following transfer function was obtained for its definition:

$$W_1 = 10\frac{0.15}{0.1s + 1}\frac{0.4s + 25}{s + 25}. \tag{18}$$

The results of the simulation are given in Figs. 2, 3, 4 and 5. Figure 2 presents the results of modeling the synthesized system in the command signal's transient process in conditions of changing the moment of inertia. During modeling, the deviation in the inertial and unbalanced moments for nominal values and their changes within the limits were considered. At the same time, the system remains stable to disturbances.

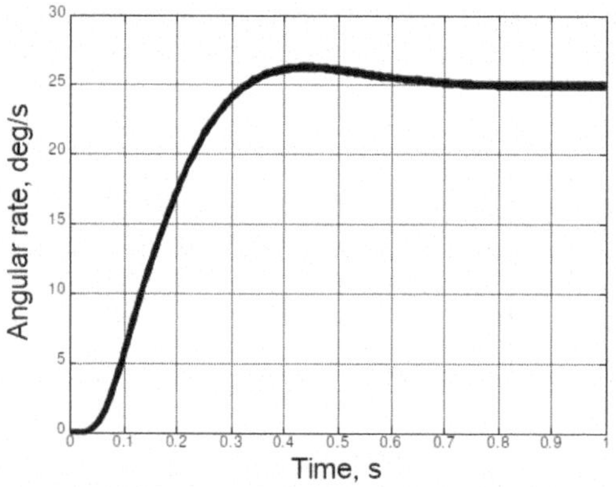

Fig. 2. The transient process on the command signal in conditions of changing the moment of inertia.

The structure of the synthesized controller after the maximum possible reduction in order takes the form

$$A = \begin{bmatrix} 0.06 & 0.4 & 0.72 & 0.53 & 0.11 \\ 0.05 & -0.95 & 0.01 & -0.03 & 0.02 \\ -0.55 & -0.09 & 1.28 & 0.29 & 0.09 \\ 0.47 & 0.27 & -0.18 & 0.61 & -0.11 \\ -0.8 & 0.15 & 0.47 & 0.49 & 1.07 \end{bmatrix};$$

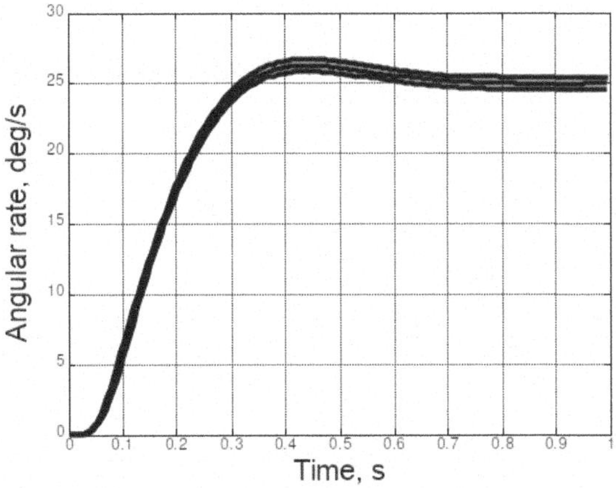

Fig. 3. The transient process on the command signal in conditions of changing the imbalance moment.

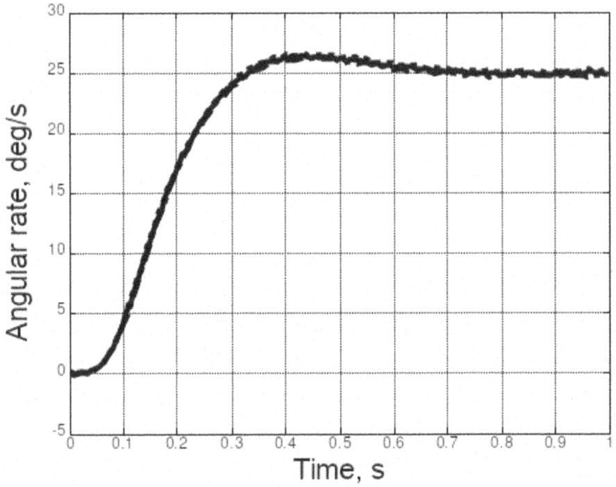

Fig. 4. The transient process on the command signal in conditions of terrain-long irregularities.

$$B = \begin{bmatrix} 0.56 & -173.4 & -52.48 \\ 0.07 & -20.41 & -6.18 \\ -0.55 & -17.22 & -3.04 \\ 1 & -106.4 & -25.5 \\ -6.54 & -34.61 & -12.39 \end{bmatrix};$$

$$C = \begin{bmatrix} -0.004 & 0.007 & 0.006 & 0.001 & -0.0004 \end{bmatrix};$$

$$D = \begin{bmatrix} 0.02 & -3.64 & -1.09 \end{bmatrix}.$$

Figure 3 shows the simulation results of the transient process on the command signal in conditions of changing the imbalance moment.

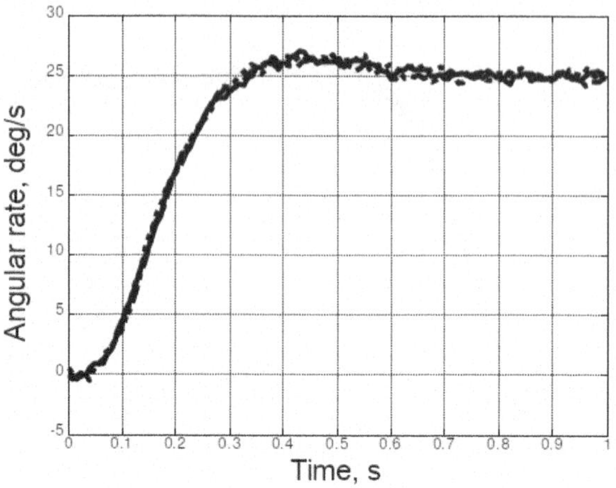

Fig. 5. The transient process on the command signal.

The modeling results of the synthesized system with the simulation of the coordinate (external) disturbances inherent in moving vehicles of the considered type have shown the high resistance of the synthesized system to external disturbances of the instrumentation assigned for operation on moving vehicles.

4 Conclusions

A new problem has been solved the H_∞-synthesis of a robust tracking system taking into account the impact of external coordinate disturbances and measurement noise.

The features of combined control are described. The sensitivity function on the disturbance is introduced. The connections between input and output signals of the generalized control systems are represented.

The compensator for the feedforward control was proposed. The matrix transfer functions for the sensitivity functions were obtained.

The efficiency of the proposed design procedure is confirmed by an example of its applied implementation as applied to robust tracking systems of information-measuring devices operated on aircraft. The weighting coefficients for the augmented control object are chosen.

The structure of the robust controller has been realized and represented in the form of a quadruple of matrices in the state-space representation. The transient processes of the synthesized system have been represented graphically. The influence of parametrical disturbances is shown in this representation.

The obtained results can be spread to moving objects of the other types.

The next research foresees the synthesis of the adaptive system and implementation of the comparative analysis between robust and adaptive systems including operational characteristics and computational burden.

References

1. Hilkert, J.M.: Inertially stabilized platform technology. IEEE Control. Syst. Mag. **1**, 26–46 (2013). https://doi.org/10.1109/MCS.2007.910256
2. Masten, M.K.: Inertially stabilized platforms for optical imaging systems. IEEE Control. Syst. Mag. **1**, 47–64 (2008). https://doi.org/10.1109/MCS.2007.910201
3. Sushchenko, O., Salyuk, O.: Approach to automated synthesis of structure for aircraft control system. In: Ostroumov, I., Zaliskyi, M. (eds.) Proceedings of the 1st International Workshop on Advances in Civil Aviation Systems Development. ACASD 2023, Lecture Notes in Networks and Systems, vol. 736, pp. 73–83. Springer, Cham (2023). https://doi.org/10.1007/978-3-031-38082-2_6
4. Kuznetsov, B.I., Nikitina, T.B., Bovdui, I.V.: Structural-parametric synthesis of rolling mills multi-motor electric drives. Electr. Eng. Electromech. **5**, 25–30 (2020). https://doi.org/10.20998/2074-272X.2020.5.04
5. Scogestad, S., Postlathwaite, I.: Multivariable Feedback Control. Analysis and Design. John Wiley & Sons, London (2005)
6. Zerrik, E.H., Castillo, O.: Stabilization of Infinite Dimensional Systems. Springer, Cham (2021)
7. Kuznetsov, B.I., Nikitina, T.B., Bovdui, I.V.: Multiobjective synthesis of two degrees of freedom nonlinear robust control by discrete continuous plant. Techn. Electrodyn. **5**, 10–14 (2020). https://doi.org/10.20998/2074-272X.2020.5.04
8. Sanfelice, R.G.: Hybrid Feedback Control. Princeton University Press, Princeton (2021)
9. Sushchenko, O.A., Shyrokyi, O.V.: H2/H∞ optimization of system for stabilization and control by line-of-sight orientation of devices operated at UAV. In: IEEE 3rd International Conference Actual Problems of Unmanned Aerial Vehicles Developments (APUAVD), pp. 235–238. IEEE, Kyiv (2015). https://doi.org/10.1109/APUAVD.2015.7346608
10. Osadchyi, S., Zozulia, V.: Synthesis of optimal multivariable robust systems of stochastic stabilization of moving objects. In: IEEE 5th International Conference Actual Problems of Unmanned Aerial Vehicles Developments (APUAVD), pp. 106–111. IEEE, Kyiv (2019). https://doi.org/10.1109/APUAVD47061.2019.8943861
11. Chikovani, V., Sushchenko, O., Tsiruk, H.: Redundant information processing techniques comparison for differential vibratory gyroscope. East.-Eur. J. Enterpr. Technol. **4**(7), 45–52 (2016). https://doi.org/10.15587/1729-4061.2016.75206
12. Osadchy, S.I., Zozulya, V.A., Bereziuk, I.A., Melnichenko, M.M.: Stabilization of the angular position of hexapod platform on board of a ship in the conditions of motions. Autom. Control. Comput. Sci. **56**(3), 221–229 (2022). https://doi.org/10.3103/S0146411622030051
13. Doyle, J., Glover, K., Khargonekas, P.: State-space solution to standard H2 and H∞ control problems. IEEE Trans. Autom. Control. **34**, 831–847 (1989). https://doi.org/10.1109/9.29425
14. Lia, Y., Huang, N.Z., Jiang, T.: Selective Maintenance Modelling and Optimization. Springer, Cham (2023)
15. Niu, S.S., Xiao, D.: Process Control. Springer, Berlin (2022)
16. Sushchenko, O.A., Bezkorovainyi, Y.N., Novytska, N.D.: Dynamic analysis of nonorthogonal redundant inertial measuring units based on MEMS-sensors. In: IEEE 38th International Conference on Electronics and Nanotechnology (ELNANO), pp. 464–469. IEEE, Kyiv (2018). https://doi.org/10.1109/ELNANO.2018.8477553
17. Voliansky, R., Sadovoi, A., Volianska, N.: Interval model of the piezoelectric drive. In: 14th International Conference on Advanced Trends in Radioelecrtronics, Telecommunications and Computer Engineering (TCSET), pp. 1–6. IEEE, Lviv-Slavske (2018). https://doi.org/10.1109/TCSET.2018.8336211

18. Sushchenko, O.A., Golitsyn, V.O.: Data processing system for altitude navigation sensor. In: IEEE 4th International Conference on Methods and Systems of Navigation and Motion Control (MSNMC), pp. 84–87. IEEE, Kyiv (2013). https://doi.org/10.1109/MSNMC.2016.7783112
19. Zhiteckii, L.S., Azarskov, V.N., Solovchuk, K.Y., Sushchenko, O.A.: Discrete-time robust steady-state control of nonlinear multivariable systems: a unified approach. IFAC Proc. Vol. **47**(3), 8140–8145 (2014). https://doi.org/10.3182/20140824-6-ZA-1003.01985
20. Le, A.T.: Adaptive Robust Control Systems. Intech Open, Vienna (2018)
21. Averyanova, Y., Larin, V., Kuzmenko, N., Ostroumov, I., Zaliskyi, M., Solomentsev, O., et al.: Turbulence detection and classification algorithm using data from AWR. In: IEEE 2nd Ukrainian Microwave Week (UkrMW), pp. 518–522. IEEE, Kyiv (2022). https://doi.org/10.1109/UkrMW58013.2022.10037172
22. Ostroumov, I., Kuzmenko, N., Bezkorovainyi, Y., Averyanova, Y., Larin, V., et al.: Relative navigation for vehicle formation movement. In: 3rd KhPI Week on Advanced Technology, pp. 1–4. IEEE, Kharkiv (2022). https://doi.org/10.1109/KhPIWeek57572.2022.9916414
23. Zaliskyi, M., Solomentsev, O., Larin, V., Averyanova, Y., Kuzmenko, N., Ostroumov, I., et al.: Model building for diagnostic variables during aviation equipment maintenance. In: IEEE 17th International Conference on Computer Sciences and Information Technologies, pp. 160–164. IEEE, Lviv (2022). https://doi.org/10.1109/CSIT56902.2022.10000556
24. Gu, D.W., Petkov, P.H., Konstantinov, M.M.: Robust Control Design with MATLAB. Springer, London (2005)
25. Holubnychyi, O., et al.: Self-organization technique with a norm transformation based filtering for sustainable infocommunications within CNS/ATM systems. In: Ostroumov, I., Zaliskyi, M. (eds.) Proceedings of the 2nd International Workshop on Advances in Civil Aviation Systems Development. ACASD 2024, Lecture Notes in Networks and Systems, vol. 992, pp. 262–278. Springer, Cham (2024). https://doi.org/10.1007/978-3-031-60196-5_20
26. Balas, G., Chiang, R., Packard, A., Safonov, M.: Robust Control Toolbox User's Guide. The Math Works Inc., Natick, MA (2008)
27. Ivashchuk, O., et al.: A configuration analysis of Ukrainian flight routes network. In: 16th International Conference on the Experience of Designing and Application of CAD Systems, pp. 6–10. IEEE, Lviv (2021). https://doi.org/10.1109/CADSM52681.2021.9385263
28. Qiu, L., Zhou, K.: Introduction to Feedback Control. Prentice Hall, New Jersey (2009)
29. Asadi, F.: State-Space Control Systems: The MATLAB/Simulink Approach. Morgan & Claypool, Sun Rafael (2021)
30. Zoppoli, R., Sanguineti, M., Gnecco, G., Parisini, T.: Neural Approximations for Optimal Control and Decision. Springer, Cham (2020)
31. Wright, S.J., Recht, B.: Optimization for Data Analysis. Cambridge University Press, Cambridge (2022)
32. Larin, V., Solomentsev, O., Zaliskyi, M., Shcherban, A., Averyanova, Y., Ostroumov, I., et al.: Prediction of the final discharge of the UAV battery based on fuzzy logic estimation of information and influencing parameters. In: 3rd KhPI Week on Advanced Technology, pp. 1–6. IEEE, Kharkiv (2022). https://doi.org/10.1109/KhPIWeek57572.2022.9916490

Chaos-Based Secured Data Transmission Scheme for Civil Aviation Applications

Roman Voliansky[1(✉)], Nina Volianska[2], and Yunifa Miftachul Arif[3]

[1] National Technical University of Ukraine "Igor Sikorsky Kyiv Polytechnic Institute", Polytechnichna Str., 37, Kyiv 03056, Ukraine
voliansky.roman@lll.kpi.ua

[2] Taras Shevchenko National University of Kyiv, Bohdan Havrylyshyn Str., 24, Kyiv 04116, Ukraine

[3] Universitas Islam Negeri Maulana Malik Ibrahim Malang, J. Gajayana Str., 50, Malang 65144, Indonesia

Abstract. Our paper is devoted to studying and designing novel chaotic systems based on known coordinate transformations from one coordinate system into another. As the initial data for such transformation, we take into account a discrete-time model for the generalized dynamical system and show its usage to solve direct and inverse dynamic problems. We consider the solution of the first one to be an equation of a system transmitter, and the last one defines the motion of the receiver subsystem. The above-mentioned transformation is based on the coordinate transformation from a Cartesian coordinate system into an angular one. The use of matrix methods allows us to define highly formalized generalized expressions that perform the above-mentioned direct and inverse coordinate transformations. These expressions and system equations can be considered state space finite difference-algebraic equations. In his case, the transformation equations can be considered as equations of some nonlinear observer. One can rewrite these equations in the finite-difference form to simplify the system analysis by substituting the observer equations into initial system ones. Our studies in solving inverse dynamic problem define several ways that one can consider as backgrounds to establish secured communication schemes. We show the use of our approach by considering the transformation of the well-known Chua circuit dynamic from Cartesian to angular coordinates and vice versa. The studies of these transformations' results prove the possibility of designing novel chaotic systems.

Keywords: Chaotic Systems · Chua Circuit · Coordinate Transformation · Secured Communications

1 Introduction

Today, secure communications in civil aviation are an important component of flight safety and integrity [1–3]. That is why the need for reliable protection of these systems from various cyber threats is becoming increasingly important [4–6].

Today, the following security issues can be identified.

I. Ostroumov et al. (Eds.): ACASD 2025, LNNS 1418, pp. 401–419, 2025.
https://doi.org/10.1007/978-3-031-91992-3_27

The first one is cybersecurity risks, which are caused by the threat of hacking, phishing, and malware attacks, which are rising in aviation communication systems because some persons may attempt to intercept or alter flight data, spoof communications, or gain unauthorized access to important information [7–9]. The next one is signal interception and jamming. Attackers use these doings to disrupt communications between aircraft and air traffic control [10–12]. The third one is related to spoofing attacks, which falsify the identities of aircraft or air traffic controllers and can cause misrouting or incorrect coordination, undermining safety [13, 14].

One can avoid the above-given threats by using various encryption and authentication methods [15–17]. These methods encrypt the transmitted messages to prevent unauthorized access to flight data, verify the identities of aircraft, and control air traffic to reduce the risk of unauthorized communication [18–21].

Due to the importance of the problem being considered, different local and international bodies provide frameworks and guidelines for secure communication in civil aviation. In the last decades, such advanced technologies as artificial intelligence [22, 23] and blockchains technologies [24, 25] are used to detect the security problems in real time. Moreover, aviation personnel need training in secure communication practices to recognize phishing attempts, identify suspicious activities, and use encryption technologies to protect communication channels [26].

Despite all-round solving the above problem in various planes and fields, its technical solution remains important. That is why we propose to use systems with chaotic dynamics at some intermediate level to perform data encoding and decoding while organizing secure information transmission channels [27–30]. It is well-known that such a class of dynamic systems is widely studied now, and various authors offer their own systems [31–34] or modify the known ones [35–37]. At the same time, the design of such systems remains unexplained. We offer to avoid this drawback by applying some coordinate transformation to a well-known chaotic system, which allows us to localize the system motions in some domain of its state space.

Our paper is organized in the following way. At first, we consider some generalized dynamic systems in the discrete-time domain. Then, we show the use of coordinate transformation to perform a system design in terms of state space equations. At third, we transform the system model in form of finite-difference equations and show the use of these equations to solve the direct and inverse dynamic problems to hide and restore some external signal. Finally, we give an example of our approach to a well-known Chua circuit.

2 Methodology

2.1 The Generalized Third-Order Dynamical System in the Discrete-Time Domain

Let us consider the generalized third-order nonlinear dynamical system

$$\frac{d}{dt}\mathbf{X} = \mathbf{F}(\mathbf{X}) + \mathbf{G}(\mathbf{U}) \tag{1}$$

where $\mathbf{X}$ is a vector of system state variables x_i and $\mathbf{F(X)}$ is a vector of system nonlinear functions, and $\mathbf{G(U)}$ is a vector of external signals

$$\mathbf{X} = (x_1\ x_2\ x_3)^T,\quad \mathbf{F(X)} = \left(f_1(\mathbf{X})\ f_2(\mathbf{X})\ f_3(\mathbf{X})\right)^T, \mathbf{U} = \left(u_1\ u_2\ u_3\right)^T,\quad \mathbf{G(U)} = \left(g_1(\mathbf{U})\ g_2(\mathbf{U})\ g_3(\mathbf{U})\right)^T. \tag{2}$$

Since a lot of various devices nowadays are implemented by using digital elements such as MCU, FPGA, and CPU we rewrite (1) into discrete-time domain by applying a finite-difference approximation to its derivative operator. In the most general case, such an approximation can be defined as follows

$$\frac{d}{dt} \approx d\left(1, z^{-1}, T\right), \tag{3}$$

where z^{-1} means that current value of system state variable is shifted in time for discretization time T.

If one substitutes (3) into (1) the following matrix system motion equation can be written down

$$d\left(1, z^{-1}, T\right)\mathbf{X} = \mathbf{F(X)} + \mathbf{G(U)}.\ \ \mathbf{X0} = \left(x_{10}\ x_{20}\ x_{30}\right)^T, \tag{4}$$

where $\mathbf{X0}$ is a system initial condition vector with components x_{i0}.

The left-hand expression in (4) depends on both current values of system state variables and their previous values. This fact makes a numerical solution (4) not clear enough. That is why we offer to solve it for vector of system current values $\mathbf{X}$

$$\mathbf{X} = d^{-1}\left(z^{-1}, T\right)(\mathbf{F(X)} + \mathbf{G(U)}), \tag{5}$$

where $d^{-1}(.)$ is operator means the solution (3) for $\mathbf{X}$.

We call (5) as discrete-time model of the considered generalized dynamical system. Contrary to the continuous-time model (1) the designed model (5) allows to define the current system state variables by using their previous values only. Such an approach makes it possible to easy implement the system (1) by using digital devices and thus to produce the considered system's multichannel output.

This model allows us to solve so called direct dynamic problem and define system (1) motions by using the information about external signals $\mathbf{U}$ and system previous position. Thus, one can consider (5) as equation for the driven data transformer which process the vector of input signals $\mathbf{U}$ into signals $\mathbf{X}$. It is clear that output signals $\mathbf{X}$ can significantly differ from the initial $\mathbf{U}$ in case of complex system dynamic which is defined by vectors $\mathbf{F(X)}$ and $\mathbf{G(U)}$ as well as initial conditions vector $\mathbf{X0}$ in (5). That is why the selecting of an appropriate above-mentioned functions allows us to use device based on (5) as some signal encryptor.

At the same time, one can use (4) to solve an inverse dynamic problem and define external signal $\mathbf{U}$ by known system current and previous positions in such a way

$$\mathbf{U} = \mathbf{G}^{-1}\left(d\left(1, z^{-1}, T\right)\mathbf{X} - \mathbf{F(X)}\right).\ \ \mathbf{X0} = \left(x_{10}\ x_{20}\ x_{30}\right)^T, \tag{6}$$

where $\mathbf{G}^{-1}(\ldots)$ is a function which inversed to $\mathbf{G(U)}$.

Thus, one can consider (6) as the equation for some signal descriptor which allows to restore data that is encrypted according to (5).

2.2 Interrelations Between the Angular and Cartesian Coordinates

It is a well-known fact that the conventional way to analyse system motions is their studying in state space, in which axes coincide with system state variables and are considered perpendicular to each other. In other words, the considered system is studied in the Cartesian coordinate system in which for each time position of system representative point A can be defined by three coordinates x_{1A}, x_{2A}, and x_{3A}, which are defined as distances from coordinate system's origin to projections of the representative point in each axis (Fig. 1).

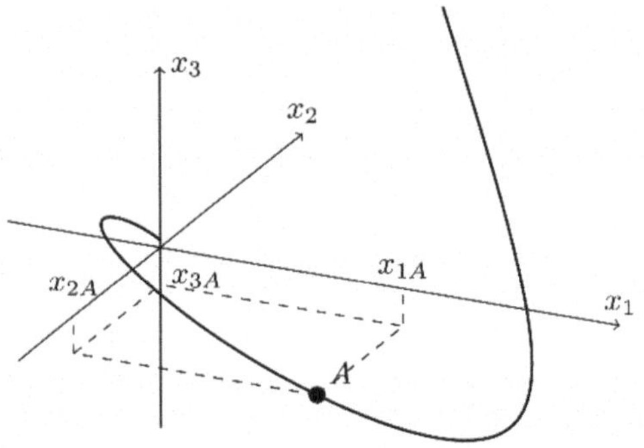

Fig. 1. System 3D phase portrait in Cartesian coordinates.

One can find this approach is not convenient in some applications because of it is necessary to define the above-mentioned projections and find corresponding distances. That is why we offer to replace an linear coordinates of representative point with its angular coordinates. System position in this case are defined by two angles α_1 and α_2 and distance ρ between the origin and representative point (Fig. 2).

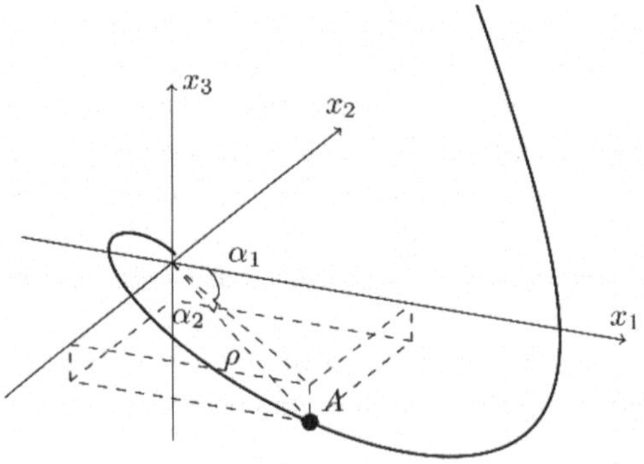

Fig. 2. System 3D phase portrait in angular coordinates.

The comparison of Figs. 1 and 2 allows us to define the following interrelations between Cartesian and angular coordinates:

$$x_{3A} = \rho \sin(\alpha_2);\, x_{2A} = \rho \cos(\alpha_2)\sin(\alpha_1);\, x_{1A} = \rho \cos(\alpha_2)\cos(\alpha_1), \tag{7}$$

In (7) we assume that all angular coordinates can take both negative and positive values and system Cartesian coordinates are considered to be inverted if its angular coordinates take negative values.

We call (7) as inverse transformation problem which allows us to define system position in the Cartesian coordinates by known angular ones. If one takes into account following angular state variable vector $\mathbf{Y}$ and inversed transformation vector $\mathbf{K}^{-1}(\mathbf{Y})$

$$\mathbf{Y} = \left(\rho\ \alpha_1\ \alpha_2\right)^T; \mathbf{K}^{-1}(\mathbf{Y}) = \left(\rho \sin(\alpha_2)\ \rho\cos(\alpha_2)\sin(\alpha_1)\ \rho\cos(\alpha_2)\cos(\alpha_1)\right)^T, \tag{8}$$

it becomes possible to generalize (7) in such a way

$$\mathbf{X} = \mathbf{K}^{-1}(\mathbf{Y}). \tag{9}$$

Solution (7) for angular coordinates allows us to formulate the direct transformation problem

$$\rho^2 = x_{1A}^2 + x_{2A}^2 + x_{3A}^2;\ \ \alpha_1 = \arctan\left(\frac{x_{2A}}{x_{1A}}\right);\ \ \alpha_2 = \arctan\left(\frac{x_{3A}}{\sqrt{x_{1A}^2 + x_{2A}^2}}\right), \tag{10}$$

which allows us to define a system position in angular coordinates by its position in Cartesian ones.

One can generalize (8) as follows

$$\mathbf{Y} = \mathbf{K}(\mathbf{X}) \tag{11}$$

by taking into account vector of angular coordinates $\mathbf{Y}$ and direct transformation vector $\mathbf{K}(\mathbf{X})$

$$\mathbf{K}(\mathbf{X}) = \left(x_{1A}^2 + x_{2A}^2 + x_{3A}^2\ \arctan\left(\frac{x_{2A}}{x_{1A}}\right)\ \arctan\left(\frac{x_{3A}}{\sqrt{x_{1A}^2+x_{2A}^2}}\right)\right)^T. \tag{12}$$

It is understood that defined in such a way vector-functions $\mathbf{K}(\mathbf{X})$ and $\mathbf{K}^{-1}(\mathbf{Y})$ are inverse to each other and following expressions become true

$$\mathbf{Y} = \mathbf{K}\left(\mathbf{K}^{-1}(\mathbf{Y})\right);\ \ \mathbf{X} = \mathbf{K}^{-1}(\mathbf{K}(\mathbf{X})). \tag{13}$$

From the control theory viewpoint one can consider (11) and (9) as a nonlinear observability matrix equation for (5) and (6). This fact allows us to write down state space equations for the considered discrete-time enctyptor and decryptor system in such a way

$$\mathbf{X} = d^{-1}\left(z^{-1}, T\right)(\mathbf{F}(\mathbf{X}) + \mathbf{G}(\mathbf{U})), \mathbf{Y} = \mathbf{K}(\mathbf{X}); \tag{14}$$

$$\mathbf{U}=\mathbf{G}^{-1}\left(d\left(1,z^{-1},T\right)\mathbf{X}-\mathbf{F}(\mathbf{X})\right),\mathbf{X}=\mathbf{K}^{-1}(\mathbf{Y}). \quad (15)$$

It is clear that (14) and (15) define motion of a wide class of nonlinear dynamical systems with nonlinear observability equations and can be used to define system dynamic in various coordinate spaces, including the generalized phase spaces which dimension is greater than three. Thus, in the most common case not only transformation (7) and (10) but much more complex ones can be considered to define the novel systems motion trajectories with unique features.

2.3 Dynamical System Models in Angular Coordinates

The main feature of the models (14) and (15) is the usage of both algebraic and finite-difference equations to define the system motions. Such a mathematical description of system dynamic is a quite complex and reduce number of known control methods to analyse and synthesize system's motions. That is why we offer to rewrite it in form of finite-difference equations only.

One can perform such a rewriting by substituting (9) into (14) and (15)

$$\mathbf{K}^{-1}(\mathbf{Y})=d^{-1}\left(z^{-1},T\right)\left(\mathbf{F}\left(\mathbf{K}^{-1}(\mathbf{Y})\right)+\mathbf{G}(\mathbf{U})\right),\mathbf{Y}=\mathbf{K}\left(\mathbf{K}^{-1}(\mathbf{Y})\right); \quad (16)$$

$$\mathbf{U}=\mathbf{G}^{-1}\left(d\left(1,z^{-1},T\right)\mathbf{K}^{-1}(\mathbf{Y})-\mathbf{F}\left(\mathbf{K}^{-1}(\mathbf{Y})\right)\right),\mathbf{K}^{-1}(\mathbf{Y})=\mathbf{K}^{-1}(\mathbf{Y}). \quad (17)$$

According to (13) the observability equations in (16) and (17) becomes an identity so only the first equations define system motions and should be considered

$$\mathbf{K}^{-1}(\mathbf{Y})=d^{-1}\left(z^{-1},T\right)\left(\mathbf{F}\left(\mathbf{K}^{-1}(\mathbf{Y})\right)+\mathbf{G}(\mathbf{U})\right); \quad (18)$$

$$\mathbf{U}=\mathbf{G}^{-1}\left(d\left(1,z^{-1},T\right)\mathbf{K}^{-1}(\mathbf{Y})-\mathbf{F}\left(\mathbf{K}^{-1}(\mathbf{Y})\right)\right). \quad (19)$$

This fact allows us to claim that the system state space equations which consist of matrix finite-difference or differential equation and algebraic one can be reduced by excluding the observability equation if one substitutes the inversed observability equation into the system motion equation. At the same time, a comparison of (18) and (14) shows that the first one is solved for some nonlinear vector-function $\mathbf{K}^{-1}(\mathbf{Y})$, so system (18) can be considered as given in the implicit form. It is clear that the numerical solution of such systems is not a trivial problem and can require the use of various Newton-based numerical methods, which usage can reduce the performance of digital systems that implement the above-given model. In the most general case, the solution (18) for $\mathbf{Y}$ can be given in such a way

$$\mathbf{Y}=\mathbf{K1}\left(d^{-1}\left(z^{-1},T\right)\left(\mathbf{F}\left(\mathbf{K}^{-1}(\mathbf{Y})\right)+\mathbf{G}(\mathbf{U})\right)\right), \quad (20)$$

where $\mathbf{K1}(.)$ means solution of (18).

We call (20) as the discrete-time model of the generalized nonlinear system in the angular coordinates. This model allows us to define the current values of system state variables by using their previous values. Structure of the model is defined by the structure of the initial system in the Cartesian coordinates and transformation function which define transformation from one coordinate system to another. This fact in the general case makes it possible to use (20) for defining system motions in various coordinate systems but not only angular one.

In a similar way

$$\mathbf{U} = \mathbf{G}^{-1}\left(d\left(1, z^{-1}, T\right)\mathbf{K}^{-1}(\mathbf{Y}) - \mathbf{F}\left(\mathbf{K}^{-1}(\mathbf{Y})\right)\right) \tag{21}$$

can be considered as the solution of system inverse dynamic problem in the angular coordinates.

3 Results and Discussions

3.1 Chua Circuit Discrete-Time Model

We show the use of our approach by designing the discrete-time model for Chua circuit which motions are defined in angular coordinates. The considered circuit schematic is shown in Fig. 3. Contrary to the classical Chua circuit we consider here the controllable circuit with one controllable voltage sources. Such a circuit allows us to form chaotic signals as functions of external single-channel one.

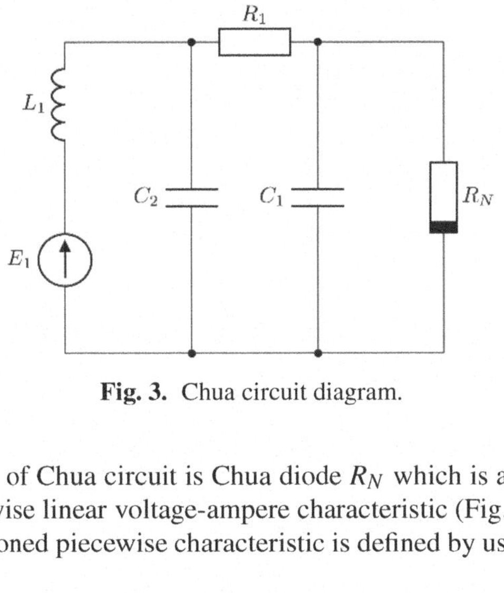

Fig. 3. Chua circuit diagram.

The main feature of Chua circuit is Chua diode R_N which is a nonlinear electronic element with piece-wise linear voltage-ampere characteristic (Fig. 4).

The above-mentioned piecewise characteristic is defined by using the simple piecewise linear function

$$I = G_{RN}(U) = \begin{cases} m_0 U + m_0 - m_1 \; if \; -U_2 < U < -U_1; \\ m_1 U \quad\quad\quad\quad\; if \; -U_1 < U < U_1; \\ m_0 U - m_0 + m_1 \; if \; U_1 < U < U_2, \end{cases} \tag{22}$$

where m_i is gain of i-th interval in piece-wise characteristic, U is an input voltage, and U_i are fracture voltages.

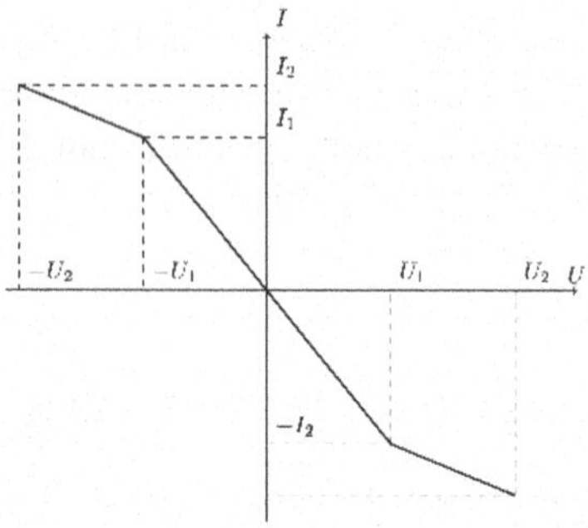

Fig. 4. Chua diode characteristic.

One can use different energy or circuit-based approaches [37] to model the device in Fig. 3. In both cases its dynamic can be defined by following equations

$$\dot{U}_{c1} = \frac{1}{C_1}\left(\frac{U_{c2} - U_{c1}}{R_1} - G_{RN}(U_{c1})\right); \tag{23}$$

$$\dot{U}_{c2} = \frac{1}{C_2}\left(\frac{U_{c1} - U_{c2}}{R_1} + I_L\right); \dot{I}_L = \frac{\mathrm{E}_1 - U_{c2}}{L_1},$$

where U_{ci} is a capacitors' voltages, R_1 is a circuit resistance, C_1 and C_2 are capacitance of circuit capacitors, I_L is a current through inductance, L_1 is circuit inductance, E_1 is the controlled energy source voltage.

Analysis of (23) shows that these equations are defined in terms of capacitor's voltages U_{c1} and U_{c2} and inductance current I_L and thus have different measurement units. It is clear that it is hard to analyse such equations because of necessary to take into account not only values of system state variables but also their physical nature. That is why, to make analysis of the considered dynamical system more flexible following dimensionless form of the Chua equations are used

$$\dot{x}_1 = a(x_2 - x_1 - g(x_1)); \tag{24}$$

$$\dot{x}_2 = x_1 - x_2 + x_3; \ \dot{x}_3 = -bx_2 + be_1,$$

$$x_1 = U_{c1}, \ x_2 = U_{c2}, \ x_3 = R_1 I_L, e_1 = E_1, \tag{25}$$

$$a = \frac{C_2}{C_1}; \ b = \frac{R_1^2 C_2}{L_1}; \ g(x) = R_1 G_{RN}(U_{c1}).$$

Equations (24) define Chua circuit dynamic in the continuous time domain and according to the approach, which is described in previous section, should be transformed into discrete-time domain. We perform such a transformation by using the simplest finite-backward difference approximation for the derivative operator

$$\dot{x} = \frac{1 - z^{-1}}{Tz^{-1}} x. \tag{26}$$

If one substitutes (26) into (24) following finite-difference equation can be written down

$$\left(1 - z^{-1}\right)x_1 = aTz^{-1}(x_2 - x_1 - g(x_1));$$
$$\left(1 - z^{-1}\right)x_2 = Tz^{-1}x_1 - Tz^{-1}x_2 + Tz^{-1}x_3; \quad (27)$$
$$\left(1 - z^{-1}\right)x_3 = -Tz^{-1}bx_2 + Tz^{-1}bx_2e_1.$$

Solution (27) for x_i allows us to rewrite Chua equation in such a way

$$x_1 = aTz^{-1}x_2 - (aT - 1)z^{-1}x_1 - aTg\left(z^{-1}x_1\right);$$
$$x_2 = Tz^{-1}x_1 - (T - 1)z^{-1}x_2 + Tz^{-1}x_3; \quad (28)$$
$$x_3 = z^{-1}x_3 - Tz^{-1}bx_2 + Tz^{-1}be_1.$$

We call (28) as Chua circuit equation in the discrete-time domain. It is clear that right-hand expression of these equations are defined by using system parameters, piecewise linear function and and previous values of its state variables. Thus, it can be easy implemented in program way by using various programmed digital devices such as FPGA, MCU, DSP, and so on.

Results of numerical sollution of (28) are shown in Figs. 5 and 6.

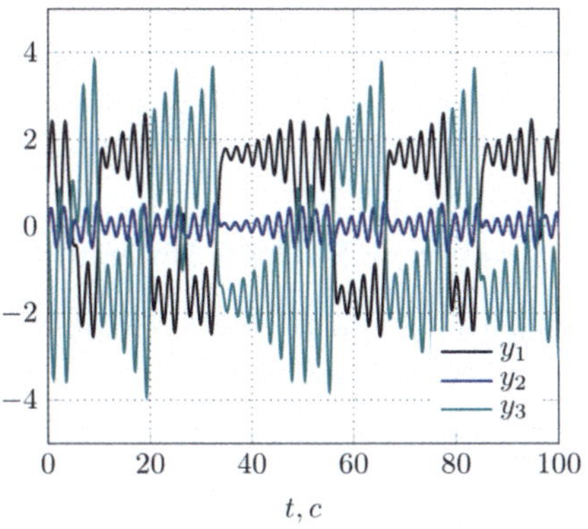

Fig. 5. Chua circuit motions.

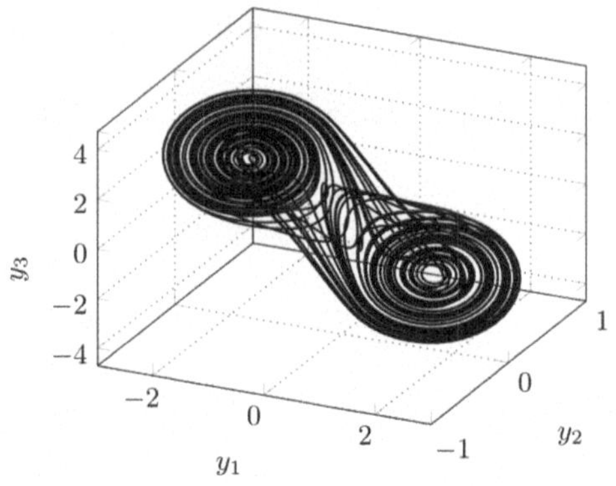

Fig. 6. Chua circuit attractor.

The given simulation results are similar to the well-known ones and prove the chaotic nature of the considered circuit.

Analysis of (24) and (27) shows the only one equation in explicit form contains external signal e_1. In the most general case, we consider this voltage source as a variable one which is controlled by some data stream to be encrypted. Other chaotic signals depend indirectly on e1 through the external signal x_3.

This fact defines only one solution of the inverse problem

$$z^{-1}e_1 = \frac{1 - z^{-1}}{Tb}x_3 + z^{-1}x_2 \tag{29}$$

and it is clear that to restore external signal e_1 it is necessary to have the current values of the state variable x_3 as well as previious values of this variable and x_2. In other words, two signals should be transmitted and received. If one of the above-mentioned signals received with error this makes it impossible to correctly restore the encrypted signal.

3.2 Transformation of Chua Circuit Model in Different Coordinate Systems

It is clear that the signals which are produced by (28) are time-depended signals that are defined by system parameters, structure, and nature. There are no any state spaces defined by these and other equations in the stage of system design. This fact raises the problem of interpreting system outputs as coordinates of representative point in some coordinate system. As it was previously-mentioned the conventional way to solve this problem is the use of Cartesian coordinates to define a state space where system motions are studied. It should be mentioned that from the such coordinate system has mathematical sense only in case of defining system motion by the canonical equations. Since these equations are defined in terms of output signal and its derivatives, orthogonal coordinate system can show phase shift angles between signals. Although, anyone is not denied to use other coordinates for system studying.

In our paper we consider two interpretations for system state variables.

Chua Circuit Model's Transformation from Cartesian Coordinates into Angular Ones. Let us assume that Chua circuit state variables are interpreted as coordinates which are given in the Cartesian space. Let us substitute (28) into (10) and solve the obtained equations for α_1, α_2, and ρ

$$\rho = z^{-1}(1-T)\rho + Tz^{-1}\Big(\rho\Big((1-a)\cos(\alpha_1)^2 + \sin(\alpha_1)(a+1)\cos(\alpha_1)\Big) \tag{30}$$

$$\times\cos(\alpha_2)^2 - \Big(\cos(\alpha_1)g(\rho cos(\alpha_2)\cos(\alpha_1))a + \sin(\alpha_1)\sin(\alpha_2)$$

$$\rho(b-1))\times\cos(\alpha_2) + e_1\sin(\alpha_2)b);$$

$$\alpha_1 = z^{-1}\alpha_1 + \frac{z^{-1}T}{z^{-1}\rho\cos\left(z^{-1}\alpha_2\right)}\Big(\Big((a+1)\cos\left(\alpha_1\right)^2 \pm a\Big)\rho\cos(\alpha_2)\Big) + \sin(\alpha_1)$$

$$\times(a-1)\cos(\alpha_1) + \rho sin(\alpha_2)\cos(\alpha_1) + g\,(\rho cos(\alpha_2)\cos(\alpha_1))\mathrm{a}\sin(\alpha_1));$$

$$\alpha_2 = z^{-1}\alpha_2 + \frac{z^{-1}T}{z^{-1}\rho}\Big(-\rho\sin(\alpha_1)(b-1)\cos(\alpha_2)^2 + \sin(\alpha_1)(a+1)\cos(\alpha_1)$$

$$\times\sin(\alpha_2) - \rho\Big(be_1 - 1 + (1-a)\cos\left(\alpha_1\right)^2\Big)\cos(\alpha_2) + +gx(\rho cos(\alpha_2)\cos(\alpha_1))asin(\alpha_2)\cos(\alpha_1) - \rho sin(\alpha_1)).$$

We call (30) as discrete-time equations of Chua circuit's motion in the 3D angular space. As one can see transformation equations from the one coordinate state into another dramatically changes system motion equations and make them more complex. At the same time the use of finite-difference form of equations allows us to define clear dependencies which can be used to define the current value of system state variables by their previous values. Moreover, the right-hand expressions in these equations are defined as linear and nonlinear combinations of system state variables and circuit parameters, which can be easily implemented in digital devices without using any cyclic construction to specify the current value of some state variable.

Initial conditions for the transformed model can be defined by substituting into (10) the initial conditions of the Chua circuit state variables in Cartesian coordinates.

Analysis of (30) shows that contrary to the initial Eqs. (28), the angular-form equations contain external signal e_1 in two of them. Furthermore, contrary to the third equation in initial discrete-time Chua circuit equations, all of angular state variables are used in these equations. This fact allows us, at first, claim about necessity to establish the three-channel communication. At second, the complex expression to restore data.

$$z^{-1}e_1 = \frac{1}{bcos\left(z^{-1}\alpha_2\right)}\Big(\Big(\alpha_2 - z^{-1}\alpha_2\Big)\rho + Tz^{-1}\Big(\rho sin(\alpha_1)(b-1)\cos(\alpha_2)^2 \tag{31}$$

$$+\rho\sin(\alpha_2)\Big(-1 + (1-a)\cos(\alpha_1)^2 + \sin(\alpha_1)(a+1)\cos(\alpha_1)\Big)\cos(\alpha_2)$$

$$-g(\rho\cos(\alpha_1)\cos(\alpha_2))\mathrm{a}\sin(\alpha_2)\cos(\alpha_1) + \rho\sin(\alpha_1)));$$

$$z^{-1}e_1 = \frac{1}{b\sin(z^{-1}\alpha_2)}((\rho - z^{-1}\rho) + Tz^{-1}(((a-1)\rho\cos(\alpha_1)^2$$
$$-(a+1)\rho\sin(\alpha_1)\cos(\alpha_1) + \rho)\cos(\alpha_2)^2 + (ag(\rho\cos(\alpha_1)\cos(\alpha_2))\cos(\alpha_1)$$
$$+\sin(\alpha_1)\sin(\alpha_2)\rho(b-1))\cos(\alpha_2))$$

makes data exchange more secured.

We call (31) as descriptor equations which are based on the solving of encryptor Eq. (30). Thus, according to (31), the external signal e_1 can be defined in two ways as solution of two different equations. We believe that this fact can be used to solve problems with checking the correctness of external signal's restoring and comparing the results of its defining which are obtained from different expressions to define the transmission errors.

Results of studies the communication system which are based on (30) and (31) are shown in Figs. 7 and 8.

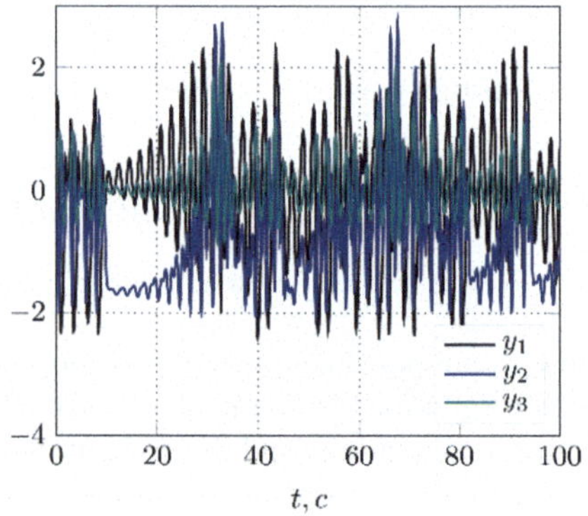

Fig. 7. Chua circuit angular motions.

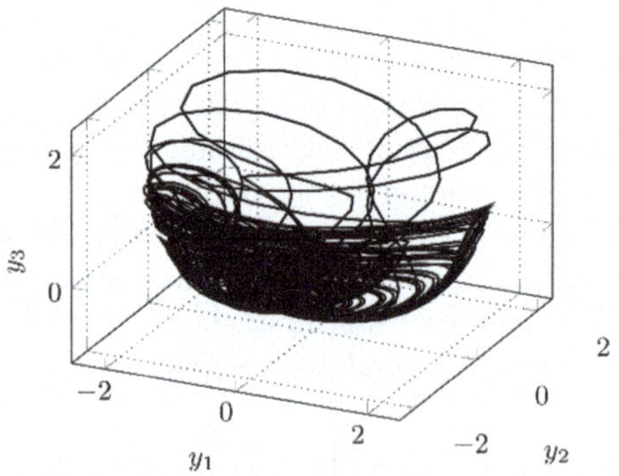

Fig. 8. Chua circuit attractor.

Analysis of given results shows the dramatically changings in system outputs from the well-known classical Chua circuit. The considered transformed system produce three outputs which have the almost same amplitude but different biases. According to the obtained results, in the angular coordinates the Chua circuit should not be considered as a conventional double-scroll chaotic system. Considered system's attractor is defined in some 3D space on a sphere which variable radius is determined by the first expression of (10).

The maximal and minimal values of this radius a defined as follows

$$\rho_{max} = \sqrt{x_{1\max}^2 + x_{2\max}^2 + x_{3\max}^2}; \quad \rho_{min} = \sqrt{x_{1\min}^2 + x_{2\min}^2 + x_{3\min}^2}. \tag{32}$$

Chua Circuit Model's Transformation from Angular Coordinates into Cartesian Ones. Let us assume that Chua circuit state variables are interpreted as coordinates which are given in the angular space as follows

$$\begin{aligned} \rho &= aTz^{-1}\alpha_1 - (aT-1)z^{-1}\rho - aTg\left(z^{-1}\rho\right); \\ \alpha_1 &= Tz^{-1}\rho - (T-1)z^{-1}\alpha_1 + Tz^{-1}\alpha_2; \\ \alpha_2 &= z^{-1}\alpha_2 - Tz^{-1}b\alpha_1 + Tz^{-1}be_1. \end{aligned} \tag{33}$$

Let us substitute (33) into (7) and solve the obtained equations for x_1, x_2, and x_3.

$$\begin{aligned} x_1 = z^{-1}x_1 &+ \frac{z^{-1}T}{\left(x_1^2+x_2^2\right)\left(x_1^2+x_2^2+x_3^2\right)}\left(-\left(x_1^2+x_2^2\right)\left(ax_1g(x_1) - ax_1\arctan\left(\frac{x_2}{x_1}\right)\right.\right. \\ &+x_2\left(x_1^2+x_2^2+x_3^2\right)\sqrt{x_1^2+x_2^2+x_3^2} - \left(x_1^2+x_2^2+x_3^2\right)\left(\left(x_2x_1^2+x_2^3\right)\right. \\ &\times\arctan\left(\frac{x_3}{\sqrt{x_1^2+x_2^2}}\right) + bx_1x_3\left(e_1 - \arctan\left(\frac{x_2}{x_1}\right)\right)\sqrt{x_1^2+x_2^2} \\ &\left.\left.\left.+\left(x_1^2+x_2^2\right)\left(ax_1 - x_2\arctan\left(\frac{x_2}{x_1}\right)\right)\right)\right)\right); \end{aligned} \tag{34}$$

$$\begin{aligned} x_2 = z^{-1}x_2 &+ \frac{z^{-1}T}{\left(x_1^2+x_2^2\right)\left(x_1^2+x_2^2+x_3^2\right)}\left(-\left(x_1^2+x_2^2\right)\left(ax_2g(x_1)\right.\right. \\ &-ax_2\arctan\left(\frac{x_2}{x_1}\right) - x_1\left(x_1^2+x_2^2+x_3^2\right)\sqrt{x_1^2+x_2^2+x_3^2} \\ &+x_1\left(x_1^2+x_2^2+x_3^2\right)\left(\left(x_1^2+x_2^2\right)\arctan\left(\frac{x_3}{\sqrt{x_1^2+x_2^2}}\right)\right. \\ &\left.\left.\left.+bx_2x_3\left(e_1 - \arctan\left(\frac{x_2}{x_1}\right)\right)\sqrt{x_1^2+x_2^2} + \left(x_1^2+x_2^2\right)\left(ax_2 + x_1\arctan\left(\frac{x_2}{x_1}\right)\right)\right)\right)\right); \end{aligned}$$

$$x_3 = z^{-1}x_3 + \frac{z^{-1}T}{x_1^2 + x_2^2 + x_3^2}\left(-ax_3\left(g(x_1) - \arctan\left(\frac{x_2}{x_1}\right)\right)\right.$$
$$\left.\times\sqrt{x_1^2 + x_2^2 + x_3^2} + (b(e_1 + \arctan\left(\frac{x_2}{x_1}\right)\right))\sqrt{x_1^2 + x_2^2 + x_3^2}\left(x_1^2 + x_2^2 + x_3^2\right)\Big).$$

We call (34) as Chua circuit equations in the Cartesian space for the case when circuit dynamic initially defined in the angular coordinates. Analysis of (34) shows that all of state variables depend each other previous values. Moreover, each equation in (34) depend on external signal e1. We use this fact to claim the possibility to form signals to transmit with desired characteristics and components. Eqs. (30) and (34) shows the use of transformation expressions (9) and (11) to define discrete-time system motion in the different state spaces. As one can see, dynamic of transformed system in these spaces are defined by equations which are differ from the initial Chua equations as well as each other.

Initial conditions for the transformed model can be defined by substituting into (7) the initial conditions of the Chua circuit state variables in angular coordinates.

Let us solve (33) for e1 to define the descriptor equations for this case

$$e_1 = \frac{-Tz^{-1}}{Tz^{-1}\sqrt{x_1^2 + x_2^2}bx_1x_3\left(x_1^2 + x_2^2 + x_3^2\right)}(((x_1^2 + x_2^2)\left(ax_1g(x_1) - \right. \tag{35}$$
$$-ax_1\arctan\left(\frac{x_2}{x_1}\right) + x_2\left(x_1^2 + x_2^2 + x_3^2\right)\sqrt{x_1^2 + x_2^2 + x_3^2} +$$
$$+\left(x_1^2 + x_2^2 + x_3^2\right)x_2\left(\left(x_1^2 + x_2^2\right)\arctan\left(\frac{x_3}{\sqrt{x_1^2 + x_2^2}}\right) - \sqrt{x_1^2 + x_2^2}bx_1x_3\right.$$
$$\left.\left.\times\arctan\left(\frac{x_2}{x_1}\right) + \left(x_1^2 + x_2^2\right)\left(ax_1 - x_2\arctan\left(\frac{x_2}{x_1}\right) + x_1 - z^{-1}x_1\right)\right)\right));$$

$$e_1 = \frac{-Tz^{-1}}{Tz^{-1}\sqrt{x_1^2 + x_2^2}bx_2x_3\left(x_1^2 + x_2^2 + x_3^2\right)}\left(\left(x_1^2 + x_2^2 + x_3^2\right)\right.$$
$$\times\left(\left(x_1^2 + x_2^2\right)\left(ax_2g(x_1) - ax_2\arctan\left(\frac{x_2}{x_1}\right) - x_1\left(x_1^2 + x_2^2 + x_3^2\right)\right.\right.$$
$$\times\sqrt{x_1^2 + x_2^2 + x_3^2} - x_1\left(\left(x_1^2 + x_2^2\right)\arctan\left(\frac{x_3}{\sqrt{x_1^2 + x_2^2}}\right) - \sqrt{x_1^2 + x_2^2}b\right.$$
$$\left.\left.\times x_2x_3\arctan\left(\frac{x_2}{x_1}\right) + \left(x_1^2 + x_2^2\right)\left(ax_2 + x_1\arctan\left(\frac{x_2}{x_1}\right) + x_2 - z^{-1}x_2\right)\right)\right);$$

$$e_1 = \frac{-Tz^{-1}}{Tz^{-1}\sqrt{x_1^2 + x_2^2}b\left(x_1^2 + x_2^2 + x_3^2\right)}((ax_3\left(g(x_1) - \arctan\left(\frac{x_2}{x_1}\right)\right)\sqrt{x_1^2 + x_2^2 + x_3^2} +$$

$$\left(x_1^2+x_2^2+x_3^2\right)\left(\arctan\left(\frac{x_2}{x_1}\right)b\sqrt{x_1^2+x_2^2}+ax_3+x_3-z^{-1}x_3\right)\Bigg).$$

If one compares the solutions of inverse dynamic problems given by (31) and (35) he finds that the use of trigonometrical functions to define coordinate transformations allows to increase the number of expressions which define the solution of inverse problem. Moreover, due to the more complex transformation from the angular coordinates into Cartesian ones, external signal can be defined by using three different expressions which are defined with different signals. Analysis of expressions (30), (31), (33), and (35) shows the presence of similar patterns in their determinations. The use of these patterns allows us to simplify the implementation of above-mentioned expressions. What is why we offer to define the arctan functions as well as square roots, squares and multiplication of different trigonometric functions one time during the calculation cycle and then use them in the decryptors and encryptors equations.

Results of studies the communication system which are based on (33) and (35) are shown in Figs. 9 and 10.

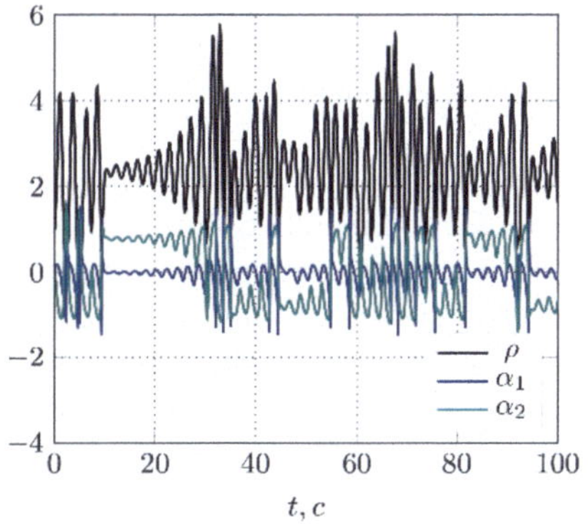

Fig. 9. Chua circuit angular motions.

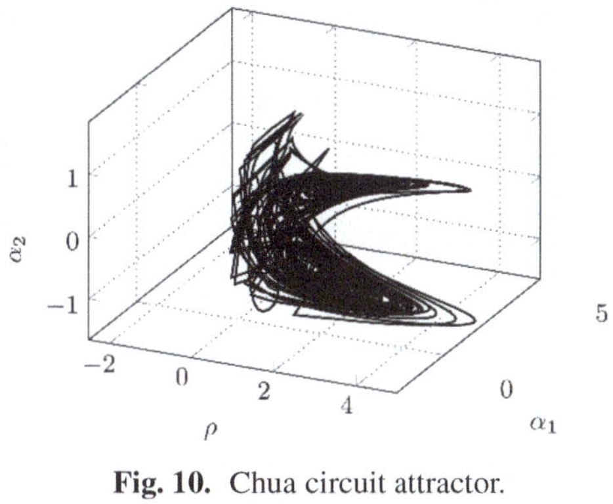

Fig. 10. Chua circuit attractor.

If one analyzes of given in Figs. 9 and 10 simulation results, he finds that one of system state variables has two equilibrium positions near ± 1. This fact forms two wings chaotic attractor which is shown in Fig. 10. It should be said that contrary to previous system the considered one can be considered as some amplifier which gain one of the outputs almost thrice in comparison with others.

4 Conclusions

The usage of the above-considered transformations allows us to define the chaotic system with a predefined attractor in the given domain of the system state space. It becomes possible to describe system trajectories by finite-difference equations, which can be easily implemented with modern FPGA/MCU devices due to the reused fragments. These fragments are defined with trigonometric functions, creating a new class of chaotic systems with harmonic periodic functions. Localization of the system in this domain forms its desired motion trajectories and determines its features.

The use of these features gives us the possibility to find the dependencies between system output and some external signal. Our approach shows that it becomes possible to solve inverse dynamic problem and define this signal by known system motions. Moreover, several such dependencies can be found. That is why we offer to use them in two ways.

At first, different dependencies for the same external signal can be used to define the transmitted signal and check its correctness. We offer to use such a case to establish a secured point-to-point communication channel by implementing a transmitter based on (30) and (34) and a receiver whose dynamic is defined by (31) or (35). One should consider a transmitter as the master device whose motion depends only on external signals. The receiver is taken as a slave device whose motions depend on the signals that are obtained from the transmitter.

Second, the use (34) and (35) defines several ways to find a signal e_1. One can use such models to establish secured point-to-multipoint communications. Here, we assume that the transmitter is implemented using (34) as a master device, and several receivers can be designed using (35). It is clear that the number of receivers depends on the space dimension where the system dynamic is transformed.

In our paper, we consider the case when initial and transformed systems have the same dimensions, but analysis of (7) and (10) shows that it is possible to consider the case when the motions of the transformed system are considered in some multidimensional state space. We see the future development of our work in design and study systems in which the order of transformed trajectory differs from the initial one.

References

1. Strohmeier, M., et al.: Assessing the impact of aviation security on cyber power. In: 2016 8th International Conference on Cyber Conflict (CyCon), pp. 223–241. Tallinn, Estonia (2016). https://doi.org/10.1109/CYCON.2016.7529437

2. Holubnychyi, O., et al.: Self-organization technique with a norm transformation based filtering for sustainable infocommunications within CNS/ATM systems. In: Ostroumov, I., Zaliskyi, M. (eds.) Proceedings of the 2nd International Workshop on Advances in Civil Aviation Systems Development. ACASD 2024, Lecture Notes in Networks and Systems, vol. 992, pp. 262–278. Springer, Cham (2024). https://doi.org/10.1007/978-3-031-60196-5_20
3. Solomentsev, O., et al.: Efficiency analysis of current repair procedures for aviation radio equipment. In: Ostroumov, I., Zaliskyi, M. (eds.) Proceedings of the 2nd International Workshop on Advances in Civil Aviation Systems Development. ACASD 2024, Lecture Notes in Networks and Systems, vol. 992, pp. 281–295. Springer, Cham (2024). https://doi.org/10.1007/978-3-031-60196-5_21
4. Strohmeier, M., et al.: On perception and reality in wireless air traffic communication security. IEEE Trans. Intell. Transp. Syst. **18**(6), 1338–1357 (2017)
5. Zaliskyi, M., et al.: Methodology for substantiating the infrastructure of aviation radio equipment repair centers. In: CEUR Workshop Proceedings, vol. 3732, pp. 136–148 (2024). https://ceur-ws.org/Vol-3732/paper11.pdf
6. Sushchenko, O., et al.: Algorithm of determining errors of gimballed inertial navigation system. In: Gervasi, O., et al. (eds.) Computational Science and Its Applications – ICCSA 2024 Workshops. ICCSA 2024 Lecture Notes in Computer Science, vol. 14816, pp. 206–218. Springer, Cham (2024). https://doi.org/10.1007/978-3-031-65223-3_14
7. Zhou, B., et al.: A method of verifying the secure communication between edge device and end device in power internet of things. In: 2020 IEEE 2nd International Conference on Civil Aviation Safety and Information Technology. ICCASIT-2020, pp. 228–231. IEEE, Weihai (2020). https://doi.org/10.1109/ICCASIT50869.2020.936855
8. Dansarie, M.: Security issues in special-purpose digital radio communication systems: a systematic review. IEEE Access. **12**, 91101–91126 (2024). https://doi.org/10.1109/ACCESS.2024.3420091
9. Mäurer, N., et al.: Efficient control-channel security for the aeronautical communications system LDACS. In: 24th International Symposium on a World of Wireless, Mobile and Multimedia Networks. WoWMoM-2023, pp. 407–412. IEEE, Boston, MA (2023). https://doi.org/10.1109/WoWMoM57956.2023.00072
10. Xin, L.: Research on computer electronic intelligent control system based on 5G wireless communication technology. In: 5th International Conference on Civil Aviation Safety and Information Technology. ICCASIT2023, pp. 47–51. IEEE, Dali (2023). https://doi.org/10.1109/ICCASIT58768.2023.10351620
11. Zhang, R., et al.: Analysis of message attacks in aviation data-link communication. IEEE Access. **6**, 455–463 (2018). https://doi.org/10.1109/ACCESS.2017.2767059
12. Mäurer, N., Bilzhause, A.: A cybersecurity architecture for the l-band digital aeronautical communications system (LDACS). In: 37th Digital Avionics Systems Conference, pp. 1–10. IEEE, London (2018). https://doi.org/10.1109/DASC.2018.8569878
13. Zhijun, W., Caiyun, W.: Security-as-a-service in big data of civil aviation. In: International Conference on Computer and Communications. ICCC-2015, pp. 240–244. IEEE, Chengdu (2015). https://doi.org/10.1109/CompComm.2015.7387574
14. Wang, R., et al.: Research on data security transmission technology of park integrated energy system based on quantum secure communication. In: 4th International Conference on Civil Aviation Safety and Information Technology. ICCASIT-2022, pp. 1418–1422. IEEE, Dali (2022). https://doi.org/10.1109/ICCASIT55263.2022.9986980
15. Mäurer, N., et al.: PMAKE: physical unclonable function-based mutual authentication key exchange scheme for digital aeronautical communications. In: International Symposium on Integrated Network Management, pp. 206–214. IEEE, Bordeaux (2021)

16. Wu, Z., et al.: Research on security authentication of air traffic management networks based on diameter. In: Cross Strait Quad-Regional Radio Science and Wireless Technology Conference, pp. 1239–1242. IEEE, Harbin (2011)
17. Zhang, Z., et al.: Trust management method of D2D communication based on RF fingerprint identification. IEEE Access. **6**, 66082–66087 (2018). https://doi.org/10.1109/ACCESS.2018.2878595
18. Kanada, N., Lu, X.: Preliminary feasibility study of quantum key distribution for future air traffic management systems. In: 15th International Symposium on Autonomous Decentralized System ISADS-203, pp. 1–4. IEEE, Mexico City (2023). https://doi.org/10.1109/ISADS56919.2023.10092072
19. Murtaza, A., et al.: IP-based space air ground information network for air traffic control communication. In: Pacific Rim Conference on Communications, Computers and Signal Processing, pp. 1–6. IEEE, Victoria (2019). https://doi.org/10.1109/PACRIM47961.2019.8985127
20. Wu, P., et al.: Design and implementation of 4A service in aviation broadband communication system. In: 43rd Digital Avionics Systems Conference, pp. 1–6. IEEE, San Diego (2024). https://doi.org/10.1109/DASC62030.2024.10748870
21. Huang, Y., et al.: Data source authentication protocol for aviation broadband communication system. In: International Wireless Communications and Mobile Computing, pp. 1242–1247. IEEE, Ayia Napa (2024). https://doi.org/10.1109/IWCMC61514.2024.10592349
22. Aydin, T., Rodosek, G.D.: Machine learning based predictive handover in unmanned aerial systems communication. In: 42nd Digital Avionics Systems Conference, vol. 3, pp. 1–5. IEEE, Barcelona (2023). https://doi.org/10.1109/DASC58513.2023.10311303
23. Averyanova, Y., et al.: Algorithm of wind-related hazards prediction for UAS flight and urban operations based on meteorological data fusion. In: CEUR Workshop Proceedings, vol. 3895, pp. 132–142 (2024). https://ceur-ws.org/Vol-3895/paper11.pdf
24. Wu, Z., et al.: Blockchain-based authentication of GNSS civil navigation message. IEEE Trans. Aerosp. Electron. Syst. **59**(4), 4380–4392 (2023). https://doi.org/10.1109/TAES.2023.3241041
25. Li, J., et al.: Analysis and future challenge of blockchain in civil aviation application. In: 6th International Conference on Computer and Communications, pp. 1742–1748. IEEE, Chengdu (2020). https://doi.org/10.1109/ICCC51575.2020.9345297
26. Znakovska, Y., et al.: The information technologies use for UAS operators' training. In: Faure, E., et al. (eds.) Information Technology for Education, Science, and Technics. ITEST 2024 Lecture Notes on Data Engineering and Communications Technologies, vol. 222, pp. 327–338. Springer, Cham (2024). https://doi.org/10.1007/978-3-031-71804-5_22
27. Chien, T.I., et al.: Design of multiple-accessing chaotic digital communication system based on Interleaved Chaotic Differential Peaks Keying (I-CDPK). In: 6th International Symposium on Communication Systems, Networks and Digital Signal Processing, pp. 638–642. IEEE, Graz (2008). https://doi.org/10.1109/CSNDSP.2008.4610717
28. Shaocheng, Q., et al.: Adaptive secure communication based on unified chaotic systems with varying parameters. In: 2009 Chinese Control and Decision Conference, pp. 18–22. IEEE, Guilin (2009). https://doi.org/10.1109/CCDC.2009.5195141
29. Wang, S.F.: Dynamical analysis of memristive unified chaotic system and its application in secure communication. IEEE Access. **6**, 66055–66061 (2018). https://doi.org/10.1109/ACCESS.2018.2878882
30. Jinlong, L., et al.: A secure communication scheme using chaotic system array. In: 34th Chinese Control Conference, pp. 407–412. IEEE, Hangzhou (2015). https://doi.org/10.1109/ChiCC.2015.7259672

31. Moufak Dukhan, A., et al.: A multi-level hybrid chaotic system with a novel damping approach to achieve variable rates. In: 27th International Telecommunication Networks and Applications Conference, pp. 1–6. IEEE, Melbourne, VIC (2017). https://doi.org/10.1109/ATNAC.2017.8215423
32. Liu, P., et al.: Dynamic analysis of novel memristor chaotic systems with influence factors. In: International Conference on Electrical, Automation and Computer Engineering, pp. 118–122. IEEE, Changchun (2023). https://doi.org/10.1109/ICEACE60673.2023.10441899
33. Rodrigues, V.H.P., et al.: Globally stable synchronization of chaotic systems based on norm observers connected in cascade. IEEE Trans Circuits Syst II Express Briefs. **63**(9), 883–887 (2016). https://doi.org/10.1109/TCSII.2016.2534798
34. Prabu, R., et al.: A new Chaotic Jerk system based only on a single quadratic equation displaying coexisting attractors. In: Third International Conference on Smart Technologies and Systems for Next Generation Computing. ICSTSN-2024, pp. 1–3. IEEE, Villupuram (2024). https://doi.org/10.1109/ICSTSN61422.2024.10670956
35. Voliansky, R., et al.: Variable-structure interval-based duffing oscillator. In: 42nd International Conference on Electronics and Nanotechnology (ELNANO), pp. 581–586. IEEE, Kyiv (2024). https://doi.org/10.1109/ELNANO63394.2024.10756860
36. Bian, H., et al.: Parameter inversion of high-dimensional chaotic systems using neural ordinary differential equations. In: Data Driven Control and Learning Systems Conference, pp. 400–405. IEEE, Kaifeng (2024). https://doi.org/10.1109/DDCLS61622.2024.10606602
37. Voliansky, R., Volianska, N.: Modified Chua's circuit in different state spaces. In: Ostroumov, I., Zaliskyi, M. (eds.) Proceedings of the 2nd International Workshop on Advances in Civil Aviation Systems Development. ACASD 2024. Lecture Notes in Networks and Systems, vol. 992, pp. 423–440. Springer, Cham (2024). https://doi.org/10.1007/978-3-031-60196-5_31

Model the Geometric Characteristics Determining of Spatial Landmarks for Survey-Comparative Navigation System

Oleksii Chuzha, Yurii Hryshchenko(✉), Olena Tachinina, and Oleksandra Bankova

National Aviation University, Liubomyra Huzara Ave., 1, Kyiv 03058, Ukraine
hryshchenko18@kai.edu.ua

Abstract. Modern survey-comparative navigation systems provide an integrated representation of the complete set of navigation data, interact with onboard navigation computers, correct other navigation systems, and serve as a crucial informational link in the "Human-Machine" system. Automatic survey-comparative navigation systems based on laser three-dimensional primary information sensors are autonomous and promising for use in modern aircraft. The implementation of survey-comparative navigation methods involves the task of recognizing ground landmarks, comparing them with reference landmarks with known positions, and establishing the coordinates of the aircraft relative to these landmarks. Therefore, the article explores the topic of determining the geometric characteristics of spatial objects based on the analysis of their three-dimensional images obtained by Light Identification, Detection and Ranging (LIDAR). Processing three-dimensional LIDAR images allows for determining the geometric dimensions and spatial structure of simple and complex volumetric landmarks, which is sufficient for solving the tasks of ground landmark recognition and the implementation of survey-comparative navigation methods.

Keywords: Survey-Comparative Navigation Methods · LIDAR · Geometric Features · Basic Elements of the Form · Solution of the Recognition Problem

1 Introduction

The flight of an aircraft consists of the translational movement of its center of mass relative to a chosen coordinate system, facilitated by means of air navigation, and the movement of the aircraft around its center of mass, which is carried out through piloting. Air navigation deals with movement of an aircraft from one point to another along specific spatio-temporal trajectories [1, 2]. The concept of air navigation encompasses three processes:

- determining the position and velocity of the aircraft in a reference coordinate system;
- comparing the measured values with the calculated ones for the given moment in time;

I. Ostroumov et al. (Eds.): ACASD 2025, LNNS 1418, pp. 420–432, 2025.
https://doi.org/10.1007/978-3-031-91992-3_28

- converting the results of the comparison into control commands that ensure the adjustment of movement parameters and the execution of the flight route to achieve the final goal.

Parameters of aircraft motion are used to formulate the trajectory movement program, are referred to as navigation parameters. Survey-comparative navigation methods involves comparing the terrain with prominent landmarks, placed on a navigation map or in navigation computer (reference image), with its real image obtained through onboard technical vision means (infrared television, radar, and others) [3]. Coordinates of aircraft location could be estimated if the reference image of the terrain matches the real one with a specified probability. Additionally, the coordinates of navigation landmarks, prominent objects, airfields, and other features can be determined.

In modern aircraft, survey-comparative navigation systems are rarely used due to the involvement of an operator in the ground landmark recognition process, and in automatic systems, due to errors associated with the low informativeness of ground landmark features.

A significant advance in the speed and memory capacity of modern navigation computers and the development of three-dimensional image formation systems allows to use specific methods of ground landmarks recognition.

2 Problem Statement

Until recently, correlation-extremal algorithms were the primary class of algorithms used for object detection and identification in survey-comparative navigation systems of aircraft. However, as demands for accuracy and reliability in detecting and recognizing more complex objects in challenging real-world conditions have increased, the limitations of this class have become more apparent.

These limitations include a high probability of errors, the need for a large number of reference images to describe multi-angle spatial objects, and instability with respect to brightness-geometric variability of images, which occurs under real registration conditions. Consequently, there has been a shift from correlation algorithms to methods and algorithms of structural image analysis.

Currently, the sequence of image processing procedures is considered according to the so-called modular paradigm [4]. This paradigm is based on extensive studies of human visual perception mechanisms, asserts that image processing relies on several sequential levels of increasing informational complexity from the "pictorial" image of objects (raster image, unstructured information) to their symbolic representation (vector and attribute data in structured form, relational structures) and should be carried out in a modular fashion through the following processing stages:

- preliminary image processing;
- primary image segmentation;
- extraction of the geometric structure of the visible field;
- determination of the relative structure and semantics of the visible scene.

The levels of processing associated with these stages are typically referred to as lower-level, mid-level, and high-level processing, respectively. While lower-level processing algorithms (simple noise filtering, histogram processing) can be considered well-developed and thoroughly studied, mid-level algorithms (segmentation) continue to be a central area of research efforts today. Significant progress has been made in recent years regarding the problems of comparing image points and fragments, feature extraction within small fragments, and high-precision 3D positioning of points, which implies appropriate modeling and calibration of sensors and their combinations, as well as the extraction of simple brightness-geometric structures like “point”, “edge”, “spot”, “straight line”, “corner”. These “primary” image features, also known as characteristic features, play a fundamental role in the construction of geometric models of objects and the development of algorithms for their extraction.

The specific choice of characteristic features and their components depends on the available computational power and the minimum required robustness of the object’s model description in terms of these features. High-level processing methods, related to “understanding images”, are still in the early stages of development in the field of computer vision and artificial intelligence. Undoubtedly, the future development of intelligent control systems largely depends on the advancement of this class of algorithms.

One of the main challenges for methods and algorithms for detecting complex objects, which distinguishes image processing methods from the well-studied theory of signal processing, is the development of object detection methods that are insensitive to various types of variability unique to images. Such specific types of variability include angle and radiometric distortions, as well as other distortions that do not reduce to probabilistic models (noise, shapes). Currently, several major algorithmic approaches and mathematical models are known, which are used in the development of practical image analysis systems. These include histogram transformations, projection analysis, linear and nonlinear image filtering, brightness and texture segmentation, correlation detection and matched filtering, the “background normalization” method, the Hough transform, the structural-linguistic approach, and several others.

Despite the achieved results, the overall state of the problem of extracting and identifying complex-structured objects in monoscopic images can be described as unsatisfactory. The task of detecting three-dimensional structures in stereoscopic images appears even more challenging. Thus, the analysis of characteristic features of flat image objects requires a large amount of software resources; additionally, the recognition of such objects is possible only based on geometric dimensions, which may change depending on the viewing angle, or on distinct contrasting features against the background. Much attention is currently being paid to the problem of high-quality landing approaches [5–8].

3 Methodology

The navigational content of survey-comparative measurement methods is determined by the appearance and quantity of landmarks. In single-landmark systems, the physical parameters of the landmark (area, geometric shape features, emission spectrum, etc.) are compared with those of measured objects. From the aircraft, the planned physical parameters of the landmark are measured, and the navigational parameters of the vector of its location relative to the navigation coordinate system are determined. A significant drawback of such systems is their low resistance to interference, as they can analyze unplanned landmarks that are physically similar to the true ones.

In multi-landmark systems, several planned landmarks are used simultaneously. These systems store information not only about the parameters of the landmarks but also about their mutual coordinates. The advantage of multi-landmark systems is their reduced dependency on the loss of some planned landmarks and interference. However, implementing such systems requires powerful electronic computing resources.

One way to increase the effectiveness of survey-comparative navigation systems (SCNS) is to perform tasks of recognizing ground landmarks to determine the aircraft's location. The features currently used in SCNS, such as true altitude and terrain changes, do not fully meet the efficiency requirements for recognition. Therefore, new features that can accurately describe volumetric figures and be easily obtained from digital images of the Earth's surface are necessary. These features can be three-dimensional geometric characteristics derived from processing digital images of ground objects.

An important step in solving the recognition problem is developing a working dictionary of features. At this stage, only those features of objects that are most informative and can be determined by available measuring devices, or specially created ones, are included in the working dictionary. Geometric features are the most suitable for recognizing ground spatial objects. However, geometric features can differ in their informativeness and extraction method.

The informativeness of geometric features can be evaluated through their study. This task can be addressed by modeling the process of forming three-dimensional images of recognized objects and processing them to extract three-dimensional geometric features for recognition. This requires developing mathematical models of the spatial object and the process of forming three-dimensional images of objects, as well as assessing the range at which information about the three-dimensional features of objects can be determined using a laser system for forming three-dimensional images.

To successfully accomplish the tasks of recognizing ground landmarks, SCNS must implement three models: a spatial object model, a three-dimensional image formation model, and a three-dimensional image analysis model for extracting geometric features of landmarks. The first stage involves considering the spatial object model.

3.1 Spatial Object Model

Spatial objects, regardless of its origin (natural or artificial), are material formations with specific dimensions and characteristic shapes. Despite the considerable diversity of these objects, they can be accurately represented using a set of material points, linear elements (straight lines, curves), simple or volumetric geometric figures (plane, parallelepiped, cylinder, sphere, cone, pyramid), or their combinations and parts (Fig. 1).

This methodology simplifies the process of object modeling for subsequent generation of their 3D images and extraction of characteristic geometric features. The spatial object model finds applications in technical vision systems, robotics, and other fields requiring the identification and analysis of objects in space.

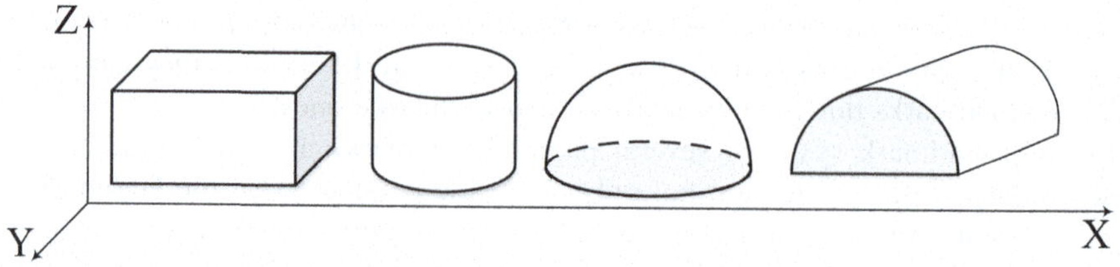

Fig. 1. Spatial object models.

A mathematical model of a spatial object for trajectory recognition systems consists of the geometric model of the object itself and basic reference elements, such as the underlying ground surface. Additionally, it is necessary to consider the model of the reflectance characteristics of the object elements and the ground surface, depending on the type of image formation sensor (radar, laser, infrared).

Several well-known methods are utilized for modeling, including parametric and non-parametric design, as well as modeling using Basic Shape Elements (BSE) [9, 10]. The implementation of algorithms for constructing images of curved surfaces using the first two methods is computationally intensive, thus limiting their practical use. A more prevalent approach for describing surfaces of spatial objects involves representing them using BSE, which allows for accurate surface construction, coordinate point transformations, computation of surface normals, and efficient utilization of computational resources (Fig. 2). Typically, simple geometric figures (planar sections) are used as BSE. However, a significant number of artificially generated objects may be sufficiently described by second-order surfaces (sphere, cylinder) or a combination of such surfaces (Fig. 3).

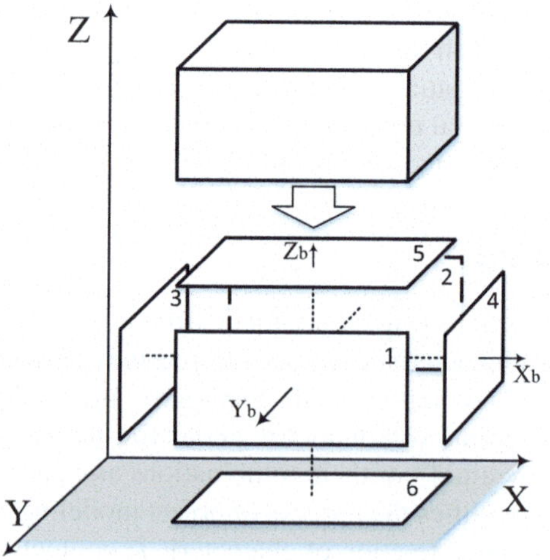

Fig. 2. Basic shape elements of parallelepiped.

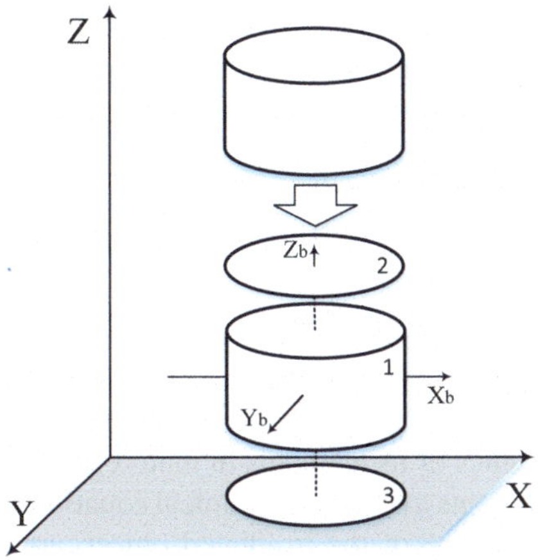

Fig. 3. Basic shape elements of cylinder.

Simple surfaces (line, plane, hemisphere) are described by second-degree equations relative to Cartesian rectangular coordinate systems. The general second-degree equation in a spatial coordinate system is expressed as [11–13]:

$$k_{11}x^2 + k_{22}y^2 + k_{33}z^2 + 2k_{12}xy + 2k_{13}xz + 2k_{23}yz$$

$$+2k_{14}x + 2k_{24}y + 2k_{34}z + k_{44} = 0. \quad (1)$$

The coefficients k_{ij} in (1) determine the shape of the surface. However, for constructing a spatial object, specifying only the values of these coefficients k_{ij} is insufficient. It is also necessary to define the boundaries of the BSE. This involves an additional six auxiliary coefficients b_{ij} that determine the boundaries of the surface relative to the reference plane (see Fig. 2).

Thus, to design a spatial object, information about sixteen coefficients is required. The expression (1) considering the new coefficients is as follows:

$$a_{11}x^2 + a_{12}y^2 + a_{13}z^2 + a_{14}xy + a_{15}xz + a_{16}yz + a_{17}x + a_{18}y + a_{19}z$$

$$+a_{110} + b_{111} + b_{112} + b_{113} + b_{114} + b_{115} + b_{116} = 0, \quad (2)$$

where $a_{11} \ldots a_{110}$ correspond to the coefficients k_{ij} in (1); auxiliary coefficients: b_{111} and b_{112} is boundaries of the shape element along the axis OX, b_{113} and b_{114} is boundaries of the shape element along the axis OY, b_{115} and b_{116} is boundaries of the shape element along the axis OZ.

If the object consists of several BSE, then each element will correspond to a separate Eq. (2), and the entire object can be characterized in the form of a coefficient matrix a_{ij} and b_{ij} with dimensions $i \times 16$, where there is the number of BSE (Fig. 2):

$$A = \left\| \begin{matrix} a_{11} \ldots a_{110} \ldots b_{111} \ldots b_{116} \\ a_{11} \ldots a_{110} \ldots b_{111} \ldots b_{116} \\ \ldots \ \ldots \ \ldots \ \ldots \ \ldots \ \ldots \ \ldots \\ a_{i1} \ldots a_{i10} \ldots b_{i11} \ldots b_{i16} \end{matrix} \right\|. \tag{3}$$

This form of recording information (3) is the most convenient for use in calculators and ensures compactness and efficient access. Thus, to obtain a model of a complex object, it is necessary to set the values of its elements in matrix A (3).

To determine the coefficients $a_{11} \ldots a_{110}$ canonical equations of second-order surfaces [14], which describe the surface of the irradiated object, are used. In this process, it is necessary to account for the change in coefficients when shifting and rotating the second-order surfaces used as BSE.

The construction of the object model can be divided into three stages.

In the first stage, the object is divided into segments, each of which can be represented by a single BSE [13]. For each segment, a "local" rectangular coordinate system $O_1X_1Y_1Z_1$ is determined (Fig. 4). Each of the selected BSEs is described by the canonical equation of a second-order surface, which can be generally represented by the expression [11]:

$$\lambda_1 x_i^2 + \lambda_2 y_i^2 + \lambda_3 z_i^2 = 0, \tag{4}$$

where λ_1, λ_2, λ_3 are coefficients that determine the shape and size of the second-order surface.

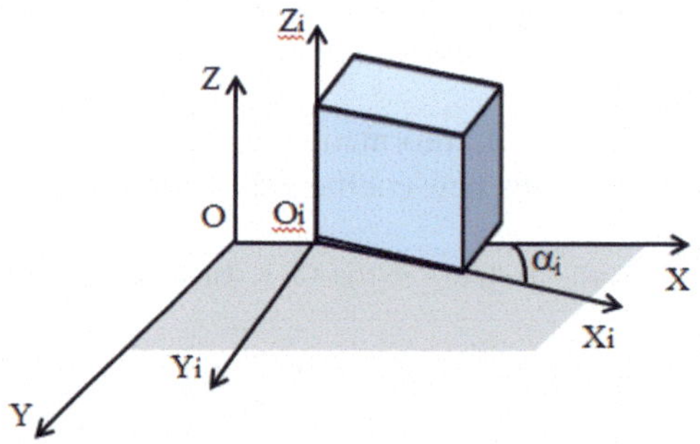

Fig. 4. BSE coordinate system ($O_iX_iY_iZ_i$).

The angles $\alpha_i, \beta_i, \gamma_i$ of the orientation of the coordinate system (CS) $O_iX_iY_iZ_i$ relative to the base coordinate system $OXYZ$ and the coordinates of the origin O_i of the system $O_iX_iY_iZ_i$ relative to the base coordinate system, which is chosen as the fixed terrestrial CS $OXYZ$ are also determined. To determine the angles α_i, β_i, $\gamma_i\alpha$ and the coordinates of the origin O_i, it is necessary to determine the orientation of the object relative to the base (geographical) coordinate system.

In the second stage, the direct calculation of the coefficients aij is performed. To do this, it is necessary to express the coordinates $O_iX_iY_iZ_i$ through the coordinates of the base CS. Using the coordinate transformation formulas, considering the simultaneous rotation and translation of the axes, known from analytical geometry [10], we obtain:

$$\begin{cases} x_i = d_{1i}(x - x_{0i}) + d_{2i}(y - y_{0i}) + d_{3i}(z - z_{0i}) \\ y_i = d_{1i}(x - x_{0i}) + b_{2i}(y - y_{0i}) + b_{3i}(z - z_{0i}) \\ x_i = c_{1i}(x - x_{0i}) + c_{2i}(y - y_{0i}) + c_{3i}(z - z_{0i}) \end{cases}, \tag{5}$$

where d_{1i}, d_{2i}, d_{3i} are direction cosines of the X axis relative to the axes X_i, Y_i, Z_i; b_{1i}, b_{2i}, b_{3i} are direction cosines of the Y axis relative to the axes X_i, Y_i, Z_i; c_{1i}, c_{2i}, c_{3i} are direction cosines of the Z axis relative to the axes X_i, Y_i, Z_i.

To determine the direction cosines, information about the orientation angles of the coordinate system $O_iX_iY_iZ_i$ relative to the base CS and the known transition matrices [9] is required. Substituting (5) into (4), after transformations and bringing the equation to the form (2), we obtain the values of the coefficients $a_{11} \ldots a_{110}$. To calculate the coefficients $b_{i11} \ldots b_{i16}$, (see Eqs. 2 and 3) which determine the minimum and maximum dimensions of the BFE relative to the X, Y, Z axes, it is necessary to use the drawings of the object and its dimensions (see Fig. 2). At the same time, it should be noted that the dimensions of the BSE are determined relative to the base CS.

On the third stage, a matrix A of computed coefficients (3) is formed, which can be stored in memory as corresponding structural data. Thus, by performing the enumerated operations, we obtain a mathematical model of the object and its surface.

To construct an image of the object, it is necessary to calculate normals to points lying on its surface. The equation of the normal to the surface represented by the i-th BSE, described by expression (2), at point (x_1, y_1, z_1) according to [14, 15] will have the form:

$$\frac{x - x_1}{a_{i1}x_1 + \frac{1}{2}a_{i4}y_1 + \frac{1}{2}a_{i5}z_1 + \frac{1}{2}a_{i7}} = \frac{y - y_1}{\frac{1}{2}a_{i4}x_1 + a_{i2}y_1 + \frac{1}{2}a_{i6}z_1 + \frac{1}{2}a_{i8}} = \frac{z - z_1}{\frac{1}{2}a_{i5}x_1 + \frac{1}{2}a_{i6}y_1 + a_{i3}z_1 + \frac{1}{2}a_{i9}}. \tag{6}$$

The mathematical model of the radar reflection characteristics of objects can vary in complexity depending on the specific task under consideration. The primary characteristic of the radar scattering ability of a surface is the scattering cross-section $l(\vec{m}, \vec{n})$ [15]. It is defined as the ratio of the brightness of the object's surface in a given direction$L(\vec{m}, \vec{n})$ to the brightness of ideal scattering L0, and is expressed as:

$$l(\vec{m}, \vec{n}) = \frac{L(\vec{m}, \vec{n})}{L_0}, \tag{7}$$

where $\vec{m}$, $\vec{n}$ – unit vectors that respectively indicate the direction of incident and scattered radiation.

The indications scattering characteristics of natural formations and artificial objects have a complex form and depend on numerous factors, which limits obtaining results in analytical form [16]. Therefore, in modeling using energy calculations, the Lambertian reflector model is often used. In this model, the scattering cross-section does not depend on the direction of incident and scattered radiation $l = \rho$, where ρ is the power reflection coefficient. Thus, the primary reflection characteristic for modeling is the power reflection coefficient.

In general, the reflection coefficients of different segments of the underlying surface and the irradiated object vary. Therefore, it is practical to represent the field of reflection coefficients as homogeneous regions, synthesizing a piecewise-homogeneous field. The formation of a piecewise-homogeneous field of reflection coefficients occurs in two stages: defining the boundaries of homogeneous regions and filling these regions with homogeneous fields of reflection coefficients. Homogeneous regions are taken as the surfaces of BSE. Therefore, the model of reflection characteristics can be expressed in the form of a matrix B (see Eq. 8), where the elements are reflection coefficients. Such a matrix takes the form:

$$B = \left\| \begin{matrix} \rho_1 \\ \rho_2 \\ \ldots \\ \rho_i \end{matrix} \right\|. \quad (8)$$

Thus, the mathematical model of the object and the underlying surface can be represented in the form of matrices, where the elements contain information about the shape of the object and the underlying surface, their dimensions, and reflection characteristics. To compute the coefficients of these matrices, it is necessary to know the dimensions of the irradiated object, its shape, and the values of the reflection coefficients of the underlying surface and the object itself.

3.2 Measuring Time Intervals

The most promising are laser systems for forming three-dimensional images, based on the LiDAR technology, which have high spatial and temporal characteristics. LiDAR is an active type distance measuring locator, which allows to implement the method of laser stereometry, which is a direct method of measuring the three-dimensional shape and size of the observed object [17].

Airborne LiDAR systems are installed on aircraft or UAVs to collect high-resolution topographic data over specific areas of interest. These systems emit laser pulses towards the ground surface and measure the time intervals it takes for the light to return to the sensor after reflecting off the target.

The model of time intervals measuring refers to the time it takes for each element of the LIDAR matrix radiation receiver (MRR) to register the signal reflected from the irradiated object.

To formulate the equation for determining these time intervals, let's refer to Fig. 5, which depicts the simplest case of probing a ground object. Here, a single range vector is characterized solely by the azimuth angle $\overrightarrow{D_0}$ relative to the fixed terrestrial coordinate system OXYZ with the objective of the LIDAR matrix radiation receiver positioned at point O.

An imaging plane is perpendicular to the distance vector $\overrightarrow{D_0}$, passing through closest to the LIDAR matrix radiation receiver (MRR) the point O_D (with coordinates $x_D y_D z_D$),

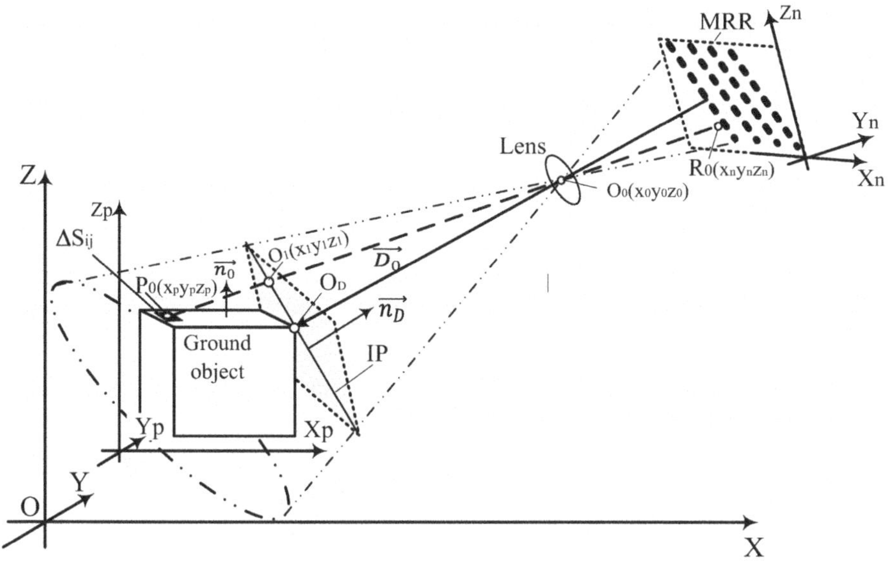

Fig. 5 The scheme to obtain time intervals from ground object.

which lies on the surface of the probed object. Central point R_0 of the MRR element is placed in coordinates $x_n y_n z_n$. Intersection point $O_1$of the line connecting points R_0 and P_0 with the imaging plane at the point with coordinates $x_1 y_1 z_1$.

Based on Fig. 5 we could determine the time interval between the beginning of radiation arrival at the LIDAR lens and the moment of registering the radiation, reflected from the segment ΔS_{ij} with the center at point P_0, by LIDAR MRR element at point R_0, as following:

$$\tau_{ij} = \frac{\Lambda_{ij} \cdot \sqrt{(x_1 - x_p)^2 + (y_1 - y_p)^2 + (z_1 - z_p)^2}}{2c} + \Delta\tau_{ij}, \tag{9}$$

where Λ_{ij} is indicator function; $\Delta\tau_{ij}$ is coefficient characterizing error in measuring time interval by MRR element (where *i* and *j* are the matrix *A* element number).

The coordinates of point O_1 are found by simultaneously solving the equations of the imaging plane and the equation of the line passing through points P_0 and O_0. The equation of the imaging plane is given by:

$$A(x - x_D) + B(y - y_D) + C(z - z_D) = 0, \tag{10}$$

where *A*, *B*, *C* are the projections of the normal vector $\overrightarrow{n_D}$ to the imaging plane (IP) onto the axes of the coordinate system *OXYZ*. Since the vectors $\overrightarrow{n_D}$ and $\overrightarrow{D_0}$ are collinear, the projections of the normal vector are determined by the specified angles that characterize the direction of the unit range vector $\overrightarrow{D_0}$ relative to the coordinate system *OXYZ*.

For the distance to the irradiated object, which significantly exceeds its dimensions, expression (9) [18]:

$$\tau_{ij} = \frac{\Lambda_{ij} \cdot \left(\sqrt{(x_0 - x_p)^2 + \left(y_0 - y_p\right)^2 + (z_0 - z_p)^2} - D_{min} \right)}{2c} + \Delta\tau_{ij}. \quad (11)$$

Therefore, to determine the time intervals, it is necessary to specify the distance to the irradiated object, the orientation angles of the distance vector $\overrightarrow{n_0}$ relative to the terrestrial coordinate system *OXYZ*, the coordinates of points P_0 and O_D, as well as the random variable $\Delta\tau_{ij}$.

3.3 Recognition Task for a Spatial Landmark

The criterion for assessing the accuracy of determining coordinates in the survey-comparative navigation system can be considered as the probability of correctly solving the recognition task $\overline{W}$ for a ground navigation landmark with known coordinates:

$$\overline{W} = \frac{n_t}{n_\Sigma}, \quad (12)$$

where n_t is number of correctly identified landmarks out of the entire set of objects; n_Σ is total number of experiments conducted to identify the landmarks.

Based on the spatial object model, a set of objects with different geometric and reflective properties can be formed. To solve navigation tasks, it is necessary to identify landmarks among these objects.

Geometric features provide a comprehensive description of the spatial object. Therefore, as features, it is proposed to use the volume V of the object and its average height $\overline{H}$. Analysis of these features is recommended using the criterion of the probability of correctly solving the recognition task $\overline{W}$.

Research results indicate that achieving $\overline{W} = 0.7$ or higher depends on the type of features and sensor characteristics [15]. Simultaneously using three-dimensional features enhances the probability of detecting ground landmarks based on the volume and average height $V \cup \overline{H}$ (Figs. 6 and 7).

The orientation angle of the object relative to the observation point and its shape also play a crucial role. The probability $\overline{W} = 0,7$ or higher for correctly solving the recognition task is achieved within the orientation angles of $70\,^\circ \leq \chi\,^\circ \leq 30^\circ$ for objects shaped like a parallelepiped. For objects shaped like a vertical second-order surface (vertical cylinder, hemisphere), this probability remains nearly unchanged (Fig. 7).

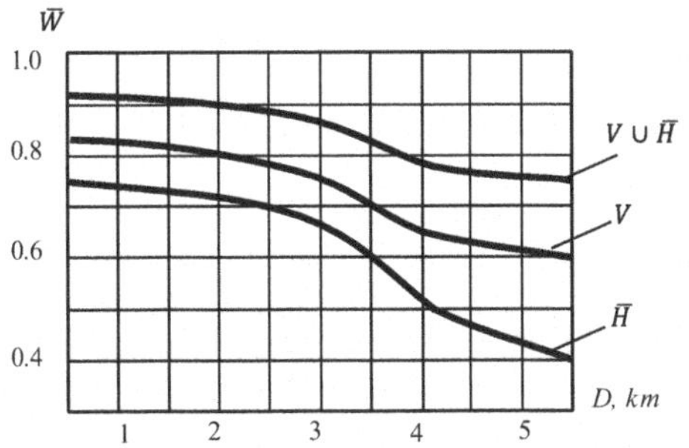

Fig. 6. Results of recognition task, depending on the type of geometric features.

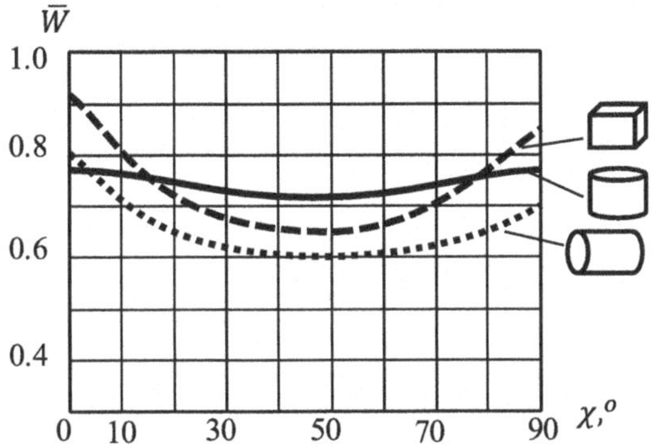

Fig. 7. Results of recognition task, depending on the type of geometric features and the object's orientation angles relative to the observation point.

Therefore, the mathematical model of spatial object allows for calculating their geometric characteristics with a specified probability, which is essential for forming recognition criteria in implementing survey-comparative navigation methods [19].

4 Conclusions

Based on the mathematical models of a spatial object and measuring time intervals, we can distinguish the following characteristics of the object: dimensions, shape, and reflective properties. Reflective properties of the object are variable parameters that depend on the angles of illumination, weather conditions, and the structure of the object. Therefore, using reflective properties as recognition features for landmarks is not advisable. It is more appropriate to use dimensions and shape of the object as recognition features for landmark detection in survey-comparative navigation systems.

In summary, understanding and utilizing the orientation angle and shape of objects are crucial in developing effective navigation systems that rely on survey-comparative methods, ensuring reliable recognition criteria based on mathematical models of spatial object and measuring time intervals.

References

1. Pleter, O.: Air Navigation: Fundamentals, Systems, and Flight Trajectory Management. Springer, Cham (2024)
2. Ostroumov, I., Ivannikova, V., Zaliskyi, M., Kuzmenko, N.: Impact analysis of Russian-Ukrainian war on airspace. J. Air Transp. Manag. **124**, 102742 (2025)
3. Mukhina, M.P., Seden, I.V.: Analysis of modern correlation extreme navigation systems. Electron. Control Syst. **1**(39), 95–101 (2014). https://doi.org/10.18372/1990-5548.39.7343
4. Marr, D.: Vision: A Computational Investigation into the Human Representation and Processing of Visual Information. MIT Press, Cambridge, MA (2010)
5. Hryshchenko, V., Romanenko, V., Hryshchenko, Y.: Quality of piloting during the approach. In: Ostroumov, I., Zaliskyi, M. (eds.) Proceedings of the International Workshop on Advances in Civil Aviation Systems Development. ACASD 2023. LNNS, vol. 736, pp. 260–270. Springer, Cham (2023). https://doi.org/10.1007/978-3-031-38082-2_20
6. Hryshchenko, Y., Chuzha, O., Romanenko, O., Bankova, O.: Alert system notifying the crew of the necessity for a go-around. In: Ostroumov, I., Zaliskyi, M. (eds.) Proceedings of the 2nd International Workshop on Advances in Civil Aviation Systems Development. ACASD 2024. LNNS, vol. 992, pp. 190–202. Springer, Cham (2024). https://doi.org/10.1007/978-3-031-60196-5
7. Ostroumov, I.V., Kuzmenko, N.S.: Accuracy assessment of aircraft positioning by multiple radio navigational aids. Telecommun. Radio Eng. **77**(8), 705–715 (2018). https://doi.org/10.1615/TelecomRadEng.v77.i8.40
8. Solomentsev, O., Zaliskyi, M.: Correlated failures analysis in navigation system. In: 5th International Conference on Methods and Systems of Navigation and Motion Control, pp. 41–44. IEEE, Kyiv (2018). https://doi.org/10.1109/MSNMC.2018.8576306
9. Faux, I., Pratt, M.: Computational Geometry for Design and Manufacture. Ellis Horwood, Chichester (1979)
10. Roberts, L.: Machine perception of three-dimensional solids. In: Tippett, J.T., et al. (ed.) Optical and Electro-Optical Information Processing (1965)
11. Buldygin, V., Alekseeva, I., Haidey, V., Dykhovychnyi, O., Konovalova, N., Fedorova, L.: Linear algebra and analytical geometry: textbook, TViMS (2011)
12. Yampolskyi, O., Shuhailo, O.: Canonical Theory of Surfaces of the Second Order. Study Guide. Kharkiv National University, Kharkiv (2020)
13. Chuzha, O., Pazyura, N.: Romanenko, V.: 3D model of landmarks for autonomous navigation of unmanned aerial vehicles. Electron. Control Syst. **3**(53), 2152, 112–119 (2017). https://doi.org/10.18372/1990-5548.53.12152
14. Hofmann-Wellenhof, B., Legat, K., Wieser, M.: Navigation. Principles of Positioning and Guidance. Springer Science & Business Media, York (2003)
15. Mikheenko, L.: Radiometric Measurements. Study Guide. NTU KPI, Kyiv (2011)
16. Bohren, C., Huffman, D.: Absorption and Scattering of Light by Small Particles. John Wiley & Sons, New York (2008)
17. Richmond, R.D., Cain, S.C.: Direct-Detection LADAR Systems. SPII, Washington (2010)
18. Chuzha, O., Romanenko, V.: Differential analysis of 3D sensor images of survey-comparative navigation system. Electron. Control Syst. **3**(57), 3226 (2018). https://doi.org/10.18372/1990-5548.57.13226
19. Chuzha, O., Pazyura, N., Romanenko, V.: Model of spatial landmarks for survey-comparative methods navigation of UAVs. In: IEEE 4th International Conference on Actual Problems of Unmanned Aerial Vehicles Developments, pp. 32–36. IEEE, Kyiv (2017). https://doi.org/10.1109/APUAVD.2017.8308770

Human Factor in Civil Aviation

Optimization of Cross-Professional Interaction "Pilot-Air Traffic Controller" During Collaborative Practical Training

Tetiana Shmelova[1], Yuliya Sikirda[2](✉), Maxim Yatsko[3], and Volodymyr Kolotusha[1]

[1] National Aviation University, Liubomyra Huzara Ave., 1, Kyiv 03058, Ukraine
[2] Ukrainian State Flight Academy, Stepana Chobanu Str., 1, Kropyvnytskyi 25005, Ukraine
sikirdayuliya@ukr.net
[3] SmartLynx Airlines "Mazrudas", Marupes novads 2167, Latvia

Abstract. The problem of cross-professional interaction between the pilot and air traffic controller (ATCO) is crucial to ensuring air traffic safety. In maintaining flight safety, it is important to organize high-quality collaborative decision-making (CDM) between the flight crew, ATCO, and other stakeholders based on their operational reliable communication in real-time. Therefore, the elaboration of CDM models for optimization of cross-professional interaction "pilot-ATCO" during their collaborative practical training is an actual problem. Based on the situation-based approach the structure of collaborative practical training of the pilots and ATCOs in a common educational environment is defined. The scheme of CDM by aviation specialists in an emergency to provide timely, competent, and reasonable recommendations of ATCO and other operators to the pilot under various disturbing actions is developed. The individual and CDM matrices for the different aviation specialists in the case of aircraft collision with a bird during takeoff are built. The optimal alternative for flight completion follows the operators' individual and collaborative decisions based on the Wald, Laplace, and Hurwitz criteria is determined. The consistency between the pilot and ATCO estimates of the objective factors in the emergency "Collision with a bird during takeoff" based on Spearman's coefficient is verified. Designed in the future based on big data analysis/machine learning tools an Intelligent System of Collaborative Decision-Making during the Training (IS CDM-T) for the collective practical training of the aviation specialists will allow for enlarging the efficiency of the air navigation system operators' cross-professional interaction in emergencies.

Keywords: Bird Collision · Decision-Making Matrix · Hurwitz Criterion · Laplace Criterion · Spearman's Coefficient · Takeoff · Uncertainty · Wald Criterion

I. Ostroumov et al. (Eds.): ACASD 2025, LNNS 1418, pp. 435–454, 2025.
https://doi.org/10.1007/978-3-031-91992-3_29

1 Introduction

1.1 Introduction to the Problem

Air traveling is the safest mode of transportation, and 2023 was the year with the best safety record in the history of flights. This is evidenced by data from the International Air Transport Association (IATA) [1, 2]. The annual safety report confirmed that on average, a person has to fly daily for 103,239 years until a lethal accident can occur. As per IATA [2], there was just one fatality in 2023 – a local flight crash in Nepal that killed 68 passengers and four crewmembers. However, it was a turboprop aircraft, not a jet, so the case was not included in the statistics. It is noted that even with a 17% increase in aircraft traffic, the accident rate has decreased. In 2023, there was one accident for every 1.26 million flights, compared to the 5-year average of one accident for every 880,000 flights. In 2023, there were also no passenger aircraft hull losses, i.e., when an aircraft sustained damage that could not be restored.

Problems remain at critical stages of flight, particularly during landing and takeoff, as more than half of all aviation accidents between 2014 and 2023 occurred at these stages. The factors leading to these accidents involve various issues ranging from aircraft configuration management to aircraft navigation, effective communication with air traffic controllers (ATCOs), mission-critical functions associated with crew resource management (CRM) practices, and key competencies. A large number of accidents in 2023 were also caused by non-compliance to standard operating procedures (SOPs), inappropriate runway contact, inadequate hands-on skills, unfavorable meteorological conditions, aircraft failure, and absence of monitoring and cross-checking. The number of aviation events by categories in 2023 is presented in Fig. 1: Landing Gear Failure (9); Ground Damage (5); Tail Strike (5); Hard Landing (4); Runway Excursion (2); Outside Airport Landing/Ditching (2); In-Flight Damage (1); Other End State (1); and catastrophe due to Loss of Control In-Flight (1) with 72 deaths [2].

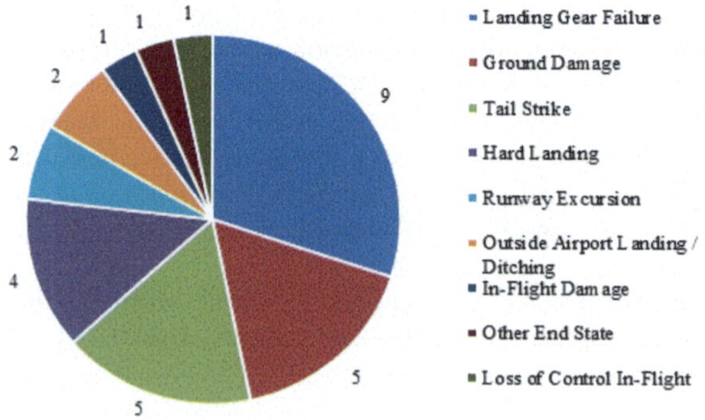

Fig. 1. The number of aviation events by categories in 2023.

Statistically, up to 80% of all aviation accidents can be traced back to human error. Pilot error is believed to be the cause of 53% of accidents, followed by technical malfunctions (21%) and meteorological hazards (11%) [3–5]. Despite the best training, when aircrews undergo comprehensive training on simulators, people tend to revert to previous patterns of behavior. Training is helpful, but muscle memory takes time to develop, which can lead to catastrophic unforeseen effects, especially when flying a high-maneuverable aircraft.

The National Transportation Safety Board (NTSB) dataset, which contains records of every reported accident from 2008, lists more than 13 thousand accidents with at least one staff-related issue. These accidents include more than 20 thousand human resources-related issues, with an average rate of 1.54 per accident [6]. On the other side, it is essential to note that there are continuously some other factors at play when pilot error happens. The aircraft problems, environmental issues, and organizational concerns are a couple of examples.

The aviation industry should cooperate to provide safety improvements through technological progress, enhanced professional training, efficient in-flight collaborative decision-making (CDM), and the use of CRM. In the meantime, promoting a positive safety culture is important.

1.2 Motivation

The present-day aviation transport system is rapidly evolving towards globalization, integration, and compatibility of its components [7–9]. The effectiveness of the system greatly depends on the quality training of competent and well-qualified aviation specialists [10–12]. Considering the International Civil Aviation Organization (ICAO) forecast data within Next Generation of Aviation Professionals (NGAP) Initiative [13, 14], the aviation transport system by 2030 will require many high-skilled professionals: approximately 160,000 pilots, 40,000 ATCOs, and 360,000 engineers. Currently, cross-professional interaction, which is an important core competence for aviation professionals to develop and make collaborative decisions, is underdeveloped, and specialists are trained separately. The National Transport Strategy of Ukraine for the period up to 2030 identifies the lack of compliance of the transport training system with modern innovative requirements as a problematic issue in the field of professional education [15]. These facts motivate the development of an intelligent system for collaborative practical training of aviation specialists (pilots, ATCOs, drone operators, maintenance engineers, flight dispatchers, handling agents, search and rescue personnel, other ground staff, etc.) in a joint educational platform based on the concept of "CDM-Education" [16].

1.3 Contribution

This research contributes to the enhancement of professional interaction between aviation professionals during the common practical training based on the use of non-stochastic, stochastic, and deterministic models of CDM as part of a specialized Intelligent Integrated Training System (IITS).

1.4 The Organization of the Paper

The paper comprises seven sections. The first section is an introduction. The second section concentrates on the analysis of related works and the state of the art. The third section considers the situation-based approach in collaborative practical training of pilots and ATCOs, and the fourth – describes the illustrative example of CDM in the emergency "Collision with a bird during takeoff". The fifth section presents results and discussions. The sixth section is a conclusion. The seventh section describes the future research directions.

2 Literature Review and Problem Statement

The problem of cross-professional interaction between pilots and ATCOs is crucial to ensuring air traffic safety. Notwithstanding improvements in the design and equipment of aircraft and the widespread use of modern automated air traffic control (ATC) systems with artificial intelligence elements, humans are still the main link in the "pilot-ATCO" interaction chain and have a considerable impact on flight safety. In maintaining flight safety, it is important to organize high-quality CDM between the flight crew, ATCO, and other stakeholders based on their operational reliable communication, considering the results of analyzing large amounts of information in real-time [8, 9, 11].

The activities of pilots and ATCOs belong to the complex types of human-operator activities, which have the following general features [17, 18]: significant psychophysiological stress, even when working in simple air and meteorological conditions; professional activity is often carried out in conditions of high intensity and time pressure against the background of great responsibility for flight safety; permanent (despite significant automation of ATC and flight processes) readiness for an adequate response in case of deviation from the "planned flight pattern"; regular and "rigorous" checks for compliance with the profession; teamwork: "pilot – aircraft crew", "ATCO – ATC work shift", "pilot – ATCO – flight support services – departmental ATC authorities", "ATCO – air traffic safety electronics personnel (ATSEP)", "ATCO – flight dispatcher (F/D)", etc.

Before starting their professional training, future aviation specialists have to undergo a multi-level aptitude test: demonstrate an appropriate level of general educational competencies, certain medical indicators, professionally required qualities, the requisite skill level in English, etc. [19, 20]. Given the rather strict requirements for future aviation professionals, the professional training system should guarantee and cost-effectively provide initial training of selected candidates with the required set of competencies to perform complex types of operator activities.

The process of co-development of decisions by aviation personnel is a key element of the NGAP programme [12]. For example, the results of a joint training programme on operational incidents between Deutsche Flugsicherung, a German air navigation service provider, and Deutsche Lufthansa, the leading national air carrier in Germany [21], confirmed that CDM contributes to the improvement of situational awareness of pilots and ATCOs during flights and ATC.

Ukraine International Airlines [22] also undoubtedly recognizes the importance of bilateral professional awareness of the activities of pilots and ATCOs. Introductory flights of ATCOs in the cockpit are a common practice all over the world [17, 18]. Such flights as part of an aircraft crew should be performed at least once every 3 years. However, it is also important for pilots to know how ATCOs work, to see the real situation at the workplace, and to understand the specifics of the tasks of those specialists who guarantee their safety in the sky while working on the ground. There is a practice of familiarizing pilots with the work of ATCOs directly at their workplaces in the controller's room [22].

The authors have presented novel ways to improve and achieve flexible CDM in the process of the professional performance and practical training of aviation specialists (machine learning and artificial neural network for pre-simulation training [23]; intelligent system for CDM support during simulation and real flights [16, 24–28]) based on the decision-making (DM) methods they have developed (Table 1, Fig. 2).

Table 1. Methods of CDM by aviation specialists.

Name of the method	An example of an emergency is considered	Aircraft type	Flight route
Method for integrating deterministic, stochastic, and non-stochastic DM models in certainty, risk, and uncertainty [24]	Failure of one engine on a twin-engine aircraft	Boeing 737–800	Lviv – Kharkiv
Method of subjective-objective CDM based on individual and collaborative DM models [24]	Failure of one engine on a twin-engine aircraft	Boeing 737–800	Lviv – Kharkiv
Method of CDM based on the priority of the Hurvitz criteria [25]	Engine failure during takeoff due to bird strike	Antonov 148–100A	Kharkiv – Lviv
Method for managing the situation's development by using the integration of individual and collaborative DM models [26]	Landing gear failure on takeoff	Generalized type	$A_1 - A_2$
Method of CDM based on the priority of the factors influencing DM [27]	Cargo fire during the climb	Boeing 737–900ER	Najaf – Madinah
Method of CDM in education "CDM-E" [16]	Engine failure during the training route flight	One-engine light propeller aircraft	Kozelets' – Kyiv (Antonov)
Method of CDM based on a multi-step DM model [28]	Loss of thrust on both engines due to falling into a cloud of volcanic ash with subsequent ditching	Cessna 310	Napoli – Palermo

The importance of CDM processes is explained by the fact that flight safety, timely and competent response to non-standard and emergency situations, and the final efficiency of all in-flight stakeholders rely on the accuracy, adequacy, and speed of the actions of the "pilot-ATCO" team. CDM processes are the core element of their professional activities, as mistakes made at the DM stage can have implications that may not be corrected at subsequent stages of the action.

The professional training of aviation specialists is currently not sufficiently advanced for effective co-development and making optimal decisions in emergencies. Therefore,

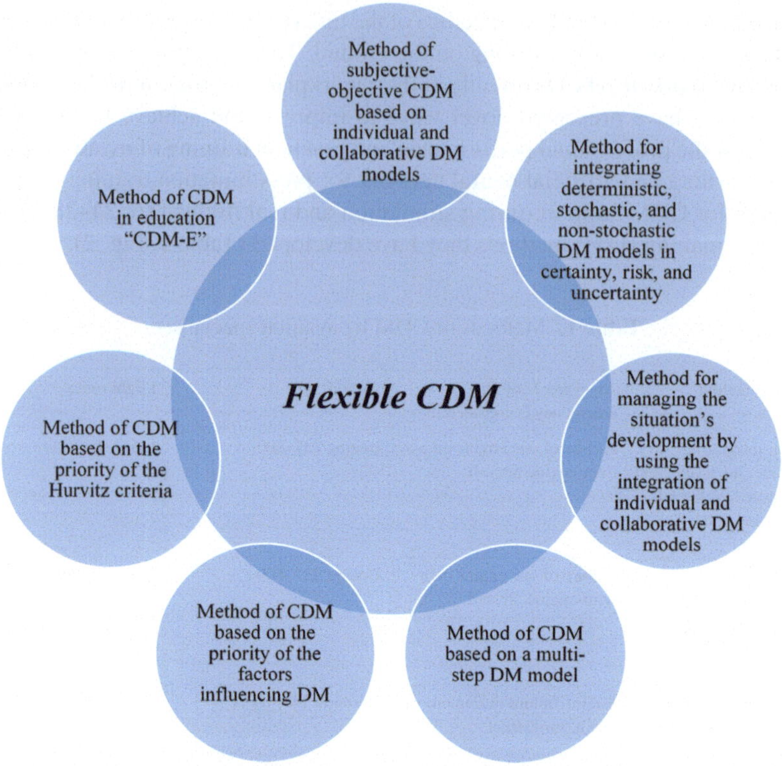

Fig. 2. Methods of CDM by the aviation specialists.

the elaboration of CDM models for optimization of cross-professional interaction "pilot-ATCO" during their collaborative practical training is an actual problem.

The purposes of this work are:

- Development of the situation-based approach in collaborative practical training of the pilots and ATCOs.
- Building the integrated CDM models for the pilot, ATCO, ATSEP, and F/D in the emergency "Collision with a bird during takeoff" and verification of their consistency using Spearman's coefficient.

3 Situation-Based Approach in Collaborative Practical Training of Pilots and Air Traffic Controllers

DM is a decisive aspect of the air traffic "pilot-ATCO" team's activities, which integrates their professional competencies, psychophysiological capabilities, and ability to work effectively in different conditions. The main active "players" in the air navigation system at all stages of the flight (without diminishing the great importance of the activities of other specialists in the air transport system) are the pilot and the ATCO. They are the only ones who understand best what decisions need to be made and implemented at a certain time, under certain circumstances, and in a certain place. Table 2 shows a description of the elements of pilot and ATCO behavior that they must demonstrate in the basic professional competence "Problem solving and decision-making" [19, 20].

Table 2. Elements of pilot and ATCO behavior in terms of basic professional competence "Problem solving and decision-making".

Pilot	ATCO
Description of the pilot competence: identify potential problems and minimize their impact. Apply appropriate DM processes.	Description of the ATCO competence: identify and apply DM processes for hazards and associated risks.
Element (indicator) of the pilot's behavior to be assessed: 1. Searches for adequate information from relevant sources. 2. Determines and verifies what caused deviations from the planned flight course and why. 3. Monitors, reviews, and corrects decisions as necessary. 4. Identifies and considers appropriate options. 5. Applies appropriate and timely decision-making methods. 6. Identifies, assesses, and resolves threats and errors promptly. 7. Works diligently to eliminate problems, while maintaining a focus on flight safety. 8. Adapts when faced with situations for which there is no appropriate algorithm of actions and instructions. 9. Demonstrates self-control when faced with an unexpected event.	Element (indicator) of the ATCO's behavior to be assessed: 1. Considers existing rules and operational procedures when determining potential solutions to the problem. 2. Utilizes appropriate tools and queries relevant systems as required to assist in identifying feasible solutions to a problem. 3. Realizes the relevant solution to the problem. 4. Determines which situations have the highest priority. 5. Organizes the execution of tasks in the appropriate order of priority. 6. Applies an appropriate strategy to minimize the impact of identified threats. 7. Ensures work in the presence of problems without reducing the level of flight safety.

The data presented on the example of the competency "Problem solving and decision-making" illustrate the similarity of the required professional competencies and certain indicators of the behavior of a pilot and ATCO. This confirms the expediency of their collaborative professional training in a common educational environment.

The practical experience of the authors in the field of professional training of pilots and ATCOs shows that the method of situation-oriented training allows the most effective, in terms of costs and benefits, to transform theoretical knowledge into practical skills in the "working" environment [16].

In the context of aviation operators, the method of situation-based training is as follows:

1. There is no single answer to the situation, but several competing "what if" alternatives.
2. Training is not focused on obtaining declarative ready-made knowledge, but on its development, with a guaranteed conviction of why it is necessary.
3. As a result of training according to this method, not only purposefully formed knowledge appears, but also practical skills.

The main advantages of the situation-oriented method in the professional training of future pilots and ATCOs:

1. Significant professional practicability: students work on real professional situations relevant to their future work.

2. Development of critical thinking and DM skills: practicing real-life situations helps to develop problem-solving skills, and make decisions in the face of risks and uncertainty of information.
3. Improvement of communication skills: the situation-oriented method of training includes teamwork, in which the interchange of individual solutions to a problem is transformed into CDM by the team.
4. Flexibility in training: searching for the most optimal solution under conditions of time pressure, risk, and uncertainty of information encourages creativity and non-standard solutions to solve the problem.

The purpose of control is to develop by the operator-pilot (control element) such optimal actions Y_{opt} to ensure that the aircraft (control object) performs the specified flight plan Y_s, provided that timely, competent, and reasonable recommendations of ATCO and other air navigation system operators Y_r under various disturbing actions Y_d, i.e., changes in the dynamic air situation or flight situation (normal, complicated, difficult, emergency, and catastrophic), using the feedback channel Y_{fb}, and minimizing the misalignment $Y = Y_a - Y_s$. The development of effective solutions is possible if the parameters of deviations of the actual values of the control object from the specified flight plan are known (Fig. 3).

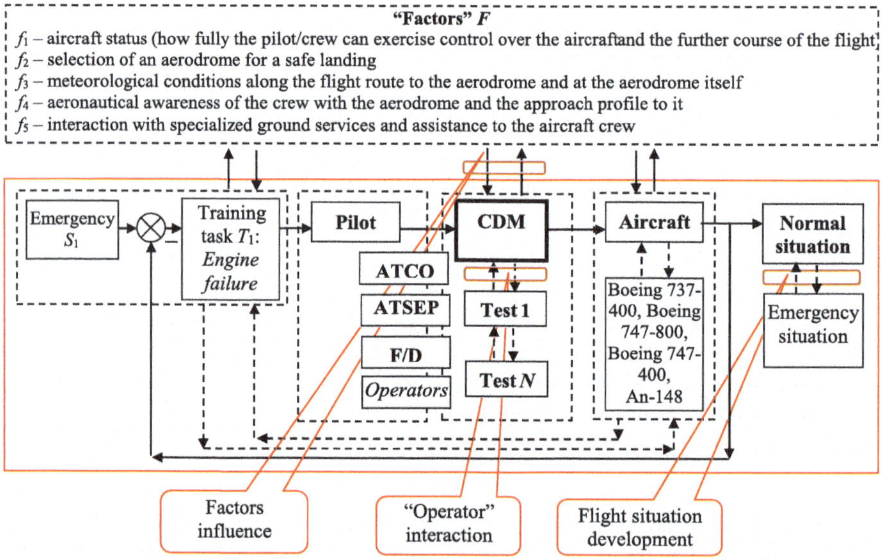

Fig. 3. The scheme of CDM by the aviation specialists in an emergency.

Factors that influence CDM by aviation specialists [24] (Fig. 3):

- f_1 – fuel supply capacity on board.
- f_2 – remoteness of the alternate aerodromes.
- f_3 – runway specifications.
- f_4 – meteorological characteristics of the alternate aerodromes.
- f_5 – lighting system for the approach.
- f_6 – accessible system of approach.
- f_7 – accessible navigation aids.
- f_8 – aircraft specifications.
- f_9 – establishment of communication (radio connection).
- f_{10} – air traffic density, etc.

The best thing for future pilots and air traffic controllers is to obtain professional-oriented knowledge with its further transformation into practical actions under the guidance of an instructor [16]. The instructor plays a key role in the implementation of the situation-oriented method. Virtual modeling as part of a group/team under the supervision of an instructor is used as a practical action. An example of the formation of the necessary professional competencies by solving "real" work situations in a single learning environment to increase the effectiveness of cross-professional integration of future pilots and air traffic controllers is as follows: the instructor develops a simulation model of a certain situation that has occurred/could occur in a real professional environment and which reflects the range of necessary knowledge and skills that students should acquire – master the necessary competencies under the supervision of the instructor.

The structural distribution of time ($T = 90$ min) according to the method of situation-oriented training is as follows:

1. Briefing – up to 10 min. Introduction. The instructor provides a general overview of the situational training and emphasizes the objectives of the training.
2. Study of the theoretical material – up to 30 min. Students, under the supervision of the instructor, receive educational material in multimedia form regarding the rules, calculations, procedures, and algorithms of actions related to the situation.
3. Individual work – up to 20 min. Students independently analyze the situation, make calculations, and develop individual solutions.
4. Work in a group/team – up to 20 min. Students are united in a group/team to discuss their proposals and develop a collective solution.
5. Debriefing "Flight analysis" – up to 10 min. The instructor provides feedback: reviews the decisions made, the correct algorithms of actions, emphasizes the practicality of the results and their significance for students, and gives examples from the professional activities of specialists as confirmation. Evaluate overall group/team and individual performance.

The actual allocation of time may depend on the simulation model of the specific situation, the competencies to be mastered, as well as the learning styles of the students and the role of the instructor.

4 The Illustrative Example of Collaborative Decision-Making (CDM) in the Emergency "Collision with a Bird during Takeoff"

4.1 Theoretical Part: Briefing-Preparation and Study of the Training Material

Shot Information About Emergency. ICAO records about 6000 collisions between aircraft and birds annually, with 95% of collisions occurring below 2000 ft. (600 m), where the aircraft is at heights of potentially the largest concentration of birds for a long time [29]. An aircraft at a low altitude has a lower speed and, accordingly, reduced aerodynamic reserves, which makes it difficult to perform complex maneuvers (turns), and the crew has limited time to stabilize the flight and make decisions to compensate for the emergency. In case of failure of one of the engines, asymmetric thrust creates an additional workload when flying the aircraft, against the background of the fact that the takeoff and landing stages are the most difficult – the crew performs many work operations at a fast pace, which in itself requires a high level of concentration, accuracy, and efficiency. During the takeoff and landing phases, there is usually significant air traffic in the aerodrome area, which requires clear coordination with the ATC authority and compliance with the instructions and guidelines of the ATCO.

The most dangerous places for aircraft to collide with birds are engines and windshields. A bird hitting a turbojet engine can cause partial or even complete failure due to mechanical damage. If the windshield is damaged, it may be cracked or fractured, leading to rapid decompression; restriction of the view from the cockpit, which will make it difficult to maneuver, especially during takeoff or landing.

Training Situation Conditions. Route of heavy aircraft Boeing 737–800 from aerodrome N to aerodrome M with alternate aerodrome K (Fig. 4). The weather at the aerodromes complies with the minimum CAT 1 (at least the following criteria: visibility 800 m, visibility range on the runway 550 m, decision height 200 ft. (60 m)). Wintertime, precipitation, braking action is average, and the temperature is closes to 0°.

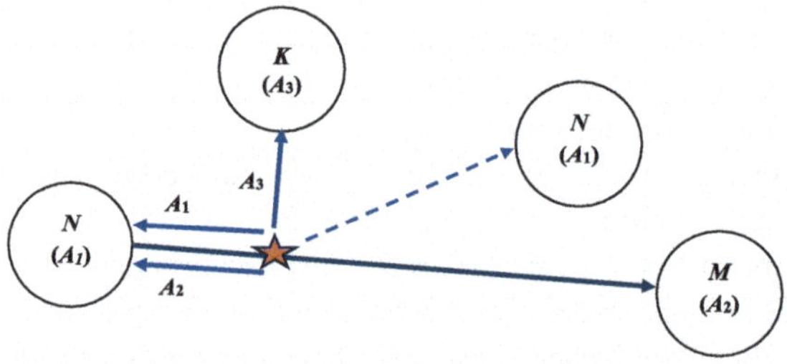

Fig. 4. Route from aerodrome N to aerodrome M with alternate aerodrome K.

The aircraft took off with a Maximum Takeoff Weight (MTOW) of 79,015 kg. Maximum Landing Weight (MLW) is 66,360 kg. There is no emergency fuel jettison system.

Place of damage due to a collision with a bird – engine; consequence – engine failure; height $H = 200$ m; safe takeoff speed V_2 is reached.

Alternate solutions for the flight completion:

- Return to the departure aerodrome N after fuel consumption (A_1).
- Return to the departure aerodrome N without fuel consumption (A_2).
- Flight and landing at the nearest suitable aerodrome – alternate aerodrome K (A_3).

Factors influencing DM for each operator:

- $\{f\}$ – factors that are taken into consideration by the operator O_1 (pilot).
- $\{l\}$ – factors that are taken into consideration by the operator O_2 (ATCO).
- $\{\lambda\}$ – factors that are taken into consideration by the operator O_3 (ATSEP).
- $\{\theta\}$ – factors that are taken into consideration by the operator O_4 (F/D).

When selecting an optimal alternate, operators are governed by the common factors (f_i, l_i, λ_i, θ_i):

- $f_1, l_1, \lambda_1, \theta_1$ – aircraft specifications.
- $f_2, l_2, \lambda_2, \theta_2$ – runway specifications.
- $f_3, l_3, \lambda_3, \theta_3$ – meteorological characteristics of the alternate aerodromes.
- $f_4, l_4, \lambda_4, \theta_4$ – accessible system of approach, navigation aids.
- $f_5, l_5, \lambda_5, \theta_5$ – establishment of communication, available interaction, and assistance to the aircraft.

In such a critical situation, a quick, balanced decision must be made with the priority of the passengers, crew, and aircraft safety.

4.2 Pilot Decision-Making – Individual and Group (in Crew) Work

The actions of the pilot of an aircraft, in particular the Boeing 737-800, in non-standard situations are subject to clear algorithmization and are regulated by the Flight Operations Manual (FOM) and checklists for various emergencies, including a bird collision:

1. Main actions of the crew (example):
    - The aircraft commander (who took off) continues piloting.
    - The co-pilot performs actions according to the checklist.
2. Procedure for the checklist "Engine Failure/Damage".
    (a) Maintain aircraft control:
        - The pilot maintains a straight flight.
        - Climb angle within 10–12° to ensure optimal climb of the aircraft on one engine.
        - Indicated Airspeed (IAS) must be within the safe takeoff speed V_2, which defines the minimum speed needed for the aircraft to continue climbing even with one engine failure. It depends on the aerodynamic characteristics, mass, air pressure, temperature, air density, mass of the aircraft, etc. Depending on the conditions, V_2 can be in the range of 213–240 km/h.

    (b) Identify and verify the problem:

- Instrumentation check to confirm engine failure or damage.
- Engine failure is declared.

(c) Confirm engine failure:

- Engine failure is performed according to the checklist "Engine Fire/Severe Damage" or "Engine Failure".
- The fuel supply to the faulty engine is stopped.

(d) Climb to a safe height.

3. The crew announces the stage of distress:
   (a) The crew reports the situation on board to the ATCO ("MAYDAY, MAYDAY, MAYDAY") in case of an emergency.
   (b) The crew shall report the following key information:

- Type of aircraft.
- Identification index.
- Nature of the problem (failure of the right engine due to a bird strike).
- Height and direction of flight (height 200 m, pilot continues to climb in a straight line with a heading of 350°).
- Requesting information necessary for the crew to make a decision or report their actions (flying over the aerodrome to consume fuel, returning to the departure aerodrome, heading to the nearest aerodrome suitable for a safe landing).

In assessing the various factors that influence the results of the flight completion, a nominal rating scale is used, where the best outcome is rated "5", and the worst outcome is rated "1". Optimal solutions are obtained with the help of the Expert Judgment Method (EJM) and based on the Wald (W), Laplace (L), and Hurwitz (H) criteria [24]. The main priority is flight safety, taking into account all possible factors. Table 3 shows a matrix of possible solutions from the pilot's point of view.

Table 3. The individual DM matrix for the pilot.

Alternatives	Factors that influence DM/Outcomes					Solutions				
	f_1	f_2	f_3	f_4	f_5	W	L	$H, \alpha = 0.3$	$H, \alpha = 0.5$	$H, \alpha = 0.8$
	Pilot (O_1)									
A_1	3	4	3	5	5	3	4.0	3.6	4.0	4.6
A_2	2	4	3	5	5	2	3.8	2.9	3.5	4.4
A_3	4	4	5	4	4	**4**	**4.2**	**4.3**	**4.5**	**4.8**

The optimal alternative for the flight completion follows the pilot's individual decision based on the criteria of the Wald (W), Laplace (L), and Hurwitz (H) with the different coefficients of optimism-pessimism ($\alpha = \{0.3; 0.5; 0.8\}$) is the flight and landing at the alternate aerodrome K (A_3).

4.3 Decision-Making by Air Traffic Controller and Other Aviation Specialists – Individual and Group (in Duty Shift) Work

In case of emergencies or unforeseen circumstances, ATCOs should adhere to the ASSIST principle as closely as possible and follow the "Standard Action Charts for ATC Specialists in Emergency and Contingency Situations" to the fullest extent possible to provide maximum assistance to the aircraft crew in distress and obtain the necessary important information for its further transmission to emergency services [30].

ATCO actions in the emergency "Collision with a bird during takeoff":

1. Clarify whether the pilot can control the aircraft.
2. Provide a long pre-landing runway if requested by the crew.
3. Check the runway (if the collision occurred after takeoff).
4. If necessary, inform the pilot of the aerodrome details as soon as possible: operating runway, length, surface, height, instrument landing system (ILS), and navigation aids frequencies.
5. Allow the pilot time to perform the necessary operations by the FOM for the approach.
6. Organize (if necessary) the lead of the emergency aircraft.

The individual DM matrices for the ATCO and other aviation specialists who can be involved in emergencies (for example, ATSEP and F/D) are built analogically with the pilot's DM matrix (Table 4).

Table 4. The individual DM matrices for the ATCO, ATSEP, and F/D (step 1).

Alternatives	Factors that influence DM/Outcomes					Solutions				
	f_1	f_2	f_3	f_4	f_5	W	L	$H, \alpha = 0.3$	$H, \alpha = 0.5$	$H, \alpha = 0.8$
	ATCO (O_2)									
A_1	3	5	4	5	5	**3**	**4.4**	**3.6**	**4.0**	**4.6**
A_2	2	5	4	5	5	2	4.2	2.9	3.5	4.4
A_3	3	2	4	3	4	2	3.2	2.6	3.0	3.6
	ATSEP (O_3)									
A_1	3	4	3	5	5	**3**	4.0	**3.6**	**4.0**	**4.6**
A_2	2	4	3	5	5	2	3.8	2.9	3.5	4.4
A_3	4	4	5	4	4	4	**4.2**	4.3	4.5	4.8
	F/D (O_4)									
A_1	4	4	3	5	5	3	4.2	3.6	4.0	4.6
A_2	2	4	3	5	5	2	3.8	2.9	3.5	4.4
A_3	4	4	5	4	4	**4**	**4.2**	**4.3**	**4.5**	**4.8**

The optimal alternative for the flight completion follows the ATCO's individual decision (step 1) based on the criteria of the Wald (W), Laplace (L), and Hurwitz (H) with the different coefficients of optimism-pessimism ($\alpha = \{0.3; 0.5; 0.8\}$) is the return to the departure aerodrome N after fuel consumption (A_1); follows the ATSEP's individual decision based on the criteria of the Wald (W) and Hurwitz (H) – is the return to the departure aerodrome N after fuel consumption (A_1), but based on the Laplace (L) criterion – the flight and landing at the alternate aerodrome K (A_3); follows the F/D's individual decision based on all criteria – the flight and landing at the alternate aerodrome K (A_3).

To determine the common opinion of aviation specialists (pilot (O_1), ATCO (O_2), ATSEP (O_3), and F/D (O_4)), based on the individual solutions of the operators (subjective factors), the collective DM matrix is constructed (Table 5).

Table 5. The collective matrix for the pilot, ATCO, ATSEP, and F/D (step 1).

Alternatives	Operators/Solutions				
	O_1	O_2	O_3	O_4	W
A_1	3	3	3	3	**3**
A_2	2	2	2	2	2
A_3	4	2	4	4	2

The optimal CDM follows the pilot's, ATCO's, ATSEP's, and F/D's collective decisions by the Wald (W) criterion is the return to the departure aerodrome N after fuel consumption (A_1).

Consider the option if the ATCO takes into account not only the ASSIST requirements but also the aircraft specifications (Table 6), which will affect the result of the optimal collective decision (Table 7).

Table 6. The individual DM matrix for the ATCO (step 2).

Alternatives	Factors/Outcomes					Solutions				
	f_1	f_2	f_3	f_4	f_5	W	L	$H, \alpha = 0.3$	$H, \alpha = 0.5$	$H, \alpha = 0.8$
	ATCO (O_2)									
A_1	2	5	4	5	5	2	4.2	2.9	3.5	4.4
A_2	2	5	4	5	5	2	4.2	2.9	3.5	4.4
A_3	5	3	5	5	4	3	**4.4**	**3.6**	**4.0**	**4.6**

Table 7. The collective matrix for the pilot, ATCO, ATSEP, and F/D (step 2).

Alternatives	Operators/Solutions														
	O_1	O_2	O_3	O_4	W	O_1	O_2	O_3	O_4	L	O_1	O_2	O_3	O_4	$H, \beta = 0.5$
A_1	3	2	3	3	2	4.0	4.2	4.0	4.2	4.1	4.0	3.5	4.0	4.0	3.75
A_2	2	2	2	2	2	3.8	4.2	3.8	3.8	3.9	3.5	3.5	3.5	3.5	3.5
A_3	4	3	4	4	**3**	4.2	4.4	4.2	4.2	**4.25**	4.5	4.0	4.5	4.5	**4.25**

The calculations with the help of table processor MS Excel and the programming language Python showed that if ATCO takes into account not only ASSIST requirements but also the aircraft specifications in its individual decisions (step 2), then both its individual and collective choice of all operators will be the flight and landing at the nearest suitable aerodrome – alternate aerodrome K (A_3).

4.4 Debriefing "Flight analysis" – Evaluation of the Collective Decision-Making (CDM) by Aviation Specialists

During the debriefing, the training leader can draw attention to the different views of specialists, taking into account their specialization in dealing with the emergency, and show practical examples of the significance of the situation. The instructor evaluates the overall group/team and individual performance; and if necessary, provides references to additional material.

Verify the consistency between the pilot and ATCO estimates of the objective factors that influence the choice of the flight and landing at the nearest suitable aerodrome – alternate aerodrome K (A_3) at step 1 and step 2 with the help of rank correlation coefficient (Spearman's coefficient) (Table 8, Fig. 5):

$$r_s = 1 - \frac{6\sum_{I=1}^{n}(x-y)^2}{n\left(n^2-1\right)}, \quad (1)$$

where x are the estimates by the pilot; y are the estimates by the ATCO; n is a number of factors.

Table 8. Calculation of the Spearman's coefficient.

Estimates	Factors that influence DM/Outcomes					Sum
	f_1	f_2	f_3	f_4	f_5	
Step 1	The flight and landing at the nearest suitable aerodrome (A_3)					
x	**4**	**4**	**5**	**4**	**4**	
y	3	2	4	3	4	
$x-y$	1	2	1	1	0	
$(x-y)^2$	1	4	1	1	0	*7*
r_s	***0.65***					
Step 2	The flight and landing at the nearest suitable aerodrome (A_3)					
x	**4**	**4**	**5**	**4**	**4**	
y	**5**	**3**	**5**	**5**	**4**	
$x-y$	−1	1	0	−1	0	
$(x-y)^2$	1	1	0	1	0	*3*
r_s	***0.85***					

The Spearman's coefficient at step 2 (when ATCO takes into account not only ASSIST requirements but also the aircraft specifications) increased which indicates that consistency between the pilot and ATCO estimates of the objective factors influencing the choice of the flight and landing at the nearest suitable aerodrome – alternate aerodrome K (A_3) became stronger.

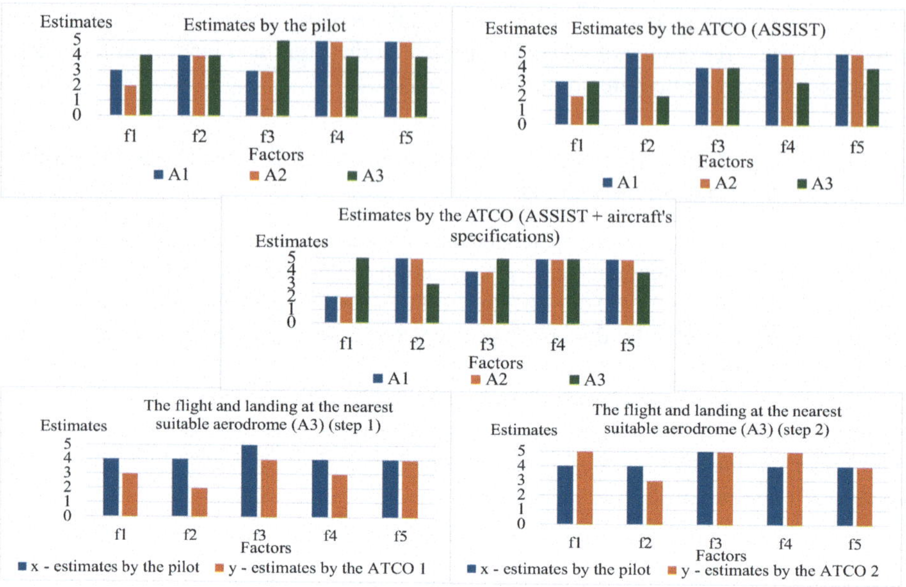

Fig. 5. Estimates of the objective factors that influence DM in the emergency by the pilot and ATCO.

5 Results and Discussions

The situation-based approach in collaborative practical training of the pilots and ATCOs is proposed. The similarity of the elements of pilot and ATCO behavior in the basic professional competence "Problem solving and decision-making" is demonstrated. This confirms the expediency of their collaborative professional training in a common educational environment.

The structural distribution of time ($T = 90$ min) according to the method of situation-oriented training is defined as: briefing – up to 10 min; study of the theoretical material – up to 30 min; individual work – up to 20 min; work in a group/team – up to 20 min; debriefing "Flight analysis" – up to 10 min.

A training example of CDM in the emergency "Collision with a bird during takeoff" at height $H = 200$ m (safe takeoff speed V_2 is reached) during the flight of heavy aircraft Boeing 737–800 from aerodrome N to aerodrome M with alternate aerodrome K is presented. The alternatives of the flight completions are considered: return to the departure aerodrome N after fuel consumption (A_1); return to the departure aerodrome N without fuel consumption (A_2); flight and landing at the nearest suitable aerodrome – alternate aerodrome K (A_3). The actions of the pilot and ATCO in the emergency "Collision with a bird during takeoff" are discussed.

Five common objective factors influencing CDM by four operators (pilot, ATCO, ATSEP, and F/D) in the case of aircraft collision with a bird during takeoff are determined: aircraft specifications; runway specifications; meteorological characteristics of the alternate aerodromes; accessible system of approach, navigation aids; establishment of communication, available interaction and assistance to the aircraft.

The optimal alternative for the flight completion follows the pilot's individual decision based on the criteria of the Wald (W), Laplace (L), and Hurwitz (H) with the different coefficients of optimism-pessimism ($\alpha = \{0.3; 0.5; 0.8\}$) is the flight and landing at the alternate aerodrome K (A_3).

The optimal alternative for the flight completion follows the ATCO's individual decision (step 1) based on the criteria of the Wald (W), Laplace (L), and Hurwitz (H) is the return to the departure aerodrome N after fuel consumption (A_1); follows the ATSEP's individual decision based on the criteria of the Wald (W) and Hurwitz (H) – is the return to the departure aerodrome N after fuel consumption (A_1), but based on the Laplace (L) criterion – the flight and landing at the alternate aerodrome K (A_3); follows the F/D's individual decision based on all criteria – the flight and landing at the alternate aerodrome K (A_3).

The optimal CDM follows the pilot's, ATCO's, ATSEP's, and F/D's collective decisions by the Wald (W) criterion is the return to the departure aerodrome N after fuel consumption (A_1).

The calculations with the help of table processor MS Excel and the programming language Python showed that if ATCO takes into account not only ASSIST requirements but also the aircraft specifications in its individual decisions (step 2), then both its individual and collective choice of all operators will be the flight and landing at the nearest suitable aerodrome – alternate aerodrome K (A_3).

The Spearman's coefficient at step 2 (when ATCO takes into account not only ASSIST requirements but also the aircraft specifications) increased which indicates that consistency between the pilot and ATCO estimates of the objective factors influencing the choice of the flight and landing at the nearest suitable aerodrome – alternate aerodrome K (A_3) became stronger.

6 Conclusions

The problem of cross-professional interaction between pilots and ATCOs is crucial to ensuring air traffic safety. Notwithstanding the widespread use of artificial intelligence elements, humans are still the main link in the "pilot-ATCO" interaction chain. It is important to organize high-quality CDM between all stakeholders based on their operational reliable communication in real-time. Nevertheless, the professional training of aviation specialists is currently not sufficiently advanced for effective co-development and making optimal decisions in emergencies. Therefore, the elaboration of CDM models for optimization of cross-professional interaction "pilot-ATCO" during their collaborative practical training is an actual problem.

Based on the situation-based approach the structure of collaborative practical training of the pilots and ATCOs in a common educational environment is defined as: briefing; study of the theoretical material; individual work; work in a group/team; debriefing "Flight analysis".

The scheme of CDM by aviation specialists in an emergency to provide timely, competent, and reasonable recommendations of ATCO and other operators to the pilot under various disturbing actions, to ensure that the aircraft performs the specified flight plan, is developed.

The individual and collaborative DM matrices for the pilot, ATCO, ATSEP, and F/D in the case of aircraft collision with a bird during takeoff are built. The optimal alternative for flight completion follows the operators' individual and collaborative decisions based on the Wald, Laplace, and Hurwitz (with the different coefficients of optimism-pessimism) criteria is determined.

The consistency between the pilot and ATCO estimates of the objective factors that influence the decision to complete the flight in the emergency "Collision with a bird during takeoff" based on Spearman's coefficient is verified.

For operational DM, it is proposed to process data using the Python programming language in the COLAB environment, an artificial neural network (ANN) is developed to predict the development of a non-standard situation on the training example of the emergency "Collision with a bird during takeoff".

7 Future Scope

Designed in the future based on the big data analysis/machine learning tools an Intelligent System of Collaborative Decision-Making during the Training (IS CDM-T) for the collective practical training of the aviation specialists (pilot, drone operator, ATCO, ATSEP, F/D, ground handler, engineer, etc.) will allow enlarging the efficiency of the air navigation system operators' cross-professional interaction in emergencies.

References

1. IATA Annual Review 2024. International Air Transport Association, Montreal, Canada (2024)
2. IATA Annual Safety Report Executive Summary and Safety Overview. 60th edn. International Air Transport Association, Montreal, Canada (2024)
3. Muñoz-Marrón, D.: Human factors in aviation: CRM (Crew Resource Management). Psychol. Papers. **39**(3), 191–199 (2018). https://doi.org/10.23923/pap.psicol2018.2870
4. Wise, J.A., Hopkin, V.D., Garland, D.J. (eds.): Handbook of aviation human factors, 2nd edn. CRC Press, Boca Raton (2016). https://doi.org/10.1201/b10401
5. PANISH | SHEA | RAVIPUDI^{LLP}. Aviation and Plane Crash Statistics. https://www.panish.law/aviation_accident_statistics.html. Last accessed 6 Dec 2024
6. National Transportation Safety Board. https://www.ntsb.gov/safety/Pages/default.aspx. Last accessed 08 Dec 2024
7. Global Air Navigation Plan 2016-2030. Doc. 9750. 5th edn. International Civil Aviation Organization, Montreal, Canada (2016)
8. Manual on Collaborative Air Traffic Flow Management (ATFM). Doc. 9971. 3rd edn. International Civil Aviation Organization, Montreal, Canada (2018)
9. A-CDM Concept of Operations. Airservices Australia, Canberra, ACT, Australia (2014)
10. Airport CDM Implementation: Manual. EUROCONTROL, Brussels, Belgium (2017)
11. Airport-Collaborative Decision Making: IATA Recommendations. International Air Transport Association, Montreal, Canada (2018)
12. International Civil Aviation Organization. Next Generation of Aviation Professionals (NGAP) Programme. https://www.icao.int/secretariat/CapacityDevelopmentImplementation/ngap/Pages/default.aspx. Last accessed 9 Dec 2024

13. International Civil Aviation Organization. ICAO study reveals strong demand for qualified aviation personnel up to 2030. https://www.icao.int/Newsroom/Pages/icao-study-reveals-strong-demand-for-qualified-aviation-personnel-up-to-2030.aspx. Last accessed 10 Dec 2024
14. Gill, O.: Europe needs 6,000 new pilots a year for next two decades, warns Boeing. The Telegraph. https://www.telegraph.co.uk/business/2022/07/27/europe-needs-6000-new-pilots-year-next-two-decades-warns-boeing/. Last accessed 10 Dec 2024
15. Ministry of Infrastructure of Ukraine. National Transport Strategy of Ukraine 2030. https://mtu.gov.ua/files/for_investors/230118/National%20Transport%20Strategy%20of%20Ukraine.pdf. Last accessed 10 Dec 2024
16. Shmelova, T., Sikirda, Y., Yatsko, M., Kolotusha, V.: Intelligent integrated training system for the aviation specialists "Collaborative Decision-Making – Education" (CDM-E). CEUR Workshop Proc. **3538**, 168–180 (2023)
17. Maldonado A.C.: Air traffic controllers ensure pilot safety. SHAW AIR FORCE BASE. https://www.shaw.af.mil/News/Article-Display/Article/1669647/air-traffic-controllers-ensure-pilot-safety/. Last accessed 11 Dec 2024
18. Martin, S.: Why pilots should visit air traffic control. Boldmethod. https://www.boldmethod.com/blog/article/2015/01/why-pilots-should-visit-atc/. Last accessed 11 Dec 2024
19. Manual on Air Traffic Controller Competency-based Training and Assessment. Doc. 10056. 1st edn. International Civil Aviation Organization, Montreal, Canada (2017)
20. Training. Doc 9868. 3rd edn. International Civil Aviation Organization, Montreal, Canada (2020)
21. Lufthansa aviation training. https://www.lufthansa-aviation-training.com/en/home. Last accessed 11 Dec 2024
22. It is important for pilots to know how an air traffic controller works. Wings: All about Ukrainian Aviation. www.wing.com.ua/content/view/19708/55/. Last accessed 11 Dec 2024
23. Shmelova, T., Sikirda, Y., Rizun, N., Lazorenko, V., Kharchenko, V.: Machine learning and text analysis in an artificial intelligent system for the training of air traffic controllers. Chapter 1. In: Shmelova, T., Sikirda, Y., Rizun, N., Kucherov, D., Dergachov, K. (eds.) Automated Systems in the Aviation and Aerospace Industries, pp. 1–50. IGI Global, Hershey (2019). https://doi.org/10.4018/978-1-5225-7709-6.ch001
24. Shmelova, T., Yatsko, M., Sikirda, Y.: Collaborative-factor models of decision making by operators of the Air Navigation System in conflict or emergency situations. Commun. Comput. Info. Sci. (CCIS). **1635**, 391–409 (2022)
25. Shmelova, T., Sikirda, Y., Yatsko, M., Kasatkin, M.: Collective models of the aviation human-operators in emergency for intelligent decision support system. CEUR Workshop Proc. **3156**, 160–174 (2022)
26. Shmelova, T., Sikirda, Y., Yatsko, M., Marienkov, I., Sahun, Y.: Collaborative decision-making models in flight emergency "Landing gear failure on takeoff". CEUR Workshop Proc. **3373**, 15–33 (2023)
27. Shmelova, T., Sikirda, Y., Yatsko, M.: Nondeterministic collaborative decision-making models in flight emergencies based on the factors' priority. In: Ostroumov, I., Zaliskyi, M. (eds.) Proceedings of the International Workshop on Advances in Civil Aviation Systems Development. ACASD 2023 Lecture Notes in Networks and Systems, vol. 736, pp. 283–300. Springer, Cham (2023). https://doi.org/10.1007/978-3-031-38082-2_22
28. Shmelova, T., Sikirda, Y., Yatsko, M., Stratonov, V.: Integration of deterministic, stochastic, and non-stochastic models to obtain an optimal collaborative decision in the flight emergency. In: Ostroumov, I., Zaliskyi, M. (eds.) Proceedings of the 2nd International Workshop on Advances in Civil Aviation Systems Development. ACASD 2024 Lecture Notes in Networks and Systems, vol. 992, pp. 205–223. Springer, Cham (2024). https://doi.org/10.1007/978-3-031-60196-5_17

29. Airport Services Manual. Part 3. Wildlife Control and Reduction. Doc 9137. 5th edn. International Civil Aviation Organization, Montreal, Canada (2020)
30. Durgut, M.: Emergency Procedures for Air Traffic Controllers. AviationFile. https://www.aviationfile.com/emergency-procedures-for-air-traffic-controllers. Last accessed 15 Dec 2024

Collaborative Decision-Making During the Rescue Operation by the Unmanned Aerial Vehicle Group

Tetiana Shmelova[1], Illia Marienkov[1(✉)], Stepan Simchenko[1], and Peter Dischler[2]

[1] National Aviation University, Liubomyra Huzara Ave., 1, Kyiv 03058, Ukraine
il.marienkov@gmail.com

[2] ISGlobal, University of Barcelona, Gran Via de les Corts Catalanes, 585, Barcelona 08007, Spain

Abstract. Rescue operations, particularly during natural disasters like floods, often require the rapid deployment of Unmanned Aerial Vehicles (UAVs) to assist in search, coordination, and logistics. However, effectively managing a group of UAVs in such dynamic environments presents several challenges, including decision-making under uncertainty, optimizing resource allocation, and ensuring seamless coordination among autonomous agents. This paper examines the complexities of multi-agent systems in UAV-based rescue operations and highlights the need for adaptive decision-making algorithms to address environmental changes and resource limitations. The key to this approach is balancing the UAVs' coverage capabilities, energy consumption, and real-time operational adjustments using collaborative decision-making methods. The paper also explores the specific roles of different UAV types – reconnaissance, transport, communication, and monitoring – in achieving efficient mission execution. A multi-level coordination system is proposed, where UAVs autonomously distribute tasks, adapt to new information and ensure continuous communication. This framework aims to enhance the reliability and scalability of UAV group operations while maintaining flexibility in the face of rapidly changing conditions. By improving these systems, rescue operations can be carried out more effectively, even in remote or challenging environments, reducing the risks to rescuers and victims.

Keywords: Group of UAVs · Collaborative Decision-Making · Monitoring · Search and Rescue Operations · Case of the Flood

1 Introduction

1.1 Introduction to the Problem

When carrying out rescue operations using a group of Unmanned Aerial Vehicles (UAV), there's a complex decision-making problem that arises in conditions of uncertainty. In some cases, if and where the use of a single UAV is ineffective, it is advisable to use

I. Ostroumov et al. (Eds.): ACASD 2025, LNNS 1418, pp. 455–472, 2025.
https://doi.org/10.1007/978-3-031-91992-3_30

a group flight of UAVs for aerial photography (photography of large areas, monitoring forest fires, patrolling territories, search, and rescue), transportation and movement of goods, and inspection of large areas [1, 2]. It is advisable to use various UAVs to perform multi-purpose tasks in a multi-agent system [3, 4]. It is proposed to organize group flights of various types of UAVs to simultaneously perform various tasks, such as monitoring, conducting search and rescue operations, and delivering urgent cargo (transportation). This issue is characterized by the need to optimize across multiple criteria while facing dynamic changes in the environment [5]. A key aspect is balancing the effectiveness of mission execution with resource costs. Decision-making (DM) needs to find a compromise between maximizing the coverage of the search area and the limited capabilities of each UAV, considering energy and time constraints.

The problem is further complicated by the stochastic nature of the environment, where conditions can change unpredictably. This requires the creation of adaptive DM algorithms that can quickly respond to situational changes and adjust plans without losing overall operational effectiveness [1, 6]. The main problem is the absence of optimization of the UAV group solutions and operations depending on the task's purposes under uncertainty. Therefore, the main problem is the lack of optimization of group decisions and the operation of UAVs depending on the task goal in conditions of DM uncertainty. According to the current ICAO requirements, the use of collaborative decision-making (CDM) models is relevant for the improving effectiveness of solutions [7]. The system's multi-agent nature adds an additional dimension to the DM issue. Ensuring effective coordination among autonomous agents, optimal task distribution, and conflict avoidance during execution is essential. Data uncertainty and the environment's limited observability create a problem of incomplete information when making decisions. The system must account for possible measurement errors and assess risks associated with each decision.

Overall, the problem falls into the class of NP-hard optimization tasks with multiple constraints in a dynamic environment. Effectively solving it requires the development of specialized methods that combine elements of decision theory, multi-agent systems, and adaptive management for obtaining solutions for optimizing CDM when organizing a group flight using heterogeneous UAVs to perform multi-purpose tasks [7, 8].

1.2 Motivation

Today, CDM by a group of UAVs during rescue operations is becoming critically important due to the increasing complexity and scale of emergencies. Natural disasters, technological accidents, and military conflicts create situations where traditional methods of conducting rescue operations may be ineffective or too risky for rescuers [2, 3]. The use of a group of UAVs capable of CDM significantly enhances the capabilities of search and rescue operations [1, 2]. Due to distributed DM, the UAV group can effectively cover large areas, adapt to changing environmental conditions, and coordinate their actions without constant operator intervention. In situations with limited time and resources, CDM among UAVs ensures optimal use of available opportunities. The development of artificial intelligence (AI) and informational technologies creates new opportunities for implementing effective CDM algorithms. This increases the effectiveness of the UAV group, reduces the workload on operators, and enhances the reliability of conducting rescue operations.

In the condition of global challenges, such as climate change and the rising frequency of natural disasters, the ability of UAV groups to CDM becomes a key factor in the effectiveness of rescue operations. This allows for the rapid deployment of rescue missions in remote and hard-to-reach areas, where traditional methods may be ineffective [2, 4]. The economic aspect also plays an important role – CDM optimizes resource use, reduces costs of conducting operations, and enhances their efficiency. This is especially important in the context of limited funding for rescue services [5, 6, 9, 10].

Modern scientific research in the field of UAV applications demonstrates significant interest in the problem of organization of rescue operations. The main research directions focus on developing coordination algorithms for actions among autonomous agents in dynamic environments, with particular emphasis on methods for optimizing the distribution of tasks and resources among UAV group members [11–13].

A significant amount of attention in scientific work is dedicated to the issue of DM under uncertainty during search and rescue operations, and DM in an emergency during UAV's flights such as bad weather conditions [7, 8]. Effective CDM in an emergency needs information about the characteristics of the situation, the goal of the mission and target tasks of UAVs, and the characteristics of landing aerodromes/places (Fig. 1).

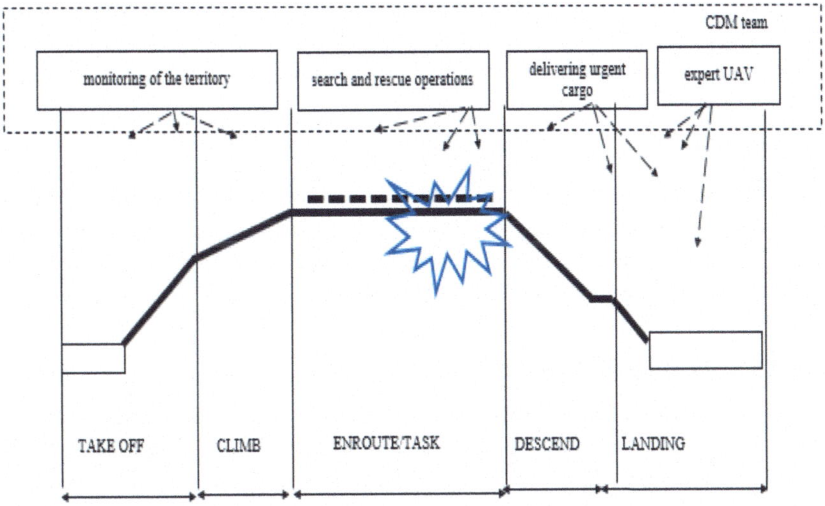

Fig. 1. Emergency in flight of UAVs.

Several methods have been developed that use fuzzy logic and Bayesian networks to assess the situation and choose optimal strategies for UAV group behavior, integrating deterministic, stochastic, and non-stochastic DM models in certainty, risk, and uncertainty [7, 8].

A separate area of research focuses on ensuring reliable communication between UAVs during CDM. The developed data exchange protocols allow the coordination of the group to be maintained even with partial loss of communication between individual agents [3, 7, 9]. Current scientific literature thoroughly examines methods for adaptive role distribution among group members. This approach is based on the use of multi-agent systems with the capability for dynamic reconfiguration of the group structure depending on mission execution conditions [9].

However, despite the substantial volume of research, several unresolved issues remain. In particular, the optimization of energy consumption in group management is insufficiently studied, as are methods for ensuring system reliability and DM algorithms under counteraction and severe weather conditions. The problem of scalability in group management algorithms requires special attention. Most existing solutions are effective only for small groups of UAVs and lose their effectiveness as the number of agents increases. This significantly limits the applicability of such systems in large-scale rescue operations [12, 13]. The integration of UAV group management systems with other components of rescue operations remains a pertinent issue. Effective methods for coordinating the actions of autonomous groups of UAVs with ground rescue services and emergency management systems need to be developed [8, 14]. Thus, the issue of CDM by a group of UAVs during rescue operations remains relevant and requires further research [15–17]. Special attention should be paid to the development of scalable, energy-efficient, and reliable algorithms for group control that can operate in real conditions during rescue missions [18–20].

The authors propose considering a comprehensive solution to target task "rescue mission" using heterogeneous UAVs for multi-purpose target tasks (terrain monitoring, object search and detection, and urgent delivery of necessary cargo to rescue victims) and according to the optimization of CDM in the condition of uncertainty [8, 21].

1.3 Contribution

This research contributes to the enhancement of unmanned aerial vehicle (UAV) operations by developing an intelligent model. The proposed approach leverages non-stochastic, stochastic, and deterministic models to improve UAV deployment efficiency within an Intelligent Integrated Training System (IITS).

1.4 The Organization of the Paper

The paper comprises seven sections. The first section is an introduction. The second section focuses on the literature review and problem statement. The third section discusses the unmanned aerial vehicle network, including reconnaissance, transport, communication, and additional monitoring aspects. The fourth section presents the coordination principles of unmanned aerial vehicles. The fifth section provides an illustrative example of CDM during a flood, divided into theoretical and practical parts. The sixth section presents the conclusions. The seventh section outlines future research directions.

2 Literature Review and Problem Statement

There is significant experience in using UAV groups for flood rescue operations in different countries around the world. In China, in Henan Province, in 2021, a group of 12 UAVs was successfully deployed, covering an area of over 100 km^2 and helping to rescue more than 200 people. The UAVs provided mapping of flooded areas, delivered rescue equipment and searched for people using thermal cameras [14, 22].

In Germany, in the Air River Valley, a group of 8 specialized UAVs carried out 24-hour monitoring of flooded areas. The main advantage of the German approach was the use of advanced 3D mapping technologies and automatic people detection systems, which significantly increased the efficiency of rescue operations [15, 23].

The benefits of UAVs include performing tasks that would otherwise pose a risk to humans and efficiently handling multiple tasks during a single mission [24–26]. The primary focus in UAV development has been on multitasking (multipurpose use). However, a single vehicle cannot complete a large volume of planned tasks within a limited time frame. Hence, employing a group of UAVs of various types for multifunctional purposes is recommended. Several scenarios for collective (group) operations of autonomous UAVs are discussed in [27–29], and additional advantages of UAV groups over individual UAVs are highlighted in [24, 25, 30].

Based on the analysis of rescue operations using UAV groups, several key areas can be identified for improving the DM system. First, it is necessary to improve the autonomy of DM by each UAV. Currently, systems are too dependent on central control, which causes delays and reduces the efficiency of operations. The introduction of machine learning algorithms and local situation analysis systems will allow each UAV to make decisions independently and adjust its route depending on the circumstances [16, 22]. The second important aspect is improving group coordination between UAVs. Existing systems have problems with coordinating actions and effectively distributing tasks between UAVs. The implementation of dynamic role distribution protocols and swarm behavior algorithms will significantly increase the efficiency of group operations. The development of automatic task redistribution systems and the optimization of collision avoidance algorithms are essential [8, 9, 17, 20].

Data processing and DM also require significant improvement. Current systems are characterized by limited information processing speed and insufficient accuracy of situation analysis. The implementation of distributed data processing systems and the use of neural networks for situation analysis will allow for an increase in the speed and quality of DM. It is also important to create effective systems for combining data from various sensors [9, 18].

System adaptability is another critical aspect that needs improvement. Existing systems demonstrate insufficient flexibility when changing conditions and limited ability to integrate different missions. The development of adaptive control algorithms and the implementation of dynamic replanning systems may increase the efficiency of operations in conditions of changing and association of missions. Special attention should be paid to the creation of mechanisms for rapid response to unforeseen situations [19].

Implementing these improvements will create a more efficient and reliable DM system for UAV teams during rescue operations. This, in turn, will increase the effectiveness of rescue operations and increase the number of lives saved during floods and other emergencies.

The group flights with various types of UAVs are organized to simultaneously perform multiple tasks, such as monitoring, reconnaissance, conducting search and rescue operations, and delivering urgent cargo (transportation), and communication.

3 Unmanned Aerial Vehicle Network

3.1 Unmanned Aerial Vehicle for Reconnaissance

In disaster situations, particularly floods, reconnaissance UAVs play a critical role in conducting rescue operations. This specialized unit is equipped with advanced technology to carry out comprehensive monitoring of the flooded area and provide rescuers with up-to-date information about the situation on the ground [5].

The main task of the reconnaissance UAV is to create a detailed picture of the disaster zone in real-time. The UAV is equipped with a high-resolution multispectral camera that operates in both visible and infrared ranges. This allows for effective detection of individuals in need of assistance, even in challenging weather conditions or at night. Thermal sensors can capture the heat signatures of people through tree foliage and lightweight structures, significantly enhancing the search capabilities for those affected. The UAV is also equipped with a lidar system that allows for the creation of accurate three-dimensional maps of flooded areas. This technology makes it possible to determine the depth of flooding in different locations, identify submerged buildings and structures, and assess the condition of critical infrastructure like bridges, dams, and roads. Based on this data, rescuers can plan optimal evacuation routes and identify priority zones for rescue operations [5, 6, 11].

A key feature of the reconnaissance UAV's design is its high resilience to adverse weather conditions. The aircraft can operate in strong winds and rainfall, which are typical during floods [1, 14, 15]. A waterproof body and special protective coatings for the electronic components ensure reliable operation in high humidity conditions. The stabilization system, using state-of-the-art gyroscopes and accelerometers, allows for clear images to be captured even in challenging weather. An important aspect of the reconnaissance UAV's operation is its ability to make autonomous decisions. The built-in artificial intelligence system analyzes the acquired data in real-time and automatically identifies potentially dangerous situations that require immediate response. When people in danger are detected, the system instantly transmits their coordinates to rescue services and other UAVs in the group, allowing for prompt organization of assistance.

To ensure prolonged operation, the UAV is equipped with high-capacity batteries and an energy consumption optimization system. Depending on the tasks at hand and weather conditions, the aircraft can stay in the air for up to 4 h. If necessary, rapid battery replacement is possible to continue the mission without long breaks in surveillance. The data transmission system of the reconnaissance UAV uses secure communication channels with high bandwidth. This allows high-quality video, thermal images, and 3D terrain maps to be transmitted to the command center in real time. All information is automatically archived and can be used for further analysis and planning of rescue operations [6]. During its operation, the reconnaissance UAV constantly interacts with other UAVs in the group, exchanging data and coordinating actions [20]. The authors proposed various configurations of groups for performing missions, determining the leader in the UAV group – the central UAV – repeater using graph theory. Such interaction helps to optimize the process of searching for victims and ensures maximum coverage of the territory during the rescue operation (Fig. 2).

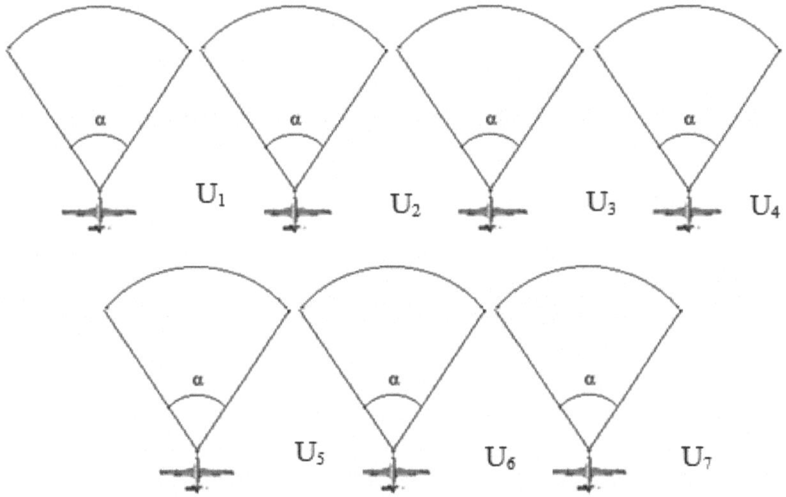

Fig. 2. Presentation of a group UAVs in a connected network proceeding aerial photography.

3.2 Transport Unmanned Aerial Vehicle

In the system of group interaction of UAVs, the transport UAV plays a critically important role in logistics for the rescue operation. This specialized UAV is designed for the fast and reliable delivery of vital supplies to disaster zones, where traditional transportation methods may be complicated or impossible. Transport UAV is built based on a reinforced frame made of composite materials, which provides an optimal strength-to-weight ratio. This configuration allows for the reliable transportation of bulky loads weighing up to 20 kg, including medications, drinking water, food, and rescue equipment. Special attention in transport UAVs has been given to the loading and unloading system, which features a universal cargo compartment with modular fastenings, allowing for quick adaptation to different types of cargo. The system can operate in various modes: from a soft landing with a complete stop of the UAV to dynamic cargo drops from a low altitude [15, 18].

The navigation equipment of the transport UAV includes a redundant geolocation system that simultaneously uses signals from multiple satellite systems (GPS, GLONASS, Galileo) [25]. This ensures high positioning accuracy even in conditions of limited satellite visibility or the presence of obstacles. Additionally, an inertial navigation system is installed, which helps maintain course accuracy during the temporary loss of satellite signal. For visual orientation in space, the UAV is equipped with a compact high-resolution video camera with an image stabilization system. The software of the transport UAV is based on a modern autopilot system that ensures fully autonomous execution of flight tasks. The system continuously analyzes telemetry data and can automatically adjust the route in unforeseen situations [7, 31]. The energy supply for the UAV is based on high-capacity lithium-polymer batteries with an active monitoring system to track the battery's status. The energy reserve is calculated to carry out a typical mission, considering the maximum load and the necessary reserve for returning to base.

The transport UAV is fully integrated into the overall group interaction system of UAVs. It receives delivery target coordinates from the reconnaissance UAV, maintains constant communication through the communication UAV, and can coordinate its actions with a backup UAV in case it needs to transfer cargo or distribute task.

3.3 Communication Unmanned Aerial Vehicle

The relay UAV is a specialized UAV primarily designed to extend the communication range between the base station and target UAVs or other entities. Its main function is to receive a signal from the transmitter, amplify it, and then pass it on to the receiver, which significantly expands the coverage area [15, 18, 19].

To ensure stable operation, the UAV is equipped with a comprehensive positioning system. The main GPS module works alongside a backup, guaranteeing accurate positioning even if one of the devices fails. In the event of a complete GPS signal loss, the UAV can navigate using an installed FPV camera that streams video back to the base station. UAV control can be performed in two modes. In automatic mode, the onboard computer with autopilot executes a pre-programmed mission – it maintains an optimal position for signal relay, adjusts altitude and position to ensure the best quality of communication. If necessary, instant switching to manual control is possible, which is particularly important in non-standard situations or when encountering obstacles [15, 18, 22, 25].

The autopilot software handles all automatic functions. It includes a mission planning system that allows you to set routes, hold position points, and establish desired flight altitude. A key feature is the constant monitoring of signal quality and automatic adjustment of the UAV's position to optimize relay [8]. The system incorporates various safety mechanisms. When the UAV's battery is running low, it automatically returns to the launch point. If it loses communication with the base station, an emergency landing protocol is activated. The FPV camera is used not only for visual monitoring but also as a backup orientation system in case of GPS failure [16–18, 25].

Before each flight, a comprehensive check of all systems is conducted, including compass calibration, testing both GPS modules, and checking the quality of the signal relay. During operation, the operator continuously monitors the UAV's telemetry, battery level, and the quality of the relayed signal [18, 25].

3.4 Additional Unmanned Aerial Vehicle Monitoring

The additional UAV monitoring provides constant oversight of the developing emergencies. The main purpose of such a UAV is to collect and transmit up-to-date information about the state of flooded areas. Equipped with a specialized optical system, including a thermal camera and a high-resolution camera, the UAV is capable of monitoring both during the day and at night. The thermal imaging equipment is particularly effective for detecting people in need of assistance in poor visibility conditions or hard-to-reach areas [19].

The UAV software includes an automatic object recognition system, which allows for the identification of people, vehicles, and damaged infrastructure. The data collected in real time is transmitted to the rescue operation control center, where it is analyzed to make quick decisions regarding the deployment of rescue teams.

Special attention is given to the UAV's resilience against tough weather conditions. The craft is designed to be water-resistant and provides enhanced stability in strong winds. The power supply system is designed for long operational periods, which is crucial for continuous situation monitoring [22, 25].

During the monitoring process, the UAV carries out several key tasks:

- Mapping flooded areas while determining the depth of submersion.
- Detecting people who need evacuation.
- Assessing the state of critical infrastructure (bridges, dams, power plants).
- Identifying safe routes for rescue teams.
- Predicting flood development based on collected data.

The information gathered is used to create up-to-date flood maps, which are constantly updated and help coordinate the actions of rescue services. This allows for the optimization of resource use and enhances the overall efficiency of the rescue operation [15, 18].

4 Unmanned Aerial Vehicle Coordination Principle

The coordination of different types of UAVs in a unified system is based on a clear division of roles and well-coordinated interaction among all components. The central element is a command center, which carries out the overall management and coordination of the actions of all UAVs (Fig. 3).

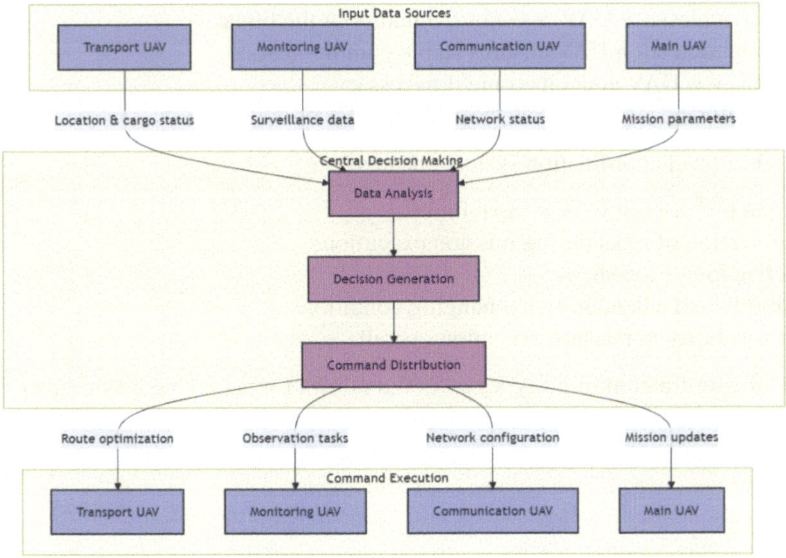

Fig. 3. Structure of Unmanned Aerial Vehicle network.

The reconnaissance UAV serves as the vanguard of the system. It is the first to explore the area, determine safe routes, and identify potential threats or obstacles. The information it collects forms the basis for planning the routes of other UAVs. Its high maneuverability and specialized reconnaissance equipment allow for rapid assessment of the situation and adaptation of mission plans according to the conditions identified.

The transport UAV operates based on data received from the reconnaissance UAV. Its routes are plotted considering the identified safe corridors and risk zones. The control center coordinates its actions with other UAVs to avoid trajectory overlaps and ensure safe delivery of cargo. If necessary, its route can be adjusted in real-time based on new intelligence data. The communication UAV occupies a strategic position in ensuring stable communication among all elements of the system. It constantly monitors signal quality and automatically adjusts its position to optimize coverage. If other UAVs move over significant distances, it can change its position to maintain reliable communication with all participants in the operation. An additional monitoring UAV provides overall oversight of the area and mission execution control. It coordinates its actions with the reconnaissance UAV, complementing its data with a more detailed analysis of specific areas of interest. Its information is used for timely adjustments to the actions of all other UAVs and to optimize their interaction.

The coordination algorithm includes:

- Continuous data exchange among all UAVs via the communication UAV.
- Automatic route adjustments to avoid collisions.
- Task prioritization and resource allocation.
- Synchronization of actions for maximum operational efficiency.
- Automatic mission replanning in case of emergencies.

In the event of critical situations, the system has a defined action protocol:

- The reconnaissance UAV assesses the nature of the threat.
- The communication UAV ensures stable communication.
- The transport UAV alters its route if necessary.
- The monitoring UAV controls the development of the situation.

Such a multi-level coordination system ensures:

- Optimal use of each UAV's capabilities.
- Minimization of risks during mission execution.
- Effective use of resources.
- Flexibility and adaptability to changing conditions.
- High reliability in mission execution overall.

Successful coordination of all types of UAVs allows for achieving a synergistic effect, where the capabilities of the system as a whole exceed the sum of the capabilities of its individual components.

5 The Illustrative Example of Collaborative Decision-Making During a Flood

5.1 Theoretical Part

Carrying out a rescue operation during a flood using a UAV group requires clear organization and centralized management. To effectively complete the tasks, the group consists of four specialized UAVs: a main UAV coordinator, a communication UAV, a transport UAV, and a monitoring UAV. Each of them has specific functions and areas of responsibility.

During the preparation phase, the UAV group is deployed considering the optimal operating altitudes for each UAV. The main UAV occupies a central position at an altitude of 100–150 m, providing the best overview of the operation area. The communication UAV is positioned at 80–100 m to create a reliable communication hub. The monitoring UAV operates at 50–70 m, which is ideal for detailed surveying of the area. The transport UAV remains on standby at the base, ready to carry out delivery tasks for necessary supplies.

The management organization is built on a clear hierarchical structure, where the main UAV acts as the coordination center. All decisions regarding task allocation, setting priorities, and adjusting plans are made at the level of the main UAV. Other group members regularly send updates on their status, including battery level, technical parameters, current location, and task completion status. The communication UAV ensures continuous communication between all participants in the operation [9].

Initial reconnaissance of the area is carried out by the monitoring UAV along a predefined route. The collected data is transmitted via the communication UAV to the main UAV, which analyzes it and creates an updated flood map. Based on this information, flood zones, water levels, potentially hazardous areas, and possible locations of victims are identified.

When victims are discovered, the next stage of the operation is activated. The main UAV analyzes the situation and prioritizes the rescue by evaluating the number of victims, accessibility, threat levels, and available resources. Based on this analysis, detailed instructions for the transport UAV are created, specifying the target coordinates, optimal route, flight altitude, and required rescue equipment.

The coordination of the group's actions is carried out in real-time. The main UAV continuously updates tasks for each group member, considering changes in the situation. A regular status information exchange is established every 30 s, which includes: current coordinates, battery charge level, task status, and technical condition of the device. In case of emergencies or changes in operational conditions, immediate adjustments to plans are made.

The safety of the operation is ensured by a set of procedures and protocols. The main safety elements include constant distance monitoring between UAVs, designated waiting zones for each device, clear action protocols in case of loss of communication, and the presence of backup return routes to the base. Special attention is given to monitoring weather conditions and the technical status of the devices [10].

The completion of the operation occurs in stages. First, the transport UAV returns to base, followed by the monitoring UAV, then the communication UAV, and lastly, the main UAV. Once the UAVs have landed, data collection and archiving commence, followed by the preparation of a detailed report on completed tasks. Subsequently, the group's performance is analyzed, and potential improvements for future operations are identified.

This centralized approach to managing the UAV group during a rescue operation ensures maximum efficiency in utilizing available resources, quick DM, and minimizing risks for all participants in the operation.

5.2 Practical Part

Alternative strategies for decision making in uncertainty by participants-operators (O) of UAVs (Fig. 4):

- O_1 is the main UAV coordinator;
- O_2 is communication UAV;
- O_3 is transport UAV;
- O_4 is monitoring UAV.

The alternative strategies for the operators (formula (1)): A_1 is to continue performing the task; A_2 is to finish completing the task; A_3 is to the departure; A_4 is to the destination.

$$A = \{A_1, A_2, A_3, A_4\}. \tag{1}$$

The factors (formula (2)) that influence the DM process, specifically:

- λ_1 is the number of casualties;
- λ_2 is ease of access;
- λ_3 is threat level;
- λ_4 are available resources;
- λ_5 is battery level of other UAVs;
- λ_6 are current coordinates of other UAVs;
- λ_7 is technical condition of other UAVs.

$$\lambda = \{\lambda_1, \lambda_2, \ldots, \lambda_7\}. \tag{2}$$

To select the optimal solution, one of the four criteria is applied (Wald, Laplace, Hurwicz, Savage).

Wald's criterion. The optimal solution according to Wald's criterion is determined by the rule (formula (3)):

$$A^* = \max_{A_i}\left\{\min_{B_j} u_{ij}(A_i, B_j)\right\}. \tag{3}$$

The Laplace Criterion. The optimal solution according to the Laplace criterion is determined by the rule (formula (4)):

$$A^* = \max_{A_i}\left\{\frac{1}{n}\sum\nolimits_{i=1}^{N} u_{ij}(A_i, B_j)\right\}. \tag{4}$$

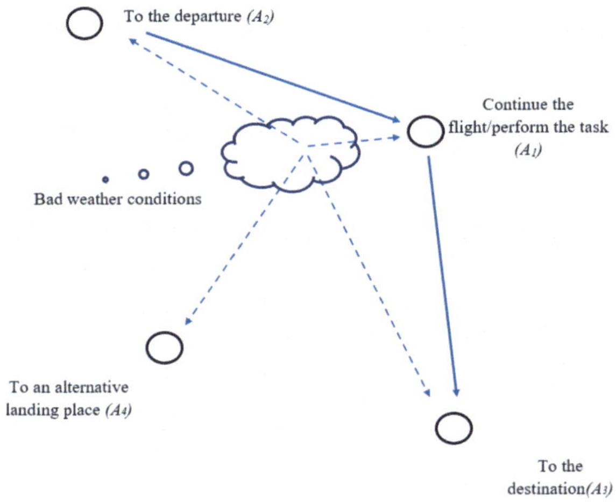

Fig. 4. Schema of an emergency in a UAV flight.

Hurwitz Criterion. The optimal solution according to the Hurwitz criterion is determined by the rule (formula (5)):

$$A^* = \max_{A_i}\left\{\alpha \; max_{B_j} \; u_{ij}(A_i, B_j) + (1-\alpha)\min_{B_j} u_{ij}\left(A_i, B_j\right)\right\}. \tag{5}$$

The optimism-pessimism coefficient α will be chosen to be 0.5.

Savage's criterion. The optimal decision according to Savage's criterion is determined by the rule (formulas (6), (7)):

$$A^* = \min_{B_j}\max_{A_i} r_{ij}(A_i, B_j), \tag{6}$$

$$r_{ij}(A_i, B_j) = \Delta_{A_i} = \max_{B_k} u_{ij}(A_i, B_j) - u_{ij}(A_i, B_j). \tag{7}$$

It's necessary to form an intermediate loss matrix r_{ij}. To obtain this matrix, you need to find the maximum element in each row of the output matrix. Then, by subtracting this element from all other elements in each row, we will get a new loss matrix.

Decision matrix for the first operator (Table 1), all values of which are determined by the expert judgment method (EJM). According to the Wald, Laplace, Hurwicz, and Savage criteria, the optimal solution is A_1 (to continue with the task).

Decision matrix for the second operator (Table 2), all values of which are determined by the EJM. For the second operator, according to the Wald's, Laplace's, and Savage's criteria, the optimal solution is A_1 (to continue with the task). According to the Hurwicz's criteria, the optimal solution is A_2 (finish completing the task).

Decision matrix for the third operator (Table 3), all values of which are determined by EJM. For the third operator, based on Wald's, Laplace's, Hurwicz's criteria, the optimal decision is A_1 (to continue with the task). According to the Savage's criteria, the optimal solution is A_4 (to the destination).

Table 1. Decision Matrix O_1 (the main UAV coordinator).

	A	λ_1	λ_2	λ_3	λ_4	λ_5	λ_6	λ_7	**W**	*L*	*H*	*S*
A_1	Continue performing the task	6	3	7	5	8	4	8	**3**	**5.85**	**5.5**	**5**
A_2	Finish completing the task	6	2	1	6	5	7	2	1	4.14	4	6
A_3	To the departure	7	3	3	5	2	6	1	1	3.85	4	6
A_4	To the destination	8	1	3	9	4	3	6	1	4.85	5	8

Table 2. Decision Matrix O_2 (communication UAV).

	A	λ_1	λ_2	λ_3	λ_4	λ_5	λ_6	λ_7	**W**	*L*	*H*	*S*
A_1	Continue performing the task	3	3	6	7	8	5	6	**3**	**5.43**	5.5	**5**
A_2	Finish completing the task	9	6	6	4	4	3	4	3	5.14	**6**	6
A_3	To the departure	5	1	2	8	5	7	3	1	4.43	4.5	7
A_4	To the destination	6	4	3	4	7	1	2	1	3.86	4	6

Table 3. Decision Matrix O_3 (transport UAV).

	A	λ_1	λ_2	λ_3	λ_4	λ_5	λ_6	λ_7	**W**	*L*	*H*	*S*
A_1	Continue performing the task	5	5	6	3	9	5	3	**3**	**5.14**	**6**	6
A_2	Finish completing the task	8	1	2	4	8	3	2	1	4	4.5	7
A_3	To the departure	5	1	2	3	7	5	2	1	3.57	4	6
A_4	To the destination	6	1	3	5	4	1	2	1	3	3.5	**5**

Decision matrix for the fourth operator (Table 4), with all values determined by EJM. For the fourth operator, according to the Wald, Laplace, Hurwitz criterion, the optimal solution is A_1 (to continue with the task). According to the Savage's criteria, the optimal solution is A_2 (finish completing the task) and A_3 (to the departure).

Table 4. Decision Matrix O_4 (monitoring UAV).

	A	λ_1	λ_2	λ_3	λ_4	λ_5	λ_6	λ_7	**W**	*L*	*H*	*S*
A_1	Continue performing the task	6	2	7	5	4	5	9	**2**	**5.43**	**5.5**	7
A_2	Finish completing the task	4	6	6	5	2	1	5	1	3.71	3.5	**5**
A_3	To the departure	2	3	5	7	3	4	2	2	3.71	4.5	**5**
A_4	To the destination	3	3	6	7	5	1	8	1	4.71	4.5	7

Table 5 presents the collective matrix of DM under uncertainty for all operators. The optimal solution according to the Wald and Laplace criterion is A_1 (to continue with the task).

Table 5. Collaborative decision-making matrix for all operators.

	A	O_1	O_2	O_3	O_4	**W**	O_1	O_2	O_3	O_4	*L*
A_1	Continue performing the task	3	3	3	2	**2**	5.85	5.43	5.14	5.43	**5.46**
A_2	Finish completing the task	1	3	1	1	1	4.14	5.14	4	3.71	4.25
A_3	To the departure	1	1	1	2	1	3.85	4.43	3.57	3.71	3.9
A_4	To the destination	1	1	1	1	1	4.85	3.86	3	4.71	4.1

For a joint decision in a CDM system the optimal strategy is chosen based on the coordinated evaluation of identical priority factors. This ensures that all stakeholders align on common objectives and that the decision reflects a balanced consideration of risks, opportunities, and shared goals. The integration of these factors promotes consistency, fairness, and the efficiency of the DM process, enhancing the overall effectiveness of the system.

6 Conclusions

As a result of the study of the problem of CDM by a UAV group during rescue operations, it was found that this technology is of critical importance for increasing the efficiency and safety of rescue missions. The use of distributed DM algorithms allows to significantly expand the capabilities of search and rescue operations and optimize the use of available resources. It was found that modern methods of UAV group control are based on a combination of different approaches, including multi-agent systems, fuzzy logic and optimization methods. This allows us to create flexible and adaptive systems that can function effectively in conditions of uncertainty and dynamic changes in the environment. Analysis of existing solutions showed the presence of certain limitations and problems that require further research. In particular, the issues of scalability of algorithms, optimization of energy consumption and ensuring the reliability of communications between members of the UAV group remain relevant.

The results of the study demonstrate the need to develop new approaches to integrating UAV group control systems with existing emergency management systems. This will allow us to create a single effective emergency response system. It was determined that the further development of CDM technologies by a UAV group should consider the requirements for energy efficiency, reliability and safety of operations. Future research should focus on developing algorithms capable of working effectively in conditions of limited resources and difficult weather conditions.

From a practical point of view, the implementation of the developed methods and algorithms will allow to increase the efficiency of rescue operations, reduce risks for rescuers and optimize the use of resources. This is of particular importance in the context of the growing frequency and scale of emergency situations.

In general, the conducted study confirms the prospects for the further development of CDM technologies by a UAV group and the need for their active implementation in the practice of conducting rescue operations.

7 Future Scope

Further research will be aimed at creating an Intelligent System of Collaborative decision-making (IS CDM) during the performance of humanitarian missions using UAVs to minimize risks and maximize safety when delivering humanitarian cargo.

In UAV-assisted rescue missions during natural disasters, determining who requires urgent assistance is a major challenge. Triage prioritization, widely used in emergency medical response, can be adapted to optimize using UAV or UAVs group. By integrating real-time data processing and AI-driven models, UAV groups can classify and prioritize victims based on severity, accessibility, and resource availability [31].

The IS CDM utilizes big data analysis and machine learning to enhance UAV-assisted humanitarian operations. This system combines AI-based automation with human oversight, ensuring adaptable responses. UAVs collect real-time environmental and victim data through sensors, while AI-driven analytics process this information to identify priority rescue targets and adjust responses dynamically. Machine learning models predict distress severity and allocate UAV resources efficiently. Reconnaissance UAVs scan affected areas, AI-driven triage models assign priority scores, transport UAVs deliver supplies or conduct evacuations, and communication UAVs facilitate seamless data exchange. The IS CDM system continuously refines UAV task allocation using reinforcement learning and predictive analytics.

Effective Integration of the DM models (DM under certainty, risk, uncertainty) optimizes UAV resource allocation [7, 20, 21]: (a) critical cases receive immediate UAV evacuation, (b) urgent cases get supply deliveries and alerts, (c) stable cases are monitored, (d) unreachable cases are reassessed later. By integrating IS CDM, UAV-assisted rescue missions become more precise, efficient, and adaptable, maximizing survival rates and ensuring optimal resource use. Future research will focus on enhancing AI-human collaboration, refining intervention protocols, and ensuring ethical, scalable, and accountable UAV operations in humanitarian missions. To effectively manage artificial intelligence decisions, hybrid intelligent systems that integrate machine and human competencies with environmental influences are necessary.

References

1. Iqbal, U., Riaz, M., Zhao, J., Barthelemy, J., Perez, P.: Drones for flood monitoring, mapping and detection: a bibliometric review. Drones. **7**, 32 (2023). https://doi.org/10.3390/drones7010032
2. Murphy, R.R., Tadokoro, S., Nardi, D.: Search and rescue robotics. In: Springer Handbook of Robotics, pp. 1151–1174 (2023). https://doi.org/10.1007/978-3-540-30301-5_51

3. Guan, Y., Zou, S., Peng, H., Ni, W., Sun, Y., Gao, H.: Cooperative UAV trajectory design for disaster area emergency communications: a multiagent PPO method. IEEE Internet Things J. **11**(5), 8848–8859 (2024). https://doi.org/10.1109/JIOT.2023.3320796
4. Ghauri, S.A., Sarfraz, M., Qamar, R.A., Sohail, M.F., Khan, S.A.: A Review of multi-UAV task allocation algorithms for a search and rescue scenario. J. Sens. Actuator Netw. **13**(5), 47 (2024). https://doi.org/10.3390/jsan13050047
5. Półka, M., Ptak, S., Kuziora, Ł.: The use of UAV's for search and rescue operations. Procedia Eng. **192**, 748–752 (2017). https://doi.org/10.1016/j.proeng.2017.06.129
6. Griffin, G.F.: The use of unmanned aerial vehicles for disaster management. Geomatica. **68**, 265–281 (2014). https://doi.org/10.5623/cig2014-402
7. Shmelova, T., Sikirda, Y., Yatsko, M.: Nondeterministic collaborative decision-making models in flight emergencies based on the factors' priority. In: Ostroumov, I., Zaliskyi, M. (eds.) Proceedings of the International Workshop on Advances in Civil Aviation Systems Development. ACASD 2023 Lecture Notes in Networks and Systems, vol. 736, pp. 283–300. Springer, Cham (2023). https://doi.org/10.1007/978-3-031-38082-2_22
8. Shmelova, T., Kucherov, D., Simchenko, S.: Organization of group flight using heterogeneous UAVs for multi-purpose target tasks. In: 7th International Conference Actual Problems of Unmanned Aerial Vehicles Developments, pp. 169–174. IEEE, Kyiv (2024). https://doi.org/10.1109/APUAVD64488.2024.10765885
9. Kucherov, D.P., Jiang, G., Liu, H., Fu, M.: UAV group control protocol with adaptive consensus. In: 7th International Conference Actual Problems of Unmanned Aerial Vehicles Developments, pp. 169–174. IEEE, Kyiv (2024). https://doi.org/10.1002/acs.3868
10. Employment of group 3/4/5 organic/non organic UAS tactical pocket guide. https://info.publicintelligence.net/JFCOM-UAS-PocketGuide.pdf. Last accessed 11 Dec 2024
11. Shah, A.F.M.S.: Architecture of emergency communication systems in disasters through UAVs in 5G and beyond. Drones. **7**(1), 25 (2023). https://doi.org/10.3390/drones7010025
12. Karamanou, A., Dreliosi, G.-C., Papadimatos, D., Hahlakis, A.: Supporting search and rescue operations with UAVs. In: 5th International Conference on Civil Protection & New Technology, pp. 63–75. IEEE, Kozani (2018). https://doi.org/10.6084/m9.figshare.8986412
13. Braverman, A.: Unmanned aerial systems (UAS) in urban search and rescue-methodology, capacity development, and integration. J. Emerg. Manag. **19**(1), 33–38 (2021). https://doi.org/10.5055/jem.0496
14. Manandhar, B., Cui, S., Wang, L., Shrestha, S.: Post-flood resilience assessment of July 2021 flood in Western Germany and China. Land. **12** (2023). https://doi.org/10.3390/land12030625
15. Goudarzi, S., Soleymani, S., Anisi, H., Ciuonzo, D., Kama, N., Abdullah, S., et al.: Real-time and intelligent flood prediction using UAV-assisted wireless sensor network. Comput. Mater. Contin. **70** (2021). https://doi.org/10.32604/cmc.2022.019550
16. Peksa, J., Mamchur, D.: A review on the state of the art in copter drones and flight control systems. Sensors. **24**, 3349 (2024). https://doi.org/10.3390/s24113349
17. Arnold, R., Osinski, M., Reddy, C., Lowey, A.: Reinforcement learning for collaborative search and rescue using unmanned aircraft system swarms. In: International Symposium on Technologies for Homeland Security, pp. 1–6. IEEE, Boston (2022). https://doi.org/10.1109/HST56032.2022.10025435
18. Papyan, N., Kulhandjian, M., Kulhandjian, H., Aslanyan, L.: AI-based drone assisted human rescue in disaster environments: challenges and opportunities. Pattern Recognit. Image Anal. **34**, 169–186 (2024). https://doi.org/10.1134/S1054661824010152
19. Zhang, P., Chen, G., Li, Y., Dong, W.: Agile formation control of drone flocking enhanced with active vision-based relative localization. IEEE Robot. Autom. Lett. **7**, 6359–6366 (2022). https://doi.org/10.48550/arXiv.2108.05505

20. Shmelova, T., Sikirda, Y., Yatsko, M., Stratonov, V.: Integration of deterministic, stochastic, and non-stochastic models to obtain an optimal collaborative decision in the flight emergency. In: Ostroumov, I., Zaliskyi, M. (eds.) Proceedings of the 2nd International Workshop on Advances in Civil Aviation Systems Development. ACASD 2024 Lecture Notes in Networks and Systems, vol. 992, pp. 205–223. Springer, Cham (2024). https://doi.org/10.1007/978-3-031-60196-5_17
21. Shmelova, T., Sikirda, Y., Yatsko, M., Marienkov, I., Sahun, Y.: Collaborative decision-making models in flight emergency "Landing gear failure on takeoff". CEUR Workshop Proc. **3373**, 15–33 (2023)
22. China's high-tech companies aid Henan flood rescues with drones and satellites. https://www.globaltimes.cn/page/202107/1229309.shtml. Last accessed 11 Dec 2024
23. Unmanned Aircraft Systems Traffic Management (UTM) – A Common Framework with Core Principles for Global Harmonization, 3ed edn., Montreal, Canada (2017)
24. Sładkowski, A., Kamiński, W.: Chapter 3: Using unmanned aerial vehicles to solve some civil problems. In: Cases on Modern Computer Systems in Aviation, pp. 52–127. IGI Global, Hershey (2019). https://doi.org/10.4018/978-1-5225-7588-7.ch003
25. Austin, R.: Unmanned aircraft systems: UAVs design, development, and deployment. Wiley, London (2010). https://doi.org/10.1002/9780470664797
26. World Civil Unmanned Aerial Systems market profile & forecast. Teal Group Corporation, Fairfax, USA (2019)
27. Bondarev, D., Djafarzade, R., Kozub, A.: The effectiveness of unmanned aerial vehicles group flights. Inf. Process. Syst. Sci Period. **9** (2014)
28. Shmelova, T., Bondarev, D.: Chapter 8: Automated system of controlling Unmanned Aerial Vehicles group flight. In: Unmanned Aerial Vehicles in civilian logistics and supply chain management, pp. 208–242. IGI Global, Hershey (2019). https://doi.org/10.4018/978-1-5225-7900-7.ch008
29. Meenu, C., et al.: Fault tolerance-based routing approach for WMN. In: International Conference on Computer and Computational Sciences (ICCCS), pp. 177–182. IEEE, Greater Noida (2015). https://doi.org/10.1109/ICCACS.2015.7361345
30. Shmelova, T., Yatsko, M., Sierostanov, I., Kolotusha, V.: Chapter 49: Artificial intelligence methods and applications in aviation. In: Handbook of Research on AI Methods and Applications in Computer Engineering, pp. 108–140. IGI Global, Hershey (2023). https://doi.org/10.4018/978-1-6684-6937-8
31. Shmelova, T., Smolanka, V., Sikirda, Y., Sechko, O.: Real-time monitoring and diagnostics of the person's emotional state and decision-making in extreme situations for healthcare. Decis. Mak. Anal. **2**(1), 11–32 (2024). https://doi.org/10.55976/dma.22024121911-32

Unmanned Aircraft Systems

UAS Communication Links and Protocols Analysis to Ensure Wind Information Obtaining and Exchange

Yuliya Averyanova[1(✉)], Oleksii Sushych[1], Irina Pryhodko[1], and Yevheniia Znakovska[1,2]

[1] National Aviation University, Liubomyra Huzara Ave., 1, Kyiv 03058, Ukraine
ayua@kai.edu.ua

[2] Interregional Academy of Personnel Management, Frometivska Str., 2, Kyiv 03039, Ukraine

Abstract. Operative meteorological information plays an important role for safety and efficiency of aircraft flights including the flights of unmanned aerial vehicles. At the same time, the UAVs can serve as instruments to obtain actual information about atmospheric state and weather phenomena in the area of flight. The collected real-time information can be shared and used to provide operative meteorological support of flights and UAV-based missions. In this paper we consider the modern communication links and protocols to provide weather-related information and further transmit or exchange between UAV operators or wider customers of weather information. An overview of the modern communication links and protocols was done. The detailed study to extract wind information was done using the MAVLink messaging protocol as it is open and one of the most popular and is supported by many unmanned aerial vehicle control platforms. It was shown that considered instruments allow to extract of information about wind speed and wind direction and match this information with horizontal and vertical coordinates. The obtained information can be shared between the interested UAV operators used in autopilots of UAVs for flight trajectory correction.

Keywords: UAV · Meteorological Measurements · Wind · Wind Components · Communication Links · Communication Protocols · Mavlink Messaging Protocol

1 Introduction

Unmanned aerial vehicles (UAVs) are actively integrated into many fields of human activity [1] including industrial applications and use for private purposes [2, 3]. Many tasks of civilian and industrial applications are performed at rather low altitudes and can be significantly affected by the weather, atmospheric conditions, and terrain peculiarities. The weather can directly influence the characteristics of UAVs [4] and restrict the functionality of UAVs. Also, the atmospheric conditions can influence the navigation and communication systems and different sensors. In [5, 6] the impacts that weather-related phenomena can make on communication are divided into four groups. These

I. Ostroumov et al. (Eds.): ACASD 2025, LNNS 1418, pp. 475–485, 2025.
https://doi.org/10.1007/978-3-031-91992-3_31

include signal attenuation, signal path bending, dispersion, and signal-to-noise (SNR) degradation. The mentioned factors can influence the GPS navigation as well [7–9]. The influence of weather and atmospheric parameters on the unmanned aircraft system (UAS) operation proves the importance of understanding the following items:

- the consequence of the weather influence on any of the UAS components and
- the way of obtaining operative meteorological information for in-time decision-making to provide safe and effective UAV-based missions.

In this context, it is important to note that UAV and UAV operators from one side are the customers of meteorological information. It is highly recommended for UAV pilots to be familiar with current and forecasted weather conditions to begin flight and UAV-based missions. On another side, the UAV can be used as an instrument for obtaining actual information about weather and atmospheric conditions. It is known that one of the industrial applications of UAV is Meteorology [10]. At the same time, the interest is to study the possibility of multifunctional use of UAVs. Under such a concept the UAV performing a particular mission simultaneously makes some meteorological measurements and transmits information to the operator. Such an approach can benefit into economic side of UAV-based missions performing two tasks at the same time and into flight safety. Flight safety benefits are expected due to the availability of operative information about the current state of the atmosphere and the presence of meteorological phenomena that can influence UAS operation.

It was our motivation to study the modern communication links and protocols to provide weather-related information and transmission or exchange between interested operators or wider weather information customers.

1.1 Motivation

Taking into account the importance of operative meteorological information for UAV flight safety and efficient mission realization as well as the possibilities to use UAV as a platform for meteorological sensors additionally to primary mission and technological ability for operative exchange with information, it was our motivation to study the modern communication links and protocols to provide weather-related information obtaining and transmission or exchange between the interested operators or wider range of weather information customers.

2 Short Analysis of the Latest Research

The advantages of weather monitoring using drones are provided (probably not limited) by possibility [11–13]:

- to measure directly the wide spectrum of atmospheric parameters using different sensors;
- to study the vertical and horizontal fields of meteorological elements;
- to make measurements in hard-to-reach areas;
- relatively low cost of measurements, especially when measurements are combined with other UAV-based tasks;

- to obtain real-time data;
- to obtain high-resolution data to study microscale phenomena [10].

The methods to measure wind parameters using drones and estimating drag and thrust are presented and discussed in [14]. The method to estimate wind using GPS and orientation sensors is given in [15].

Taking into account the technological ability to measure atmospheric parameters and to obtain information about weather phenomena it is also important to study the communication channels and protocols to exchange real-time information. In the paper [16], the ZigBee protocol for meteorological data sharing and exchange in urban environments and radio noises was studied. The aeronautical infocommunications under dynamically changing conditions is discussed in [17]. In this paper, the learning technique for sustainable infocommunications is proposed. Paper [18, 19] is devoted to the problem of secure communication between UAVs. Nowadays this becomes of high importance as the increase of security attacks on the background of rising commercial drone market [20].

The overview of modern standard links and protocols is discussed in Sect. 3 of this study.

3 Communication Links and Protocols Overview

Command and control of unmanned aircraft systems are provided by special communication lines and protocols, which transfer information between the Ground Control Station (GSC) and the unmanned aerial vehicle (UAV) (see Fig. 1).

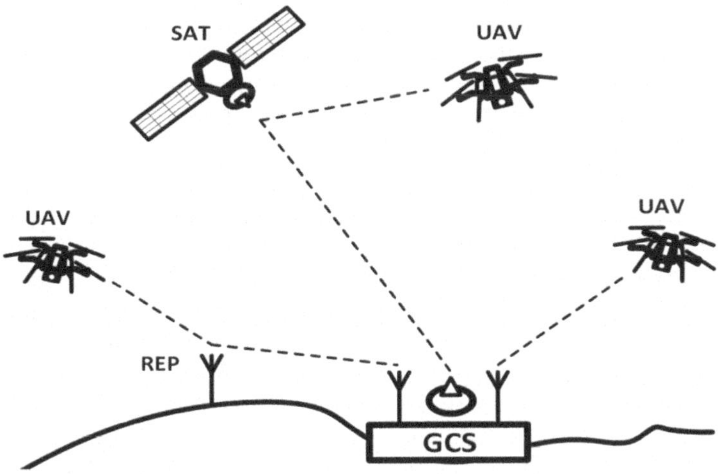

Fig. 1. Communication lines for command and control of UAV.

Communication Links. Unmanned aircraft system missions can be either controlled or fully autonomous. The controlled missions require the provision of the communication link between the unmanned aircraft system and the ground control station.

This communication link can be named as the UAV Command and Control (C2) link [21]. However, even for autonomous missions, it is desirable to have a bidirectional communication link – UAV to ground control station.

The C2 bidirectional wireless control and monitoring line allows the commands transmission from the ground control station operator to the UAS. The telecommunications equipment of the UAS provides the telemetry transmission in the direction of the ground control station and other elements of the system via the C2 line (it includes other UAVs, repeaters). The communication lines are also used to transmit information from the payload of the UAS.

In this way, the UAS makes the information exchange with operators, other UAVs, manned aircraft pilots, and air traffic control systems to ensure flight safety. This type of communication is called Controlled and Non-Payload Communication (CNPC).

At the same time, the unmanned aircraft system exchanges information related to the mission – photographs, videos, and data for end users. This is accomplished by using payload communication. However, the technical requirements for the CNPC communication line and the communication line with the payload differ significantly. Also, and it is of particular note that the CNPC communications link is the crucial to flight safety.

The radio frequency spectrum of the CNPC link and the payload link is allocated by international and national legal aspects of radio frequency spectrum regulation and can be operated both in the range of licensed frequencies and in the range where a license is not required [22].

The most commonly used frequencies for Visual Line of Sight (VLOS) flights are the frequencies in the range 433 MHz, 850–965 MHz (control panel), 2.4 GHz and 5.8 GHz (control, video transmission), 1575 MHz and 1227 MHz (satellite navigation) [10]. The use of this frequency ranges ensures the straight-line propagation of radio waves and allows the use of small-sized antennas and low-power telecommunications equipment, which can be placed even on the light unmanned aircraft systems. The availability and accuracy of Global Navigation Satellite System (GNSS) estimation can be found in [23].

The operation of UAS flights beyond the visual line of sight (BVLOS) is based on the use of repeaters. The satellites or ground-based equipment can be used as the repeaters (see Fig. 1).

The use of satellite transponders is associated with additional costs and increased complexity of UAV telecommunications equipment. If geostationary satellites are used, there will be significant delays in information transmission. It can reach several seconds in some cases. However, in the case of long-term missions over large distances, there is no alternative to satellite communication channels. Modern satellite telecommunications equipment is a fairly compact and relatively low-power solution. As an example, Iridium and Inmarsat [24] satellite communication systems can be used as ICAO-recommended mobile satellite communication systems (Aeronautical Mobile Satellite Service - AMSS). It should also be taken into account that providing a high-speed CNPC communication line and especially a high-speed payload communication line is associated with a significant increase in the cost and complexity of on-board UAV telecommunications equipment.

The use of ground-based repeaters can be implemented using broadband mobile communication systems (Broadband Mobile Communication - BMC). Modern 4G/LTE 5G telecommunication systems provide the necessary bandwidth for both the CNPC communication line and the payload communication line [22]. However, the use of mobile communication networks, along with clear advantages, also has some quite significant limitations. The main disadvantages include the limited operating area of mobile communication systems and the reliability of channels that provide communication between the unmanned aircraft system and the ground control station.

Communication Protocols. Data transmission through communication lines between the UAV and the ground control station is implemented using standardized data transmission protocols. Today, quite a large number of communication protocols have been developed and put into practice. Some protocols are proprietary, some are open source. Open-source protocols have some advantages with the appropriate level of support from the development team.

The main advantages include:

- Low or no financial burden;
- Reliability, because an error in open-source code is easier to detect;
- Long-term support of open-source code by the developer community;
- Security is implemented through open-source code to check for vulnerabilities.

Also, in favor of open-source software is the development of open software as an alternative to proprietary software.

Quite several communication protocols have been developed for unmanned aircraft systems. The most popular protocols are Cyphal, DroneLink, and MAVLink [25, 26].

Cyphal is an open-source protocol and technology for distributed computing and real-time communications for manned and unmanned aircraft, spacecraft, robots, and cars [25].

DroneLink is a cross-platform mobile and web application for automating UAV flights, mapping flights, setting waypoints, and creating missions [26].

MAVLink (Micro Air Vehicle Link) is a very lightweight cross-platform messaging protocol for communication with unmanned aerial vehicles and between on-board UAV components [27]. This protocol is one of the most popular and is supported by many unmanned aerial vehicle control platforms, such as Ardupilot and PX4. These two open-source autopilot systems are supported by a global developer community and control all types of vehicles, from UAVs to land vehicles and underwater vehicles, and are also supported by many unmanned aircraft systems [28, 29].

The MAVLink protocol is used for real-time communication, this means that MAVLink messages are quite compact, do not require a communication channel, and can be transmitted by various low-speed communication systems. The use of a digital signature and checksum verification guarantees the integrity of the transmitted information. It should be noted that the MAVLink protocol transmits messages without encryption, and this makes it easy to read data and intercept control. Due to the characteristics described above, the MAVLink protocol has become widely used to provide communication between ground control stations and UAVs [27].

ArduPilot supports the MAVLink protocol for communication with ground control stations, such as Mission Planner [29]. ArduPilot, using the Kalman filter, allows estimating wind parameters at the location of a multi-copter unmanned aircraft system without an additional airspeed sensor [30].

4 Wind Information Retrieval and Exchange

Unmanned aircraft systems including UAVs performing flight missions provide the necessary data. This data allows obtaining objective information about the movement of air masses over the earth's surface. This horizontal air movement is known as wind. The calculated value of wind speed and direction is transmitted using the MAVLink protocol in the WIND message and can be viewed in real-time at the Mission Planner control ground station as can be seen in Fig. 2 [30, 31].

Fig. 2. Calculated wind speed and direction.

The ArduPilot Kalman filter estimates wind parameters. These parameters are recorded by the open-source platform in the VWN and VWE fields in the onboard log file messages. The estimated wind parameters are shown in Fig. 3 [30].

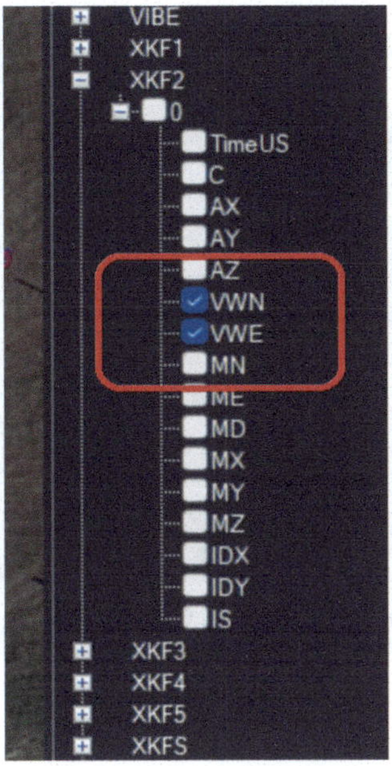

Fig. 3. Estimation of wind parameters using the ArduPilot Kalman filter.

The Kalman filter iterations are divided into two phases: extrapolation and correction [32]. During extrapolation, the filter obtains a priori estimate of the system state $\hat{x}_{k|k-1}$ for the current step based on the final state estimate from the previous step. In the correction phase, the a priori extrapolation is supplemented with the corresponding current measurements to correct the estimate. The corrected estimate is called as a posteriori state estimate, or simply the state estimate $\hat{x}_{k|k}$. Correction of the previously obtained state extrapolation can be represented with formula (updated system state estimate) [32]:

$$\hat{x}_{k|k} = \hat{x}_{k|k-1} + K_k \tilde{y}_k, \tag{1}$$

where $\hat{x}_{k|k-1}$ is the vector of state extrapolation; K_k is the optimal Kalman matrix of amplification coefficients; $\widetilde{y}$ is the deviation of the measurements from expectations.

The WIND MAVLink message contains information about wind direction, and horizontal and vertical wind speed [31].

Data analysis of wind assessment obtained via telemetry channels or recorded in onboard log file messages can be performed using the standard graphical analysis tools of the Mission Planner ground control station. An example of the results of flight data analysis is shown in Fig. 4. This also can be done using other software, for example, MAVExplorer [29, 30].

MAVExplorer is an interactive, community-supported log file data analysis tool for the graphical presentation of information. MAVExplorer supports MAVLink telemetry logs and allows one to obtain both defined graphs and a graph from any data from log files using the command GRAPH in the terminal [33].

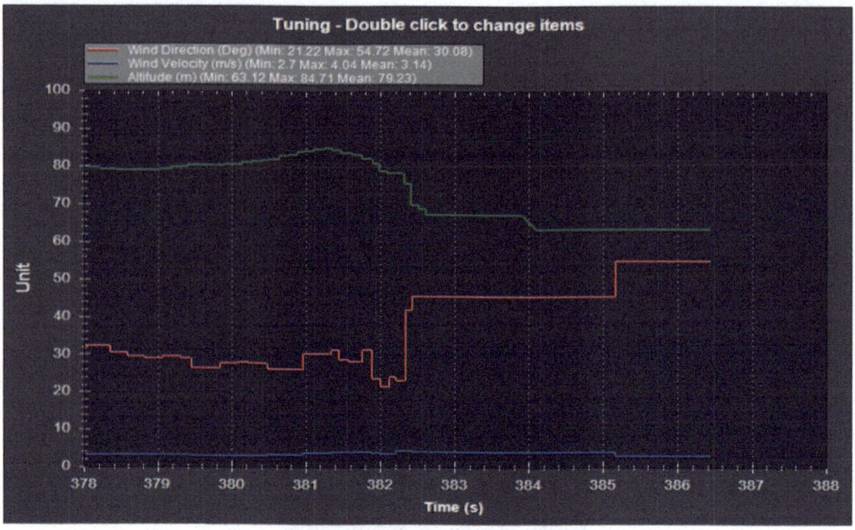

Fig. 4. Graphical analysis tools of the Mission Planner.

For the experiment, telemetry data from real flights of unmanned aircraft systems were used, and then the analysis of the obtained log file data was done. Data on changes in horizontal (red curve) and vertical (green curve) speed obtained using MAVExplorer are shown in Fig. 5.

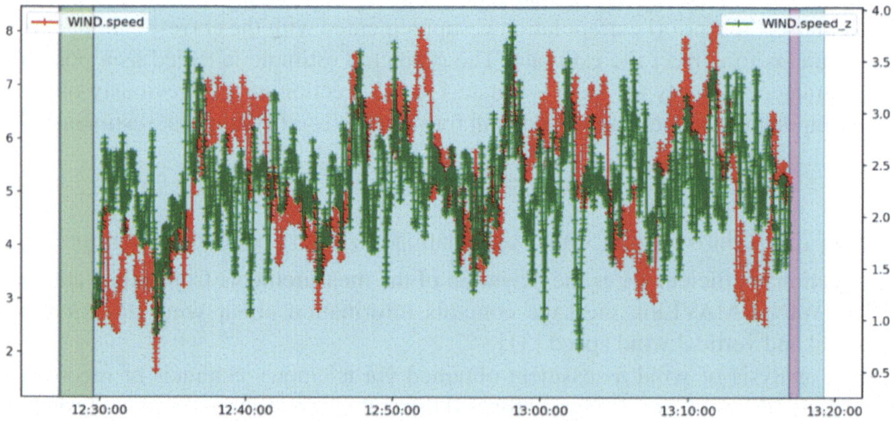

Fig. 5. Changes in horizontal and vertical speed.

The data on changes in vertical speed (green curve) and altitude (red line given in mm) obtained using MAVExplorer are shown in Fig. 6.

The considered instruments allow to extract of information about wind speed and wind direction and match this information with horizontal and vertical coordinates. Thus, the wind fields along the flight trajectory can be obtained.

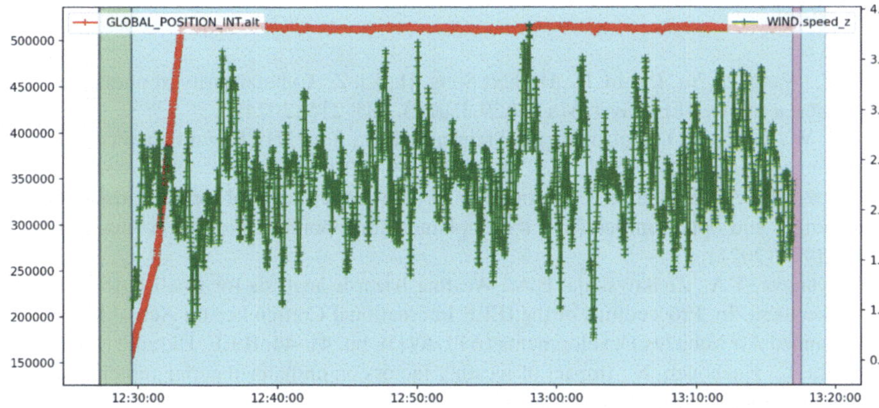

Fig. 6. Vertical speed and altitude changes (altitude in mm).

The obtained operative information can be used by UAV operators when decision-making before and during the flight as well as used by autopilots for in-flight trajectory correction or optimization. The proposed approaches and algorithm of trajectory correction can be found in [34–36].

5 Conclusions

This paper studied modern communication technologies to provide weather-related information. We were focusing on wind measurements and considering the protocols that allow to make measurements without any additional airspeed sensors. We have analyzed the ability of MAVLink protocol and Ardupilot control platforms.

In this paper, we presented the results of wind assessment obtained via telemetry channels or recorded in onboard log file messages using the standard graphical analysis tools of the Mission Planner ground control station.

Also, we demonstrated the analysis of possibilities to extract real-time information about wind using the MAVExplorer.

The analysis of the obtained results shows the possibility of obtaining real-time information about wind velocity along the flight trajectory of UAVs. This information is important as wind and wind-related phenomena are highly dynamic and it can be quite difficult to their characteristics and match with the exact time and place. In turn, the significant variations of wind components can be important information for flight planning or flight trajectory correction.

Using the considered data links and protocols, the information about the wind field can be shared between the interested UAV operators used in autopilots of UAVs for flight trajectory correction. Also, the considered approach of obtaining and sharing operative weather information can be used as a component of global weather observation and study. It is also can be used for local atmospheric parameters and microscale phenomena for different branches of people's activity including aviation, UAV-based missions, and urban flying, the internet of things, smart cities, and agriculture.

References

1. Yu, Z., Wang, Z., Yu, J., Liu, D., Herbert Song, H., Li, Z.: Cybersecurity of unmanned aerial vehicles: a survey. IEEE Syst. Mag. **9**(39, Part 2), 185–215 (2024)
2. Raza, W., Renkoff, O., Ogirimah, O., Stansbury, R.S., Song, H.: Flight data recorders: past, present, and future. IEEE Syst. Mag. **9**(39, Part 2), 120–145 (2024)
3. Mourtzis, D., Angelopoulos, J., Panopoulos, N.: UAVs for industrial applications: identifying challenges and opportunities from the implementation point of view. Procedia Manuf. **55**, 183–1905 (2021)
4. Averyanova, Y.A., Znakovskaja, E.A.: Weather hazards analysis for small UASs durability enhancement. In: Proceedings of the IEEE International Conference on Actual Problems of Unmanned Air Vehicles Developments (APUAVD), pp. 41–44. IEEE, Ukraine (2021)
5. Mishra, L., Kaabouch, N.: Impact of weather factors on unmanned aerial vehicles' wireless communications. Future Internet. **17**(1), 27 (2025)
6. Merzlikin, A., Pavlikov, V., Ruzhentsev, N., Sobkolov, A., Tsopa, O., Zhyla, S.: Peculiarities of microwaves attenuation statistics for Ukrainian region. In: 2020 IEEE Ukrainian Microwave Week UkrMW Proceedings, pp. 761–764, Kharkiv, Ukraine (2020)
7. Kutsenko, O., Ilnytska, S., Konin, V.: Investigation of the residual tropospheric error influence on the coordinate determination accuracy in a satellite landing system. Aviation. **22**(4), 156–165 (2018)
8. Solomentsev, O., et al.: Efficiency analysis of current repair procedures for aviation radio equipment. In: Ostroumov, I., Zaliskyi, M. (eds.) Proceedings of the 2nd International Workshop on Advances in Civil Aviation Systems Development. ACASD 2024 Lecture Notes in Networks and Systems, vol. 992, pp. 281–295. Springer, Cham (2024). https://doi.org/10.1007/978-3-031-60196-5_21
9. Pishko, Y.: The effect of absolute humidity on GPS-positioning accuracy. Geod. Cartogr. Aer. Photogr. **87**(87), 30–35 (2018)
10. Mourgelas, C., Micha, E., Chatzistavrakis, E., Voyiatzis, I.: Classification of unmanned aerial vehicles in meteorology: a survey. Environ. Sci. Proc. **26**(1), 135 (2023)
11. Advanced Weather Drones. https://www.meteomatics.com/en/meteodrones-weather-drones. Last accessed 19 Jan 2025
12. Hervo, M., Gonzague, R., Giovanni, M., Weusthoff, T., Haefele, A.: Evaluation of an automatic meteorological drone based on a 6-months measurement campaign. (2023). https://doi.org/10.20944/preprints202307.2043.v1
13. Meier, K., Hann, R., Skaloud, J., Garreau, A.: Wind estimation with multirotor UAVs. Atmosphere. **13**(4), 551 (2022)
14. Compere, M.D., Adkins, K.A., Muthu Krishnan, A.: Go with the flow: estimating wind using uncrewed aircraft. Drones. **7**(9), 564 (2023)
15. Sziroczak, D., Rohacs, D., Rohacs, J.: Review of using small UAV-based meteorological measurements for road weather management. Prog. Aerosp. Sci. **134**, 105 (2022)
16. Shershen, B., Ivanytskyi, M., Averyanova, Y., Znakovska, Y.: Study of ZigBee-based communication to share operative weather information between UASs in the frame of smart cities concept. In: Ortiz-Rodriguez, F., et al. (eds.) Electronic Governance with Emerging Technologies. EGETC 2024 Communications in Computer and Information Science, vol. 2245, pp. 1–15. Springer, Cham (2025)
17. Holubnychyi, O., et al.: Self-organization technique with a norm transformation based filtering for sustainable infocommunications within CNS/ATM systems. In: Ostroumov, I., Zaliskyi, M. (eds.) Proceedings of the 2nd International Workshop on Advances in Civil Aviation Systems Development. ACASD 2024 Lecture Notes in Networks and Systems, vol. 992, pp. 262–278. Springer, Cham (2024). https://doi.org/10.1007/978-3-031-60196-5_20

18. Ko, Y., Kim, J., Duguma, D.G., Astillo, P.V., You, I., Pau, G.: Drone secure communication protocol for future sensitive applications in military zone. Sensors. **21**(6), 2057 (2021). https://doi.org/10.3390/s21062057
19. Averyanova, Y., Larin, V., Kuzmenko, N., Ostroumov, I., Zaliskyi, M., Solomentsev, O., et al.: Turbulence detection and classification algorithm using data from AWR. In: IEEE 2nd Ukrainian Microwave Week (UkrMW), pp. 518–522. IEEE, Ukraine (2022). https://doi.org/10.1109/UkrMW58013.2022.10037172
20. Dergachov, K., et al.: GPS usage analysis for angular orientation practical tasks solving. In: 2022 IEEE International Conference on Problems of Infocommunications. Science and Technology, pp. 187–192. IEEE, Kharkiv (2022). https://doi.org/10.1109/PICST57299.2022.10238629
21. Aeronautical Telecommunications. Communication Systems and Procedures Relating to Remotely Piloted Aircraft Systems C2 Link. Annex 10, Volume VI, ICAO
22. Radio Regulations. https://www.itu.int/pub/R-REG-RR-2024. Last accessed 19 Jan 2025
23. Ivashchuk, O., et al.: A configuration analysis of Ukrainian flight routes network. In: 16th International Conference on the Experience of Designing and Application of CAD Systems, pp. 6–10. IEEE, Lviv (2021). https://doi.org/10.1109/CADSM52681.2021.9385263
24. Manual on the Aeronautical Mobile Satellite (Route) Service (Doc 9925). https://store.icao.int/en/manual-on-the-aeronautical-mobile-satellite-route-service-doc-9925
25. Cyphal. https://opencyphal.org. Last accessed 19 Jan 2025
26. Drone Flight Control. https://www.dronelink.com. Last accessed 19 Jan 2025
27. MAVLink Developer Guide. https://mavlink.io/en. Last accessed 19 Jan 2025
28. Ardupilot. https://ardupilot.org. Last accessed 19 Jan 2025
29. Mission Planner. https://ardupilot.org/planner. Last accessed 19 Jan 2025
30. Windspeed Estimation and Baro Compensation. https://ardupilot.org/copter/docs/airspeed-estimation.html. Last accessed 19 Jan 2025
31. Dialect: ArduPilotMega. https://mavlink.io/en/messages/ardupilotmega.html#WIND. Last accessed 19 Jan 2025
32. Hamilton, J.: The Kalman filter. In: Time Series Analysis, pp. 372–408. Princeton University Press (1994)
33. Using MAVExplorer for Log Analysis. https://ardupilot.org/dev/docs/using-mavexplorer-for-log-analysis.html. Last accessed 19 Jan 2025
34. Averyanova, Y.A., Rudiakova, A.N., Yanovsky, F.J.: Aircraft trajectories correction using operative meteorological radar information. In: Proceedings of the International Radar Symposium (IRS), pp. 256–259. IEEE, Poland (2020)
35. Ivanytskyi, M., Znakovska, Y., Averyanova, Y.: UAS flight trajectory optimization algorithm based on operative meteorological information. CEUR Workshop Proc. **3426**, 287–297 (2023)
36. Ivanytskyi, M., Averyanova, Y., Znakovska, Y., Shershen, B.: Aircraft flight path correction during the flight execution phase using the operative information on weather hazards. In: Proceedings of the 2nd International Workshop on Advances in Civil Aviation Systems Development. ACASD 2024 Lecture Notes in Networks and Systems 992 LNNS, pp. 31–41 (2024)

Assessment of the Impact of Variable Mass of an Unmanned Aerial Vehicle on Flight Range and Endurance

Andreii Hnashuk[1], Valentina Konovaliuk[2], Gennadiy Yun[1], and Kristina Marintseva[3](✉)

[1] Private Enterprise "Secret-Service", Baggovutivska Str., 4(10), Kyiv 04107, Ukraine
[2] National Aviation University, Liubomyra Huzara Ave., 1, Kyiv 03058, Ukraine
[3] Kingston University London, Roehampton Vale Campus, Friars Ave., London SW15 3DW, UK
k.marintseva@kingston.ac.uk

Abstract. This study examines the impact of variable mass on the flight range and endurance of unmanned aerial vehicles (UAVs), focusing on their potential for delivering specialised cargo. Key applications include medical supply transport, agricultural logistics, and humanitarian missions, highlighting the role of UAVs like the Agrobee in overcoming geographical and logistical challenges. By integrating theoretical principles, aerodynamic analysis, and practical design considerations, the paper explores how variable mass dynamics – arising from fuel consumption and mid-flight cargo drops – affect UAV performance. The research provides analytical models and operational insights for optimising UAV efficiency, bridging the gap between theoretical ideals and real-world constraints. Additionally, the study evaluates the practical implications of variable mass on UAV stability, flight endurance, and energy consumption. It examines payload management strategies, and mission planning considerations, essential for ensuring operational safety and efficiency. UAVs' capability to dynamically adjust their weight mid-flight through controlled cargo drops presents opportunities for enhanced mission flexibility and cost-effective logistics. However, challenges such as aerodynamic stability, cargo release mechanisms, and weather dependencies must be carefully addressed. This research contributes to the development of improved UAV operational frameworks, promoting more sustainable and efficient logistics solutions in various industries.

Keywords: Variable Mass · Range · Endurance · Unmanned Aerial Vehicle (UAV) · Special Cargo · Range · Flight Time · Agrobee · Logistics · Variable Mass · Transport Operation

1 Introduction

Specialised cargo encompasses a wide range of goods that require specific handling, urgent delivery, or adherence to unique environmental requirements. This broad category encompasses items such as medical equipment, perishable goods, high-precision devices,

I. Ostroumov et al. (Eds.): ACASD 2025, LNNS 1418, pp. 486–497, 2025.
https://doi.org/10.1007/978-3-031-91992-3_32

and other materials, each requiring tailored approaches to ensure their effective, safe and secure transportation.

The transportation of specialised cargo poses numerous challenges, particularly for goods with strict storage conditions or those destined for remote or inaccessible locations. Time sensitivity, compliance with regulatory requirements, and the need for precise delivery add complexity to an already demanding process. Operational challenges include managing unpredictable weather patterns, overcoming compliance obstacles, and maintaining the condition of sensitive cargo en route.

The advent of unmanned aerial vehicles (UAVs), commonly known as drones, has altered conventional freight transport and become a pivotal innovation in the delivery of specialised goods. These items encompass essential medical supplies and devices used for environmental assessments. UAVs have emerged as a breakthrough technology in specialised cargo logistics, offering solutions to key transportation challenges. Their speed and precision allow for the rapid delivery of goods to hard-to-reach locations, while their manoeuvrability makes them ideal for handling delicate and high-priority shipments [1–3].

The healthcare sector has witnessed significant advancements in the use of UAVs for the transportation of critically important medical supplies [4, 5]. Zipline's use of UAVs in Rwanda and Tanzania has exemplified how drones can revolutionize healthcare logistics. Their ability to deliver critical medical supplies swiftly to remote areas has proven essential in life-saving interventions and has established UAVs as a viable solution for medical supply chains [6, 7].

Equipped with refrigeration technology, specialized UAVs facilitate the rapid and controlled transport of perishable items. This is particularly beneficial in agriculture, where maintaining produce freshness during transit is crucial for supply chain efficiency. At the same time, postal services in various countries, including Australia, the UK, Switzerland, and Ukraine are experimenting with UAVs to enhance delivery operations, assessing their feasibility and economic impact [8].

The logistics industry has been revolutionised by UAVs, which now play a vital role in the efficient transport of specialised cargo. Their speed, precision, and adaptability allow them to tackle complex logistical challenges effectively. As UAV technology advances, its applications in cargo transportation will continue to expand, leading to more sustainable and optimised delivery solutions across various sectors.

2 Cargo Drop Methods

In UAV transport operations, precise cargo release is vital, especially when serving remote locations or navigating difficult landscape [9, 10]. Selecting an airdrop method – whether using fixed-wing or rotorcraft UAVs – relies on considerations like runway access and the need for parachute systems. Advanced technologies like the Joint Precision Airdrop System (JPADS) [11] play a vital role in ensuring cargo safety and drop accuracy. In this article, we consider fixed-wing UAV airdrop as the key method for subsequent operations, assuming the availability of a network of airfields with suitable runway quality in the region under consideration. This method involves transporting cargo in containers secured within the UAV's fuselage. The decision between airfield-based and parachute-based airdrop methods, including advanced systems like JPADS, is crucial for maintaining the safety and accuracy of cargo delivery.

This section discusses two main methods: airfield-based and parachute-based airdrops.

Parachute airdrop [10, 11] is a manoeuvre that accounts for wind conditions, cargo weight, and the altitude at which the parachute canopy is deployed. The parachute provides a stable and controlled descent trajectory, making it suitable for various scenarios, from humanitarian transport operations to commercial cargo delivery. Parachute airdrop is especially effective for transporting medical supplies or equipment to specific locations. In regions with complex landscapes or limited accessibility, parachute drops enable cargo to reach destinations that would otherwise be challenging to access. The JPADS system enhances parachute-based airdrops by integrating GPS guidance technology, offering improved accuracy, real-time tracking, and navigation, which allows for meticulous planning and execution of airdrop operations.

Airfield-based airdrop involves transporting cargo in containers secured within the UAV's fuselage. Upon arrival at the destination airfield, the parcels are unloaded through a hatch designed into the UAV's structure. Securely fastening cargo within the container is essential to prevent displacement during flight [12].

The choice between airfield-based and parachute-based airdrops depends on the specific requirements of the transport operation. The most effective approach is dictated by considerations like cargo specifications, environmental factors, and delivery accuracy needs. In some cases, combining both methods can help overcome various operational challenges.

The characteristics and applications of each method enable informed decision-making based on the unique requirements of cargo transportation. As UAV technologies continue to evolve, these methods, including innovations like JPADS, play a pivotal role in expanding the scope and efficiency of airborne cargo delivery.

3 The Impact of Variable Mass on UAV Flight

The fundamental principles governing UAV flight dynamics, particularly for fully autonomous vehicles (FAV-UAV) designed for urgent and specialised cargo delivery, are well established in aerodynamics. Lift, drag, thrust, and weight are key forces that determine the behaviour of UAVs in the air [13, 14]. The lift force Y is mathematically expressed by the lift equation:

$$Y = \frac{1}{2} C_Y \rho V^2 S,$$

where C_Y is the lift coefficient, ρ is the air density, V is the velocity, and S is the wing area.

Simultaneously, aerodynamic drag, which resists the UAV's forward motion, is given by the drag equation:

$$D = \frac{1}{2} C_D \rho V^2 S,$$

where C_D is the drag coefficient. Thrust, the force propelling the UAV forward, counteracts drag. When studying the impact of variable mass, the weight of the UAV becomes a dynamic factor. The balance between these forces, combined with weight (the gravitational force acting on the UAV, W), dictates UAV motion. As fuel (or fuel and chemicals for agricultural UAVs) is consumed during flight, the UAV's mass decreases, affecting its centre of gravity and aerodynamic characteristics. Understanding the impact of variable mass is crucial, as it introduces dynamism into the system [15].

As fuel is expended during flight, its effect on the UAV's centre of gravity and aerodynamics becomes more pronounced. A unique feature of this principle is the ability to drop cargo mid-flight, adding another layer of dynamism. Suppose the initial mass of the UAV (m) is 650 kg (350 kg for the UAV and 300 kg for the cargo), with the gravitational acceleration $g \approx 9.8\ m/s^2$. If 300 kg of cargo is dropped mid-operation (m_c), the weight calculations are as follows:

$$W = mg = 650 \times 9.8 \approx 6370N,$$

$$\Delta W = (m - m_c)g = (650 - 300) \times 9.8 \approx 3430N.$$

The efficiency coefficient E can be calculated as:

$$E = \frac{W}{\Delta W} = \frac{6370}{3430} \approx 1.86.$$

Thus, dropping 300 kg of cargo increases the UAV's efficiency by approximately 86%. This demonstrates how the efficiency coefficient increases with the release of substantial cargo, emphasising the strategic use of variable mass dynamics to optimise UAV performance in scenarios requiring urgent cargo delivery.

Reducing the weight of a UAV contributes to fuel efficiency and performance optimisation during transport operations. Understanding flight principles is crucial for optimising UAV performance, particularly considering the impact of variable mass. According to Greenwood [16], UAVs in practice can carry a payload greater than the manufacturer's stated capacity. However, the maximum payload must be calculated with respect to the engines' maximum thrust. Flight characteristics, including flight time and stability, deteriorate significantly when the maximum payload capacity is reached.

Flight time depends on various factors, including payload, fuel quantity, atmospheric temperature, and wind conditions. Prolonged flight time enhances overall operational endurance, a critical factor in scenarios where extended airborne presence is mandatory.

In summary, understanding flight principles and accounting for the impact of variable mass are fundamental to optimising UAV performance. Predictive models derived from aerodynamic data enhance operational accuracy in the design process of UAV aircraft configurations.

4 Range and Endurance of Flight

The flight range (L) is a critical parameter (Fig. 1) influenced by various factors:

$$L = L_G + L_{CR} + L_L, \tag{1}$$

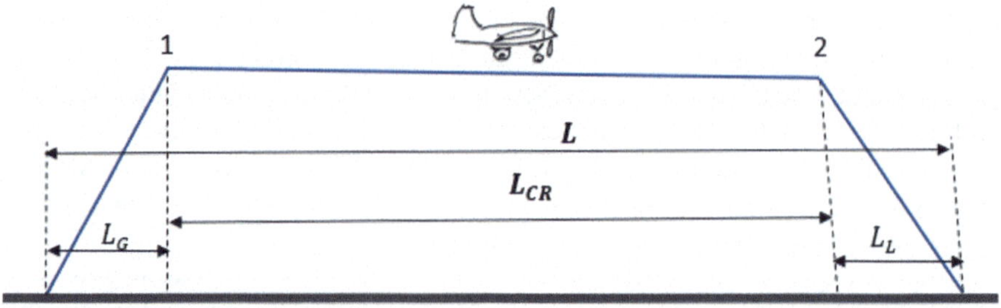

Fig. 1. The flight profile of a UAV in an aircraft configuration.

where L_G represents the distance during climb, L_{CR} is the cruising distance, and L_L denotes the landing distance.

The cruise phase is characterised by constant speed and altitude, during which fuel is consumed. However, not all onboard fuel is used during cruising. Fuel is allocated for operations such as engine start, taxiing, take-off, climb, descent, and landing. A portion is reserved for contingencies, forming the available fuel reserve $m_{f.\ res.}$. Two fundamental concepts underpin flight range estimation: technical range and practical range.

The technical range denotes the distance covered under ideal conditions, such as a calm atmosphere, with the complete consumption of all available fuel. In contrast, the practical range refers to the horizontal distance travelled using the operationally available fuel, excluding the aviation fuel reserve (AFR). According to airworthiness standards, $m_{f.AFR}$ must ensure the aircraft can sustain maximum climb speed $V_{gh}(k = k_{max})$ for 30–45 min. The typical AFR standard is approximately 5–10% of the total fuel.

The hourly fuel consumption is calculated as:

$$q_h = C_{PR}\frac{mg}{k}, \tag{2}$$

where C_{PR} is the specific fuel consumption coefficient, m is the total mass of the UAV, g is the acceleration due to gravity ($\approx$9.8 m/ s^2), and k is the rate of climb. Fuel consumption per kilometer is given by:

$$q_{km} = C_{PR}\frac{mg}{3.6V \cdot k}, \tag{3}$$

where V is the cruising speed of the UAV.

The differential equations for mass flow rate are [13]:

$$\frac{dm}{dL} = -q_{km}, \tag{4}$$

$$\frac{dm}{dt} = -q_h. \tag{5}$$

Integrating Eqs. (4) and (5) yields:

$$L_{CR} = -\int_{m_1}^{m_2}\frac{dm}{q_{km}} = \frac{3.6}{g}\int_{m_2}^{m_1}\left(\frac{V \cdot k}{C_{PR}}\right)\frac{dm}{m}, \ \text{[km]} \tag{6}$$

$$T_{CR} = -\int_{m_1}^{m_2} \frac{dm}{q_h} = \frac{1}{g} \int_{m_2}^{m_1} \left(\frac{k}{C_{PR}} \right) \frac{dm}{m}, \text{ [hours]}. \tag{7}$$

The integrand in Eq. (6) $\varphi = \left(\frac{V \cdot k}{C_{PR}} \right)$ is referred to as the range parameter:

- m_0: take-off mass of the UAV;
- $m_{t.\,g.}$: fuel mass required for engine start, taxiing, and take-off;
- m_{cru}: fuel mass consumed during cruising flight;
- $m_{f.\,AFR}$: fuel mass for altitude gain at maximum speed V_{gh};
- $m_1 = m_0 - (m_{t.\,g.} + m_{f.\,AFR})$: the fuel mass at the beginning of the cruise flight phase (Point 1 on Fig. 2) for the technical flight range;
- $m_2 = (m_{t.\,g.} + m_{f.\,AFR} + m_1) - m_{f.\,res.}$: the fuel mass at the end of the cruise flight phase for the practical flight range.

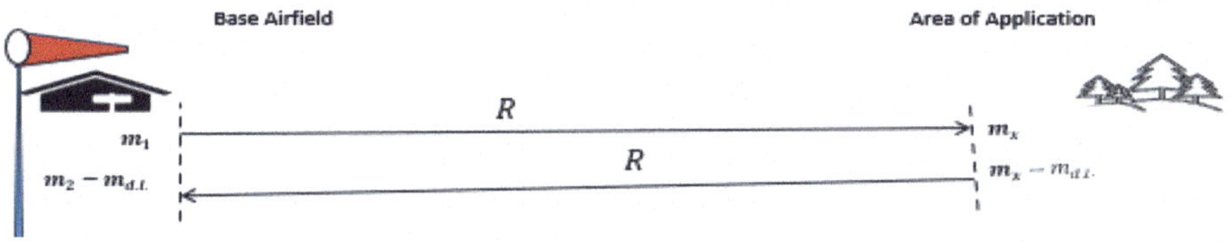

Fig. 2. UAV flight range.

A detailed examination of flight range factors is facilitated by these equations and parameters, which account for fuel consumption, reserves, and operational phases. The range parameter integrates key variables such as flight speed, climb rate, and specific fuel consumption, offering a robust framework for evaluating range performance.

Estimating the maximum range and endurance of a UAV involves complex calculations, each contributing to a comprehensive understanding of the UAV's capabilities. The primary equations for calculating the maximum cruise range ($L_{cr.\,max}$) and maximum flight endurance ($T_{cr.\,max}$) are as follows:

$$L_{cr.\max} = -\frac{3.6}{g} \int_{m_2}^{m_1} \left(\frac{V \cdot k}{C_{PR}} \right)_{max} \frac{dm}{m},$$

$$T_{ct.\max} = -\frac{1}{g} \int_{m_2}^{m_1} \left(\frac{k}{C_{PR}} \right)_{max} \frac{dm}{m}.$$

In a simplified scenario, assuming that the integrand functions are independent of the flight mass m, we obtain [15]:

$$L_{cr.\max} = \frac{3.6}{g} \left(\frac{V \cdot k}{C_{PR}} \right)_{max} \ln\left(\frac{m_1}{m_2} \right), \tag{8}$$

$$T_{ct.\max} = \frac{1}{g} \left(\frac{k}{C_{PR}} \right)_{max} \ln\left(\frac{m_1}{m_2} \right). \tag{9}$$

Considering the influence of wind on flight range, with a tailwind or headwind component W, the fuel consumption per kilometer is:

$$q_{km} = \frac{q_h}{3.6 \cdot V} = C_{PR} \frac{P_n}{3.6(V \pm W)} = q_{km(W=0)} \frac{1}{1 \pm \frac{W}{V}}, \tag{10}$$

where P_n is the required thrust for steady horizontal flight, selected by the UAV operator, equal to the aerodynamic drag force of UAV.

Thus, the UAV range L_{cr} is:

$$L_{cr} = L_{cr.(W=0)} \frac{1}{1 \pm \frac{W}{V}}. \tag{11}$$

Interpretation of optimal flight conditions:

- To maximize flight range: conduct cruise flight at altitudes 2–3 km below the theoretical ceiling with a speed $V = V_{cr}$.
- To maximize endurance: operate at altitudes 1–2 km below the theoretical ceiling with a speed $V = V_{gh}$ corresponding to the maximum climb speed.

Equation (11) highlights how identifying optimal flight conditions integrates theoretical analysis with practical application. The strategic selection of altitude and speed plays a key role in maximising range, while maintaining these parameters within the constraints of maximum climb speed is essential for endurance. These considerations bridge the divide between idealised flight models and real-world operational demands. Ultimately, the equations capture the UAV's performance range, balancing theoretical potential with real-world complexities like variable winds.

4.1 UAV Maximum Range

In UAV operations that require returning to the base airfield after completing a mission in the designated area (Fig. 3), the flight range parameter R becomes a crucial factor. Like range, the operational radius has both technical and practical aspects.

The formula for maximum cruise range is expressed as:

$$L_{cr.\max} = \frac{3.6}{g} \left(\frac{V \cdot k}{C_{PR}} \right)_{max} \ln\left(\frac{m_1}{m_2} \right) = \frac{3.6}{g} \left(\frac{V \cdot k}{C_{PR}} \right)_{max} \ln\left(\frac{m_x - m_{d.l.}}{m_2 - m_{d.l.}} \right), \tag{12}$$

where: m_x is the UAV mass at the point of application, $m_{d.\,l.}$ is the mass of the payload dropped at the point of application.

Equivalently, in terms of payload mass, Eq. (12) becomes:

$$\ln\left(\frac{m_1}{m_x} \right) = \ln\left(\frac{m_x - m_{d.l.}}{m_2 - m_{d.l.}} \right). \tag{13}$$

Solving for m_x [15]:

$$m_x = \frac{m_{d.l.}}{2} + \sqrt{\left(\frac{m_{d.l.}}{2} \right)^2 + m_1(m_2 - m_{d.l.})}. \tag{14}$$

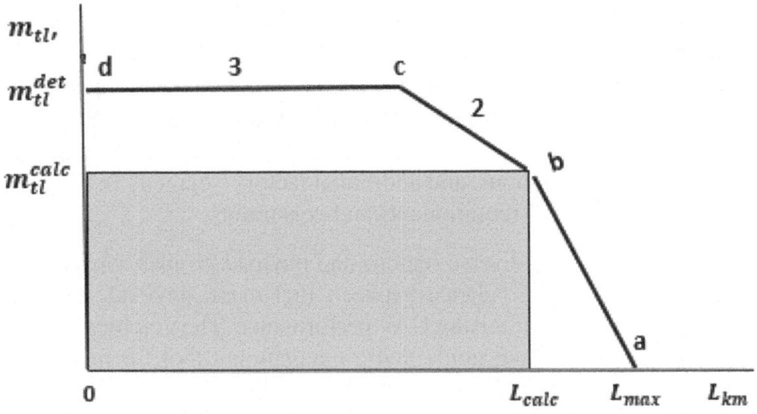

Fig. 3. Impact of aircraft payload on flight range.

Substituting m_x into Eq. (12), the range R is determined as:

$$R = \frac{3.6}{g}\left(\frac{V \cdot k}{C_{PR}}\right)_{max} \ln\left(\frac{1}{\frac{m_{d.l.}}{2}} + \left(\frac{m_{d.l.}}{2}\right)^2 + m_1(m_2 - m_{d.l.})\right). \tag{15}$$

In the presence of constant wind with a component W along the flight path, the range decreases regardless of wind direction. Practically, this means that the effective range of a UAV must carefully account for wind conditions:

$$R = R_{w=0}\left(1 - \left(\frac{W}{V}\right)^2\right). \tag{16}$$

The wind's influence, represented in Eq. (14), adds a pragmatic aspect to the concept of effective range. Accounting for wind becomes crucial in operational planning, as it directly impacts the UAV's performance envelope.

In real-world applications where UAV operations necessitate a return to the base airfield after completing designated tasks, the derived formulas offer a thorough analysis of the key variables and parameters. These factors play a crucial role in assessing mission feasibility and formulating strategies to enhance UAV performance and efficiency. The performance of UAVs is intricately tied to the balance between their technical and practical flight range – two key metrics underlying operational planning.

These considerations are crucial when determining the feasibility of TO and explores the interdependence of these metrics and their dynamic relationship with the target payload mass $m_{t.l.}$. Figure 3 illustrates the impact of payload on aircraft flight range, offering mission planners a deeper understanding of key considerations. These factors are essential for assessing operational feasibility and crafting detailed strategies to optimise UAV performance.

1. *Zero Payload:* When the target payload mass $m_{tl} = 0$, the flight range is at its maximum ($L = L_{max}$). However, introducing a non-zero payload while maintaining constant fuel mass ($m_f = const$) leads to a monotonic decrease in flight range (segment 1 in

Fig. 3). This scenario corresponds to point "b" on the diagram, which determines the calculated flight range (L_{calc}).
2. *Payload Increase*: Further increases in target payload (m_{tl}) can only be achieved by reducing fuel mass (m_f), resulting in a steeper decline in range (segment 2). A critical payload threshold (m_{tl}^{det}) emerges, determined by variables including compartment capacity, operational limitations, and additional factors (segment 3). The shaded area reflects specific operational requirements and constraints.

UAV missions typically cover diverse regions and payload profiles, each affecting load conditions. Understanding the balance between fuel mass, payload, and operational constraints is paramount for optimising UAV performance. The graphical representation in Fig. 3 serves as both a visual guide and a navigational tool for mission planners, bridging theoretical concepts and practical UAV operations.

These nuances underscore the dynamic nature of UAV payload management and its direct influence on operational range.

5 UAV Agrobee Example

In practice, there are two main methods for developing a new UAV. The first is the classical approach: preliminary design → final design → prototype production → ground and flight testing → mass production → operation. The second method involves converting an existing manned aircraft into an unmanned one. A comparison of these two methods in terms of cost-efficiency, weight efficiency, and implementation time has shown that conversion is effective when there is a large fleet of infrequently used aircraft or when their operation involves significant risks for pilots, commercial payloads, and the aircraft itself in the event of an accident. For the design of the UAV Agrobee, the developers opted for the first method [17].

The UAV Agrobee, designed and manufactured in Ukraine, is a multipurpose UAV platform intended for various civilian applications, including agriculture and the transportation of specialised cargo [18, 19]. Featuring a robust construction and advanced avionics, this UAV surpasses conventional boundaries, delivering exceptional performance. The Agrobee can be adapted for a wide range of civilian applications, including but not limited to:

- in precision agriculture, the Agrobee can be adapted for crop monitoring, plant protection, and the efficient distribution of agricultural resources, promoting sustainable and optimised farming practice [17, 18];
- it is capable of delivering 24 standard cargo containers (322 × 314 ×308 mm) using either parachute drops or landing, even on unpaved strips as short as 400 m. For safety, the containers are secured with cargo nets;
- the Agrobee can be easily modified to undertake a wide range of tasks, including geological surveys, border patrols, radio monitoring, reconnaissance missions, rabies control, and more.

Technical specifications of Agrobee are presented in Table 1.

Precise and flexible "door-to-door" cargo delivery mechanisms enable the Agrobee to tackle unique challenges in transporting specialised cargo to remote and hard-to-reach

Table 1. Technical specifications of Agrobee.

Specification	Value	Notes
Take-off weight	650 kg	
Empty weight	135 kg	
Fuel tank capacity	35 L (fuselage), 40–70 L (wing)	
Wingspan	8.6 m	
Length	5.2 m	
Height (on the ground)	2.3 m	
Wing area	10.4 sq. m	
Wing chord	1.3 m	
Fuselage	Welded thin-walled stainless steel tubes	Clad with sheet duralumin
Engine (Rotax 912ULS/S)	Power: 100 hp	Serial UL: 260 hp
Fuel consumption	0.3 L/km	
Propeller	Ø1.8 m	Carbon composite
Maximum payload	300 kg	
Maximum flight altitude	4500 m	
Cruise speed	For agricultural use: 100–140 km/h	For transport operations: with 912ULS/S: 150–170 km/h; with UL260: 200–240 km/h
Maximum speed	210 km/h	
Stall speed	45 km/h	
Rate of climb	7.5 m/s	
Take-off speed	70 km/h	
Landing speed	80 km/h	
Overload	+4/−2	Calculated
Take-off distance	320 m	
Runway length required	350–400 m	
Autopilot	Pixhawk-4 or Micro-Pilot	1 unit
Manual control system	16-channel Taranis X9D	1 unit
Communication channel range	7 W	
Forward-facing cameras	2 units	
Telemetry channel (RFD 900X)	1 unit	

regions. The Agrobee can land (or water land when equipped with floats) on prepared sites, with cargo securely stored in the fuselage compartment.

In addition to standard landing methods, it offers a parachute drop alternative, ensuring efficient cargo deployment in both urbanised and isolated areas. The UAV's controlled descent mechanism ensures secure delivery of goods, even in complex environments. This makes the Agrobee a crucial asset in humanitarian logistics, allowing it to transport emergency supplies, including food, water, and medical assistance, to disaster-stricken or war-torn areas. Additionally, in remote locations with insufficient healthcare facilities, it facilitates the transportation of specialised medical equipment and supports telemedicine initiatives to improve healthcare outreach.

6 Conclusions

The study of flight aerodynamics and the impact of variable mass can bridge theoretical principles with possible applications in the future flight test of Agrobee – a specialised UAV developed for transporting special cargo. The analysis of UAV payload impact on range has revealed the complex interplay between payload capacity and operational performance.

Agrobeees robust technical capabilities and adaptability make it an ideal choice for transporting specialised cargo. Its ability to operate under varied conditions and handle different cargo profiles showcases its reliability in overcoming logistical challenges.

The presented graphical analyses provide visual guidance for operational planners. Optimal flight conditions, accounting for factors such as altitude and speed, establish a foundation for maximising both range and endurance. These considerations are critical for achieving efficient operational objectives. In real-world scenarios, unpredictable weather can affect UAV performance, necessitating additional considerations during operational planning.

The operational range and adaptability of UAVs are also influenced by regulatory frameworks, which may evolve over time. Operational planners must stay informed of regulatory changes to ensure compliance and adjust their strategies accordingly. Based on defined theoretical frameworks and assumptions, the models presented in this study may not fully account for variations observed in real-world operations.

References

1. Shikhara, H., Sawalmeh, A.H., Al-Fuqaha, A., Dou, Z., Almaita, E., Khalil, I., Guizani, M.: Unmanned Aerial Vehicles (UAVs): a survey on civil applications and fundamental research challenges. IEEE Access. **7**, 48572–48634 (2019)
2. Yun, G., Pederiy, Y., Gnashuk, A.: The advantages of plant protection using UAV of aircraft configuration and UAV technology. AEROjournal. **14**(2), 13–18 (2019)
3. PwC.: Global Market for Commercial Applications of Drone Technology Valued at Over 127bn. https://press.pwc.com/. Last accessed 10 Jan 2025
4. Marintseva, K., Yun, G., Vasilenko, I.: Delivery of special cargoes using unmanned aerial vehicles. In: Unmanned Aerial Vehicles in Civilian Logistics and Supply Chain Management, pp. 33–63. IGI Global, Hershey (2019)

5. Zipline's drones are delivering medical supplies and PPE in North Carolina. https://www.theverge.com/2020/5/27/21270351/zipline-drones-novant-health-medical-center-hospital-supplies-ppe. Last accessed 10 Jan
6. Kremer, P., et al.: An impact assessment of the use of aerial logistics to improve access to vaccines in the Western-North Region of Ghana. Vaccine. **41**(36), 5245–5252 (2023)
7. Moneygall, Ireland.: Drone delivery startup Manna adapted to the coronavirus pandemic. https://www.businessinsider.com/drone-delivery-startup-manna-adapted-to-coronavirus-pandemic-2020-5?r=US&IR=T. Last accessed 10 Jan 2025
8. BBC: For the first time in Ukraine, mail was delivered by UAV. https://www.bbc.com/ukrainian/news-58015578. Last accessed 10 Jan 2025
9. Yakimenko, O.: Autonomous parachute-based precision delivery systems. In: Unmanned Aircraft Systems, p. 29. NAU, Ukraine (2016)
10. Benney, R., et al.: The joint precision airdrop system advanced concept technology demonstration. In: 18th AIAA Aerodynamic Decelerator Systems Technology Conference and Seminar, pp. 1–13. American Institute of Aeronautics and Astronautics, Reston (2005)
11. Aviation Trail.: The Joint Precision Air Drop System (JPADS). https://www.aviationtrailinc.org/post/new-exhibit-opens-at-parachute-museum. Last accessed 10 Jan 2025
12. Bimicon GmbH.: Development of 3D Models of Cargo Containers. https://dsgners.ru/graphic-design/5959-razrabotka-3d-modeley-gruzovyih-konteynerov-i-cgi-rolikov. Last accessed 10 Jan 2025
13. Anderson, J.D., Bowden, M.L.: Introduction to Flight. McGraw-Hill Higher Education, New York (2005)
14. Toren Beek, E., Wittenberg, H.: Flight physics: essentials of aeronautical disciplines and technology, with historical notes. Springer Science & Business Media (2009)
15. Balabanov, O.V., Pashintsev, V.T.: Investigation of approximate optimal cruising flight modes of an airplane of variable mass. Sci. Notes TSSA. **20**(3), 52–60 (1989)
16. Greenwood, W.W., Lynch, J.P., Zekkos, D.: Applications of UAVs in civil infrastructure. J. Infrastruct. Syst. **25**(2), 04019002 (2019)
17. Belan, V., Gnashuk, A., Yun, G.: The problem of manned aircraft conversion into unmanned aerial vehicles. Proc. Natl. Avia. Univ. **2**, 67–74 (2018)
18. Gnashuk, A.: Agricultural unmanned aircraft with a high payload. In: Festival of innovation projects Sikorsky challenges, KPI, Project 56 (2021)
19. Yinka-Banjo, C., Ajayi, O.: Sky-farmers: applications of unmanned aerial vehicles (UAV) in agriculture. In: Dekoulis, G. (ed.) Autonomous Vehicles, pp. 107–128. IntechOpen, London (2019). https://doi.org/10.5772/intechopen.89488

Method for Correcting Unreliable Readings of a GNSS Receiver When Using a Cost-Efficient Autopilot

Vitalii Larin(✉) and Bohdan Blazhei

National Aviation University, Liubomyra Huzara Ave., 1, Kyiv 03058, Ukraine
vjlarin@gmail.com

Abstract. This paper addresses a critical vulnerability in low-cost UAV autopilot systems in case of GNSS data unavailability, which can result from external interference or intentional spoofing. In the paper, we use a method that compensates for erroneous GNSS readings by substituting them with corrected signals derived from an external Inertial Navigation System (INS) combined with high-precision barometric altimeter measurements. The proposed approach integrates the inertial unit's angular velocity and acceleration data using an Extended Kalman Filter (EKF3) to generate accurate flight parameters that mimic the standard GNSS output, thereby ensuring reliable navigation. To validate this method, a series of field experiments were conducted using a VTOL UAV platform equipped with a HEX Cube Orange+ flight controller. During the test, an interference generator simulated false GNSS signals, causing significant deviations in altitude and position readings from the built-in GNSS receiver. In contrast, the INS-based system maintained a stable and accurate flight profile, as confirmed by telemetry data processed in MATLAB. Comparative analyses of flight altitude, groundspeed, and positional data clearly demonstrate the proposed method's effectiveness in mitigating the impact of GNSS jamming. This work contributes to enhancing the safety and reliability of UAV operations in environments prone to GNSS interference.

Keywords: Aeronavigation · Altimetry · Positioning · INS · Ardupilot · EKF

1 Introduction

Only a few decades ago, Unmanned Aerial Vehicle (UAV) technology became highly available to civilian and scientific research communities that dedicated effort and time to making open-source hardware and software. Ardupilot grew up from powering small aircraft and quadcopters to running big jet air-craft, helicopters, rovers, blimps, and copters. Project scale gathered attention from developers all over the world, which made possible combinations of different types of aircraft, to benefit from both worlds, a great example is Vertical Take-Off and Landing (VTOL) aircraft which have the advantage of quadcopter and aircraft also greatly improving the safety and reliability of UAVs. Safety is one of the greatest cornerstones of modern UAV technology. One of the most easily accessible autopilot software is Ardupilot, which is capable of running on a wide variety

I. Ostroumov et al. (Eds.): ACASD 2025, LNNS 1418, pp. 498–508, 2025.
https://doi.org/10.1007/978-3-031-91992-3_33

of hardware, starting from almost barebone STM32 chip with InvenSense MPU-6000 and bosch BMP280 up to 64-bit quad-core ARM Cortex-A72 processor and external Attitude Head-ing Reference System (AHRS) such as vectornav or inertiallabs. This flexibility and ease of access allows practically anyone to build UAVs not going too deep into software, potentially posing a risk in modern congested airspace and populated areas. In this paper, the authors show how safety could be tampered with and a possible solution to prevent accidental and malicious manipulation of Ardupilot-based UAVs.

2 Related Works

Large UAV of the HALE (High-altitude long endurance) and MALE (Medium-altitude long endurance) types, due to its large size and sufficient maximum take-off weight, have the identical set of onboard navigation equipment that used in manned aircraft. Authors of numerous scientific studies pay attention to improving various aspects of the operation of standard navigation equipment, taking into account its use for performing automatic flights.

A promising direction in aircraft navigation is a positioning method based on the use of communication network signals, which can be applied to aircraft flying at low altitudes. The study [1] defines the necessary conditions for the successful implementation of such a method of navigation of fast-moving objects.

A standard onboard device for aircraft is an aviation weather radar, with the help of which in the article [2] a solution was proposed to avoid the problem of increasing information awareness about meteorological phenomena through the developed algorithm for identifying turbulences of varying intensity.

Awareness of the aircraft flight altitude is a very determining factor for achieving the required level of aviation safety. Low-altitude radio altimeters are standard equipment for aircraft. The development of civil digital communication systems and increased traffic has created a new challenge for the use of the aircraft radio altimeter method due to the overlap of frequency ranges of communication systems and radio altimeters [3–5], which can significantly increase the risk of flight safety [6, 7]. As a result of this problem, there is a need to develop new models of radio altimeters. Structural approaches to the development of new radio altimeters are proposed in [8, 9]. The development of new devices for radio altimetry requires new tools and new development models. Proposed software-implemented interference models for radio altimeters, which were obtained by statistical modelling using the Monte Carlo method [10]. These models are needed for testing experimental prototypes of new radio altimeters.

In the paper [11] for predicting the deterioration of aviation equipment through comparative analysis, it was determined that the two-segment quadratic regression model is the most suitable due to the minimum standard deviation and the best predictive properties. The hypothesis about increasing the probability of making the right decision was been proved by the authors of [12], they proposed to use hierarchical monitoring system developed to support the life cycle of aviation equipment.

For small aircraft, especially UAV, the use of civil aircrafts' standard on-board equipment is impossible due to their large size, significant power consumption, and cost. Therefore, micromechanical devices are used as navigation equipment for this type of aircraft.

A new structure of an inductive sensor with a non-orthogonal arrangement of the constructive axes of this sensor has been proposed in [13]. Since small aircraft are used to solve various tasks of technological processes of enterprises, much attention has been paid to the possibilities of using navigation devices not only to perform their main task - determining the location of the aircraft, but also to solve related tasks. Signals from an external source of electromagnetic interference from a power line to create a local orientation system [14].

For safe flight of UAV, their autopilots must take into account many factors. One of the important factors for electric UAVs is the supply of electricity generated by the autonomous power source. The rate of discharge of the battery in flight also depends on external factors - temperature of environment, wind force. Paper [15] proposed an economical simple algorithm based on the developed rules of fuzzy logic, which allows taking into account the listed above factors for predicting the time of discharge of the battery.

Correct execution of a flight task often requires correction of the autopilot using additional measuring devices. The authors of [16] proposed a method for checking the compliance of the shooting mode by the angle of inclination of the device with the data corrected by the Slant range system, based on an assessment of the geometry of images obtained during a flight in opposite directions.

Authors [17] in the article discusses the vulnerability of UAV navigation systems to GNSS (Global Navigation Satellite System) spoofing and proposes a modification to the EKF (Extended Kalman Filter) algorithm in Ardupilot to filter out unreliable GNSS data. This modification allows the UAV to rely solely on INS (inertial navigation system) data when discrepancies are detected, enhancing flight safety and reliability. The article [18] presents a method to detect GNSS spoofing using the abnormal energy in the Q channel correlators. The proposed method is validated through simulations and field tests, showing improved detection performance compared to traditional signal quality monitoring methods, especially in overpowered and dynamic scenarios. The study highlights the robustness and effectiveness of the Q energy detector in practical applications, making it a valuable tool for enhancing GNSS security. The article [19] proposes a statistical test to detect GNSS spoofing attacks using multiple receivers on a rigid body. The test is based on the generalized likelihood ratio test (GLRT) and checks the consistency between observed range measurements and known satellite and receiver geometry. The study [20] demonstrates that increasing the number of GNSS receivers improves the detection of spoofing attacks.

The article [21] proposes a method to detect and locate GNSS spoofing attacks using low-cost commercial GNSS boards and antennas. The articles [22–24] presents a method for detecting GNSS spoofing by combining multiple metrics while maintaining a fixed maximum false alert probability. The framework uses the Generalized Likelihood Ratio Test (GLRT) to integrate various metrics, achieving a significant reduction in missed detection probability compared to traditional methods.

The article [25] presents a method for identifying and mitigating GNSS spoofing attacks using independent and simultaneous measurements from INS. The proposed method uses an extended Kalman Filter to detect discrepancies between estimated and measured ranges and range rates, enhancing the reliability of GNSS-aided INS for launch vehicle navigation. The article [26] presents a method to detect GPS spoofing by combining multiple signal quality monitoring (SQM) metrics using an OR fusion rule. The article [27] proposes an anti-spoofing method for integrated GPS and INS navigation. The method uses error propagation characteristics and prediction estimation in the Kalman filter to detect spoofing, improving the accuracy and robustness of navigation without needing additional equipment.

3 Methodology

The peculiarity of the software design responsible for processing Ardupilot navigation data is that the main priority device for determining navigation parameters is the GNSS signal receiver. GNSS receivers are known to be vulnerable to various external interference signals. Such external influences lead to distortions in the GNSS signal receiver readings. The consequence of such an impact is a false determination of navigation parameters.

Let us represent the vector of aircraft location coordinates in the NED (North East Down) navigation system as $P_N^{sat}, P_E^{sat}, P_D^{sat}$, the third parameter is the value of the flight altitude and for convenience we will denote it as $P_D^{sat} = h^{sat}$. The vector constantly updates its values with a step of k. The following expression shows a simplified mathematical description of the process of determining the aircraft coordinates based on calculations performed according to the readings of the receiver of signals of GNSS (for this we indicated the superscript sat).

$$\begin{bmatrix} P_N^{sat} \\ P_E^{sat} \\ h^{sat} \end{bmatrix}_{k+1} = \begin{bmatrix} P_N^{sat} \\ P_E^{sat} \\ h^{sat} \end{bmatrix}_k + \begin{bmatrix} \varepsilon_N^{sat} \\ \varepsilon_E^{sat} \\ \varepsilon_h^{sat} \end{bmatrix}. \tag{1}$$

The expression (1) contains a vector of errors in determining each parameter, traditionally characteristic of satellite signals: $\varepsilon_N^{sat}, \varepsilon_E^{sat}, \varepsilon_h^{sat}$.

Elimination of various distortions at determining navigation parameters is performed by the autopilot software code, which implements a specific algorithm of a type of Kalman filter. Strapdown implementation of a EKF used to provide aiding using data sources such as:

- 3D velocity measurements from GNSS,
- 2D position from GNSS,
- Height measurement from GNSS with a barometric compensation,
- 3-axis body fixed magnetic flux measurements,
- laser range finder measurements (if available),
- True airspeed measurement (if available),
- Ground optical flow measurements (if available).

Delta angles of the IMU and delta velocities are assumed to be simple integrals of the corresponding angular rates and axial accelerations, with no coning or sculling compensation applied to them before input to the filter [28, 29].

At the same time, the built-in Kalman filtering algorithm in the Ardupilot autopilot has certain drawbacks associated with too high a degree of trust in the GNSS data source without any data validation algorithm, which makes the algorithm vulnerable to interference or malicious attacks, which does not allow detecting and eliminating false GNSS signals, even when they clearly appear as measurement misses. Particularly dangerous is additive random interference when determining the flight altitude based on false readings from the GNSS receiver. This false signal leads to the appearance of erroneous flight altitude readings, the value of which can be increased almost twofold. In this case, Eq. (1) will take the following form:

$$\begin{bmatrix} P_N^{sat} \\ P_E^{sat} \\ h^{sat} \end{bmatrix}_{k+1} = \begin{bmatrix} P_N^{sat} \\ P_E^{sat} \\ h^{sat} \end{bmatrix}_k + \begin{bmatrix} \varepsilon_N^{false} \\ \varepsilon_E^{false} \\ \varepsilon_h^{false} \end{bmatrix}, \tag{2}$$

where ε_N^{false}, ε_E^{false}, ε_h^{false} are components of the random error vector whose magnitude may be an order of magnitude greater than the magnitude ε_N^{sat}, ε_E^{sat}, ε_h^{sat}.

Also, the following constraints should be done:

$$\varepsilon_N^{false} \gg \varepsilon_N^{sat}, \varepsilon_E^{false} \gg \varepsilon_E^{sat}, \varepsilon_h^{false} \gg \varepsilon_h^{sat}. \tag{3}$$

Exposure to such interference for some time will result in incorrect calculations of the autopilot navigation solution since the autopilot will implement control actions aimed at returning the aircraft to the altitude specified by the flight mission. This means that the aircraft will descend from the true flight altitude to a dangerously low altitude, a flight at which will most likely lead to a collision with objects, trees, or directly with the ground.

Thus, it is necessary to send a signal from an external source to the Ardupilot autopilot in such a way that the autopilot perceives this signal as a signal from an external GNSS signal receiver, but with correct readings. That is, connect a device between the GNSS receiver and the autopilot that emulates a GNSS signal corrected from interference and impacts as an output signal.

We propose to use the combined navigation unit INS-BU as an external source of "correct" signals. This unit contains an extended version of the EKF3, which allows algorithmically ignoring false signals from the GNSS receiver and using signals from another sensor - a high-precision barometric altimeter - to determine the flight altitude instead.

In this case, the output signal of the INS-BU navigation unit provides for the output of data on the flight altitude, without specifying from what specific type of measuring device they were received. Thus, when connecting the INS-BU unit to the Ardupilot autopilot, the internal algorithms of the latter perceive the external signal as a GNSS system signal. Rewriting expression (1) taking into account the above, we change the sat index to "sat". Also, since the coordinates will be determined by the inertial unit, we will write the corresponding components of the error vector for the errors of the geographic coordinates (ε_N^{ins}, ε_E^{ins}), and the error component in measuring altitude as the error of the barometric altimeter, which is part of the inertial unit $\left(\varepsilon_h^{baro}\right)$:

$$\begin{bmatrix} P_N^{sat} \\ P_E^{sat} \\ h^{sat} \end{bmatrix}_{k+1} = \begin{bmatrix} P_N^{"sat"} \\ P_E^{"sat"} \\ h^{"sat"} \end{bmatrix}_k + \begin{bmatrix} \varepsilon_N^{ins} \\ \varepsilon_E^{ins} \\ \varepsilon_h^{baro} \end{bmatrix}. \tag{4}$$

4 Results of Experimental Verification of the Proposed Method

To test the efficiency of the proposed method for reducing the vulnerability of the autopilot GNSS receiver to powerful interference, a full-scale experimental study was conducted. VTOL UAV-based platform was selected for experiment study (Fig. 1).

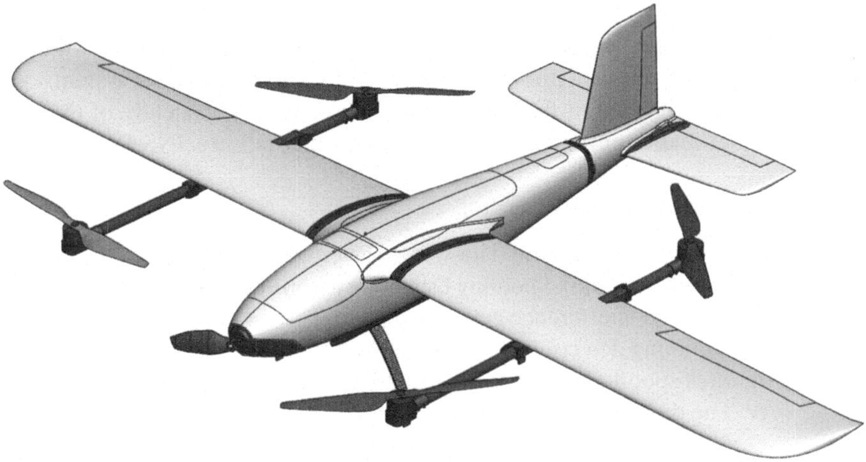

Fig. 1. UAV VTOL platform.

The following equipment was installed on board the UAV:

1. Flight controller: HEX Cube Orange+;
2. Power system: Mauch PowerCube 3V3;
3. Telemetry radio: Analog Devices AD9361;
4. INS: InertialLabs INS-BU;
5. GNSS receiver: Tualcom Tulaj 4100.

The test flight was during the day under the following environmental conditions:

- Air temperature: +2 °C;
- Relative humidity: 725 mmHg;
- Wind direction: South-southeast (SSE);
- Wind speed: 9 m/s.

The route plan, composed using Mission Planner, involved the UAV taking off to an altitude of 480 m MSL, then flying along waypoints in fixed-wing mode. During the flight along these waypoints, independent researchers activated an interference generator, after which the UAV returned to the starting point and landed on the designated landing pad, also in VTOL mode.

To simulate a false GNSS satellite signal during the flight, the interference generator was activated. During the experimental flight, remote pilots monitoring the ground control station display observed no change in the UAV's planned flight trajectory, both before and after the interference generator was activated. This observation preliminarily indicated the successful operability of the proposed method.

After decoding the telemetry data (flight log), the following data were obtained, in particular:

- Data on flight altitude, was acquired independently of the GNSS receiver signal and from the barometric altimeter of the InertialLabs INS-BU navigation system.
- Data on the true airspeed of the aircraft, also obtained independently of the GNSS receiver signal and from the inertial unit of the InertialLabs INS-BU navigation system.
- Data from the spoofing signal detector.

For ease of further processing, these data were converted into the MATLAB modeling environment format. The U-Blox Spoofing Detection identifies 12-minute mark of strong influence when activated interference generator began. The spoofing signal reached level 2 that indicating jamming detection up to 15 min, and from 17 to 20 min the signal level 3, which is signifies spoofing. U-Blox Spoofing Detection algorithm was activated on board the UAV. The data in Fig. 2 shows the readings of altitude by INS and GNSS.

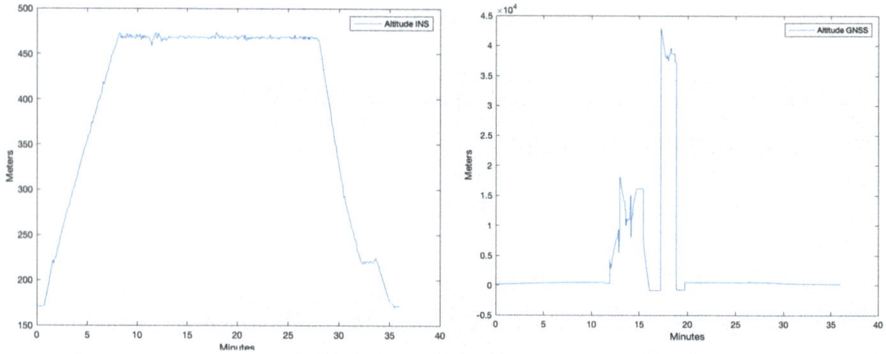

Fig. 2. Altitudes from INS and GNSS.

In the GNSS altitude graph, it is clearly seen that UAV actually flying at 480 m along the route – rapidly increase altitude up to 17 km at 12 min and fluctuated around 15 km until 15 min. Then, from 17 to 20 min the altitude readings increased again up to 40–43 km. Readings of the barometric altimeter from InertialLabs INS-BU navigation system has been fed to the autopilot. This graph represents the true altitude profile of the UAV. It can be seen that from the 8th min there is a rapid ascent from 175 m to 480 m. Then, from 8 to 29 min, the altitude remains constant, which corresponds to the UAV moving at a fixed altitude along the route. From 29 to 36 min, a rapid descent in altitude is observed as the UAV switched to VTOL mode for landing.

Figure 3 gives groundspeeds from INS and GNSS. A ground speed of the UAV initially is 12–24 m/s and increased rapidly to 200–300 m/s. Also during 15–20 min the ground speed fluctuated chaotically between 200 and 380 m/s.

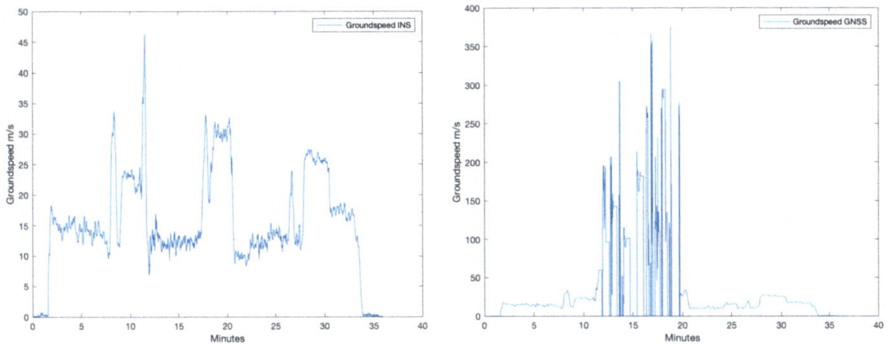

Fig. 3. Groundspeeds from INS and GNSS.

It is evident that the UAV, at an actual flight altitude of 480 m above sea level, the GNSS receiver receives an interference signal to measure altitude of 45 km. The moment of a short-term increase in altitude in Fig. 2 (Altitude from GNSS) coincides with Fig. 3, clearly showing the moment when an incorrect signal was sent and how the GNSS receiver responded to it. Also, Fig. 3 (groundspeed INS and groundspeed GNSS) shows a change in ground speed when affecting GNSS. At an actual ground speed within 12–24 m/s, jumps of up to 380 m/s are visible during the interference effect.

The impact of interference on the GNSS signal receiver also leads to receiving erroneous data on the aircraft coordinates. In this case, the autopilot will suddenly record a significant deviation from the planned flight route while the aircraft is actually on this route. Then the autopilot's response will be the development of a control action aimed at returning the aircraft to the "route" line. The remote pilot and the air traffic controller will record a violation of the flight assignment in the form of a sharp change in the aircraft's course. In this case, the risk of a conflict situation - a collision with another aircraft - increases significantly. To prevent a possible collision, the air traffic service will have to urgently make adjustments to the flight assignments of those aircraft whose courses are in the zone of dangerous proximity to an uncontrolled aircraft, the autopilot of which is guided by erroneous readings from the GNSS signal receiver.

To prevent the autopilot from generating a control action as a response to an error in the GNSS signal receiver, the authors proposed using the same algorithm that was used to obtain correct data on the flight altitude. Only in this case, instead of data from the GNSS signal receiver, data obtained on the basis of calculations of the inertial unit readings of angular velocities and accelerations were used to determine the flight coordinates.

Figure 4 shows the data from the GNSS receiver (left plot) that were subjected to external influence. Right plot in Fig. 4 shows coordinates of the UAV obtained after the InertialLabs external INS. The influence of the interference on the GNSS receiver by 2D coordinates of longitude and latitude, as well as the influence on the altitude Fig. 2 reading are clearly visible.

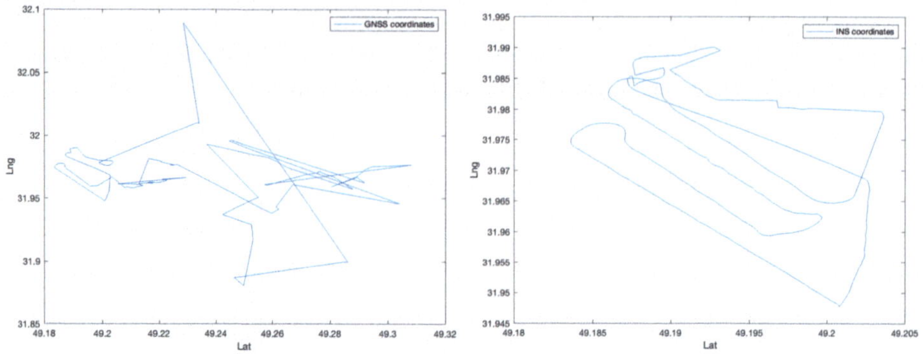

Fig. 4. Coordinates of UAV from INS and GNSS.

5 Conclusions

Low-cost common autopilots, as shown by the example of the Ardupilot based autopilot, which are produced in large amount and are widely distributed, have a significant drawback - a significant dependence of the generation of control actions on the readings of the GNSS signal receiver, which cannot be eliminated by changing its internal parameters. Ignoring this fact can lead to incidents and accidents, since this type of autopilot is widely used by operators of UAV.

The method is proposed for compensating for erroneous readings of the GNSS signal receiver by replacing the signal of the built-in receiver with signals of a so-called "external" GNSS receiver. The correct signals are actually generated by the inertial measuring unit and the barometric altimeter. The results of the mathematical model-ling clearly demonstrate the generation of an adequate navigation solution when the block of the condition for switching to an alternative data source is satisfied, in contrast to the solution obtained from significantly distorted data of the built-in GNSS signal receiver model.

References

1. Ostroumov, I., Kuzmenko, N., Bezkorovainyi, Y., Averyanova, Y., Sushchenko, O.: Relative navigation for vehicle formation movement. In: 3rd KhPI Week on Advanced Technology, KhPI Week, pp. 10–13. IEEE, Kharkiv (2022). https://doi.org/10.1109/KhPIWeek57572.2022.9916414
2. Averyanova, Y., Larin, V., Kuzmenko, N., Ostroumov, I., Zaliskyi, M., Solomentsev, O.: Turbulence detection and classification algorithm using data from AWR. In: 2nd Ukrainian Microwave Week, UkrMW, pp. 518–522. IEEE, Ukraine (2022). https://doi.org/10.1109/UkrMW58013.2022.10037172
3. Special Airworthiness Information Bulletin: AIR-21-18/FAA Aviation Safety (2021)
4. Interference Risk to Radar Altimeters from Planned 5G Telecommunication Systems/ RTCA Webinar (2020). https://www.rtca.org/wp-content/uploads/2020/12/Slides-5G-Interference-Risk-to-Radar-Altimeters.pdf. Last accessed 15 Feb 2025

5. Operational and technical characteristics and protection criteria of radio altimeters utilizing the band 4 200–4 400 MHz. ITU Recommendation ITU-R M.2059-0. https://www.itu.int/dms_pubrec/itu-r/rec/m/R-REC-M.2059-0-201402-I!!PDF-E.pdf. Last accessed 15 Feb 2025
6. Szpakowska-Peas, E.: CRW 13 The new generation radio altimeter: the concept, design, and tests. Aircr. Eng. Aerosp. Technol. **92**, 1413–1419 (2020)
7. Taylor, J.: Aircraft operation of radio altimeters. In: 24th Meeting of Working Group F of Aeronautical Communications Panel (ACP), pp. 17–21, Paris, France (2011)
8. Next Generation Radar Altimeter Testing/Eastern OptX. https://www.microwavejournal.com/articles/18895-next-generation-radar-altimeter-testing. Last accessed 15 Feb 2025
9. Isgandarov, I.A., Babayeva, N.H.: Analyzes of modern problems in radio altimeters system and its solution methods. Bull. Civil Avia. Acad. **1**(24), 17–22 (2022)
10. Larin, V., Chichikalo, N., Larina, E.: Model of reflected radio altimeter signal based on Monte Carlo statistical modeling. In: Proceedings of the International Workshop on Advances in Civil Aviation Systems Development Lecture Notes in Networks and Systems, vol. 736, pp. 141–151 (2023)
11. Zaliskyi, M., Solomentsev, O., Larin, V., Averyanova, Y., Kuzmenko, N., Ostroumov, I.: Model building for diagnostic variables during aviation equipment maintenance. In: 17th International Conference on Computer Sciences and Information Technologies, CSIT, pp. 160–164. IEEE (2022). https://doi.org/10.1109/CSIT56902.2022.10000556
12. Solomentsev, O., Zaliskyi, M., Sushchenko, O., Bezkorovainyi, Y., Averyanova, Y., Ostroumov, I.: Data processing through the lifecycle of aviation radio equipment. In: 17th International Conference on Computer Sciences and Information Technologies, CSIT, pp. 146–151. IEEE (2022). https://doi.org/10.1109/CSIT56902.2022.10000844
13. Sushchenko, O., Bezkorovainyi, Y., Solomentsev, O., Kuzmenko, N., Averyanova, Y., Zaliskyi, M.: Airborne sensor for measuring components of terrestrial magnetic field. In: 41st International Conference on Electronics and Nanotechnology, ELNANO. pp. 687–691. Kyiv, Ukraine (2022)
14. Sushchenko, O., Bezkorovainyi, Y., Golitsyn, V., Kuzmenko, N., Averyanova, Y., Zaliskyi, M.: Integration of MEMS inertial and magnetic field sensors for tracking power lines. In: XVIII International Conference on the Perspective Technologies and Methods in MEMS Design, MEMSTECH, pp. 33–36. IEEE, Lviv (2022). https://doi.org/10.1109/MEMSTECH55132.2022.10002907
15. Larin, V., Solomentsev, O., Zaliskyi, M., Shcherban, A., Averyanova, Y., Ostroumov, I.: Prediction of the final discharge of the UAV battery based on fuzzy logic estimation of information and influencing parameters. In: 3rd KhPI Week on Advanced Technology, KhPI Week. pp. 44–49. Kharkiv, Ukraine (2022). https://doi.org/10.1109/KhPIWeek57572.2022.9916490
16. Pasichnyk, N., Komarchuk, D., Opryshko, O., Shvorov, S., Kiktev, N.: Methodology for software assessment of the conformity of atmospheric correction from the UAV's zenith sensor. In: 6th International Conference on Actual Problems of UAV Development, (APUAVD), pp. 179–183. IEEE, Kyiv (2021)
17. Blazhei, B., Larin, V., Kuzmenko, N.: Software detection and denying false GNSS data on open-source UAV autopilot. CEUR Workshop Proc. **3732**, 186–196 (2024)
18. Wang, J., Tang, X., Ma, P., Wu, J., Ma, C., Sun, G.: GNSS spoofing detection using Q channel energy. Remote Sens. **15**, 5337 (2023)
19. Kalantari, A., Larsson, E.G.: Statistical test for GNSS spoofing attack detection by using multiple receivers on a rigid body. EURASIP J. Adv. Signal Process. **2020**(8), 2–16 (2020). https://doi.org/10.1186/s13634-020-0663-z
20. Kuzmenko, N., Ostroumov, I., Bezkorovainyi, Y., Averyanova, Y., Larin, V., Sushchenko, O., et al.: Airplane flight phase identification using maximum posterior probability method. In:

3rd International Conference on System Analysis & Intelligent Computing (SAIC), pp. 1–5, Kyiv, Ukraine (2022)
21. Mao, P., Yuan, H., Chen, X., Gong, Y., Li, S., Li, R., Luo, R., Zhao, G., Fu, C., Xu, J.: A GNSS spoofing detection and direction-finding method based on low-cost commercial board components. Remote Sens. **15**(2781) (2023)
22. Rothmaier, F., Chen, Y.H., Lo, S., Walter, T.: A framework for GNSS spoofing detection through combinations of metrics. IEEE Trans. Aerosp. Electron. Syst. **57**(6), 3633–3647 (2021). https://doi.org/10.1109/TAES.2021.3082673
23. Rothmaier, F., Taleghani, L., Chen, Y.-H., Lo, S., Walter, T.: GNSS spoofing detection through metric combinations: calibration and application of a general framework. J. Inst. Navig. **68**(2), 243–258 (2021). https://doi.org/10.1002/navi.420
24. Priseborough.: GitHub priseborough/InertialNav: Inertial Navigation Filter. GitHub. (n.d.). https://github.com/priseborough/InertialNav. Last accessed 15 Feb 2025
25. Vinoj, V., Lalu, V.: INS aided spoofing detection of high dynamic GNSS receiver or launch vehicle applications: a loosely coupled approach. Adv. Space Res. **74**(13), 2814–2829 (2024). https://doi.org/10.1016/j.asr.2024.03.018
26. Benachenhou, K., Bencheikh, M.L.: Detection of global positioning system spoofing using fusion of signal quality monitoring metrics. Comput. Electr. Eng. **92**, 107159 (2021). https://doi.org/10.1016/j.compeleceng.2021.107159
27. Liang, W., Li, K.: A dynamic calibration and compensation method for the asynchronous time between two inertial navigation systems. IEEE Sensors J. **21**(8), 10091–10101 (2021). https://doi.org/10.1109/JSEN.2021.3049122
28. Wenwei, L., Li, K., Qihang, L.: Anti-spoofing Kalman filter for GPS/rotational INS integration. Measurement. **193**(8), 110962 (2022). https://doi.org/10.1016/j.measurement.2022.110962
29. Liu, Y., Li, S., Fu, Q., Liu, Z., Zhou, Q.: Analysis of Kalman filter innovation based GNSS spoofing detection method for INS/GNSS integrated navigation system. IEEE Sensors J. **19**(13), 5167–5178 (2019). https://doi.org/10.1109/JSEN.2019.2902178

Short Packets Transmission Modelling in UAV Systems

Volodymyr Kharchenko, Andrii Grekhov(✉), and Vasyl Kondratiuk

National Aviation University, Liubomyra Huzara Ave., 1, Kyiv 03058, Ukraine
grekhovam@gmail.com

Abstract. Ultra-reliable and low-latency transmission will be a key feature of the next generation of wireless communications and will be based on the transmission of short packets. To model short packets transmission in mobile relay systems using Unmanned Aerial Vehicles (UAVs), a two-way channel with peer-to-peer protocol is considered. Using mobile relay to transmit data requires reliable communication. For this, it is necessary to explore the possibilities of combining the Ground Control Station (GCS) with a UAV for effective interaction with intensive data traffic. This work focuses on modelling information transmission, estimating packet latency and communication channel throughput for different packet sizes, data rates and Bit Error Rates (BERs). UAV short packets transmission via a mobile drone was first simulated using NetCracker software. The practical value of the data obtained is the ability to estimate the functioning of the communication channel, detect problems at the beginning of designing reliable mobile relay channels, which will reduce time, capital investments and provide scalability for new projects. The results obtained allow us to predict the exchange of information in mobile relay systems using UAVs.

Keywords: UAV · Mobile Relay · Short Packets · Latency · Throughput

1 Introduction

Future wireless systems will need to support new types of traffic that use short packets to carry traffic generated by Unmanned Aerial Vehicles (UAV) sensors and guidance devices. Ultra-reliable communications are required for the UAV to communicate with the Aerial Relay Drone and wirelessly coordinate within the swarm, which is mission critical connection. New applications have been created in which small packets carry important information, received with low latency and ultra-high reliability. Packets are short and may contain correlated data. In this case, reliability must be high to cope with critical events.

Current wireless systems are not designed to transmit short packets because they are based on the assumption that control information is small compared to the actual payload information. Therefore, new theoretical principles governing the transmission of short packets are needed. A fundamentally different approach to communication systems is required than that used in modern high-data-rate systems [1, 2].

I. Ostroumov et al. (Eds.): ACASD 2025, LNNS 1418, pp. 509–519, 2025.
https://doi.org/10.1007/978-3-031-91992-3_34

Ultra-Reliable and Low-Latency Communications (URLLC) is a new type of communication that is being developed by Third Generation Partner Project (3GPP) for 5G networks, Industry 4.0 and the Internet of Things. URLLC aims to provide extremely high levels of reliability and low latency to support mission-critical applications [1].

The high mobility of drones and UAVs creates unique opportunities to improve communication efficiency. Currently, there is not enough research in this area. This paper discusses the modelling of a mobile control relay channel from Ground Control Station (GCS) to the UAV. However, transmitting control signals from the operator to the UAV poses a communication challenge because such connections are subject to stringent latency and reliability requirements for mission-critical functions [2].

Relay is widely used in land and sea communication systems to improve capacity, reliability and range. Most repeaters in terrestrial systems are deployed in fixed locations. In this study, we use an Aerial Relay Drone-enabled mobile relay strategy, which performs particularly well when packet transmission latency is low. We simulate data traffic in two-way channel with peer-to-peer protocol for the GCS, which is connected through the Aerial Relay Droned to the UAV. To analyze the performance of such communication channel we have done the following:

- a channel model was built to simulate data exchange using NetCracker software;
- short packets transmission for a UAV via a mobile drone was first simulated using NetCracker software;
- short packets latency and throughput were evaluated for different transect sizes and data rates;
- the results are of practical interest, since they allow us to predict the channel performance under critical conditions.

The article is organized as follows. Section 2 is devoted to related work. Section 3 presents the channel architecture and the model parameters. Section 4 gives calculation method and data traffic. Results are discussed in Sect. 5. Description of possible equipment for building air relay and Remote UAV Side is proposed in Sect. 6 and Conclusions are given at the end.

2 Related Works

The article [1] provides the latest UAV communication technologies and examines task modules, antennas, resource processing platforms, and network architectures. Machine learning and path planning are explored to improve existing drone communication methods. Encryption and optimization methods to ensure long-lasting and secure communications are discussed. Applications of UAV networks for navigation and surveillance, URLLC, edge computing, and artificial intelligence work are covered. The complex interaction between UAVs, modern cellular communications and the Internet of Things is analyzed. The need for additional research in the areas of drone-to-drone and drone-to-device communications is identified.

Specification of technologies in standards organizations, support of operator business development and ecosystem development require a holistic view of URLLC. Study [2] includes a review of URLLC use cases that have been developed in standards development organizations (SDOs) and industry consortia.

The paper [3] presents an overview of the 3GPP 5G system architectural and security enhancements to support URLLC services based on standardization in 3GPP. Supporting high reliability through redundant transmission, supporting low latency during handover, enhancing session continuity during user mobility, monitoring quality, packet latency, automatic service recovery after handoff, and security aspects are covered.

URLLC in wireless networks is reviewed in [4], where basic concepts and potential applications are considered. Current research on URLLC at the physical, link, and network layers is presented, and potential research topics and problems are identified.

A systematic study of practical URLLC procedures in the Internet of Things (IoT) was carried out in review [5]. The selected methods are divided into four categories: structure-based, diversity-based, metaheuristic-based, and channel state-based. The advantages and disadvantages of using URLLC in an IoT network are outlined.

The review [6] provides an overview of the integration of UAVs into cellular networks. In particular, types of consumer UAVs, interference issues, standardization solutions, serving air users with existing ground base stations, capabilities of flying repeaters and UAV-based base stations, prototype test benches, new rules for commercial use of UAVs, and cellular security in using UAVs.

The review [7] examines air-to-ground, ground-to-ground, and air-to-air channels for UAV communications in various cases. Recommendations are provided for managing the UAV communication link budget, taking into account losses and channel attenuation effects. The gains from transmit/receive diversity and spatial multiplexing when using multiple antennas are analyzed. Issues for modeling UAV communication channels are discussed.

The article [8] provides guidance on the use of UAVs in wireless communications. 3D UAV deployment, performance analysis, channel modeling and energy efficiency are studied. Open problems and research directions in UAV communications are presented. Optimization theory, machine learning, stochastic geometry, transportation theory, and game theory are covered. The manual contains recommendations for the analysis, optimization and design of wireless communication systems based on UAVs.

The paper [9] models a UAV-based relay system for delivering command information in accordance with URLLC requirements. The optimization problem is solved based on the proposed iterative algorithm. The figures show the dependences of the approximate probability of decoding error on the number of iterations for different heights and the probability of decoding error on the length of blocks. The size of the command signal packet is Lbits, the transmission of which must be completed within Tmax seconds. The total length of the block will be M = B Tmax, where B is the system bandwidth. The simulation results confirm that the proposed algorithm outperforms existing algorithms.

The article [10] reviews recent advances in URLLC and describes methodologies tailored to their unique characteristics. Tools for modeling and optimizing URLLC-related problems at the network level are demonstrated. This work can be considered as a first step in defining a scalable URLLC design that considers latency, reliability, packet size, network architecture, and topology.

The study [11] explored the potential of using UAVs to support URLLC using better line-of-sight communication quality. The latency, reliability and availability of the UAV communication systems network are considered. Simulations have shown that increasing

UAV density can significantly reduce overall throughput. In urban areas, the use of a single wireless ground-to-air channel is not sufficient to ensure the quality of service and availability of the URLLC network.

Ultra-reliable, low-latency transmission is a key feature of the next generations of wireless communications and will be based on the transmission of short packets. Paper [12] investigated a two-way UAV control system in which two sources communicate with each other using a UAV repeater. Exact expressions for the average block error rate are obtained and a solution is proposed to maximize system throughput by optimizing the packet length.

A secure communication system using short packets and including relay using a UAV in the presence of a ground-based passive listening device is proposed in article [13]. The goal is to improve the throughput of the secure system by designing the encoding block length and UAV trajectory. Numerical results show that the proposed design provides significant performance improvement.

The paper [14] studied the impact of packet size on loss rate and latency characteristics in a real-time wireless application. An analytical model of the dependence of packet length on delay characteristics has been developed. It is shown that proper design of packetization schemes can significantly improve the utilization of channel resources when streaming delay-sensitive information.

The article [15] investigated the effect of the number of packets on delivery time. It is shown that latency is a nonlinear function of the number of hops, message size, and number of packets. Appropriate selection of the number of packets can be used to minimize latency. The optimal number of packets has been found to minimize the delay. They showed how the optimal number of packets depends on the message size.

3 Channel Architecture

Multiple drones can form a network of relay nodes to create a long-distance communication system to directly control the First Person View (FPV) drone. Maintaining a stable communication link is a particularly challenging issue for FPV drones, as the relative motion of all drones must be constantly adjusted to control a reliable connection. Guaranteeing line of sight throughout the entire communication channel of the FPV drone is ensured by artificial intelligence by adapting repeaters to changing conditions.

As the first step in studying the relay communication channel of the FPV drone with the operator, the simplest model presented in Fig. 1. The purpose of such a communication system is to transmit and receive data using a drone, which is a relay node that provides unobstructed line of sight.

Network model is designed using Professional NetCracker 4.1 software. Model characteristics are given in Table 1. The network model "Ground Control Station – Aerial Relay Drone – UAV" is shown in Fig. 1 and contains Aerial Relay Drone (50 km from the GCS) and the low-altitude UAV (50 km from the Aerial Relay Drone).

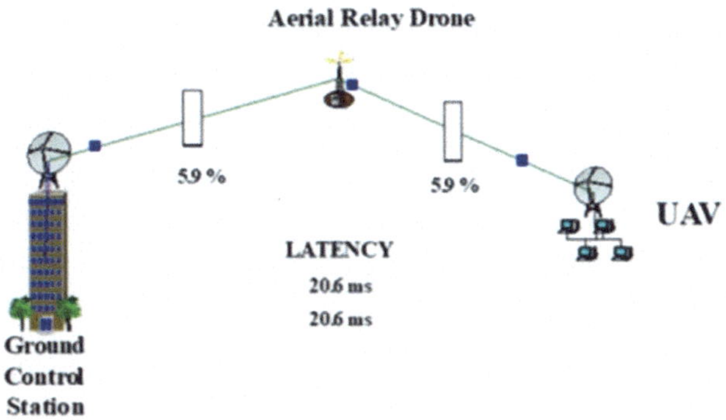

Fig. 1. Mobile control using Aerial Relay Drone.

Table 1. Model characteristics.

Characteristics → Structure elements ↓	Bandwidth (Mbps)	Length (km)
Ground Control Station (GCS)		–
Common Data Server	1000	–
Antenna – Server link	1.544	10^{-2}
Antenna	1000	–
"GCS – Aerial Relay Drone" wireless link	1.544	50
Aerial Relay Drone	1000	–
(Packet Latency – 0 ms, Packet Fail Chance – 0)		
"Aerial Relay Drone – UAV" wireless link	1.544	50
UAV		
Antenna	1000	–
"Antenna – Server link"	1.544	10^{-2}
Server	1000	–

4 Calculation Method

NetCracker realize simulation using mathematical equations and is written in Java EE. Parameters are divided into internal (which are obtained by mathematical modelling) and external (on which the internal characteristics depend). Channel throughput and packets latency are internal parameters. External parameters included the Transaction Size (TS), the Time between Transactions (TBT), the Bit Errors Rate (BER) and the Throughput in the channel. During the simulation the internal parameters are calculated using the specified external parameters. Latency refers to the time it takes for a packet to travel from a source device to a destination end device. In our case, from the GCS to the UAV.

By the throughput of the communication channel "Aerial Relay Drone – UAV" we mean the amount of data successfully transmitted over a certain period of time, measured in bits per second.

A traffic with Local Area Network peer-to-peer protocol is specified for the created model (TBT = 1 s). The model parameters were simulated taking into account the Const probability distribution law ($\varpi(x)$ = Const, $\varpi(t)$ = Const) as statistical distributions for the TS and the TBT. Formulas for the average length of the transmitted packets, the average time interval between two adjacent packets, and the average packet travel time (Latency) are given by following formulas (1), (2), (3) and (4).

The average length of the transmitted packets is determined as the mathematical expectation:

$$TS = \int_{-\infty}^{\infty} x\omega(x)dx, \tag{1}$$

and the average time interval between two adjacent packets:

$$TBT = \int_{-\infty}^{\infty} t\omega(t)dt. \tag{2}$$

Based on the values of TS and TBT parameters, the Average Utilization (AU) on the communication link is determined by the formula:

$$AU = TS/TBT = TS/\mu, \tag{3}$$

where $\mu = 1/TBT$ is the packet transmission rate over the communication channel.

The average packet travel time (Latency) over the communication channel is determined by the formula:

$$TT = \frac{TS}{AU_{link}}. \tag{4}$$

5 Results

The high mobility of drones and UAVs creates unique opportunities to improve communication efficiency. This paper discusses the modelling of a mobile control relay channel from a ground control station to a UAV. However, transmitting control signals from the operator to a UAV poses a communication challenge because such connections are subject to stringent latency and reliability requirements for mission-critical functions.

Figure 2 demonstrates the dependence of the Latency on the TS parameter and the role of bit errors in the Aerial Relay Drone-UAV channel. From the data shown in this figure, it is clear that the Latency is a nonlinear function of message size. In the absence of bit errors, the Latency is practically the same for all considered packet sizes. The appearance of bit errors in the Aerial Relay Drone-UAV channel dramatically increases the Latency at TS > 100 bits.

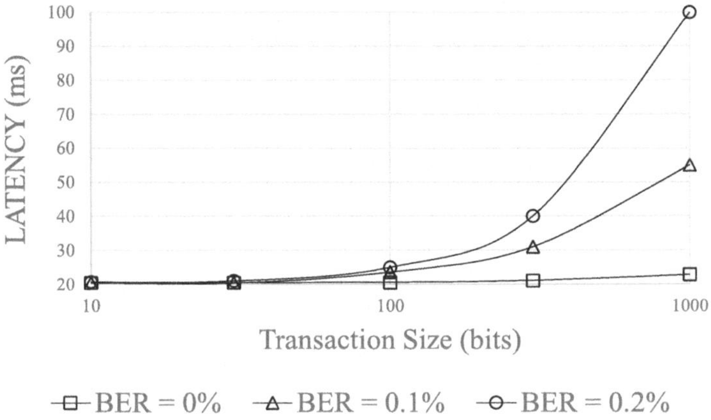

Fig. 2. Impact of packet size on end-to-end latency for different bit error levels.

The article [15] studied the effect of packet size and bit errors on packet delay based on a theoretical approach and showed that the Latency is a nonlinear function of message size, and bit error rate. During the theoretical modeling, packets of size 1–10 Kbyte were considered, which had data rate of 1 Mbit/s. This article assumed that all channels have the same bandwidth, which is not actually implemented. As a result, the delay equation used in the simulations using the NS3 simulator was not accurate. Nevertheless, the qualitative conclusions of the authors [15, 16] confirm our results.

Message delivery delay is a critical parameter in URLLC. Figure 3 shows the latency dependences on the packet sizes for different data rates (T1 = 1.544 Mbit/s, T3 = 44.736 Mbit/s). For TS $\leq$ 100 bits, the latency changes slightly for both data rates, but increases with packet size. Thus, the latency in message delivery is determined not only by the data transfer rate, but also by the size of the transactions. From the data in Fig. 3 it follows that for small packets the latency is approximately the same.

Figure 4 shows the dependences of the throughput (average load) of the "Aerial Relay Drone-UAV" wireless link when exchanging data between the GCS and the UAV on the transaction size for various BERs in this link. Increasing the transaction size results in a significant increase in throughput for TS > 300 bits and virtually zero for 10 bits $\leq$ TS $\leq$ 100 bits. Increasing the BER from 0% to 0.2% results in a ~ 5x increase in throughput for TS = 1000 bits. This circumstance plays a significant role in the transmission of intelligence content.

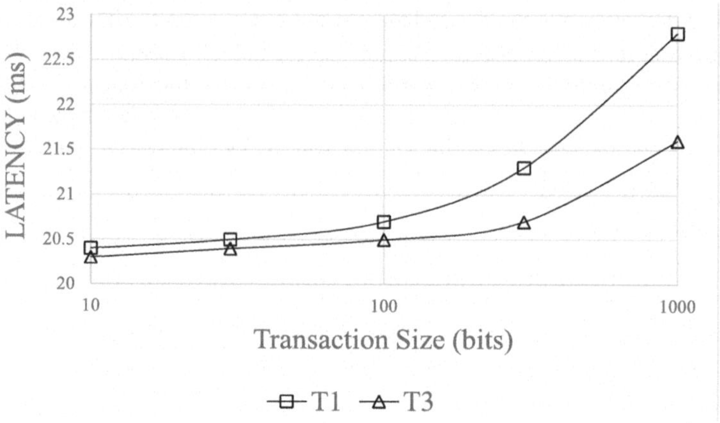

Fig. 3. Impact of packet size on end-to-end latency for different data rates.

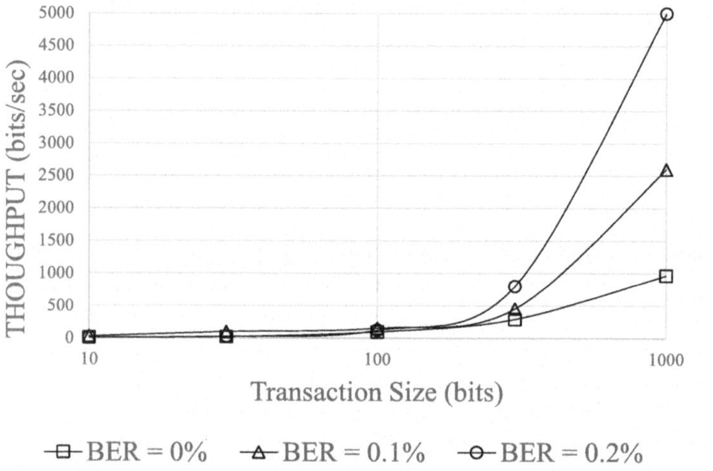

Fig. 4. Impact of packet size on throughput for different bit error levels.

The authors [14] derived analytical relationships between packet length and latency characteristics and validated the theoretical analysis through experiments in an ad hoc network using WLAN technologies. The simulation was carried out with data rate of 1.2 Mbit/s during 100 s for packet sizes of 30, 700 and 1400 bytes. From these results, it is concluded that low average latency corresponds to low packet size. Relative packet transit times ranged from 0 ms to 50 ms, excluding isolated bursts associated with packet losses. The difference between the curves for each packet size is very small. Relative one-way trip times become approximately the same for all three packets approximately in the interval of 15–20 ms. Figure 4 of paper [14] clearly shows the difference between latencies for packets of different sizes.

6 Description of Possible Equipment for Building Air Relay and Remote UAV Side

To organize retransmission in the air and provide communication with remote UAVs, it is necessary to consider the equipment for the ground station, the air repeater and the equipment on board the UAV. Let us consider the equipment for each part of the system.

6.1 Communication Channel Block Diagram

Channel block diagram is represented in Fig. 5.

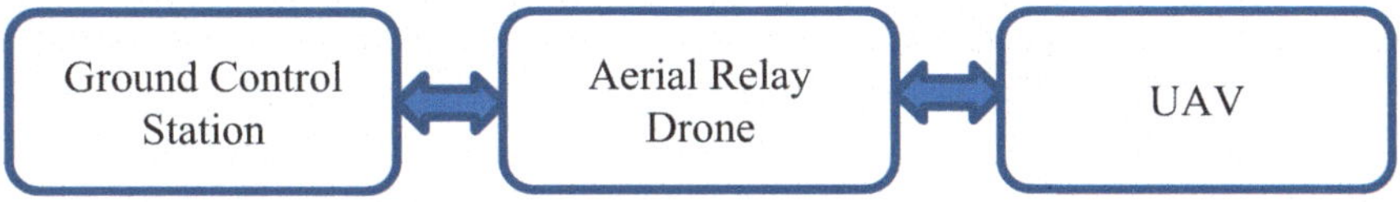

Fig. 5. Channel block diagram.

Return channel is for telemetry/video feedback, and forward channel is for control commands and mission data. Components:

1. Ground Control Station (GCS). Transmits control and monitoring data to Aerial Relay Drone. Primary communication channel: LTE/5G or dedicated RF channel.
2. Aerial Relay Drone: Serves as a data repeater between the GCS and UAV. Transmits data from the GCS to UAV via a two-way channel. Equipped with signal amplifiers and directional antennas.
3. UAV (end devices): Receive data from Aerial Relay Drone and send back telemetry.

6.2 Equipment for Remote UAV Side

The remote UAV may have the following equipment that provides two-way communication with the repeater and contains the following components:

1. Radio modules: RFD900X (915 MHz) for long-range telemetry transmission; Microhard pMDDL2450 for high-speed data transmission.
2. Antennas: Conventional dipole antennas for basic communications; Panel antennas, such as Laird Technologies.
3. On-board computers: NVIDIA Jetson TX2 or Raspberry Pi Compute Module.
4. Camera modules and sensors: For video stream transmission: Sony Alpha or DJI Zenmuse; For data collection: thermal imagers (Flir Vue) or lidars (Velodyne VLP-16).
5. Power systems: Modular batteries from Tattu or MaxAmps, providing 20–40 min of flight; Backup batteries for communication in critical situations.
6. Software: QGroundControl or ArduPilot for control; MavLink for telemetry and mission control.

6.3 Aerial Relay Drone Equipment

The airborne relay acts as an intermediary between the ground station and remote UAV, providing an extension of the communication zone, and may contain the following components and equipment:

- Airborne relay selection parameters: Flight altitude: 300–800 m to increase the communication radius. Payload: Minimum 5–10 kg for installing communication equipment. Flight time: Minimum 3–5 h to ensure stable communication coverage or hybrid systems with fuel cells for continuous operation. Using hybrid engines or solar panels to increase battery life. Trajectory: Orbital trajectory around the coverage area. Weather conditions: resistance to wind load up to 15 m/s, operating temperature from $-20\ ^\circ$ C to $+50\ ^\circ$ C. Equipment sealing from rain and dust (IP65 and above).
- Equipment for the repeater: Volansi VOLY 50: 9 kg payload, 8 h of battery life. DJI Matrice 300 RTK: 2.7 kg payload, 55 min of battery life.
- Radio modules/transceivers: Silvus StreamCaster 4200: MIMO support, 400 MHz–6 GHz ranges; Ubiquiti RocketDish for high-frequency links.
- Antennas: Equipment with circular or directional polarization, such as TAOGLAS Antennas for wideband operation; Phased array antennas with automatic beamforming support.
- Batteries: Li-ion batteries or hybrid systems with fuel cells, such as Intelligent Energy for increased flight time.
- Communication protocols: LTE/5G for connection to the ground station; Wi-Fi 6 or MANET for communication with remote UAVs.
- Additional equipment: Precise positioning mechanisms (RTK, GNSS); Payload for data transmission, such as video streams or telemetry.

7 Conclusions

The requirements for operating and targeting remote UAVs are increasing, especially in wartime. Providing control over long distances and transmitting data at high speeds places high demands on ultra-reliable and low-latency communications.

The range of the UAV's line-of-sight channels is limited by the non-linearity of the operator's ground station amplifier, fading, noise and interference. The use of mobile relay for UAVs improves the capabilities of navigation and flight over long distances. Our approach provides a simple method for predicting latency and throughput in UAV mobile relay links.

Real-time implementation of reliable mobile relay links with UAVs is a challenging task. In such cases, data transmission modes in which the communication channel with the UAV may be closed are of particular interest. There is no information about such critical situations in the literature. The original model of the UAV communication channel with a mobile repeater was constructed using NetCracker software. This work is the first study with estimating packet latency and communication channel throughput for different packet sizes, data rates and BERs.

The practical value of the data obtained is the ability to predict the functioning of the communication channel, detect problems at the beginning of designing reliable mobile relay channels, which will reduce time, capital investments and provide scalability for new projects. It is clear that developers of drone communication systems will not be able to do without such preliminary calculations.

The considered wireless communication channel can be used to communicate with a swarm of UAVs, which will allow you to perform tasks that are more complex.

References

1. Sharma, A., Vanjani, P., Paliwal, N., Basnayaka, C.M., Jayakody, D.N., Wang, H.C., Muthuchidambaranathan, P.: Communication and networking technologies for UAVs: a survey. J. Netw. Comput. Appl. **168**, 102739 (2020)
2. Service requirements for cyber-physical control applications in vertical domains (3GPP TS 22.104 version 16.5.0 Release 16). ETSI TS 122 104 V16.5.0 (2020-09)
3. Yoshizawa, T., Sheeba, M., Kunz, A.: Overview of 5G URLLC system and security aspects in 3GPP. In: 2019 IEEE Conference on Standards for Communications and Networking (CSCN) (2019)
4. Feng, D.Q., Lai, L.F., Luo, J.J., et al.: Ultra-reliable and low-latency communications: applications, opportunities and challenges. Sci. China Inf. Sci. **64**(2), 120301 (2021)
5. Sefati, S.S., Halunga, S.: Ultra-reliability and low-latency communications on the internet of things based on 5G network: literature review, classification, and future research view. Trans. Emerg. Telecommun. Technol. **34**(6), 1–38 (2023)
6. Fotouhi, A., Qiang, H., Ding, M., Hassan, M., Giordano, L.G., Garcia-Rodriguez, A., Yuan, J.: Survey on UAV cellular communications: practical aspects, standardization advancements, regulation, and security challenges. IEEE Commun Surv Tutor. **21**(4), 3417–3442 (2019)
7. Yan, C., Fu, L., Zhang, J., Wang, J.: A comprehensive survey on UAV communication channel modeling. IEEE Access. **7**(1), 107769–107792 (2019)
8. Mozaffari, M., Saad, W., Bennis, M., Nam, Y.H., Debbah, M.: A tutorial on UAVs for wireless networks: applications, challenges, and open problems. IEEE Commun Surv Tutor. **1**, 1 (2019)
9. Cunhua, P., Hong, R., Yansha, D., Maged, E., Arumugam, N.: Joint blocklength and location optimization for URLLC-enabled UAV relay systems. IEEE Commun. Lett. **23**(3), 498–501 (2019)
10. Bennis, M., Debbah, M., Poor, H.: Ultrareliable and low-latency wireless communication: tail, risk, and scale. Proc. IEEE. **106**(10), 1834–1853 (2018)
11. She, C., Liu, C., Quek, T., Yang, C., Li, Y.: UAV-assisted uplink transmission for ultra-reliable and low-latency communications. In: 2018 IEEE International Conference on Communications Workshops (ICC Workshops), pp. 1–6, Kansas City, MO, USA (2018). https://doi.org/10.1109/ICCW.2018.8403626
12. Tien-Tung, N., Anh-Vinh, N., Tan-Loc, N., Van-Hung, L.: Two-way UAV-aided systems with short packet communications. Comput. Netw. **235**(12), 109957 (2023)
13. Mamaghani, M., Zhou, X., Yang, N., Swindlehurst, A.: Secure short-packet transmission with aerial relaying: blocklength and trajectory co-design. arXiv:2310.05142v1. (2023)
14. Korhonen, J., Wang, Y.: Effect of packet size on loss rate and delay in wireless links. IEEE Commun. Soc. **3**, 1608–1613 (2005)
15. Uttam, R.: An introspection on impact of bit-error and framing on packet delay. In: 2016 International Conference on Computer, Electrical & Communication Engineering (ICCECE), pp. 1–5, Kolkata, India (2016). https://doi.org/10.1109/ICCECE.2016.8009552
16. Kharchenko, V., Grekhov, A., Kondratiuk, V., Kutsenko, O.: Investigation of factors affecting data transfer in space-air-ground integrated network. In: Ostroumov, I., Zaliskyi, M. (eds.) Proceedings of the International Workshop on Advances in Civil Aviation Systems Development. ACASD 2023 Lecture Notes in Networks and Systems, vol. 736, pp. 45–59. Springer, Cham (2023). https://doi.org/10.1007/978-3-031-38082-2_4

Management of Unmanned Aerial Vehicles Swarm: An Overview

Dmytro Krytskyi(✉)

National Aerospace University KhAI, Vadyma Manka Str., 17, Kharkiv 61070, Ukraine
d.krickiy@khai.edu

Abstract. The article considers various types of interaction of a Unmanned Aerial Vehicle (UAV) swarm (flocks, groups, swarms, hives, clusters), its capabilities for performing missions, features of such formations, advantages and disadvantages of its use. Variants of tactical use of UAVs swarm based on the use of various formations are considered. A description of swarm management methods is given, such as Markov Decision Processes, Q-Learning method, Policy Gradient Algorithms for the implementation of building a formation and performing the goal set before the swarm, its differences and advantages in performing various types of tasks are considered. Models of behavior during the execution of a task when using a UAVs swarm are described, taking into account control based on the above methods: Flocking, Select Nearest, Double Swarm, Flock, Rabbit, Assign Nearest. The results of the simulation of the use UAVs swarm to perform particular tasks and move behind the leader are given.

Keywords: UAVs Swarm · Markov chains · Q-Learning Method · Policy Gradient Algorithms

1 Introduction

Unmanned aerial vehicles (UAVs) play a key role in the future, providing support for various services for humanity. Although UAVs play major roll in defense and military purposes, also UAVs helps in a wide range of scientific and applied problems [1–3]. Every day, the issues of using UAVs in groups, swarms [4, 5] and other types of joint use are being addressed more and more intensively. This increases the speed of task completion, and sometimes leads to a multi-effect, when the overall efficiency of the result increases according to several indicators. But the joint use of UAVs varies greatly depending on the selected type of interaction: group, swarm, hive, network group, column of UAVs, segmented group (the most common types of joint interaction of UAVs in the literature) [1, 6, 7].

The main differences between the formations lie in the UAV control system, principles of organization and interaction within the group itself.

Swarm. This is one of the most complex formations, where each UAV operates autonomously, following simple rules of interaction with neighboring UAVs. There is no fixed leader in the swarm, and each UAV uses information about the location of

I. Ostroumov et al. (Eds.): ACASD 2025, LNNS 1418, pp. 520–529, 2025.
https://doi.org/10.1007/978-3-031-91992-3_35

neighboring UAVs to coordinate its actions [8–10]. The main characteristics of such a formation are: the use of decentralized control (each UAV receives data from its neighbors, and not from a single control center), the possibility of scalability of the formation (adding or removing UAVs does not disrupt the structure and stability of the swarm), self-organization (the swarm can adapt to changes in the environment or task without external commands). Such a formation is characterized by the use of artificial intelligence methods (machine learning, multi-threaded interaction) for control. Therefore, each UAV must be equipped with a set of sensors to determine the distance, detect obstacles, analyze the environment. Communication between UAVs must be in real time, and there must also be a constant analysis of the location, using GPS or local positioning systems, for accurate navigation.

UAV Swarm (Hive). A hive is a formation in which UAVs closely interact and perform tasks as a single unit, often using a single algorithm or a given route. Unlike a swarm, a hive can be more organized and have a specific structure or main goals that they achieve collectively [11]. Unlike a swarm, in a hive, coordination is carried out around a common algorithm (all UAVs obey a common instruction or task), while the hive can form different structures (for example, linear or radial), while maintaining a common goal. It should also be noted that a hive is more resilient to the loss of even a large number of UAVs (the focus on completing the goal will be achieved in any situation). This approach is characterized by the use of centralized control, but the sets of sensors will be similar to the sets of a swarm, the difference will be that the hive must use machine learning to improve mutual coordination and adaptation to different environments, and the communication equipment must be designed for data exchange between UAVs and a centralized hub.

Cluster. A cluster is a formation in which several UAVs are united to perform a common task, but each UAV can perform a unique role [12–14]. A cluster often has a leader or central UAV that coordinates the actions of the others. This type of interaction is implemented through the use of centralized or semi-centralized control (the leader or several key UAVs manage the coordination), to increase efficiency, each UAV can do its part of the work depending on its purpose. Such a formation is characterized by control from a single source (leader or server), coordination of tasks and routes, lidars, ultrasonic sensors, cameras and infrared sensors are used for navigation, communication is carried out in real time. When using a cluster formation, it is possible to achieve maximum presence in the air of a swarm of UAVs.

Networked Formation. The network formation implies that UAVs are united in a network, where each UAV can transmit information to another, as well as process and retransmit data [15]. This formation is characterized by the use of continuous data exchange (each UAV serves as a node in the network, allowing information to be transmitted over long distances). When using this formation, constant synchronization and re-checking of each UAV for validity is necessary; in case of failure, the route is built using another UAV. This approach is best used to organize communication over large areas. Therefore, this formation is characterized by the use of a mesh network (each UAV serves as a node, creating a stable network with data retransmission between nodes, AODV (Ad hoc On-Demand Distance Vector) protocols are used: for dynamic selection of the optimal transmission route. MANET: to adapt the network to changing conditions and the movement of nodes. UAVs must be equipped with powerful on-board systems, since preliminary analysis and filtering of information occurs on board, which reduces the load on the network.

Formation Flight. A column is a more organized formation in which UAVs follow a clear sequence or form a predetermined geometric shape [16]. This formation is characterized by UAVs following a given route in a certain sequence, with each UAV having its own fixed place in the formation. The distance between UAVs in the column is equal and constant, to maintain the shape. The "leader-follower" algorithm is mainly used to follow the leader [17].

Segmented teams are a structure in which UAVs are organized into subgroups, each of which performs a separate part of the work within the overall task. Each segment works relatively autonomously but coordinates its actions with other segments [18]. In such a formation, it is convenient to replace subgroups without disrupting the work of the entire team. This formation is characterized by the use of dynamic distribution of tasks between segments based on their status and capabilities.

2 Tactical Options for Using a UAVs Swarm

The following tactics of using a UAVs swarm are distinguished (Fig. 1).

Close (Swarm). Each UAV acts according to its own rules, but within the framework of a common strategy. Although the UAVs do not follow a single leader, they still interact and perform a common task, which allows for a certain level of organization in the chaotic conditions of the mission.

Mass (Hive). Several UAVs are combined into groups to perform a task (joint work of UAVs according to a single algorithm). They act synchronously, following certain routes, which gives them numerical superiority and allows them to achieve greater results, for example, during observation or synchronous actions. Here, each UAV, although autonomous, still works as a team, which allows it to remain effective when performing large missions (a large number of UAVs are used simultaneously).

Maneuvering (Network Formation). In network formations, UAVs can coordinate its actions by transmitting data to each other and adapting routes in real time, which allows for saturating the required areas. This allows for maneuvers, such as flanking an enemy. In this formation, UAVs can be divided into groups that perform different tasks, but constantly interact and adjust their actions to achieve a common goal.

Segmented Teams. Dividing the swarm into groups that perform different tasks; one group can scan, and another - deliver equipment. The key here is the constant exchange of information between subgroups, which ensures coordination and efficiency in the execution of complex missions. This organization allows for several tasks to be performed simultaneously, providing high flexibility in mission execution.

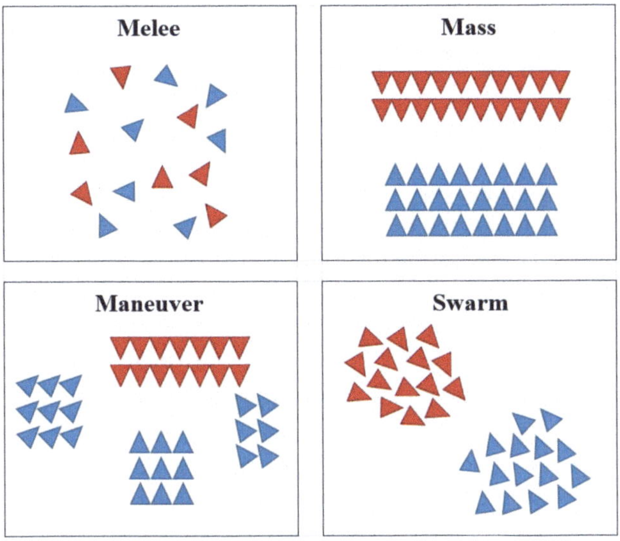

Fig. 1. UAV Swarm tactics.

3 Methods for Implementing Swarm Management

Markov Decision Processes. In order to use such swarm tactics, it is possible to use Markov decision processes (MDP). In MDP, the world is divided into "states" and each state represents a possible configuration of the environment (Fig. 2). A state is a position in space, a completed functional task, an opportunity to use an available tool, etc. In each situation, the agent has several options for behavior. Each action changes the environment and causes a transition from one state to another. These transitions are stochastic and are determined by the transition function T (s, a, s'). The transition function T (s, a, s') determines the probability that the agent is in state s and will move to state s' if it performs action a. The goal of the agent in MDP is to maximize the "reward" it receives for completing a task. Reward is a signal from the environment to the superior that indicates the agent's success or failure in completing a task. Reward is provided either at the end of the task (in case of success or failure) or during the execution of the task, depending on the type of task.

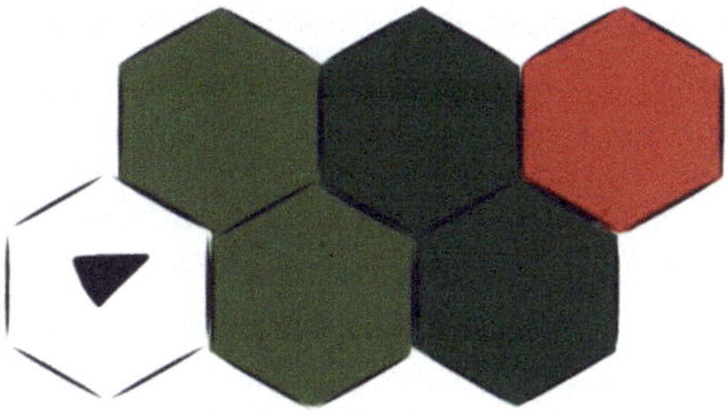

Fig. 2. Progress of the agent towards the goal taking into account the reward.

Formal description:

1. Environment: everything that exists in the world and can be used to solve a given problem.
2. State: a momentary description of the state of the agent and the environment (the agent's location, properties, surrounding objects). The state space is the set of all possible states. The current state is the state at a given time. The goal state is the state in which the task is completed or the problem is solved.
3. The agent is an entity that acts in the environment to change its state.
4. Action: the steps the agent takes to move from one state to another. Actions change the environment or the agent's position in the environment, bringing it closer to the desired outcome.

Rewards provide an estimate of the value of each state. States that bring positive rewards or are close to them are assigned a higher value than states that are far from positive rewards. If the values of all possible neighboring states are known, the agent chooses an action to move to the state with the highest value. The value of this state is denoted by V(s) and is calculated using the Bellman formula:

$$V(s) = \max_a \left(R(s, a) + \gamma \sum_{s'} T\left(s, a, s'\right) V\left(s'\right) \right), \quad (1)$$

where $V(s)$ is the value of the current state; s is current state; a is performed action; $s^{'}$ is following state; $T(s, a, s^{'})$ is transition function, the probability of transition from state s to state $s^{'}$ if action a is performed; $V(s^{'})$ is the value of the next state, is calculated recursively; γ is a discount factor from 0 to 1, the amount by which the future reward is reduced, a higher γ increases the value of the future reward, and a lower γ focuses on the reward in the present and near future; $R(s, a)$ is the reward function, the number of rewards an agent receives for an action in state s; $\max_a$ is maximize $V(s)$ by choosing the action a that brings the highest reward and value for the state s.

Bellman's formula states that the value of a state is equal to the reward for the most profitable behavior in the current state plus the discounted sum of future rewards for moving to the next state. The equation recursively computes $V(s)$ by adding up all possible future states. γ is less than 1, so the total converges to a given value. When the value of each state is known, the agent decides to perform an action that leads to the next possible state with the highest value, moves to the target state, and completes the task. The formula recursively calculates $V(s)$ by adding up all possible future states; since γ is less than 1, the sum converges to the target value. When the value of each state is known, the agent decides on the action that will lead it to the next possible state with the highest value, moves to the target state, and completes the task.

Q-Learning Method. Q-learning is a method that allows an agent to determine the optimal strategy by calculating and storing values for each state. The value of each state is calculated based on the reward function and the transition function, and all states must be known. In the case where the functions are unknown, the agent learns experimentally (performing a random action and storing the result). After conducting experiments, the agent chooses actions that lead to states with the highest value until it reaches the goal or rewarding state. This approach helps in learning agents when there is no information, but there are limitations: if the transition function is unknown, the agent cannot accurately predict which state each action will lead to. This makes it difficult to choose the optimal action. To overcome this problem, another approach to reinforcement learning is used - instead of calculating state values, the agent calculates the Q value for each state-action pair. This solution allows you to evaluate the effectiveness of each action in a particular state, even without knowing the transition function. The formula for calculating Q values takes into account the current reward and the expected reward for future actions, which allows the agent to learn from experience:

$$Q(s, a) = \sum_{s'} T(s, a, s')(R(s, a) + \gamma V(s')), \tag{2}$$

and then determines the best action by:

$$a_{best} = \text{argmax}_a Q(s, a). \tag{3}$$

The process of exploring the environment and analyzing the reward is assessed by the value of the state-action pairs. The need for an exact transition function ($T(s, a, s')$) is not mandatory, in this case the agent independently selects the data. The more actions the agent performs, the more rewards it receives, which allows it to more accurately assess the value for each state-action pair. Subsequently, the agent selects actions that allow it to achieve the set goal with the greatest effect.

Policy Gradient Algorithms. The policy $\pi(a|s, \theta)$ is a mapping between a set of states s and parameters θ and the probability that the agent will perform action a. For example, if the agent is in state A, the probability that it will perform action 1 is 60%, the probability that it will perform action 2 is 25%, and the probability that it will perform action 3 is 15%. The parameter θ is a weight that determines the probability of performing each action in each state. At the beginning of training, the parameters are initialized randomly. As the environment is explored, the most optimal parameters (the set of parameters that map states to actions that most effectively perform the task) are selected. $J(\theta)$ is the direction that provides optimal performance of the policy. This is the performance measure used to update the policy. This approach is called gradient descent:

$$\theta_{t+1} = \theta_t + \alpha \nabla J(\theta_t), \tag{4}$$

$J(\theta)$ can be further decomposed into a value function of the state-action pair and the current policy:

$$\theta_{t+1} = \theta_t + \alpha \frac{\hat{Q}(\mathrm{s, a}) \nabla \pi_{\theta t}(a|s)}{\pi_\theta(a|s)}, \tag{5}$$

where θ_{t+1} is updated parameters; θ_t is current parameters; α is programmer-controlled hyper-parameter that determines how large of a step the policy moves, usually starts large and decreases as training continues in order to finetune policy; $\hat{Q}(\mathrm{s, a})$ is estimated value of the state-action pair, estimation of how good this action is; $\nabla \pi_{\theta t}(a| s)$ is gradient of the current policy, direction that moves towards a better policy; $\pi_\theta(a| s)$ is probability that the action will be taken in the given state, how often the policy is being updated.

4 Modeling the Behavior of a UAV Swarm

Based on the above and the encountered variants of tactical use of a UAV swarm, we can distinguish six of the most common tactics: "flock", choosing the closest, distribution among the closest, "rabbit", circular capture, swarm (Fig. 3).

A "flock" act implies the movement of the swarm to the target regardless of any obstacles (external influence on the swarm is not taken into account, the swarm continues to move regardless of anything). When using the flock tactic with the choice of the closest, each UAV chooses as a task the one that is closer to its location, and the criteria on the basis of which the UAV from the swarm chooses this task are usually specified. The following can be used as such criteria: the distance to the target is less than 20 m, the distance between the center of the swarm and the closest target is less than the distance between the obstacles and the closest target, etc. The tactic of choosing the closest is the choice of the closest target. Rabbit tactics involve using a single UAV as a distraction. Swarm tactics involve dividing the swarm into groups or clusters to perform tasks. Circle tactics involve attempting to encircle a necessary area to effectively complete a task.

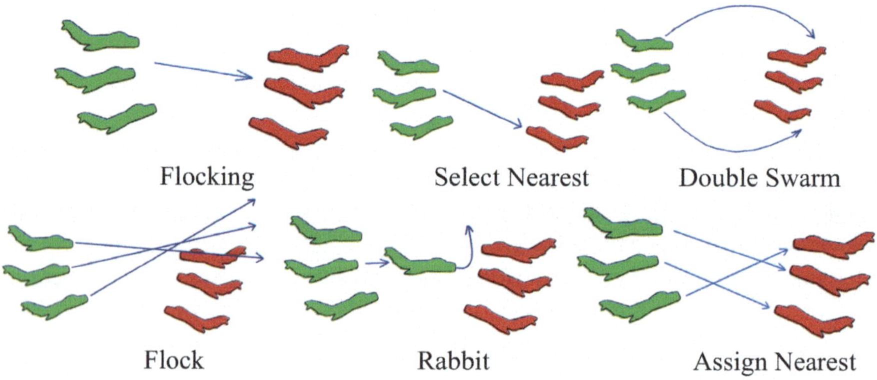

Fig. 3. Swarm tactics.

So, when modeling missions in particular territory, the area of the territory to be considered for the selected period of time was chosen as a parameter of monitoring efficiency. For this purpose, the simulation model realized the function of "coloring" the territory located in the ellipsoid of the local neighborhood for the observed UAVs (Fig. 4). After the end of the monitoring time (set by the user before the experiment), the area of the "colored" territory is calculated.

For modeling the movement towards a given object, it was found that the greatest effect is expected when using UAVs in the form of self-organizing groups (flocks, swarms, clusters, orders, etc.) with a variable degree of autonomy - in time and by control functions - from remote control to fully automated behavior.

For modeling the movement of a UAV swarm, several UAV models were placed on the stage and its original positions were adjusted. After adding the Ros interaction script and the control program modeling subscripts, the UAV team behavior simulation was launched (Fig. 5).

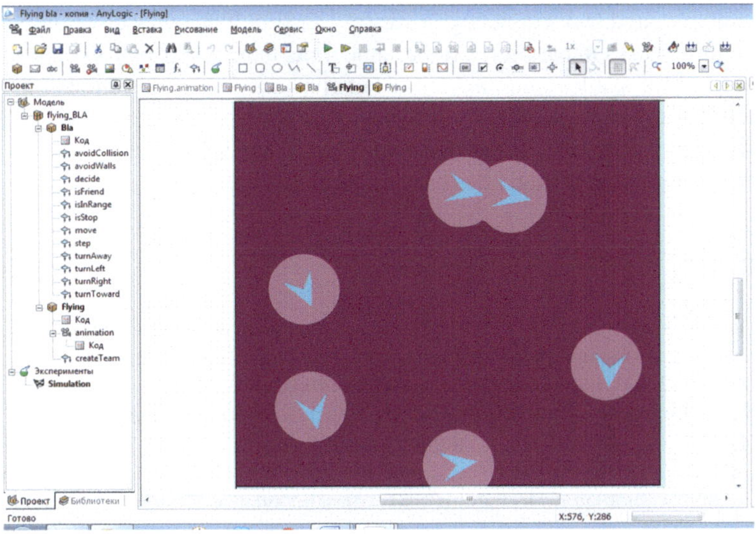

Fig. 4. Simulation model.

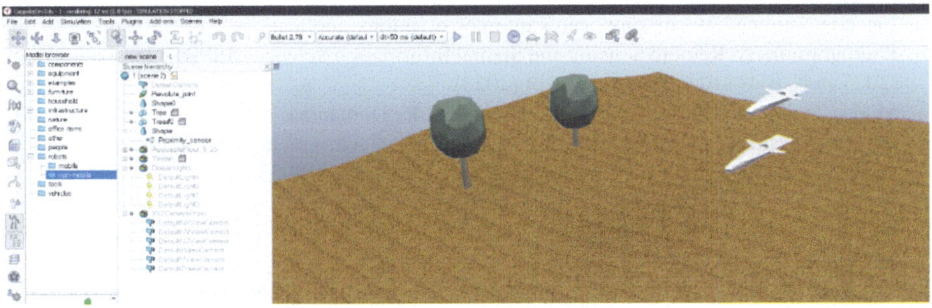

Fig. 5. Model of visualization of UAV flight according to the "master-slave" model.

5 Conclusions

UAV swarms are a powerful tool that opens up many new opportunities in a variety of industries. Thanks to decentralized control, UAV swarms can operate autonomously, performing tasks that require high precision, adaptability, and speed. The expansion of swarm systems and the integration of artificial intelligence enable new application scenarios that significantly improve the efficiency and safety of processes in industry, environmental protection, healthcare, and other areas. Each formation is suitable for its own tasks, determined by the level of autonomy, coordination requirements, and task complexity. For example, flocks or swarms are suitable for tasks that require flexibility and stability, while a group or column structure is more suitable for tasks with a clear hierarchy and direct control, such as perimeter protection or joint flights. Thus, the main advantages of using a swarm are scalability (the number of UAVs in a swarm can be easily increased or decreased), resilience (UAVs can compensate for the loss of several swarm members), and task efficiency (a swarm can simultaneously cover large areas or perform complex tasks in parallel). At the same time, there are also problems with using UAVs: the need for stable communication and constant data synchronization, the need for large computing power, energy consumption (duration of stay in the air). In future work, a comparative analysis of the effectiveness of using various tactics in carrying out tasks of protection, monitoring, attack, etc. will be carried out.

References

1. Mohsan, S.A.H., Othman, N.Q.H., Li, Y., et al.: Unmanned aerial vehicles (UAVs): practical aspects, applications, open challenges, security issues, and future trends. Intell. Serv. Robot. **16**, 109–137 (2023). https://doi.org/10.1007/s11370-022-00452-4
2. Hamdi, I.: Solving the cumulative capacitated vehicle routing problem with UAVs. J. Ind. Prod. Eng. **41**(4), 344–361 (2024). https://doi.org/10.1080/21681015.2024.2304570
3. Sładkowski, A., Kaminski, W.: Using unmanned aerial vehicles to solve some civil problems. In: Research Anthology on Reliability and Safety in Aviation Systems, Spacecraft, and Air Transport (2021)
4. Tahir, A., Boling, J., Haghbayan, M., Toivonen, H., Plosila, J.: Swarms of unmanned aerial vehicles - a survey. J. Ind. Inf. Integr. **16**, 106 (2019). https://doi.org/10.1016/j.jii.2019.100106
5. Sayed, O.H., Youssef, S.M., Ismail, O.M.: Swarm unmanned aerial vehicles (UAVs)-based fog computing platform supporting internet of things applications. ISPRS Annals of the Photogrammetry, Remote Sensing and Spatial Information Sciences, ISPRS Geospatial Week (2023), pp. 1041–10482, Cairo, Egypt (2023)
6. Marek, D., Paszkuta, M., Szygula, J., Biernacki, P., Domanski, A., Szczygiel, M., Krol, M., Wojciechowski, K.: Swarm of UAVs in a simulation environment-efficiency and adaptation. Appl. Sci. **14**(9), 3703 (2024). https://doi.org/10.3390/app14093703
7. Kumari, N., Lee, K., Barca, J.C., et al.: Towards reliable identification and tracking of UAVs within a swarm. J. Intell. Robot. Syst. **110**, 84 (2024). https://doi.org/10.1007/s10846-024-02115-1
8. Bu, Y., Yan, Y., Yang, Y.: Advancement challenges in UAV swarm formation control: a comprehensive review. UAVs. **8**(7), 320 (2024). https://doi.org/10.3390/UAVs8070320
9. Adoni, W., Fareedh, J., Lorenz, S., Gloaguen, R., Madriz, Y., Singh, A., Kuhne, T.: Intelligent swarm: concept, design and validation of self-organized UAVs based on leader–followers paradigm for autonomous mission planning. UAVs. **8**(10), 575 (2024). https://doi.org/10.3390/UAVs8100575
10. Gackowska-Katek, M., Cofta, P.: Explainable machine learning model of disorganisation in swarms of UAVs. Sci. Rep. **14**, 22519 (2024). https://doi.org/10.1038/s41598-024-73220-2
11. Jinkun, M., Chunmeng, Z.: An advanced cooperative multi-hive UAV swarm system for global dynamic multi-source information awareness. J. Ind. Inf. Integr. **40**, 608 (2024). https://doi.org/10.1016/j.jii.2024.100608
12. Yan, X., Chen, R., Jiang, Z.: UAV cluster mission planning strategy for area coverage tasks. Sensors. **23**(22), 9122 (2023). https://doi.org/10.3390/s23229122
13. Xia, X., Yan, Y., Li, Z., Meng, D., Xia, L., Chen, C.: A survey of UAV clustering algorithm. In: 24th International Conference on Advanced Communication Technology (ICACT), pp. 191–196, Republic of Korea (2022). https://doi.org/10.23919/ICACT53585.2022.9728826
14. Mehmood, A., Iqbal, Z., Shah, A.A., Maple, C., Lloret, J.: An intelligent cluster-based communication system for multi-unmanned aerial vehicles for searching and rescuing. Electronics. **12**(3), 607 (2023). https://doi.org/10.3390/electronics12030607

15. Kumari, N., Lee, K., Ranaweera, C., Barca, J.C.: Visually detecting UAVs in UAV swarm formations topologies. In: Ullah, A., Anwar, S., Calandra, D., Di Fuccio, R. (eds.) Proceedings of International Conference on Information Technology and Applications. ICITA 2022 Lecture Notes in Networks and Systems, vol. 839, pp. 21–30. Springer, Singapore (2024). https://doi.org/10.1007/978-981-99-8324-7_3
16. Pohudina, O., Bykov, A., Kritskiy, D., Kovalevskyi, M.: The method of flight mission formation for a group autonomous flight of unmanned aerial vehicles. In: Nechyporuk, M., Pavlikov, V., Kritskiy, D. (eds.) Integrated Computer Technologies in Mechanical Engineering – 2021. ICTM 2021 Lecture Notes in Networks and Systems, vol. 367, pp. 894–901. Springer, Cham (2022). https://doi.org/10.1007/978-3-030-94259-5_69
17. Pyvovar, M., Pohudina, O., Pohudin, A., Kritskaya, O.: Simulation of flight control of two UAVs based on the "Master-Slave" model. In: Nechyporuk, M., Pavlikov, V., Kritskiy, D. (eds.) Integrated Computer Technologies in Mechanical Engineering – 2021. ICTM 2021 Lecture Notes in Networks and Systems, vol. 367, pp. 902–907. Springer, Cham (2022). https://doi.org/10.1007/978-3-030-94259-5_70
18. Siemiatkowska, B., Stecz, W.: A framework for planning and execution of UAV swarm missions in a hostile environment. Sensors. **21**(12), 4150 (2021). https://doi.org/10.3390/s21124150

Visual Positioning of the UAV in Conditions of Impossibility of Using External Navigation Signals

Oleksii Hurtovyi, Kostiantyn Dergachov(✉), and Anatolii Yaremenko

National Aerospace University KhAI, Vadyma Manka Str., 17, Kharkiv 61070, Ukraine
k.dergachov@khai.edu

Abstract. The use of UAV is widespread both in industrial or amateur projects and adapted for many others missions over long distances. Usually, a large number of factors affect the development of a system or algorithm, and also complicates the processes of data acquisition, transmission and calculation, especially when there is no external navigation communication. This article presents an algorithm for identifying an object by a color mask, establishing the real distance and finding its spatial coordinates based on data obtained from the camera and onboard sensors of deviation angles. A description of the necessary components and functions for creating an algorithm for establishing the spatial X and Y coordinates of a detected object is also presented. Software architecture is designed in such a way that each block is autonomous and easily integrated. Taking into account the functional structure of the avionics equipment, the optimal approach to implementing the method was selected. The calculations are based on trigonometric equations and information about the state of vehicle. In addition, the algorithm was tested on a small quadcopter model with process optimization, which improved its autonomy and increased the accuracy of real-time calculations.

Keywords: Visual Navigation · GPS-Denied Environments · Computer Vision · Position Estimation · Real-Time Tracking · Sensor Fusion · Autonomous UAV Control System

1 Introduction

1.1 Relevance of the Topic

The need to accurately determine the coordinates of both landmarks and own coordinates is an urgent goal of many modern studies. Nowadays, much attention is paid to the study of the autonomous operation of Unmanned Aerial Vehicle (UAV) is becoming increasingly popular, especially the independent detection of objects, their tracking and calculation of data coming from reference systems, sensors, cameras, etc.

Among the tasks used by UAV is the complexity of calculating and implementing many tasks, which is due to a large number of uncertainty factors and external influences in the real conditions of its use.

I. Ostroumov et al. (Eds.): ACASD 2025, LNNS 1418, pp. 530–547, 2025.
https://doi.org/10.1007/978-3-031-91992-3_36

For many years, finding a landmark, tracking it and determining its parameters (coordinates, dimensions, etc.) has been a difficult problem due to the need to use significant resources and components for the appropriate data analysis and calculation. This paper presents a new approach that, with the help of on-board measurement of the distance to the fixed object in real time and based on the received data on the current angle of deviation from the axis of the starting (global) coordinates at the time of starting the engines, makes it much easier to find the X and Y coordinates of the landmark.

An algorithm is proposed that uses a basic set of functions of various libraries to detect static objects in space according to the specified color range and calculate first their distance and then their spatial coordinates. This approach provides the ability to save computer computing resources, so the range of use and price policy of this method is within the range of public access.

As proof of the system's performance, testing was conducted on a small-sized DJI Tello EDU model, and the features of the quadcopter's operation during algorithm execution, as well as the principle of interaction with it to receive and process data by the components of the built-in board, were presented [1]. Moreover, the development of the system was carried out taking into account many influencing factors and limitations of the on-board computer capabilities of the UAV model involved in the development.

1.2 Literature Review

The search for spatial coordinates of objects has quite a variety of implementation methods. Changing the current position of the landmark or object from which the data is recorded significantly affects accuracy of calculations.

One of the ways to find the position of an inconspicuous object is to create a projection using two independent centers (UAV) emitting radiowaves. Its intersection determines the position of the object in space. Kinematic design takes into account the features of the terrain (air, land, water and underwater space), and provides a display of an object moving in a Cartesian coordinate system which updates in real time [2, 3].

At work [4] localization is considered as a key task for autonomous robots. By combining a tracking camera and an RGB (red-green-blue) sensor with markers, high positioning accuracy is achieved. A dispersion filter is used to minimize errors. The system has been successfully tested on a UAV that flies autonomously indoors using an Nvidia Jetson Nano computer. This creates a strong foundation for autonomous UAV that can perform a variety of tasks without human intervention.

One of the most relevant areas of research is photogrammetry and 3D modeling. A large number of images obtained from the UAV camera are used for calculations. One of the main requirements for their creation is the overlap of the created images. With this approach, a detailed analysis of objects, their components, metric data, and orientation in space is achieved.

Calculation results when examining similar images may show a significant difference. This is caused by image fixation factors. Optical cameras have a different interpretation of the projection. Hence, some objects may be closer than others, although in the real world, this may not be the case. When working with a large number of images, there is a need to stitch them together to obtain a complete understanding of object and to avoid optical and geometric errors. Under the condition of equidistant of all details lying on each image, the greatest accuracy is achieved.

Such methods are used in agricultural mapping tasks, creation of digital flight models, and researched landmarks. At the same time, the cross-platform geographic information system QGIS is used for spatial data processing and analysis. On the basis of orthophotos, the layers of the terrain model, the flight path, and information about each stage of passage are created [5].

When exploring the unknown space, Simultaneous Localization and Mapping (SLAM) methods are implemented for the synchronous localization and tracking of the position of the aircraft and the algorithmically calculated task of building and updating map of unknown environment. Such methods do not prevent the quadcopter from drifting and error in coordinates which are increasing with time. Therefore, this approach is relevant only during a limited time and requires the presence of several cameras to establish the exact distance to the point or object. Also, it could initialize the map before localization. The last method aims to insert the first two keyframes, which define the origin of the coordinate system and its scale relative to the real world [6].

In order to prevent complex and at the same time resource-consuming experiments in real conditions, and to take into account the peculiarities of the development and testing of autonomous control algorithms, the AirSim simulator has been developed. Accelerating the development and research of projects, this system is quite accurately synchronized with the conditions that arise during real field tests. Among the key points of this development: realism of models; ability to configure scene components for testing; physical parameters of bodies and environments, as well as support for inertial measurement units (IMU), barometers, magnetometers [7], global positioning system (GPS) and other sensors with simulation of their errors and characteristics [8].

Many modern developments rely on the use of computer vision in their projects. This is one of the most relevant areas of research. Visual navigation is no exception. In the source [9] an overview of existing methods of map construction, obstacle avoidance, route planning, SLAM methods, optical flow, and GPS alternatives was conducted. Among the actual problems: autonomous avoidance of obstacles, accuracy of measurement, complexity of calculations, and their dependence on lighting.

Along with developments related to computer vision, there are projects using deep learning methods for autonomous navigation. This is how the system self-learns and calculates the value of depth of the map based on a sequential set of frames from the camera. A rather versatile study covering both integration of textural information RBG and scene geometry, and an attention mechanism method to process data at the level of individual patches while preserving spatial detail. Moreover, the application of the algorithm provides scale-consistent depth, which, with the help of regularization, minimizes errors and increases the efficiency of navigation [10].

A combination of autonomous positioning and visual technologies odometry is considered in [11]. The main difficulties in methods of fusion are distributed measurements in UAV cluster applications with a highly dynamic topology. The use of several UAV, SLAM methods, geoinformation data, the received video stream after the return of the Earth, as well as visual positioning technologies made it possible to research and create an information map with invariant features of terrestrial objects.

Autonomy UAV have been studied for a long time, but even nowadays solving the latest problems without human intervention in the execution process is relevant. There are many military and industrial tasks that have already been taken on by robotics and UAVs. Paper [12] illuminates cargo transportation strategy and trajectory planning based on ARUCO markers. These are 2D binary-encoded fiducial patterns designed to be quickly located by computer vision systems. Indeed, in this way, efficiency is achieved in terms of cargo delivery, but unfortunately, the markers are not attached to the enemy's equipment, which makes this methodology narrow-profile.

Improved interaction mechanics with UAV have also been achieved due to amateur and commercial areas of development. Hence, in [13] orientation and rapid response to flight situations without the use of GPS are described. The purpose of planning the UAV flight path for the AlphaPilot championship is based on finding the time-optimal path from the current position to the final gate. The system was successfully evaluated at the test site, where UAV successfully flew, with the gate not visible to UAV for most of the flight.

Machine learning shows significantly better performance when compared to systems that don't use it. To achieve the goal, the architecture of predicting visual attention, which is distributed over the RGB pixels of the image, is involved [14]. The results of the study showed a higher speed of learning and feedback of the system. The Kullback-Leibler difference and the Pearson correlation coefficient were used as metrics for visual forecasting.

In recent years, competition in areas of research related to unmanned technology has contributed to significant progress. Thus, the speed of movement of autonomous UAV in races has increased from 2 m/s to 22 m/s over the past 5 years, and at the same time, the quality and speed of processing information received from sensors and cameras has improved. In the study [15] methods of visual inertial odometry under difficult conditions (vibration, poor lighting), condition assessment, the architecture of a nonlinear control prediction model, machine learning, as well as approaches to minimizing failures, errors, and route travel time are presented.

Results of a comprehensive study of formation management of multi-agent robots in a dynamic environment is presented in [16], where testing took place on UAV and mobile manipulators. The availability of local and global planning of the system ensured the optimization of calculations and parameters taking into account limitations and avoiding obstacles. In addition, the construction of paths between the initial and target configuration through safe trajectories passing through static and moving obstacles and including narrow passages is considered.

Achieving stable operation of the algorithm in complex scenarios where there are significant difficulties for autonomous UAV navigation is described in [17]. The states of UAV behavior in the Markov Decision Process approach are considered, as well as the Soft Actor-Critic algorithm, which is combined with the concept of entropy and applied to balance between the maximum efficiency indicator and the randomness of the action strategy under different circumstances. The success of UAV navigation and response is due to use determine the position, velocity, or other characteristics of distant objects by analysis of pulsed laser light reflected from their surfaces, this is optical sensing technology (LiDAR data), as well as the features of the space of states and actions. That is, data on position, speed, corner dynamics, distance to the target, as well as linear and angular velocities are taken into account during calculations. Such a set of information helps adapt the UAV's behavior when avoiding obstacles and optimizing the route.

At work [18] the methods of generating the UAV flight path are designed and the comparison of the effectiveness of different path planning implementation architectures is demonstrated, as well as the memory consumption when using parallel calculations on different flight tracks. Visualization using software Airsim (open-source, cross platform simulator for UAV) and Unreal Engine (real-time 3D creation tool) contributed to the study of the performance of flight simulation algorithms. The development team presented opensource datasets for assessing the state of visual algorithms in extreme situations.

2 System Components

2.1 Camera Setup

Measuring distance to object requires getting an image from the camera and correctly reading information. To get these data, it is necessary to determine the parameters of the camera (focal length and dimensions of the output image which are height and width).

The focal length of the camera is one of the key characteristics of the optical system of camera, which can determine what size objects can have compared to reality and also set the amount of light on the matrix. This value can be obtained from the documentation of the camera being used, or after conducting your own calculations after calibrating camera using a special checkerboard.

To accurately calculate the focal length of a camera in the context of computer vision tasks, a camera calibration process is usually used. This approach requires a chessboard with known square sizes. It can be printed and glued to a flat surface. Usually, a chessboard has 9×6 squares. Next, several images (usually 10–20) of the chessboard are taken from different angles and from different distances. It is important that in all images the checkerboard is visible in its entirety. After that, code using special OpenCV functions, such as *cv2.calibrateCamera*, calculates the parameters of the camera's internal matrix, including the focal length [19].

Camera calibration process is not fast when it comes to instant reaction to events. Therefore, the presented material shows the result of development of an algorithm that works based on the previously specified width of the object and the actual parameters of the camera, which were taken from the technical documentation.

In this paper, the work on the graphical user interface (GUI) is put in the background, since the number of algorithm components can be changed during the work on the project, and the development of the interface is carried out after successful testing of the full software.

2.2 Image Processing

The main libraries used in the project are *NumPy* - a base for working with data arrays, and OpenCV, which provides tools for image and video processing and analysis. Programming is done in the high-level Python language.

NumPy (Numerical Python) is a library for working with multidimensional arrays and matrices. In the context of image and video processing. It is used to represent images in the form of pixel arrays, which allows easy data manipulation.

In digital form, an image is simply a matrix where each value is the color of a pixel. NumPy functions provide the capability to convert an image into an array (for example, using OpenCV) and perform various mathematical operations on pixels (filtering, brightness, contrast conversion, etc.). Also, the library is often used to create and process masks, filters, or other interactions with images, such as object selection, noise reduction, etc. [20].

OpenCV is a versatile opensource computer vision package used for image and video processing and object recognition. OpenCV provides powerful features for capturing, processing, and analyzing both images and video streams.

Image processing can include tasks such as color space conversion (for example, RGB to HSV (Hue-Saturation-Value), BGR to RGB, or inverse, and others), image smoothing, blurring, edge detection (for example, the Canny method), and much more. Switching from one color space to another, or swapping the sequence of channels, is a standard technical necessity due to the different color representation formats in different tools. Moreover, CV2 is capable of recognizing objects in images, including faces, geometric shapes and even tracking movements [21].

3 Functions and Workflow

3.1 Initialization Parameters and Input Values

For the smooth execution of this task, several global variables are specified at the beginning, which are used and dynamically changed during the operation of the full cycle of software. This is how the variable parameters characterizing the distance from the camera, the focal length, the number of recorded pixels, the real width of the object and some important constants, which are necessary when additional functions and conditions are involved, were created. By default, some variables are assigned zero values or None to avoid possible crashes, since most of the engine parameters will update their values when the video stream is started.

Tello quadcopter is equipped with sensors that measure and display its Height of flight, Pitch (slope or movement around transverse or lateral axis), Roll (motion around longitudinal axis), Yaw (rotation around vertical axis), Barometer, onboard Temperature, etc. Using the values obtained during the flight a corresponding function was created. Obtained functions return the yaw angle, engine activity time, battery charge value, and other parameters. To minimize computations on-board, some data collection functions have been suspended, as they are energy-consuming, which negatively affects the research process.

The yaw angle data acquisition system works in such a way that the direction of the quadcopter when first turned on before takeoff always has a zero value. It corresponds to the frontal direction of the UAV and the global Y-axis, which is the future reference system for any deviations and maneuvers.

3.2 Green Object Detection Function

Identification and processing of objects in the frame takes place in several color spaces. Most often it is RGB, BGR, HSV, binary-black, and white mask is also used. The difference between the first two color spaces in the sequence of channels occupying the corresponding priority area during their processing. This is directly related to the processes performed in the code using the OpenCV library. HSV is a color space corresponding to such parameters as hue, saturation, and brightness. These data support the precise determination of the contour sets of objects, for example, in comparison with BGR. Since each of the channels stores its own personal type of information, this provides the capability to more accurately distinguish the ranges necessary for determining the color and its contour, regardless of shadows and changes in lighting. In addition, the use of a black-and-white mask helps to establish the exact capture range of the desired color.

The exact setting of the range for each value lies within the following limits for Hue - it is from 0 to 180, Saturation and Value from 0 to 255. Since the conditions for setting several different colors and their shades can be used in the project, it is worth noting that the color ranges are quite wide can lead to incorrect fixation of the object on the video.

After determining the upper and lower limits for finding the desired color, the *inRange()* function is used, which builds a black-and-white mask. It highlights in white the middle of the found object, which corresponds to the given color, and in black - everything else outside. Finally, object is identified by the color mask.

The next step in the processing is noise removal using morphological operations. They are needed to fill possible holes and neutralize excess particles when fixing the actual dimensions of the mask. Thus, depending on the situation, the boundaries of white and black are increased or decreased for the most accurate filling and display of the contour of captured object.

The search and fixation of the largest value of the contour occurs after all the above operations. To increase the accuracy, an additional condition is applied indicating the extreme possible limits in which the contour should be fixed.

3.3 Contour Processing Function

The algorithm of *process_rect* block receives variable data with information about the color mask as input, namely its indexed pixel value of the dimensions. It is written into a tuple, which is an unchanging sequence of elements. Along with this, another parameter of type *np.ndarray* is passed to the input, which is an array that stores information about the frame and is used to avoid possible software crashes.

To prevent the appearance of artifacts, all necessary variables are first created. Next, there is a check for the presence of a color mask. In the case of its presence, a rectangle is created on the original image that covers the entire area captured by the color. It has the following basic parameters:

- center is coordinates of the centroid;
- size is dimensions of the rectangle (width, height);
- angle is angle of rotation of the rectangle.

Using these parameters and additional coding instructions, information about its highest fixed point, centroid, etc. initial scenario is established, and only then it is marked on the source image in the form of lines and points. The output parameters are the values of the coordinates of the centroid and the value of the highest captured point of the mask, which are entered into the corresponding variables.

3.4 Distance Calculation Function

The calculation function that determines the distance to the object works on the basis of an input parameter containing information about the pixel size of the fixed color mask on the display. This parameter is calculated in the corresponding blocks, where the mask is searched and its data is obtained using the *cv2.minAreaRect* instruction, which examines all possible contours in a predetermined range.

Finding both the contour value and the limits of the color range must be set manually, taking into account the artifacts that may appear when the limit values of the ranges are superimposed or when noises enter the first frames of the video stream. It may be accompanied by an incorrect display of pixels color that lie within the specified limits. This often happens at takeoff, when the camera additionally adjusts autofocus and exposure, or when moving from a less lit area to a more brightly lit area.

Thus, based on the calculated number of pixels occupied by the fixed mask, the distance between the object and the camera (D) is calculated:

$$D = \frac{Wf}{P}, \tag{1}$$

where W is the actual width of the object in real units; f is a focal length of the camera; P is the number of pixels occupied by the object in the image.

The fewer pixels an object occupies in the image, the further it is from the camera. Conversely, the more pixels, the closer the object.

The most accurate distance values are obtained, for the case of object width specified in the system. This value assumes that the size of the object in the world is known, and it is used to calculate the distance to the object using an equation based on the principle of similarity of triangles.

For objects of a special class, such as cars, airplanes, devices, or elements of the environment, an appropriate database must be created, according to which the obtained characteristics of the object are compared and identity is established using systems of technical vision and machine learning [22].

Since the creation of a database containing a large number of educational materials always entails a significant use of resources and the need for appropriate system parameters for calculation processes, when developing this approach, the research was based on the optimization of processes and availability, therefore, such methods were not used.

3.5 Principles of Work with Yaw Angle Calculation

The stage of finding coordinates X and Y to solve this task can be done in different ways. On the one hand, this is an approach through finding Δx and Δy. It is quite straightforward, does not require knowledge of angles and uses only the difference of coordinates. It can be convenient when you need to find the lengths of the sides, knowing the coordinates of the vertices. This method is based on the following equations:

$$\Delta x = | x_2 - x_1 |, \tag{2}$$

$$\Delta y = | y_2 - y_1 |, \tag{3}$$

where Δx is horizontal side; Δy is the vertical side; x, and y are the known coordinates of the vertices.

The method is fast when finding the values of the sides, but if it is necessary to calculate the angles, it uses the equation with the arctangent, which in turn can lead to results that are equal to infinity.

Instead, the method proposed in this article solves based on the known value of the hypotenuse, and the sine of the yaw angle (ψ) using the Pythagorean theorem:

$$\sin^2(\psi) + \cos^2(\psi) = 1. \tag{4}$$

Based on the on-board value of the rotation angle, a special function was developed based on the calculations of sine according to Eq. (4). Here are the conditions for obtaining value, based on the distribution of the values transmitted by the sensor [23]. So, at the input of the function, the deviation angle is given, and at the output, accordingly, the sine value will be obtained.

Figure 1 shows the quarters from which the value will be obtained according to the sign. When looking at the UAV from above before takeoff, its left and right sides of the turn are divided into ranges of values from 0 to 180, where the front and back left values have a "Minus" sign, and the front and back right side have a "Plus" sign.

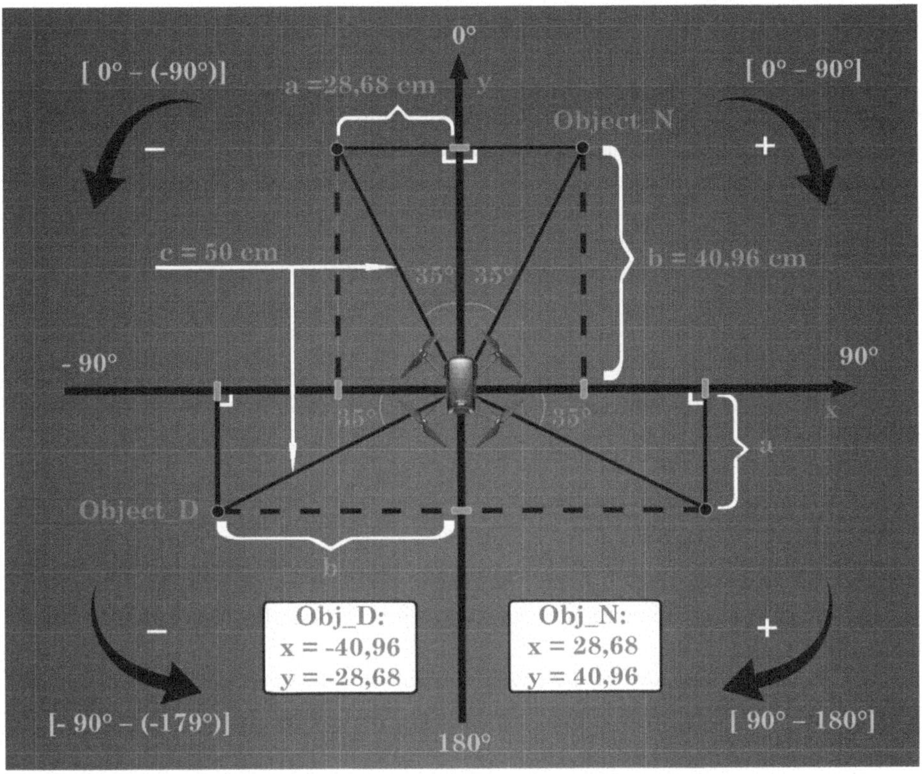

Fig. 1. DJI Tello rotation angles and changing side values.

Since trigonometric functions have periodic and symmetric properties for angles, in case of range from 0° to 90°, the quadcopter rotates in each direction by 180° and − 180° respectively. Angles ranging from 91° to 180° are calculated as follows:

$$\sin(180 - \psi) = \sin(\psi). \tag{5}$$

Calculation of equation works as a reflection of angle values.

3.6 Spatial Coordinates Calculation

The main task is a geometric problem with an initial data set of a triangle. The data set contains the length of the hypotenuse which is a distance to detected object, and yaw (the angle between the hypotenuse and the adjacent leg), on which the other unknown components are determined.

After obtaining the value of angle when turning from control axis to the landmark, as well as the length of the hypotenuse, it becomes possible to get the X and Y of the fixed object in space.

Since the object is captured using an image, there is a difference in the coordinates obtained using cv2, which works with the image using it as a 2D plane. In our study we determine the coordinates of object in 3D space, based on data obtained from onboard sensors and an image from a camera.

Having the exact value of the yaw angle and using that sum of all angles in triangle is equal to 180°, it is possible to find unknown angle. This value can be used both for the development of future calculations related to the processing of data on location of the landmark relative to the UAV, and for the construction of new and more accurate flight routes when applying spatial orientation from the involvement of a larger amount of data obtained from various objects.

The algorithm for finding spatial coordinates X and Y for a fixed object works based on trigonometry. The first leg is measured using the sine of the angle of rotation and the hypotenuse.

Let's present the problem visually. Given: Hypotenuse (D) as result of Eq. (1) and adjacent leg (*b*) and opposite leg (*b*). The yaw angle of the UAV relative to the detected object in modern time ψ_t, which is described as:

$$\psi_t = \psi_0 + \int_{t_0}^{t} \omega(\tau)d\tau, \tag{6}$$

where ψ_0 is initial yaw angle; $\omega(\tau)$ is instantaneous value of angular velocity measured by a gyroscope; $d\tau$ is time increment used in integration; t and t_0 are current and initial time respectively.

In order to find the sides of the triangle, which are the coordinates of the detected object, we use (5) to obtain value of $sin(\psi)$. Also, the results (1) and (6) were integrated into the trigonometric equation to obtain the opposite leg:

$$\mathrm{a} = \mathrm{Dsin}(\psi), \tag{7}$$

$$b = \sqrt{D^2 - a^2}. \tag{8}$$

Calculation of the length of the sides of the triangle gives the idea of the distance of each of the coordinates.

In order to run this algorithm on the DJI Tello quadcopter, a special approach was used based on certain conditions, enabling precise adjustment of the lengths according to the angle of rotation, using the method of screening out the excess. The calculated coordinates of the fixed objects (sides a and *b*) in different quarters do not have the same position relative to the yaw angle, which is carried out from the control axis. Thus, there is a difference in finding the correct coordinate value for the corresponding X or Y axis.

The algorithm uses the coordinate values that are inverted according to the quarter in which the landmark is detected when the rotation angle becomes greater than 90°. This approach simplifies calculations and increases system efficiency.

In the case of method integration for the UAV DJI Tello EDU, an additional splitting of yaw ranges from 0° to 180° in each direction was performed. Thus, 4 quarters are obtained and the impossibility of obtaining a turning angle of more than 90°. Thus, with the help of the screening out excess method, the obtained value is checked for compliance with the range in which the object is located.

4 Implementation of the Algorithm

4.1 UAV DJI Tello EDU

Expanding the range of functions using Python and testing the algorithm took place on a small model of quadcopter. DJI Tello EDU is a programmable UAV. With the help of its on-board equipment, position, angles, and engine data were obtained. The 5-megapixel camera captures the object and calculates the distance to it.

During software development, many points related to the specifics of working on this model were taken into account. Thus, the use of an excessive number of sensors negatively affects the state of the battery charge, and the large volume of software code forces UAV to skip calculation operations of some incoming frames. Considering such moments, the program code was optimized, and the set of functions was reduced to the most necessary in order to get the most accurate result with stable loads on the UAV board.

As a result of the work, software was obtained that in real time finds an object based on a color mask, calculates the distance to it, and then finds the spatial coordinates X and Y of its centroid.

4.2 The Real-Time Navigation Algorithm

The sequence of components has a clear logic, and the algorithm itself is a weaving of input and output parameters from one function to another. The development begins with the initialization of the basic values, and during testing, the received data on the system status and information about displacement in space are involved in the course of work.

Each stage of the algorithm is developed in separate blocks, for better optimization and interaction at the stages of upgrading individual elements of the system. A software block is a function with a specific set of instructions. Depending on the task, it performs image processing or calculations.

A description of the process of the X and Y coordinate research method according to the Unified Modeling Language (UML activity diagram) is presented in Figs. 2 and 3. Among the presented functions: *Green_mask*, *Process_rect*, *Get_dist*, *Get_sine*, *Location_XY*. The principle of interaction consists of the sequential acquisition of necessary values, and at the moment when they are sufficient to proceed operations.

The system works with a video stream, which is a sequential sequence of frames, each of which is processed in real-time in several stages. The sequence of functions that is indicated is also implemented by calls in an infinite «while True» loop, which will be executed while code is running. Thus, while hovering in the air, UAV waits for both new control commands to set the yaw angle and to detect the object of green color in the frame.

Image preparation before computation is necessary to obtain the most information for each frame. The higher precision of captured object size in the frame gives better accuracy of coordinates and distance measurement.

Some blocks of code can be created in separate calculation functions and used only when it is necessary to call it in the main one. A good example of such a combination of components is the function *get_sine()* and *location_XY()*. The *get_sine()* is a separate

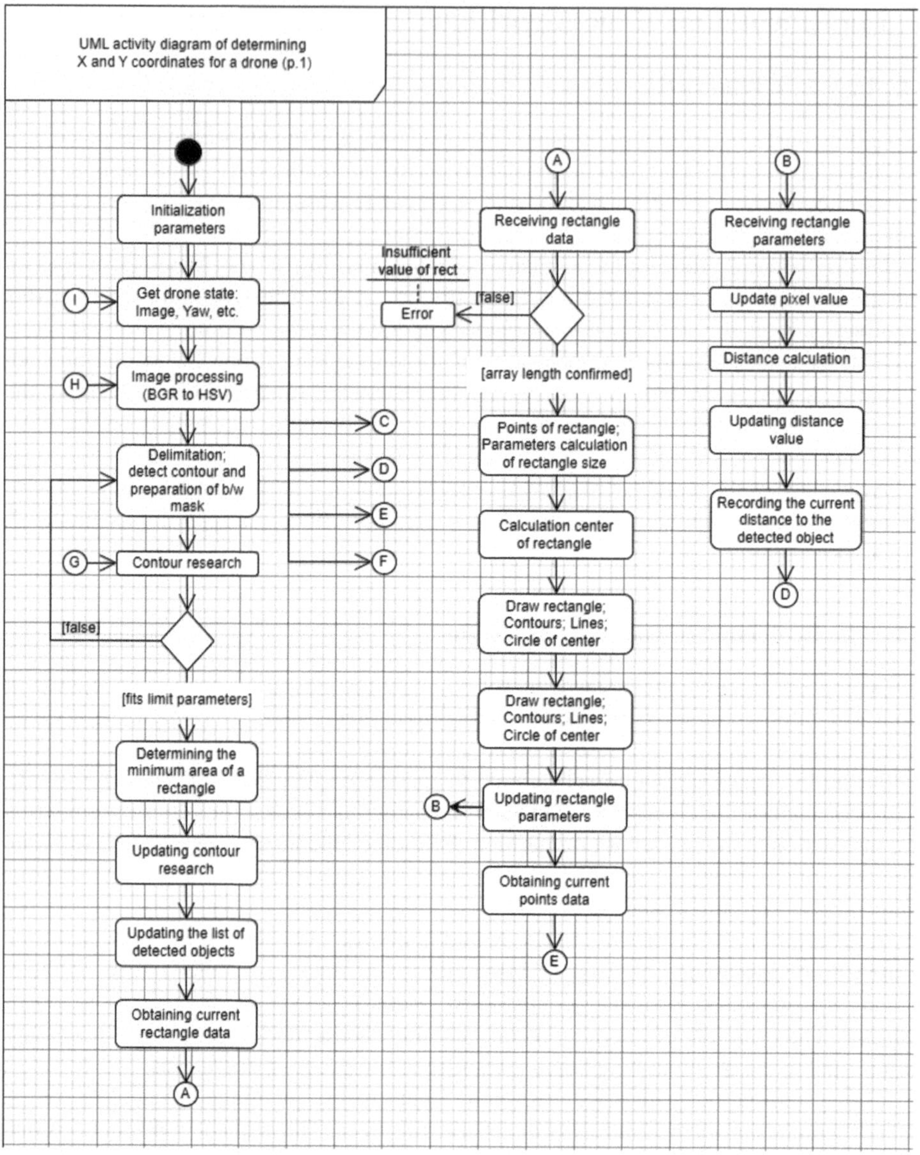

Fig. 2. UML diagram and components of the algorithm (part 1).

autonomous unit created for memory optimization and is needed in this algorithm only in the case of using the function of setting object coordinates. Moreover, this approach significantly optimizes the system [24], because when new stages are created in future development, the output parameter of the sine of the angle will be called by the name of corresponding function without copying a significant number of lines of code.

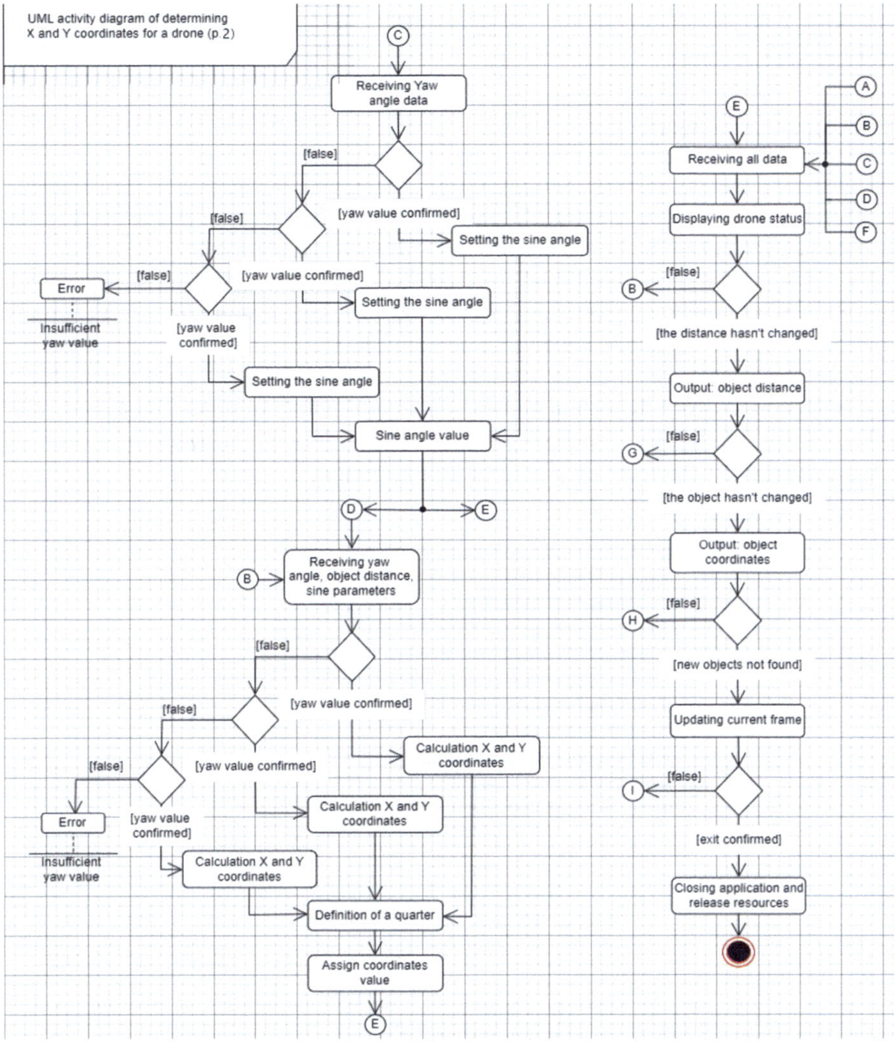

Fig. 3. UML diagram and components of the algorithm (part 2).

The similarity of some blocks is due to the selection of the best working options according to the model for testing. Determination and constant verification of the necessary parameters take place in several stages. Among the main methods: matching and screening out excess. The first method implements the analogy of values that are necessary for calculations, while the second is responsible for demonstrating the correct location of the object in space.

The study showed that in order to achieve the highest accuracy in calculations, it is necessary to use both more powerful sensor systems, including an on-board camera, and computation hardware. The difference in calculations becomes significant when it comes to working with personal computer video recording and real-time video streaming on board the UAV.

4.3 Comparative Analysis

To determine accuracy of proposed algorithm, the actual distances and coordinates of the recorded objects were calculated, taken during real measurements, as well as data sets obtained for X and Y coordinates during software calculations.

Based on the data sets, additional software code was developed that allows calculations of the programmatically obtained coordinate values universally. That is, the number of indicators can be expanded depending on the task.

Figure 4 demonstrates how the recorded values of the X and Y coordinates for the green object differ from the actual values at different points in time. The actual values for X is 35.91 sm, and for Y is 98.668 sm. The diagram demonstrates the values that were recorded every 5 s from the received video stream.

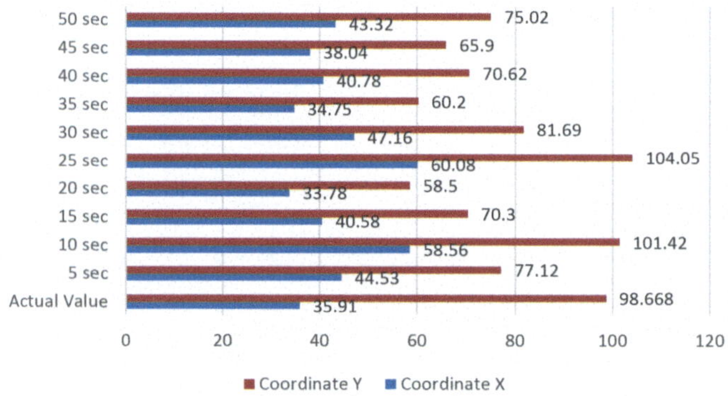

Fig. 4. Green object coordinates.

Figure 5 illustrates the absolute and relative error dependence for averaged sets of X and Y coordinates across repeated calculation cycles.

Calculations show that at the beginning of image (object) fixation, the error is quite large, due to the fact that the matrix and camera focus adapt to lighting or shaking of the UAV camera. As indicated, over time the error approaches the norm with constant updating of coordinate calculations.

Moreover, based on data analytics, a table was constructed containing indicators:

Mean absolute (MAV) and relative (MRV) errors for each set of X and Y coordinates, standard deviation (DV), sum squared deviation (SSqDev), as well as the average value of the actual coordinates (AV_ XY) and the average value of the programmatically calculated coordinates (AMV_XY) (Table 1).

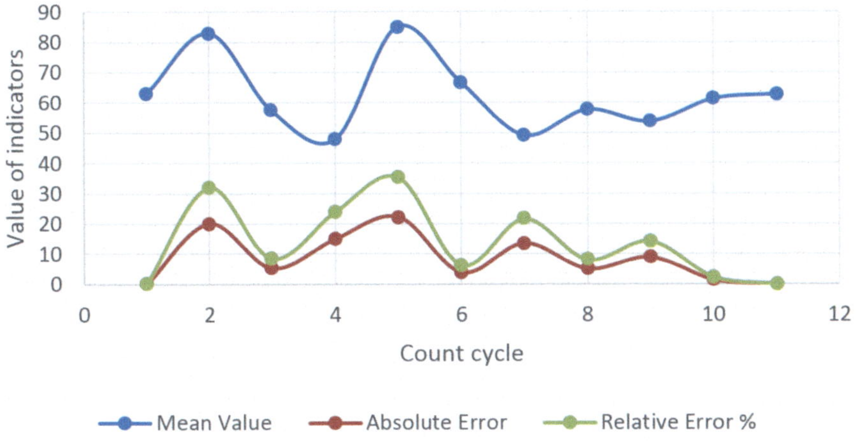

Fig. 5. Dependence of error values and averaged coordinates on time.

Table 1. Averaging, deviation and error indicators.

MAV X	MAV Y	MRV X %	MRV Y %	SSq Dev	AV_ XY	AMV_XY	DV
8.906	23.81	24.8	24.13	37.91	62.75	62.44	12.71

Thus, the results of the research shows that algorithm of establishing coordinates X and Y of object based on data obtained from the UAV camera provides approximate accuracy to the actual indicators. Of course, due to a number of external influences, artifacts may influence calculations, so the method should be used in combination with other approximation algorithms to achieve better accuracy. In addition, among the main advantages of the method is its ease of implementation, as well as transparency in comparing the obtained indicators with the actual ones. The successful operation of this algorithm permits exact establish the Z coordinates for the object detected from the camera, which opens up new opportunities in the study of visual navigation methods.

5 Conclusions

The rapid development of technologies for the development of UAV equipment and software has made it possible to widely apply object recognition methods and its calculations to solve navigation problems. Visual navigation is one of the most urgent tasks, because the accuracy of the UAV is necessary, both for the performance of the task and for its return to the starting point.

Solving the task of finding spatial coordinates X and Y provides a basis for designing a system for calculating the Z coordinate based on trigonometry, but in a different projection.

In turn, the acquisition of all coordinates of an object by visual observation opens up the prospect of establishing one's current coordinates based on the coordinates of several fixed objects. Also, in conditions of uncertainty, situations of partial or complete loss of communication with the operator, the method can be used to independently lay the return route.

Testing of such algorithms require several successive stages to pass, such as rendering, and checks to avoid possible artifacts in the code or inaccuracies in calculations. In the development process, the features of both the software components and the functionality of the UAV model are taken into account in order to optimize the processes. The principle of implementation of this method using the Tello UAV is presented for a visual demonstration that such a calculation method can be embedded in various management systems, regardless of their specifics of work and pricing policy.

References

1. Ghazi, G., Voyer, J.: Use of a DJI Tello drone as an educational platform in the field of control engineering. In: Proceedings of the Canadian Engineering Education Association Conference (CEEA 2024), Paper No. 32, pp. 1–8. (2024). https://doi.org/10.24908/pceea.2023.17061
2. Svidrak, I., Aftanaziv, I., Shevchuk, L., Strogan, O.: Determination of coordinates of unmanned aircraft by means of kinematic projection. Math. Model. Comput. **9**(2), 459–469 (2022). https://doi.org/10.23939/mmc2022.02.459
3. Popov, A., Tserne, E., Volosyuk, V., Zhyla, S., Pavlikov, V., Ruzhentsev, N.: Invariant polarization signatures for recognition of hydrometeors by airborne weather radars. In: Computational Science and Its Applications – ICCSA 2023 Lecture Notes in Computer Science, vol. 13956, pp. 201–217. Springer, Cham (2023). https://doi.org/10.1007/978-3-031-36805-9_14
4. Mraz, E., Rodina, J., Babinec, A.: Using fiducial markers to improve localization of a drone. In: 23rd International Symposium on Measurement and Control in Robotics (ISMCR), pp. 1–5, Budapest, Hungary (2020). https://doi.org/10.1109/ISMCR51255.2020.9263754
5. Barna, R., Solymosi, K., Stettner, E.: Mathematical analysis of drone flight path. J. Agric. Inform. (JAI). **10**(2), 533 (2019). https://doi.org/10.17700/jai.2019.10.2.533
6. Skoda, J., Bartak, R.: Camera-based localization and stabilization of a flying drone. In: The 28 International Florida Artificial Intelligence Research Society Conference (FLAIRS 2015), pp. 372–377 (2015). Available at: http://www.aaai.org/ocs/index.php/FLAIRS/FLAIRS15/paper/view/10398
7. Nikitina, T., Kuznetsov, B., Averyanova, Y., Sushchenko, O., Ostroumov, I.: Method for design of magnetic field active silencing system based on robust meta model. In: Shukla, S., Sayama, H., Kureethara, J.V., Mishra, D.K. (eds.) Data Science and Security. IDSCS 2023 Lecture Notes in Networks and Systems, vol. 922, pp. 103–111. Springer, Singapore (2024). https://doi.org/10.1007/978-981-97-0975-5_9
8. Shah, S., Dey, D., Lovett, C., Kapoor, A.: AirSim: high-fidelity visual and physical simulation for autonomous vehicles. In: Hutter, M., Siegwart, R. (eds.) Field and Service Robotics Springer Proceedings in Advanced Robotics, vol. 5, pp. 1–14. Springer, Cham (2018). https://doi.org/10.1007/978-3-319-67361-5_40K
9. Arafat, M., Alam, M., Moh, S.: Vision-based navigation techniques for unmanned aerial vehicles: review and challenges. Drones. **7**(2), 89 (2023). https://doi.org/10.3390/drones7020089
10. Tang, T., Du, H., Yu, X., Yang, Y.: Monocular camera-based point-goal navigation by learning depth channel and cross-modality pyramid fusion. Proc. AAAI Conf. Artif. Intell. **36**(5), 5422–5430 (2022). https://doi.org/10.1609/aaai.v36i5.20480

11. Tong, P., Yang, X., Yang, Y., Liu, W., Wu, P.: Multi-UAV collaborative absolute vision positioning and navigation: a survey and discussion. Drones. **7**(4), 261 (2023). https://doi.org/10.3390/drones7040261
12. Miranda, V., Rezende, A., Rocha, T.: Autonomous navigation system for a delivery drone. J. Control Autom. Electr. Syst. **33**, 141–155 (2022). https://doi.org/10.1007/s40313-021-00828-4
13. Foehn, P., Brescianini, D., Kaufmann, E.: AlphaPilot: autonomous drone racing. Auton. Robot. **46**, 307–320 (2022). https://doi.org/10.1007/s10514-021-10011-y
14. Pfeiffer, C., Wengeler, S., Loquercio, A., Scaramuzza, D.: Visual attention prediction improves performance of autonomous drone racing agents. PLoS One. **17**(3), e0264471 (2022). https://doi.org/10.1371/journal.pone.0264471
15. Hanover, D.: Autonomous drone racing: a survey. IEEE Trans. Robot. **40**, 3044–3067 (2024). https://doi.org/10.1109/TRO.2024.3400838
16. Alonso-Mora, J., Baker, S., Rus, D.: Multi-robot formation control and object transport in dynamic environments via constrained optimization. Int. J. Robot. Res. **36**(9), 1000–1021 (2017). https://doi.org/10.1177/0278364917719333
17. Sheng, Y., Liu, H., Li, J., Han, Q.: UAV autonomous navigation based on deep reinforcement learning in highly dynamic and high-density environments. Drones. **8**, 516 (2024). https://doi.org/10.3390/drones8090516
18. Han, Z., Wang, Z., Pan, N., Lin, Y., Xu, C., Gao, F.: Fast-racing: an open-source strong baseline for SE(3) planning in autonomous drone racing. IEEE Robot. Autom. Lett. **6**(4), 8631–8638 (2021). https://doi.org/10.1109/LRA.2021.3113976
19. Zhang, Y.J.: Camera calibration. In: 3-D Computer Vision, vol. 62, pp. 37–65 (2023). https://doi.org/10.1007/978-981-19-7580-6_2
20. Nelli, F.: The NumPy library. In: Python Data Analytics, pp. 18–147. Apress, Berkeley (2023). https://doi.org/10.1007/978-1-4842-9532-8_3
21. Rakesh, V., Chilukuri, P., Vaishnavi, P., Sreekaran, P., Sujala, P., Yadav, K.: Real time object recognition using OpenCV and Numpy in Python. In: 2023 International Conference on Innovative Data Communication Technologies and Application (ICIDCA), pp. 421–426, Uttarakhand, India (2023). https://doi.org/10.1109/ICIDCA56705.2023.10099584
22. Amjoud, A.B., Amrouch, M.: Object detection using deep learning, CNNs and vision transformers: a review. IEEE Access. **11**, 35479–35516 (2023). https://doi.org/10.1109/ACCESS.2023.3266093
23. Dergachov, K., Havrylenko, O., Pavlikov, V., Zhyla, S., Tserne, E., Volosyuk, V.: GPS usage analysis for angular orientation practical tasks solving. In: IEEE 9th International Conference on Problems of Infocommunications, Science and Technology (PIC S&T), pp. 187–192. IEEE, Kharkiv (2022). https://doi.org/10.1109/PICST57299.2022.10238629
24. Nikitina, T., Kuznetsov, B., Ruzhentsev, N., Havrylenko, O., Dergachov, K., Volosyuk, V., Shmatko, O., Popov, A.: Algorithm of robust control for multi-stand rolling mill strip based on stochastic multi-swarm multi-agent optimization. In: Shukla, S., Sayama, H., Kureethara, J.V., Mishra, D.K. (eds.) Data Science and Security. IDSCS 2023 Lecture Notes in Networks and Systems, vol. 922, pp. 247–255. Springer, Singapore (2024). https://doi.org/10.1007/978-981-97-0975-5_22

Simulation Platform for Testing and Validating of UAV Visual Guidance Algorithms

Kostiantyn Dergachov(✉), Vladyslav Dubinin, Eugene Ovdiyuk, Iryna Bychkova, and Denys Puhach

National Aerospace University KhAI, Vadyma Manka Str., 17, Kharkiv 61070, Ukraine
k.dergachov@khai.edu

Abstract. Autonomous control systems development of aerial vehicles is one of the promising directions of modern unmanned aviation. One of the complicated factor of Unmanned aerial vehicles development is the need to test the system in various environmental conditions, such as insufficient or excessive lighting, increased cloud cover, and other factors that impact the performance of control algorithms. Therefore, the development of tools for flight test simulation is a crucial scientific task to significantly enhance the efficiency of guidance system development. This paper describes a system based on Unreal Engine and the OpenCV library, which allows for the simulation of field tests for UAV guidance algorithms targeting a specified objective. This process involves synthesizing information about the virtual environment (video data from a virtual onboard camera and virtual sensors) and providing it as input signals to the control algorithm. Based on the algorithm's performance, reports are generated to evaluate its effectiveness, enabling further optimization and the elimination of deficiencies.

Keywords: Simulation System · Virtual Environment · Guidance · Autonomous UAV Control System

1 Introduction

Market of unmanned aerial vehicles (UAVs) has been experiencing incredible growth over the last 10 years. They are widely used in many different industries such as agriculture, logistics, monitoring, media, mapping, and others. The rapid development of this market requires manufacturers to offer new technical solutions to remain competitive. One such solution is automated UAV control systems for target tracking. These systems show their effectiveness in challenging conditions of UAV use, such as rescue operations in areas with no GPS or obstacles to signal transmission.

Such systems are usually quite inexpensive and simple from a technical standpoint, yet their development requires significant expenditures due to iterative development methods. In other words, each addition or modification to the system requires repetitive testing, linked to the complexity of the task and the presence of numerous factors whose overall impact on the system is hard to predict. Based on this, damage or complete destruction of the UAV during the entire development cycle is inevitable.

I. Ostroumov et al. (Eds.): ACASD 2025, LNNS 1418, pp. 548–565, 2025.
https://doi.org/10.1007/978-3-031-91992-3_37

Moreover, the total number of such tests requires a substantial amount of time and additional involvement of specialized professionals. Another complicating factor is the high demand for specific environmental conditions during testing, such as insufficient or excessive lighting, dense cloud cover, etc. [1].

Given the large number of such iterations throughout the entire development cycle and the significant material costs associated with testing, it becomes practical to develop a simulation system. Such a system can replace the need for real-world testing with its virtual equivalent until the final product testing stage.

Therefore, the development of flight-testing simulation tools is a critically important scientific task for significantly enhancing the efficiency of autonomous control system development for UAVs and substantially reducing its cost. One possible solution to this task is the creation of a system for simulating a virtual environment, in which target tracking algorithms will be tested [2]. There are many UAV flight simulators software available, however AirSim and Liftoff are the most frequently used.

AirSim is an open-source system based on the Unreal Engine engine and was developed for working with artificial intelligence. It offers a detailed 12-kilometer-long city street map. The system includes sensor simulation for devices such as a barometer, gyroscope, accelerometer, magnetometer, and GPS [3]. Additionally, the system generates a depth map of objects and an object segmentation map. An important advantage is the ability to adjust weather conditions for flights.

Liftoff is built on Unreal Engine and is used for pilot training. The system includes a set of detailed flight tracks and a large variety of drone models. The program provides a highly accurate physical model of drones. In addition, the simulator allows users to complete tracks for speed and race mode. It supports a wide range of controllers for operation.

2 Requirements Specification

The primary objective of the system is to optimize the development of UAV control algorithms by replacing the traditional testing approach with a virtual analog. The system should fully implement the "brain in a vat" [4] approach, in which the control system's sensors receive signals that simulate real operating conditions. The system should also track feedback from the test object and respond validly by altering the simulation signals affecting it.

The following requirements for the simulation system are identified:

- Generation of visual data about the surrounding environment (simulation of an onboard camera).
- Generation of additional information (simulation of onboard sensors).
- Control of the conditions under which tests are conducted.
- Presence of a controlled UAV model with accurate response to signals from autonomous control algorithms.
- Simulation of a moving target.
- Evaluation of the results of the autonomous control algorithm's performance.

3 Choice of Tools

Due to the main source of information for the test object is the onboard camera, it can be concluded that the simulation system should generate a highly realistic visual representation of the surrounding environment and changes in the state of objects within it. Moreover, the system should provide an accurate physical simulation to implement the UAV model and generate data from onboard sensors.

The OpenCV [5] library was selected for the implementation and processing of video streams. This decision was driven by the availability of a comprehensive API and numerous ready-to-use efficient algorithms. Moreover, the library has a vast knowledge base and an active community that regularly contributes to it. It is also free.

Based on these needs, the decision was made to use Unreal Engine [6]. It is a powerful, flexible tool that facilitates the creation of a detailed environment with a realistic physical model, offering cross-platform capabilities and various control interfaces. Additionally, it features an active community and comprehensive documentation and is also available free of charge.

C++ was selected as the development language as it is the foundational language of Unreal Engine and is compatible with OpenCV.

Python was selected as the language for writing autonomous UAV control algorithms due to its simplicity, the availability of the OpenCV library, and the ease of integration with Unreal Engine. Python does not require additional compilation, making it a convenient tool for writing and testing new algorithms.

4 Platform Architecture

The interaction between the system modules is presented in Fig. 1.

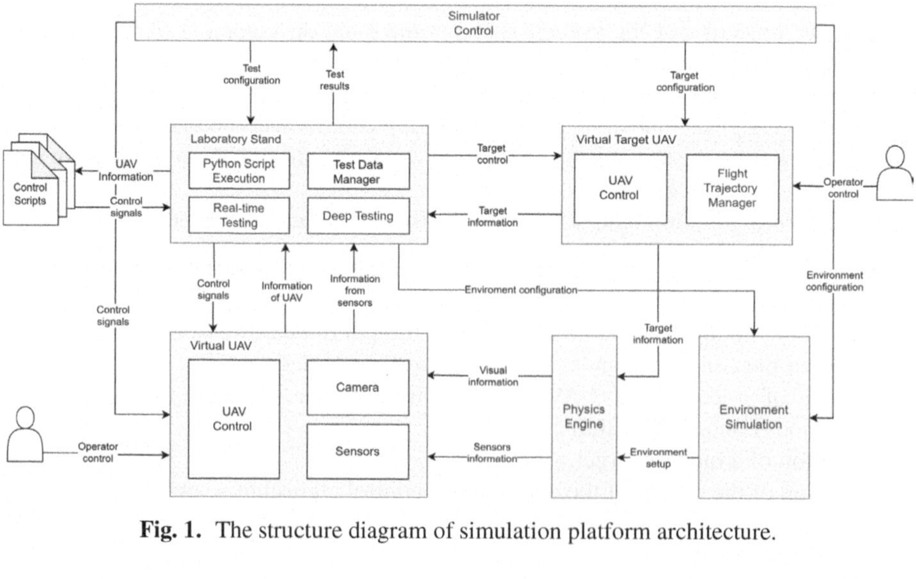

Fig. 1. The structure diagram of simulation platform architecture.

The simulation system is a sophisticated structure that consists of several interconnected blocks [6]:

- Environment Simulation Block. Creates a virtual three-dimensional space for flights and performs all fundamental environmental calculations.
- Virtual UAV Block. Describes the physical model of the UAV.
- Virtual Target Block. Responsible for simulating the behavior of a UAV serving as a virtual target.
- Laboratory Stand Block. Responsible for evaluating the effectiveness of the autonomous control algorithm.
- Simulator Control Block. Responsible for initiating appropriate scenarios, such as recording the target flight trajectory, testing, and configuring the system settings.

4.1 Environment Simulation

The environment simulation module is responsible for managing the surroundings that serve as the environment for the virtual UAV [7].

A wide range of tools enables the detailed customization of factors such as lighting, time of day, star brightness, cloudiness, fog, and more. These factors significantly influence the resulting image from the drone's virtual camera and make it easy to achieve various conditions for testing algorithms [8]. Examples of environments configured with different settings are presented in Fig. 2.

Fig. 2. Different environment simulation settings.

Another important task of this module is the generation of various types of landscapes. The landscape is an integral component of the UAV camera image and serves as a significant complicating factor for target recognition due to the presence of additional objects.

The system supports the use of different types of landscapes for various testing conditions. For low-altitude flight tests, landscape generation is based on sets of static 3D assets, including buildings and vegetation. For high-altitude tests, the landscape consists of textures based on high-resolution aerial photographs of the ground taken from flight altitudes [9]. Examples of different landscape types are presented in Fig. 3.

Landscape generation is performed by creating a virtual grid and filling its cells with instances of object sets, where each set has a width and length corresponding to a single grid cell.

Fig. 3. Different landscape generation settings.

To calculate the potential size of the map, the center of the map is determined. This point corresponds to the x and y coordinates of the controlled drone camera's position and a constant z coordinate set to 0. This positioning ensures the map's center follows the drone's movement while consistently remaining at ground level, regardless of its movement. From this point, the distance to the visible horizon is calculated. This distance defines the radius of the future landscape.

The drone's camera has a wide field of view, a long visibility range, and a high movement speed during flight. Rendering a high-quality landscape under such conditions requires creating an enormous number of objects. This process is highly resource-intensive, so the algorithm requires additional optimization. To address this issue, a method called frustum culling was used [10]. This approach ensures that objects outside the camera's field of view are not generated. It significantly improves performance by avoiding computations for 3D objects that are not visible.

To implement this approach, the portion of the circular map corresponding to the UAV camera's field of view was isolated. This portion represents the grid sector that must be populated with objects.

As the UAV and its camera move, the position of this sector shifts accordingly. In each frame, objects that move out of the sector are destroyed, while missing objects are created to fill the empty grid cells. Objects that remain visible between frames are left unchanged. The results of this algorithm are presented in Fig. 4.

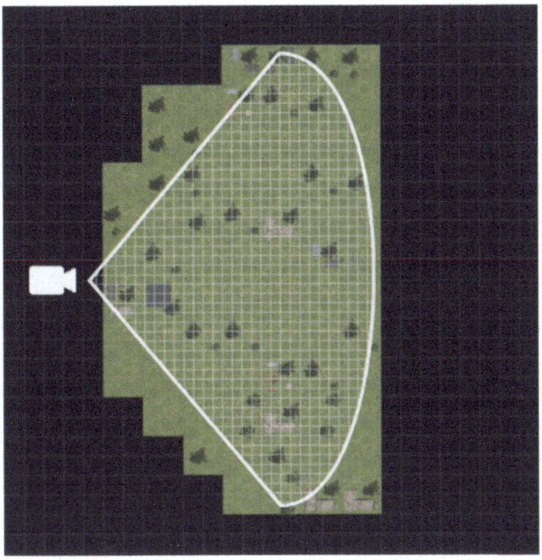

Fig. 4. Result of landscape generation algorithm.

4.2 Virtual UAV

The virtual UAV module acts as the main link between the UAV control system and the simulation system. Essentially, this module fulfills two key functions: simulating the onboard camera and sensors of the UAV and controlling the UAV in the virtual environment [11]. Its structure is described in Fig. 5.

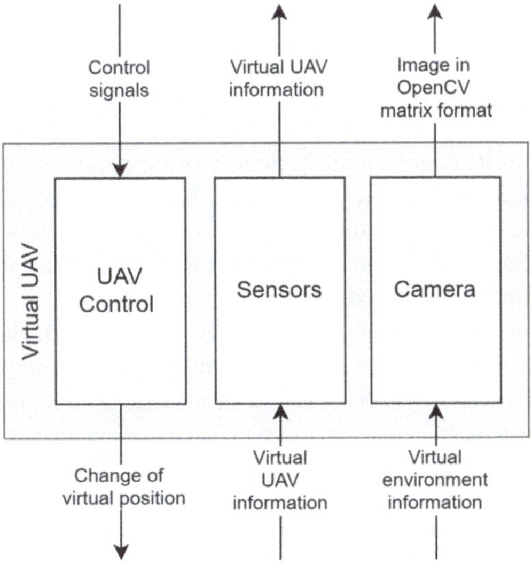

Fig. 5. General diagram of the "Virtual UAV" block.

UAV Control. The Control Module is responsible for receiving control signals and converting them into UAV movement within the virtual environment. During each simulation cycle, the module reads input data and processes it, considering the UAV's current position, speed, and movement vector in space. The result of this processing is the UAV's movement in three-dimensional space within the simulation cycle [12].

Let's examine the working principle of this module in more detail. The Control Module operates based on the following input parameters:

- Thrust – a numerical value of the applied thrust.
- Yaw – a numerical value representing movement along the yaw axis.
- Pitch – a numerical value representing movement along the pitch axis.
- Roll – a numerical value representing movement along the roll axis.

During each simulation cycle, the Control Module performs calculations of the UAV's position based on these parameters in virtual space [13].

Virtual Camera. The images from the camera are the primary source of information for the UAV control system regarding the current position of the aircraft and the target in space. Replacing this data with test data generated by the simulation system is a key point in the system's implementation.

In simplification, the primary task of this block is to solve a chain of the following tasks: obtaining information from the virtual camera in the graphics engine, converting this data into a format compatible with the OpenCV library, and transferring it to the UAV control system [14].

The decision was made to use the SceneCaptureComponent2D tool provided by Unreal Engine as the virtual camera. Unlike a regular CameraComponent, SceneCaptureComponent2D is used to capture an image from the virtual world into a "real-time texture," rather than just displaying the result on the screen.

By texture, it refers to TextureRenderTarget2D, a tool in Unreal Engine that represents an image as a two-dimensional matrix, which can be used as a target for rendering. These render targets allow storing images that are generated on-the-fly during scene rendering. TextureRenderTarget2D allows storing images in various color formats, including but not limited to, RGBA, 8-bit, 16-bit float.

After capturing the image from the camera and transferring it to the texture, the data from TextureRenderTarget2D is converted into a format compatible with the OpenCV library, namely two-dimensional pixel matrices cv::Mat.

For this purpose, information about the height and width of the texture is used, as well as an array of FColor structures that carry information about each individual pixel (R, G, B, A in a range from 0 to 255). This set of information is provided as parameters to the cv::Mat constructor.

This sequence of actions is repeated in every frame for the camera. The obtained image is used as the result of the camera simulation and is further transmitted to the system for further use. This process is described in Fig. 6.

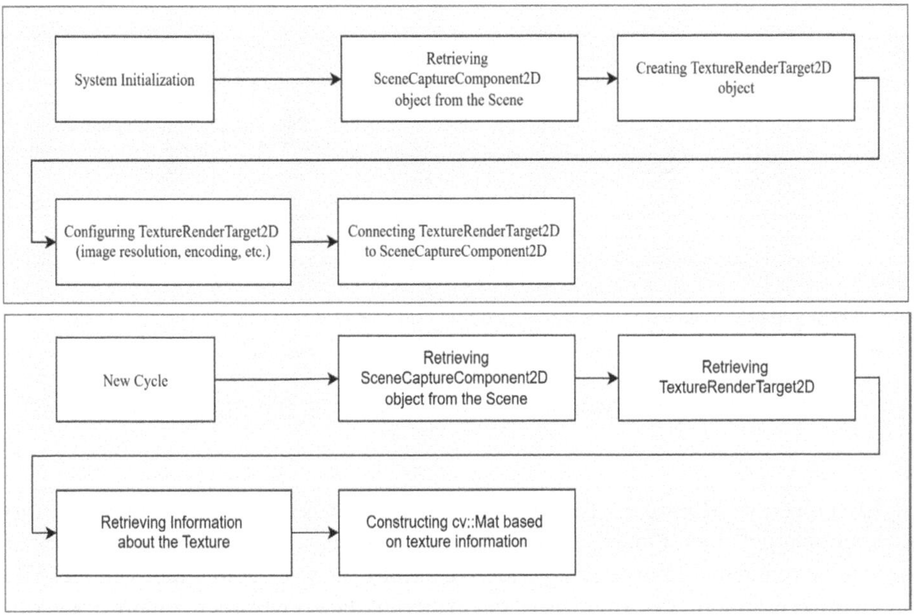

Fig. 6. Description of the Virtual Camera Workflow.

Sensors. Another important source of information about the surrounding environment for UAVs is onboard sensors. The simulation system provides the following virtual sensor analogs: accelerometer and gyroscope, magnetometer [15], barometer, and GPS positioning system. For their simulation, information from the physical engine is taken as a basis and converted into corresponding measurement units, considering the proportional relationship between the virtual UAV and its real-world counterpart [16].

To simplify the process of controlling the virtual UAV, the operator receives visual information from the onboard sensors [17].

4.3 Virtual Target

Target simulation is an integral part of the guidance algorithm testing system. The main task of this mechanism is the realistic simulation of enemy UAV movement trajectories.

The system consists of two components:

- UAV Control – a component similar to the one described in Sect. 4.1. It performs the movement of the target in space based on control signals.
- Flight Trajectory Manager - a component that enables the operator to replicate complex target maneuvers on which the testing of guidance algorithms is based.

The overall structure of the module is presented in Fig. 7.

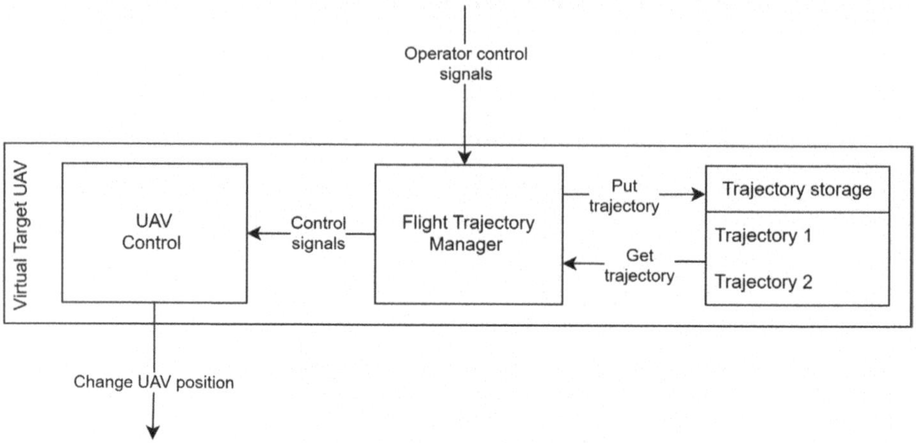

Fig. 7. General diagram of the "Virtual Target UAV" block.

Flight Trajectory Manager. The system's primary task is to create the flight trajectory of an aerial target. Target management can occur in a long-term mode, where the trajectory can be replicated in tests or for video recording, as well as in a quick mode. This mode can be used for a one-time visual evaluation of the effectiveness of target tracking algorithms. In turn, the long-term mode consists of stages of recording the trajectory and its subsequent reproduction as required.

Recording Trajectory. The simulation system is deterministic, meaning that at any given moment, based on the input data provided, it is possible to predict and reproduce its state. During the system's operation in trajectory recording mode, the operator performs the required maneuvers on a virtual UAV, which serves as the target. Each cycle of the system's operation involves recording four input control parameters (Thrust, Yaw, Pitch, Roll) and the delta time. These parameters are stored in the trajectory storage and can be used later to replicate the flight.

Reproducing Trajectory. In this process, the flight trajectory of the aerial target is replicated using previously recorded trajectories. Each cycle of the system reads the control signals that were transmitted to the UAV and the delta time. Taking this parameter into account avoids the problem of reproduction at different FPS values. If sufficient time has elapsed since the last signal transmission, new control signals will be sent to the control block, resulting in a change in the target's position. The operation process of the Flight Trajectory Manager is provided in Fig. 8.

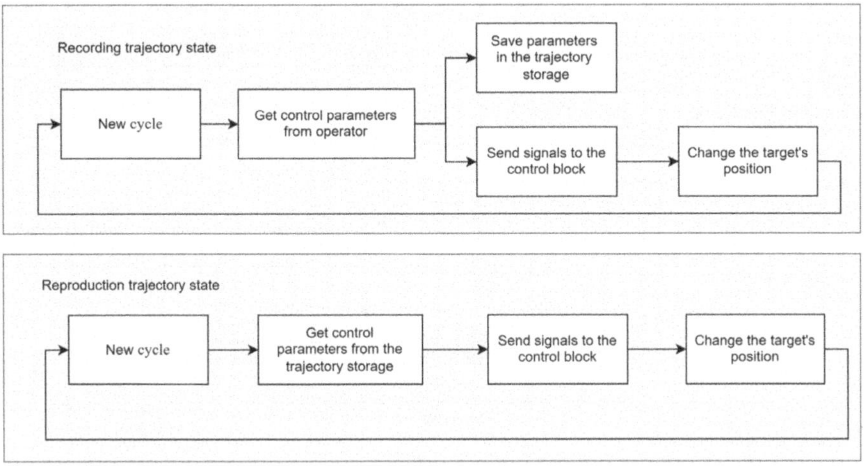

Fig. 8. General diagram of the recording and reproducing trajectory processes.

4.4 Laboratory Stand

This is a system responsible for conducting tests on UAV control algorithms, comparing them, and providing reports on their performance. The system has several modules that assist in preparing data for testing and conducting various types of tests. More details are provided below [18].

Test Data Manager. Utilized for preparing test data. During its operation, the target follows a pre-recorded trajectory while the operator controls the UAV pursuing the target in real time. At this time, the system records the video stream from the UAV's camera and telemetry information at each simulation cycle. This data will later be used for deep testing of the algorithms [19].

Python Script Execution. The module is responsible for integrating user algorithms with the control system. Its task is to initialize scripts written in Python, format the input information as expected by the scripts, execute them, and return the result [20].

Real-Time Testing. Utilized for evaluating the algorithm's performance under various conditions. The algorithm, analyzing information from the camera and sensors, sends control signals to the UAV and thereby adjusts its position in virtual space. During this time, the operator manages the target, changing its position and behavior [21]. The operating principle of the system is presented in Fig. 9.

In this mode, the system uses a split screen, where the target UAV is displayed on the top part of the screen, and the chasing UAV is on the bottom. This approach enables a rapid evaluation of the algorithm's effectiveness. This interface is presented in Fig. 10.

Deep Testing. Utilized for comparing algorithms or for evaluating the effectiveness of changes made to them. This is accomplished by generating a test data set that includes a video stream from a UAV camera and recording its telemetry. This data is used to establish identical testing conditions for different algorithms. Additionally, based on this data, information about the benchmark performance of the algorithms is created,

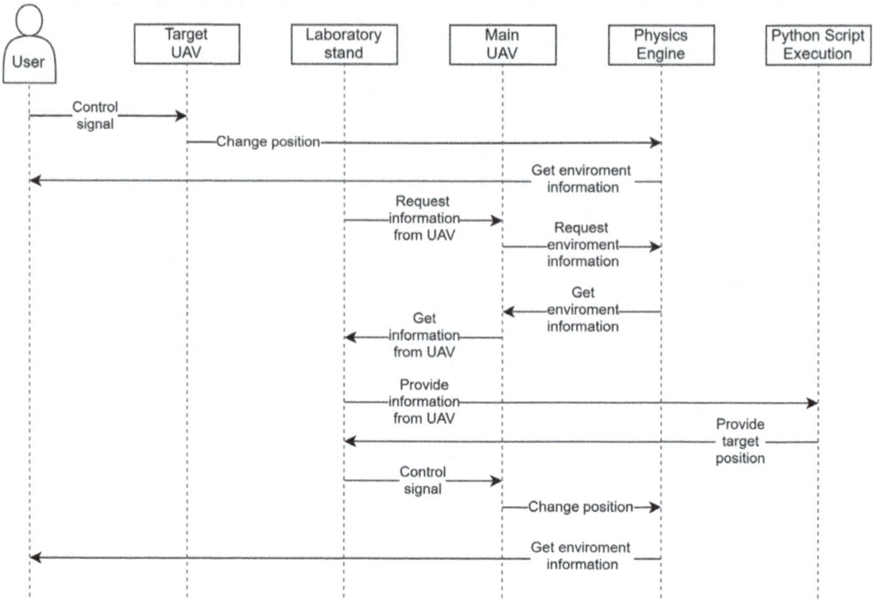

Fig. 9. General diagram of singe iteration in Real-time Testing processes.

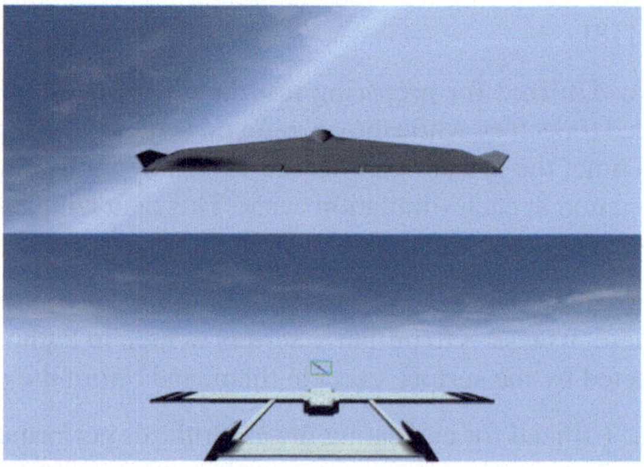

Fig. 10. Real-time testing interface. At the top is the UAV controlled by the operator, and at the bottom is the UAV controlled by the algorithm.

which will subsequently be used to assess their effectiveness [22]. For this purpose, information about the actual location of the target is added to each video frame. An example of the interface is presented in Fig. 11.

This test data is fed into the algorithms, simulating information about the surrounding environment. Based on this, the algorithm analyzes the data and determines the potential location of the target. The results are then compared to benchmark data, and a report on the algorithm's effectiveness is generated. The system's operation is depicted in Fig. 12.

Fig. 11. The process of creating information about the current position of the UAV.

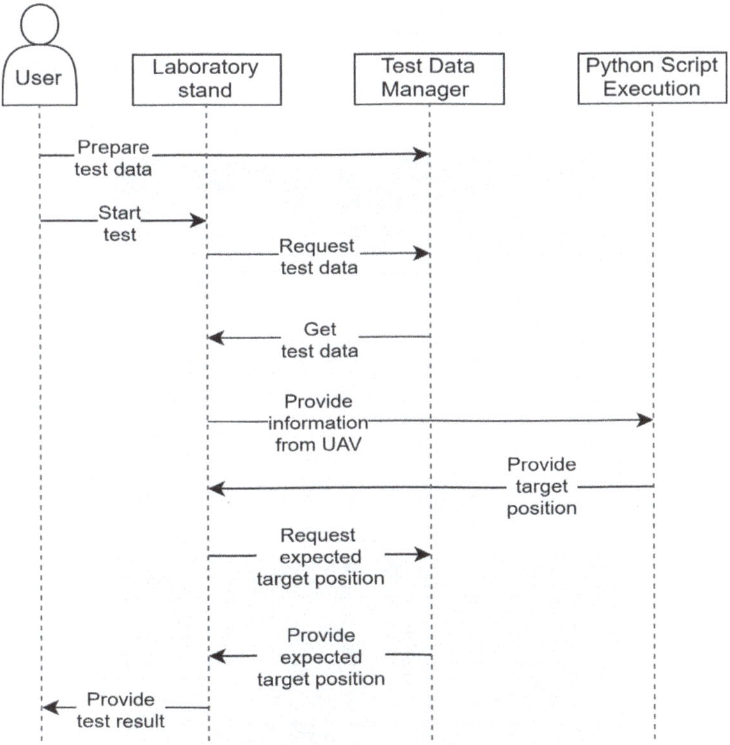

Fig. 12. General diagram of Deep Testing processes.

The system provides the following statistics: target position accuracy, frame processing speed, the percentage of total time spent on successful target tracking, and the speed at which the target is located after being lost.

5 Optimization of the Algorithm Based on Testing Results

As previously mentioned, the main task of the system is to optimize the algorithm development process for automated target tracking. This is accomplished through a realistic and flexible environment simulation system, efficient algorithm testing, and providing detailed reports on them. Let's look at this process in more detail.

For testing purposes, a basic target tracking algorithm was developed using the OpenCV library. The algorithm operates by cyclically shifting the region of interest within the frame and comparing its pixel matrix with the target image using the *cv2.absdiff(img1, img2)* [23] method. If the difference is less than a certain threshold value, the object is considered detected, its coordinates are returned as results, and the process is repeated for the next frame.

Prior to testing, a set of test data was generated, namely recording video streams from a UAV camera following a moving target. For this, three virtual flights were conducted under varying environmental conditions. Also, due to the simplicity of the algorithm, it was decided to exclude map generation and telemetry data from the UAV.

The next step for each frame of the test video was to identify the area where the target was located. This area will later be used in tests to assess the accuracy of the target detection algorithm as a reference result. Providing the target position process is presented in Fig. 13.

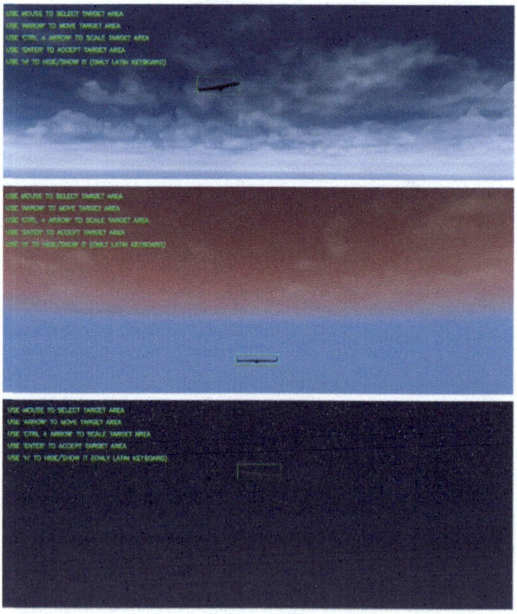

Fig. 13. Frames of target search process.

Having all the test assets and the basic algorithm, we will conduct its testing according to the following criteria: frame processing speed and target detection accuracy. The visualization of the testing process presented in Fig. 14.

Fig. 14. Visualization of the simple algorithm testing process.

As expected, the simple algorithm demonstrated relatively average results. Additional challenges like clouds and stars are quite critical for it. The best result was shown by videos in which the sky was clear and the target stood out well against the general background. Also, a weakness of the algorithm was its inability to reacquire a lost target, which made its further work ineffective. The testing results are presented in Figs. 15 and 16.

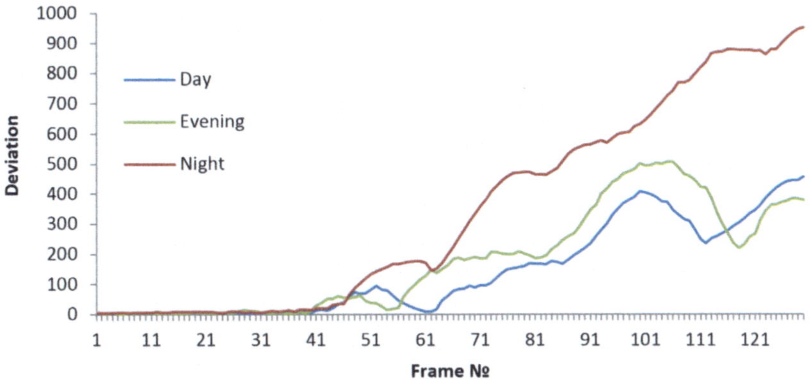

Fig. 15. The simple algorithm's deviation change. The value is calculated by the total deviation of the algorithm's output area from the target area along the x-axis and y-axis.

Based on the testing results, it can be concluded that the main problem of the current algorithm is the accumulation of errors over time. This occurs because the target being compared to its initial image while the data of its appearance are not updated.

To address this issue, it was decided to improve the algorithm by adding a mechanism for automatic re-detection of the target when the deviation increases. This is achieved by the detection of edges of the most contrast objects in the frame, for which the Canny [24] algorithm provided by the OpenCV library is used. The most contrasting edges are merged into closed morphology using the *cv2.morphologyEx* [25] method in MORPH_CLOSE mode. The largest closed area is identified the target. The test will be repeated for the updated algorithm. The testing results are presented in the graphs Figs. 17 and 18.

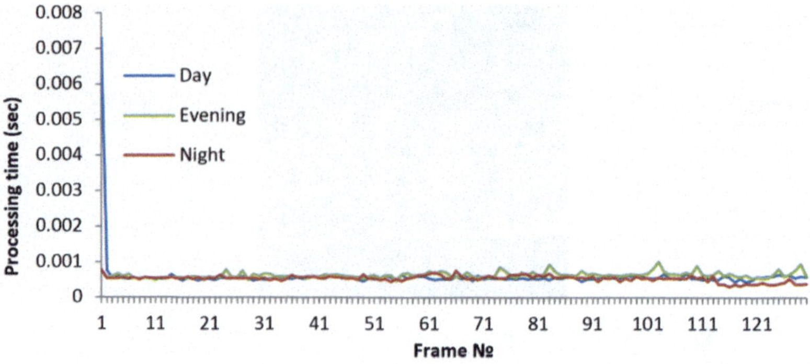

Fig. 16. Changes in the frame processing time by the simple algorithm.

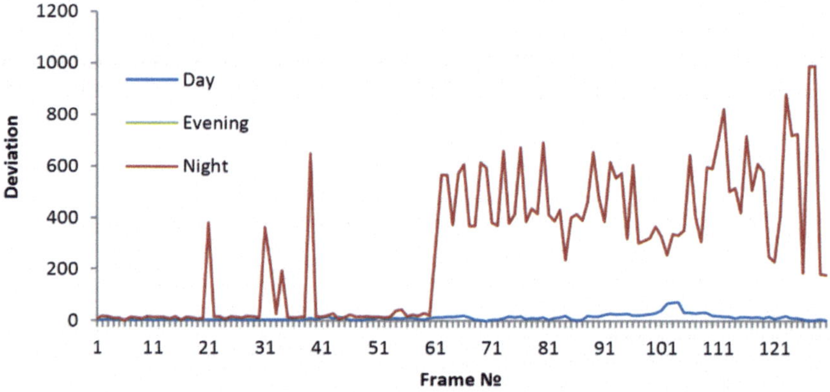

Fig. 17. The improved algorithm's deviation change. The value is calculated by the total deviation of the algorithm's output area from the target area along the x-axis and y-axis.

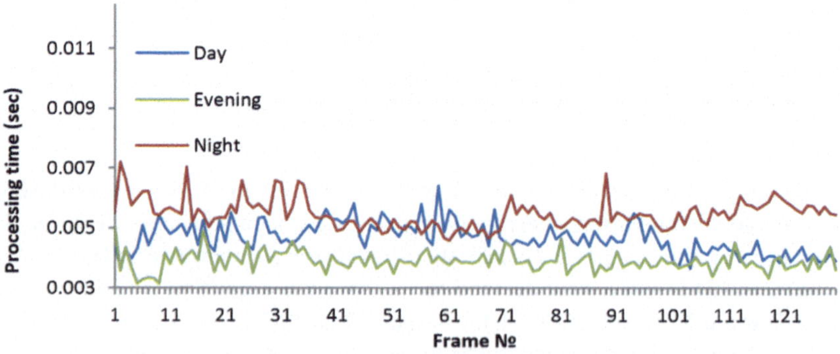

Fig. 18. Changes in the frame processing time by the improved algorithm.

Comparing the results of the two algorithms, allows to confidently conclude that the updated algorithm shows significantly better results in good visibility conditions.

Additionally, minor obstacles such as clouds do not pose a challenge for it. However, it is still not effective in tracking targets at night or under conditions with numerous bright stars. Also, the increased complexity of the algorithm's logic has led to an increase in processing time for each frame.

All this information allows to thoroughly assess the results of the work done and to identify potential paths for further improvement of the algorithm. One promising direction is improving the algorithm's efficiency in conditions of insufficient visibility or its overall performance.

6 Conclusions and Future Work

Working on automatic UAV targeting systems is a complex and resource-intensive task. The simulation system we are developing, based on Unreal Engine and OpenCV, serves as a powerful tool that significantly simplifies, accelerates, and reduces the cost of this process.

The test evaluations of the algorithms demonstrated the effectiveness of the system in development. The chosen approach of virtual simulation allowed for an accurate assessment of algorithm performance across a wide range of environmental conditions. This evaluation was conducted in isolation from external factors, such as technical issues with the drone or inconsistencies between flights. Furthermore, the process was time-efficient, eliminated the risks associated with UAV damage, and did not require additional operators.

Despite these successes, there are several potential directions for system improvement. In the future, we plan to enhance the environmental simulation by adding elements such as wind and rain. We also aim to introduce support for stereo cameras and develop tools for their calibration, potentially enabling the creation of depth maps for algorithmic applications. Additionally, we intend to increase the accuracy of sensor simulations by introducing noise and interference into the resulting signals.

References

1. Kulik, A., Dergachev, K.: Intelligent transport systems in aerospace engineering. In: Intelligent Transportation Systems – Problems and Perspectives, pp. 243–303. Springer, Cham (2015)
2. Shah, S., Dey, D., Lovett, C., Kapoor, A.: AirSim: high-fidelity visual and physical simulation for autonomous vehicles. (2017). https://doi.org/10.48550/arXiv.1705.05065
3. Dergachov, K., Havrylenko, O., Pavlikov, V., Zhyla, S., Tserne, E., Volosyuk, V.: GPS usage analysis for angular orientation practical tasks solving. In: 2022 IEEE 9th International Conference on Problems of Infocommunications, Science and Technology (PIC S&T), pp. 187–192. IEEE, Kharkiv (2022). https://doi.org/10.1109/PICST57299.2022.10238629
4. Putnam, H.: Brains in a Vat. ISBN 5-901574-09-5 (2002)
5. Open Source Computer Vision. https://docs.opencv.org/4.x/index.html. Last accessed 11 Feb 2025
6. Buck, A., Camaioni, R., Alvey, B., Anderson, D.T., Keller, J.M., Luke II, R.H., Scott, G.: Unreal engine-based photorealistic aerial data generation and unit testing of artificial intelligence algorithms. (2022). https://doi.org/10.1117/12.2618812

7. Unreal Engine 5.5 Documentation: Lighting the Environment. https://dev.epicgames.com/documentation/en-us/unreal-engine/lighting-the-environment-in-unreal-engine. Last accessed 11 Feb 2025
8. Popov, A., Tserne, E., Volosyuk, V., Zhyla, S., Pavlikov, V., Ruzhentsev, N.: Invariant polarization signatures for recognition of hydrometeors by airborne weather radars. In: Computational Science and Its Applications – ICCSA 2023 Lecture Notes in Computer Science, vol. 13956, pp. 201–217. Springer, Cham (2023). https://doi.org/10.1007/978-3-031-36805-9_14
9. Arafat, M.Y., Alam, M.M., Moh, S.: Vision-based navigation techniques for unmanned aerial vehicles: review and challenges. (2023). https://doi.org/10.3390/drones7020089
10. Unreal Engine 5.5 Documentation: Visibility and Occlusion Culling. https://dev.epicgames.com/documentation/en-us/unreal-engine/visibility-and-occlusion-culling-in-unreal-engine. Last accessed 11 Feb 2025
11. Wang, S., Chen, J., Zhang, Z., Wang, G., Tan, Y., Zheng, Y.: Construction of a virtual reality platform for UAV deep learning. (2017). https://doi.org/10.1109/CAC.2017.8243463
12. Nikitina, T., Kuznetsov, B., Ruzhentsev, N., Havrylenko, O., Dergachov, K., Volosyuk, V., Shmatko, O., Popov, A.: Algorithm of robust control for multi-stand rolling mill strip based on stochastic multi-swarm multi-agent optimization. In: Shukla, S., Sayama, H., Kureethara, J.V., Mishra, D.K. (eds.) Data Science and Security. IDSCS 2023 Lecture Notes in Networks and Systems, vol. 922, pp. 247–255. Springer, Singapore (2024). https://doi.org/10.1007/978-981-97-0975-5_22
13. Blaga, B.C., Nedevschi, S.: Employing simulators for collision-free autonomous UAV navigation. (2022). https://doi.org/10.1109/ICCP56966.2022.10053959
14. UE4: Guidebook: Integrating OpenCV into Unreal Engine 4. https://unreal.gg-labs.com/wiki-archives/ar-vr/integrating-opencv-into-unreal-engine-4. Last accessed 11 Feb 2025
15. Nikitina, T., Kuznetsov, B., Averyanova, Y., Sushchenko, O., Ostroumov, I.: Method for design of magnetic field active silencing system based on robust meta model. In: Shukla, S., Sayama, H., Kureethara, J.V., Mishra, D.K. (eds.) Data Science and Security. IDSCS 2023 Lecture Notes in Networks and Systems, vol. 922, pp. 103–111. Springer, Singapore (2024). https://doi.org/10.1007/978-981-97-0975-5_9
16. Brierley-Green, A.: Global Navigation Satellite System Fundamentals and Recent Advances in Receiver Design (2017). https://www.ieee.li/pdf/viewgraphs/gnss_fundamentals.pdf. Last accessed 11 Feb 2025
17. Yang, X., Zou, D., Pei, L., Sartori, D., Yu, W.: An efficient simulation platform for testing and validating autonomous navigation algorithms for multi-rotor UAVs based on unreal engine. (2019). https://doi.org/10.1007/978-981-13-7759-4_46
18. Yildirim, A.S., Berker, E., Kayakese, M.E.: System level test automation in UAV development. (2018). https://doi.org/10.1109/AUTEST.2018.8532551
19. Dai, X., Ke, C., Quan, Q., Cai, K.Y.: RFlySim: automatic test platform for UAV autopilot systems with FPGA-based hardware-in-the-loop simulations. Aerosp. Sci. Technol. **114**, 106727 (2021). https://doi.org/10.1016/j.ast.2021.106727
20. Unreal Python API Documentation: Integrating Python with Unreal Engine. https://dev.epicgames.com/documentation/en-us/unreal-engine/scripting-the-unreal-editor-using-python?application_version=5.4. Last accessed 11 Feb 2025
21. Sfetcu, N.: Electronic warfare and artificial intelligence. (2024). https://doi.org/10.58679/mm14430
22. Li, R., Liu, H., Lou, G., Zheng, J., Liu, X., Chen, T.: Metamorphic testing on multi-module UAV systems. In: 36th IEEE/ACM International Conference on Automated Software Engineering (ASE), pp. 1171–1173, Melbourne, Australia (2021). https://doi.org/10.1109/ASE51524.2021.9678841
23. Core module in Open Source Computer Vision. https://docs.opencv.org/4.x/javadoc/org/opencv/core/Core.html. Last accessed 11 Feb 2025

24. Canny edge detection in Open Source Computer Vision. https://docs.opencv.org/4.x/da/d22/tutorial_py_canny.html. Last accessed 11 Feb 2025
25. Imgproc Module in Open Source Computer Vision. https://docs.opencv.org/4.x/javadoc/org/opencv/imgproc/Imgproc.html. Last accessed 11 Feb 2025

Author Index

I. Ostroumov et al. (Eds.): ACASD 2025, LNNS 1418, pp. 567–568, 2025.
https://doi.org/10.1007/978-3-031-91992-3

The manufacturer's authorised representative in the EU is Springer Nature Customer Service Centre GmbH, Europaplatz 3, 69115 Heidelberg, Germany. If you have any concerns regarding our products, please contact ProductSafety@springernature.com

Printed and bound by CPI Group (UK) Ltd, Croydon, CR0 4YY
01/10/2026
02225911-0016